Molecular Exercise Phy

Fully revised and expanded, the second edition of *Molecular Exercise Physiology* offers a student-friendly introduction. It introduces a history documenting the emergence of molecular biology techniques to investigate exercise physiology, the methodology used, exercise genetics and epigenetics, and the molecular mechanisms that lead to adaptation after different types of exercise, with explicit links to outcomes in sport performance, nutrition, physical activity and clinical exercise.

Structured around key topics in sport and exercise science and featuring contributions from pioneering scientists, such as Nobel Prize winners, this edition includes new chapters based on cutting-edge research in epigenetics and muscle memory, satellite cells, exercise in cancer, at altitude, and in hot and cold climates. Chapters include learning objectives, structured guides to further reading, review questions, overviews of work by key researchers and box discussions from important pioneers in the field, making it a complete resource for any molecular exercise physiology course. The book includes cell and molecular biology laboratory methods for dissertation and research projects in molecular exercise physiology and muscle physiology.

This book is essential reading for upper-level undergraduate or postgraduate courses in cellular and molecular exercise physiology and muscle physiology. It is a valuable resource for any student with an advanced interest in exercise physiology in both sport performance and clinical settings.

Adam P. Sharples, PhD, is a Professor of Molecular Physiology and Epigenetics at the Norwegian School of Sport Sciences (NiH), Olso, Norway; an institute ranked 2nd (out of 300+) in the world for sport and exercise sciences. He investigates the underlying cellular, molecular and epigenetic mechanisms of muscle growth (hypertrophy) and wasting (atrophy) using both cell modelling and whole-body approaches. His work first demonstrated that human muscle possesses an "epigenetic memory" of exercise. He used to play professional Rugby League in the UK.

James P. Morton, PhD, is a Professor of Exercise Metabolism at Liverpool John Moores University (LJMU). His research evaluates the impact of nutrient availability on muscle metabolism during exercise and the molecular regulation of skeletal muscle adaptations to exercise training. James has also worked in a number of performance related roles across both high-performance sport and industry, working with some of the world's most high profile athletes, sports teams and institutions.

Henning Wackerhage, PhD, is a Professor and Molecular Exercise Physiologist. He is specifically interested in the molecular mechanisms by which exercise improves our fitness and health, particularly the role of the so-called Hippo proteins in skeletal muscle and the association between the proteome, metabolome, athletic performance, disease and ageing.

Molecular Exercise Physiology

An Introduction

Second Edition

Edited by
Adam P. Sharples, James P. Morton
and Henning Wackerhage

Routledge
Taylor & Francis Group

NEW YORK AND LONDON

Cover image: © kirstypargeter / Getty Images

Second edition published 2022
by Routledge
605 Third Avenue, New York, NY 10158

and by Routledge
4 Park Square, Milton Park, Abingdon, Oxon, OX14 4RN

Routledge is an imprint of the Taylor & Francis Group, an informa business

© 2022 Taylor & Francis

First edition published by Routledge 2014

Library of Congress Cataloging-in-Publication Data
Names: Sharples, Adam P. (Professor in Molecular Physiology and
Epigenetics), editor. | Morton, James P., editor. | Wackerhage, Henning, editor.
Title: Molecular exercise physiology: an introduction / edited by
Adam P. Sharples, James P. Morton, and Henning Wackerhage.
Description: Second edition. | New York, NY: Routledge, 2022. |
Includes bibliographical references and index. |
Identifiers: LCCN 2021061845 (print) | LCCN 2021061846 (ebook) |
ISBN 9781138086876 (hardback) | ISBN 9781138086883 (paperback) |
ISBN 9781315110752 (ebook)
Subjects: LCSH: Sports sciences. | Sports medicine. | Genetics. |
Sports—Physiological aspects. | Exercise—Physiological aspects.
Classification: LCC GV558 .M65 2022 (print) | LCC GV558 (ebook) |
DDC 613.7/1—dc23/eng/20211220
LC record available at https://lccn.loc.gov/2021061845
LC ebook record available at https://lccn.loc.gov/2021061846

ISBN: 978-1-138-08687-6 (hbk)
ISBN: 978-1-138-08688-3 (pbk)
ISBN: 978-1-315-11075-2 (ebk)

DOI: 10.4324/9781315110752

Typeset in Joanna MT & Frutiger
by codeMantra

Contents

Figures

Tables

Contributors

Keith Baar
University of California, Davis, USA

Claude Bouchard
Pennington Biomedical Research Center, USA

Jatin G. Burniston
Liverpool John Moores University, UK

Kenneth A. Dyar
Institute for Diabetes and Cancer (IDC) at Helmholtz Zentrum, München

Brendan Egan
Dublin City University, Ireland and Florida Institute for Human and Machine Cognition, USA

Javier Gonzalez
University of Bath, UK

Piotr P. Gorski
The Norwegian School of Sport Sciences, Norway

Mark Hearris
Liverpool John Moores University, Liverpool, UK

Nathan Hodson
University of Toronto, Canada

Pernille Hojman
University of Copenhagen, Denmark

Jonathan C. Jarvis
Liverpool John Moores University, UK

Neil R.W. Martin
Loughborough University, UK

Tormod S. Nilsen
The Norwegian School of Sport Sciences, Norway

Daniel J. Owens
Liverpool John Moores University, UK

Stephen Roth
University of Maryland, USA

Martin Schönfelder
TUM Department of Sport and Health Sciences, Technical University of Munich

Robert A. Seaborne
University of Copenhagen, Denmark

Claire E. Stewart
Liverpool John Moores University, UK

Daniel C. Turner
The Norwegian School of Sport Sciences, Norway

Mark Viggars
Liverpool John Moores University, UK and University of Florida, USA.

CHAPTER ONE

Introduction to molecular exercise physiology

Adam P. Sharples and Henning Wackerhage

INTRODUCTION

'Molecular exercise physiology' is a discipline within exercise physiology and a shortened version of the term 'molecular and cellular exercise physiology' which was used, amongst others, by Frank W. Booth (1), a pioneer in this area (see Box 1.1). In the first part of this chapter, we define molecular exercise physiology, distinguish it from exercise biochemistry and trace its roots in molecular biology and exercise physiology.

Origins and definition of molecular exercise physiology

We define molecular exercise physiology as 'the study of the molecular responses to exercise and underlying mechanisms that lead to physiological adaptation following exercise'. The field is focused particularly at the level of the molecule and cell, and investigates how the microscopic make-up of our cells responds to exercise, ultimately leading to adaptation at the cellular, tissue and system levels. The discipline combines the use of molecular biology wet-laboratory techniques with physiological methods to examine the molecular make-up and responses of our cells and tissues to exercise. Therefore, molecular exercise physiologists are concerned with four overarching areas of study: (1) **The role of heritable genetic traits** (variation in the genetic code found within our DNA) and their associated influence on the physiological response and adaptation to exercise. More recently with a sharper focus on the role of (2) **'epi'-genetics** (meaning 'above' genetics), where both the environmental 'stressor' of exercise and underlying genetics interact to influence molecular responses and adaptation to exercise, for example, by molecular modifications to our genetic code at the DNA level that subsequently affect how our genes are turned on and off following exercise. In terms of the molecular responses to exercise, exercise physiologists also want to understand: (3) How the environmental 'stressor' of exercise modulates the abundance and activity of **molecular 'signal-transduction' networks**, leading to the turning on or off of genes and therefore the resulting protein levels, culminating in changes/adaptation at the cellular or tissue level. Of course, these molecular mechanisms are also applicable across other exercise physiology sub-disciplines, such as sports and exercise nutrition, different environmental conditions (e.g. hot, cold conditions, at altitude and different time zones), ageing and diseases (e.g. cancer, diabetes and obesity). Therefore, molecular exercise physiology has diversified rapidly to incorporate the use of molecular biology to study exercise in aligned sport and exercise fields across performance and health-related disciplines. The majority of these regulatory and molecular mechanisms are studied in blood and skeletal muscle tissue and/or isolated satellite cells (regenerative cell found in skeletal muscle) and sometimes adipose tissue, albeit to a lesser extent than muscle tissue. This is due to the relative ease of sampling of

DOI: 10.4324/9781315110752-1

BOX 1.1 Frank W Booth: the advent of molecular biology techniques in exercise physiology

A brief history of a pioneering early scientist in molecular biology of physical inactivity and exercise. Prof. Dr. Frank W Booth was a leading pioneer in using molecular biology to study mechanisms underlying exercise responses and adaptation.

My philosophy of science has evolved under a number of teachings. A conundrum I often ponder is do men make history or does history make men? From my viewpoint, this thought-provoking phrase is a truncated tribute to Karl Marx who wrote, 'Men make their own history, but they do not make it as they please, they do not make it on their self-selected circumstances, but on their circumstances existing already, given and transmitted from the past'. Certainly, circumstances throughout my training and career have guided my philosophical evolution, but I can't deny that genes play an important role as well. I started liberal arts at Denison University with an intent to pursue law. Circumstances led me elsewhere. My biology classes gripped my attention. Serendipity had my University begin a swim team in my sophomore year (the only way I could have made the squad), and the assistant swim coach, Robert Haubrich, was also my academic biology advisor. My senior paper required for biology majors in 1964 was on adaptations to exercise. Two events occurred together: (1) My applications to medical school were not successful so I had to look at an alternative, and (2) Professor Haubrich gave me a flyer about a new exercise physiology PhD programme under Charles Tipton at the University of Iowa. I got on a train from Columbus, Ohio and went to Iowa City to view the exercise physiology programme, whose described course of study had intrigued my interest so much that I decided to visit and eventually enter in 1965, to replace my interest in medical school. Other students in the Tipton programme were James Barnard, Ken Baldwin and Ron Terjung. 'Tip', a name we respectfully called Charles Tipton, taught me to go after mechanisms and supported my desire to challenge existing dogmas and policies, as this was the period of student discontent over the Vietnam War. University life in the late 1960s was so tremulous with continual protest rallies, and the killing of student protesters at Kent State University, that University administers had to accept student challenges as acceptable on their campus (as opposed to today where minimal protest occurs over the loss of academic freedoms and civil liberties on University campus in response to forced compliance to regulations from the non-democratic, non-elected bureaucrats). Also, during the Cold War period (February 1945–August 1991), the Russians, on 4 October 1957, launched Sputnik I, an orbiting satellite that greatly accentuated the continual threat that the U.S. feared from the Soviet Union. The same rocket that launched Sputnik could send a nuclear warhead anywhere in the world in a matter of minutes (as I saw in the fictional classic 1964 movie Dr. Strangelove or: How I Learned to Stop Worrying and Love the Bomb, which showed a nuclear holocaust from an insane general, in an insanely funny comedy). In 1958, the U.S. Congress passed the National Defense Education Act, whose outgrowth funded Tip's exercise physiology programme, and my graduate student salary. Russia and the U.S. were in a space race. The first human walk on the moon in 1969 was from the U.S., which further fuelled my interest to the physiological effects of lack of gravity (physical inactivity), along with Jere Mitchell's and Bengt Saltin's classic Dallas bedrest study published in the journal Circulation in 1968, and fostered my decision to take a first post-doc at the School of Aerospace Medicine. John Holloszy's 1967 paper on the biochemical adaptations of exercise led me to my second post-doc experience to join Baldwin and Terjung at John's lab, a time during which John taught me critical thinking. After finishing my work with Holloszy, I had a choice of two faculty positions in medical schools (Wayne State in Detroit, or the new University of Texas Medical School at Houston). I selected the latter because it was the home of NASA. Soon after arriving in Houston in 1975, molecular biology was coming to the forefront. I saw the potential of explaining inactivity mechanisms in terms of genes. In the early 1980s, I was fortunate to be present when molecular biology began to appear as a tool in biology at the University of Texas Medical School in the Texas Medical Center. Baylor College of Medicine was one block from my medical school, and I listened to many seminars at Baylor during which I began to connect molecular techniques with my research in exercise and inactivity. My first grad student, Peter Watson, collaborated with Joe Stein in biochemistry and endocrinology, whose lab was just a few doors down from mine, to measure mRNAs. We used dot-blot hybridization of P32-labelled plasmids containing the skeletal muscle alpha-actin cDNA in isolated RNA on a nitrocellulose membrane and published the work in the *American Journal of Physiology* in 1984. Concurrently, in 1980, Kary Mullis invented the PCR, a method for multiplying DNA sequences *in-vitro*, and I began using this technique when it became commercialized.

I will end with where I started: Do humans make history or does history make humans perform some event? I think it is both, like gene–environment interaction determining phenotype. History and human curiosity interact to determine how well scientific understanding can explain why physical inactivity is an actual contributor to chronic disease and longevity. As Robert Frost, Pulitzer Prize for Poetry, United States Poet Laureate, ended his poem 'Stopping by Woods on a Snowy Evening':

> The woods are lovely, dark and deep.
> But I have promises to keep,
> And miles to go before I sleep,
> And miles to go before I sleep.

So my journey of life continues, as I have miles and miles to go before I sleep.
Frank Booth, University of Missouri, August, 2013

blood and the ever-increasing obtainability of skeletal muscle biopsies in human participants and/or patients under various exercise conditions. Finally, molecular exercise physiologists have also been interested in (4) **the role of satellite cells** and their role in the repair and regeneration of muscle after exercise, and more recently its role as a molecular 'communicator' in skeletal muscle.

BOOK OVERVIEW AND CHAPTER OUTLINE

After providing a short history of the evolution of molecular exercise physiology as a field (**Chapter 1**), the book will start by introducing the main methods (**Chapter 2**) employed by molecular exercise physiologists. This will provide an important understanding of the main methods that are employed in the research that is then discussed in the later theoretical chapters. **Chapter 3** will introduce sport and exercise genetics and the 'central dogma' of molecular biology, DNA to RNA to protein, and how this relates to exercise physiology. Importantly, it will cover how variations in inherited information found in our genetic code may contribute to changes in physiology that could affect exercise and physiological performance. **Chapter 4** will then discuss the genetics of muscle mass and strength, and in **Chapter 5**, we go on to cover the genetics of endurance exercise. **Chapter 6** will discuss how exercise can 'epigenetically' modify our inherited genetic code and lead to altered molecular responses involved in exercise adaptation. Because epigenetic modifications to DNA can be retained over longer periods, this chapter will also discuss the concept of epigenetic 'muscle memory' and how this paradigm might be an important consideration across the exercise and sport sciences. **Chapter 7** will introduce the 'signal transduction' theory of molecular exercise responses and how these lead to adaptation at the physiological level. **Chapters 8 and 9** will look at these molecular responses in both resistance exercise and endurance exercise, respectively. Then, the book will move on to discuss molecular exercise physiology in the context of altered nutrition (**Chapter 10**), different environmental conditions, e.g. hot and cold conditions, at altitude and in different time zones and the influence of circadian rhythms (**Chapter 11**), and with disease such as cancer (**Chapter 12**). Finally, we will address the role of satellite cells in molecular exercise physiology (**Chapter 13**).

A SHORT HISTORY OF THE EVOLUTION OF MOLECULAR EXERCISE PHYSIOLOGY AS A FIELD

How did molecular exercise physiology evolve? In exercise physiology, as in other life sciences in the 20th century, the trend was to move the frontier of research from the whole-body level and organ systems towards cells and molecules (2). This does not mean that molecular studies now replace organ systems research. Instead, the molecular research complements and often explains what occurs at the organ and whole-body levels. Molecular studies enable researchers to answer questions that previous generations of exercise physiologists were unable to answer due to the lack of suitable methods. Thus, molecular exercise physiology is an extension of, and in complement to, classical exercise physiology. The era that preceded and prepared the ground for molecular exercise physiology was the biochemistry of exercise era. While the whole-body and organ-level exercise physiologists frequently used non-invasive methods, exercise biochemists begun to use invasive measures and made wet lab research their main analytical method of study. Consequently, physiologists needed to learn wet lab methods and had to equip their laboratories with the necessary consumables and equipment. This included, for example, chemicals, pipettes, pH meters, centrifuges, spectrophotometers and microscopes. Human exercise physiologists with a skeletal muscle focus also needed to learn the muscle biopsy technique, which was introduced to the field by Jonas Bergstrom, and/or had to use animal models to derive skeletal muscle tissue. This was necessary because tissue samples, especially from skeletal muscle, were needed for subsequent biochemical or histochemical analysis (see Chapter 2 for a summary of histochemical analysis of muscle tissue).

The first use of human skeletal muscle biopsy techniques to study exercise adaptation in humans at the biochemical level was by Jonas Bergstrom in 1962, publishing the paper 'Electrolytes in Man' (3). Then in 1966, Bergstrom and Hultman were the first to measure glycogen synthesis in human muscle after exercise to exhaustion (4, 5). Bergstrom and Saltin in 1967 also assessed muscle glycogen in the quadriceps of both trained and untrained males, and after different starting nutritional backgrounds (e.g. high carbohydrate/high protein) (6, 7). This work was followed by Costill in 1971, who extended biopsy measurements for muscle glycogen to the gastrocnemius and soleus (as well as quadriceps) after exhaustive running and following multiple bouts of exercise (8, 9). Exercise biochemistry was therefore the first discipline that attempted to understand changes in some of the important biochemical parameters following acute and repeated exercise bouts. In this era, exercise biochemistry was primarily the study of metabolites and protein activity using exciting novel methodologies that are now taken for granted such as enzyme-based assays, spectrophotometric and fluorometric analyses together with histological measurements. However, even prior to this work in human muscle, Gollnick's seminal work in 1961 applied exercise biochemistry in rodents to investigate the regulation of adenosine triphosphate (ATP) and lactate dehydrogenase (LDH) activity in heart and skeletal muscle tissues following exercise (10, 11). John Holloszy also pioneered the early exercise biochemistry field by measuring biochemical enzyme activity associated with mitochondrial adaption to aerobic exercise in rats (12). Frank Booth (see Box 1.1), together with Hollozsy (13–15), extended this earlier work in considerably more detail over the ensuing decade, studying in detail the biochemical adaptation to exercise (16–19). Pertinent examples, which were probably the first *bona fide* use of molecular biology in exercise physiology, came from Watson, Stein and Booth who undertook the first work into changes of gene/messenger RNA (mRNA) expression (how much a gene turns on or off) investigating alpha-actin mRNA after muscle immobilization that simulates disuse and cessation of physical activity in the muscle (20). Even more impressively, this research used techniques for gene expression analysis that predated the future Nobel Prize-winning method of polymerase chain reaction (PCR), described below. They subsequently went on to investigate alpha-actin and cytochrome C mRNA expression following both exercise and immobilization in rats (17, 21–23). Frank Booth also undertook some of the first work on muscle protein synthesis (24) and degradation (25), later combining these with adaptation at the biochemical and gene expression levels (26, 27). It was here that early exercise biochemists laid the basis for a field that would soon emerge, leading Booth to his postulation within the literature of a specific field called 'cellular and molecular exercise physiology' (1, 28–30) (see Box 1.1). The advent of PCR to amplify RNA and DNA templates by Kerry Mullis in the late 1980s, leading to the Nobel Prize in chemistry in 1993, made gene expression analysis more sensitive, rapid, accurate and enabled higher throughput within the molecular biology field (31, 32). Importantly, Booth's lab was one of the first to simultaneously undertake direct gene transfer into the skeletal muscle of live rodents (33, 34), allowing them to alter the level of gene expression of specific genes enabling a powerful model to mechanistically ratify (or even refute) the role of individual genes in exercise adaptation. These gene 'knock-down' and gene 'knock-in' (or overexpression) models are discussed below, and the theory for these models in the context of sport and exercise covered in more detail in Chapter 3.

SPORT AND EXERCISE GENETICS

Towards the end of the biochemistry of exercise era, in the 1980s, there were also technological advances in genetic analysis. Research led by Claude Bouchard (see Box 1.2) first identified that aerobic capacity and aspects of physical fitness could be attributed, in part, by genetic inheritance (35–37). Bouchard began to genotype individuals (assess small variations in genetic code between humans) within the field of sport and exercise science between 1984 and 1989 (38, 39). These early efforts were based on red blood cell antigens, plasma and muscle protein mass and charge variants, and histocompatibility tissue antigens. Soon after this, both mitochondrial and nuclear DNA sequence variants were studied (40, 41). These studies were based on the exploration of DNA sequence variation using restriction fragment length polymorphisms, and subsequently on variable number of repeated short DNA sequences (microsatellites) combined with linkage analytical methods (42). Bouchard, together with Malina and Pérusse, wrote the first and widely appreciated textbook within the sport and exercise genetics field, entitled: *Genetics of Fitness and Physical Performance* (43). The book was a fundamental tool for academics, scientists and students within sport and exercise science as its content covered both classical and novel molecular genetic theories relevant to exercise scientists. Then, in 1998, Claude Bouchard identified that genetic heritability accounted for 47% of VO_2max in the sedentary state, an important predictor of endurance performance. A year later, he reported that the trainability of VO_2max in response to the standardized endurance training programme of the HERITAGE Family Study was also characterized by a genetic component accounting for about 50% of the individual differences in responsiveness. In both cases, there was 2.5 times more variation in VO_2max between families than within families (44, 45).

In the same year (1998), Hugh Montgomery in a paper in *Nature* reported that the 'I' polymorphism (a single mutation/variation in the genetic code) of the human ACE gene (angiotensin converting enzyme – important in regulating blood pressure and electrolyte balance) was associated with increased duration of repetitive biceps flexion after a programme of general physical training. This 'I' allele (variant form of the gene) was also increased in frequency within a population of elite mountaineers (46). His team later also showed that an increased frequency of the I allele was observed in elite 5,000-metre British runners compared with both 200 and 400–3,000-metre runners, suggesting a role for this particular genotype in endurance performance. In 2000, Alun Williams, in a study within the group of Montgomery, reported again in the journal *Nature* that the presence of this ACE allele also conferred an enhanced mechanical efficiency in trained muscle (47). After these initial studies in the early 2000s, and the final version of the human genome reported in 2003, more than 80 million common alleles had been identified, setting the stage for the wide-reaching effort to uncover the genomic basis of human variation, including the genetic polymorphisms associated with sport performance.

Bouchard and Teran Garcia were the first to use a combination of skeletal muscle gene expression profiling and DNA sequence variants to investigate human variability in response to an exercise programme (48, 49). With the development of high-throughput Single Nucleotide Polymorphism (SNP) genotyping methods, the human genome sequence and an ethnic group inventory of DNA sequence variants (from the *HapMap* and the *1000 Genomes* projects), the first Genome-wide-association-studies (GWASs) were conducted in 2005 (50). Four years later in 2009, the first GWAS was conducted that assessed 1.6 million SNPs associated with physical activity levels in 2,622 non-related individuals (51). Subsequently, DNA sequence 'microarrays' (microarrays are discussed below in this chapter and the in Methods in Chapter 2) were used to search for a wide range of known polymorphisms that could explain variations in endurance performance and trainability (52, 53). The recognition that the associations between DNA variants and human variation in exercise-related phenotypes were highly complex, and not easily explained, led to the incorporation of new technologies and complementary approaches aimed at refining our understanding of the genetics underlying human biological variability. Metabolomic profiling was used to identify specific plasma metabolites related to the benefits of regular exercise (54). A panel of about 5,000 plasma proteins were used to identify protein biomarkers of cardiorespiratory fitness in the sedentary state and its response to the HERITAGE endurance training programme (55). A bioinformatics pipeline was used to explore the underlying biology of cardiorespiratory fitness and develop a list of prioritized candidate genes (56). Nowadays, by taking advances of these important early breakthroughs and more recent advances in genomics, epigenomics, transcriptomics, proteomics, metabolomics, computational biology and bioinformatics, the field of molecular exercise biology is well positioned to investigate, in depth, the complex molecular mechanisms underlying exercise traits and their responsiveness to various exercise regimens.

There are however, several practical, methodological and ethical issues surrounding sport and exercise genetics as a field. This includes the sheer number of DNA variations now identified (e.g. 1000genomes.org), with some of these variations common, yet others rare or extremely rare, meaning large participant sample sizes are typically required to detect an association with sport- and exercise-related traits. Furthermore, while genome-wide analysis or whole genome sequencing have become financially cheaper to run as technology advances, it is still expensive when they have to be performed on hundreds and thousands of individuals. Although not as dramatic as in other fields of applications, there are ethical questions to consider when undertaking genome-wide analysis for human performance traits or an individual's ability to adapt to exercise, as it may in the process provide information on increased disease risk or provide information that may be associated with negatively perceived connotations towards an individual's genetic ability to perform well in certain performance measures or sports (57). One of the most critical ethical issues is when the acquired genetic information related to performance is used to try to identify talented children with the aim of maximizing their training and performance development opportunities (58). The theory behind exercise genetics is covered in Chapter 3, and genetic variations associated with muscle size and strength are discussed in Chapter 4. Finally, the genetic variants associated with endurance exercise are covered in detail in Chapter 5.

BOX 1.2 Claude Bouchard: from descriptive to comprehensive exercise genomics

Prof. Dr. Claude Bouchard, a pioneering father of molecular exercise physiology, was the first to conduct exercise genetics and DNA sequence studies in sport and exercise science.

I have pursued two main research tracks over the last four decades, one focused on the genetics of obesity and its morbidities, the other on the genetics of exercise traits and their response to exercise training. I have had the good fortune of being able to lead my research programmes at two highly supportive institutions, Université Laval in Québec City, Canada, until 1999, and subsequently the Pennington Biomedical Research Center, a campus of the Louisiana State University System in Baton Rouge, Louisiana. Early on, we reported that the heritability level of cardiorespiratory fitness traits and other exercise-related traits as well as their trainability was typically around 40–50% of the trait variance adjusted for the proper concomitants. Our first attempts at identifying genetic markers associated with human exercise-related phenotypes or traits were quite naïve by today's standards. They were based on polymorphisms in red blood cell proteins and enzymes as well as variations at the HLA loci. Later, skeletal muscle enzyme mass and charge variants were added to our pool of genetic markers. None of these studies yielded significant genetic predictors of performance or fitness traits. As DNA screening technologies became available, we embarked on family based genome-wide linkage screens and tested positional genomic loci and large numbers of candidate genes. There were several assumptions behind this line of research, including that there was a direct relation between a genetic variant and a phenotype, that an allelic variant of interest had a measurable effect size on the phenotype and that the regulation of gene expression was determined by a simple set of elements. However, advances in genomics and in the understanding of the architecture of the human genome and how it is regulated have shown that none of these assumptions were true.

With the advent of gene expression profiling, more potent candidate genes could be investigated for their associations with relevant traits, and the combination of transcriptomics and genomic variants generated much interest. Advances in high-throughput DNA genotyping and sequencing technologies resulted in a paradigm shift for exercise genomics. We have seen a clear trend away from single gene, correlative studies to unbiased, genome-wide mapping approaches with a focus on mechanisms and prediction. Data accumulated thus far support the global conclusion that complex exercise traits and their trainability are influenced by hundreds and likely thousands of genomic loci, with most characterized by very small effect sizes and residing in noncoding regions of the genome. Each person carries up to 5 million DNA variants, including single nucleotide variants, insertion/deletion polymorphisms, splice site disruptions, variants impacting promoter or enhancer sequences, CpG islands, miRNAs and other RNAs encoding sequences, and up to 500,000 rare variants that are generally unique to an individual or the family ancestry of that person. The central dogma of contemporary human genetics is that human beings share more than 90% of the genomic DNA sequence. More than 320 million DNA variants have been uncovered in human genome sequenced to date but only about 10% of these variants are unique to specific populations. Thus, not only do we have a common humanity but we also share the same genome and biological ancestry.

It remains an extraordinary challenge to establish a causal relationship between DNA variants and human variation for a given trait. Even after incorporation of epigenomics, transcriptomics, proteomics and metabolomics to critical information on the genomic DNA (or mtDNA) sequence, and even with the support of the most sophisticated computational biology and bioinformatics tools, the genomic signals may remain diffuse and hard to isolate. There are several potential explanations for this singular challenge. Among them, the fact that most DNA variants individually exert a very small effect, most of the time a fraction of 1% of the variance, is of critical importance. Equally relevant is also that the regulation of gene expression in any given tissue is highly heterogeneous and widely distributed among dozens of molecules and DNA regulatory sites for any given gene. Additionally, there is widespread biological redundancy which, among other potential consequences, can lead to attenuation of large excursions in the regulation of metabolic pathways. The latter favours the maintenance of homeostasis in systems that could otherwise exhibit large and perhaps sudden deviations from normality.

Despite all these challenges, progress is being made. Negative results are important to ponder as they may be truly invalidating the prevailing theory, but they may also simply be a reflection of a deficient model or present-day limitation of technology or instrumentation. The failure of early genetic studies of exercise phenotypes is a good example of these limitations. Large collaborative efforts mobilizing a wide range of complementary resources and skills have the potential to generate substantial advances in our understanding of the genomic and underlying biology of exercise-related phenotypes and their responses to various exercise prescriptions with a view to improve our understanding of their roles in performance and disease prevention. Now is not the time to be discouraged by the magnitude of the task. As scientists, we thrive when faced by challenges, the presence of the unknown or our own ignorance about what is true or not. Insightful ignorance, as suggested by Stuart Firestein (Columbia University), is a powerful driver of scientific progress. Molecular exercise physiologists have an abundance of sagacity and insight to bear on these issues.

The future of exercise genomics looks exciting, but the path forward will not be linear and will be repleted with challenges and obstacles.

Claude Bouchard, Pennington Biomedical Research Center, Baton Rouge, Louisiana, June 2021.

Non-human mammalian models of exercise adaptation kick started our understanding of signal transduction and gene regulatory networks in exercise adaptation in humans

The majority of work within cellular and molecular exercise physiology in humans is undertaken in skeletal muscle tissue due to the increasing availability of human skeletal muscle biopsies. However, non-human mammalian models, predominantly using rodents, have paved the way and continue to provide key mechanistic insights into understanding molecular adaption to exercise in humans. One of the advantages of animal experimentation is the use of gene 'knock-out' (KO), and transgenic ('knock-in' or 'overexpression') models that are unavailable in humans due to ethical and legal reasons (i.e. gene therapy). This type of Nobel Prize-winning work has been available in the laboratory since the 1980s following original pioneering work by Capecchi, Evans and Smithies for the introduction of customized genetic mutations in rodents (59–63). A pertinent example for the use of this technology includes work into the metabolic regulator, PGC1-α (Peroxisome proliferator-activated receptor gamma coactivator 1-alpha), now probably the most common gene/protein to be investigated in response to and following adaptation of skeletal muscle to aerobic exercise (molecular regulators of endurance exercise are discussed extensively in Chapter 9). Briefly, PGC-1α is a co-transcriptional regulator and activates a group of transcription factors, including nuclear respiratory factors 1 (NRF1) and 2 (NRF2), which activate mitochondrial transcription factor A (mtTFA) leading to initiation of the replication and transcription of mitochondrial DNA, and ultimately mitochondrial biogenesis (64, 65), potentially therefore allowing for improvements in oxidative capacity and performance after exercise training. Building on the work of Bruce Spiegelman's team, who were one of the first laboratories to identify PGC-1α as a crucial regulator of skeletal muscle mitochondrial biogenesis (66, 67), using a gene KO model of PGC-1α in rodents, researchers were able to demonstrate that following aerobic exercise or cold water therapy, PGC1-α was fundamental in the promotion of mitochondrial biogenesis enabling increases in oxidative metabolism (66, 67). It was also shown that an increase/overexpression of PGC1-α in rodent tissue also caused type-IIa fibres to become redder and oxidative in phenotype, akin to type-I fibres (68). Baar and Hollozy then went on to show that an acute bout of swimming in rats increased PGC-1α protein levels and NRF-1 binding to the δ-aminolevulinate synthase (δ-ALAS) gene promoter and NRF-2 binding to the cytochrome c-oxidase IV promoter (69), further confirming PGC-1α's molecular role in the signals leading to mitochondrial biogenesis in a rodent exercise model. At a similar time, it was observed that the energy sensor AMP kinase (AMPK), identified by Graham Hardie in 1996, as well as p38 MAPK directly phosphorylated PGC-1α (70) and activation of AMPK was also shown to increase the gene expression of PGC-1α in rat muscle (71). All this earlier work in rodent models paved the way for the first human study in 2003 demonstrating robust PGC-1α increases after aerobic exercise (72). Later, it was also confirmed that exercise performance was impaired following PGC1-α KO in animal models (73). Today, PGC1-α is now probably the most well-studied gene/protein after aerobic exercise interventions, cold water immersion and alterations in nutrients (particularly glycogen) in humans. Despite this, its importance relative to other, more recently discovered signalling events after aerobic exercise has been debated (74), something that is later discussed in Chapter 7. However, it is important to outline that without earlier discoveries in rodent models, the understanding of the molecular responses to aerobic exercise in human muscle would perhaps have been considerably hampered. AMPK/p38 MAPK/PGC-1α signalling as well as other important regulators of endurance exercise responses are discussed extensively in Chapter 9.

Another pertinent example for rodent gene KO and overexpression experiments was in the late 1990s. Chin, within the group of Williams paved the way for understanding regulators of calcium release from the sarcoplasmic reticulum, following muscle contraction and the subsequent regulation of the calcium-camodulin-calcineurin pathway and its control of the transcription factor NFAT that regulated the switching on and off of 'slow' or 'fast' muscle fibre genes (75). They used both gene overexpression of calcineurin in-vitro (in muscle cells in culture) that evoked the expression of slow troponin and myoglobin genes, and pharmacological inhibitors of calcineurin, cyclosporine A in-vivo (in rodent muscle), that increased the number of fast type-II fibres by 50%. Taken together, they suggested that there was a fundamental mechanism whereby increases in calcineurin promoted slow muscle fibre formation and that inhibition promoted fast fibre formation. This was later confirmed in KO mice in 2000 (76). The same year, in the Schiaffino group, it was also determined that the upstream MAPK pathway, notably ERK, was important in this fibre-type transition (77). The fibre-type transition, however, has been somewhat controversial in human exercise studies, something that will be discussed in Chapter 9.

In terms of animal models being fundamental in understanding muscle mass regulation and adaptation to resistance exercise, the double muscling phenotype with a naturally occurring myostatin mutation in cattle (78) was confirmed using myostatin KO mice (79). Myostatin is now considered one of the most important negative regulators of muscle mass and how this signalling pathway is altered with exercise in humans is discussed in detail in Chapter 8. Other models in rodents that are not possible in humans but have led to large strides forward in understanding the molecular regulation of muscle growth (hypertrophy) following exercise include synergistic ablation (also termed compensatory hypertrophy), which after removal of agonist muscles culminates in rapid loading of the antagonist muscles and robust hypertrophy, and serves as a model to investigate the molecular regulation of muscle growth after mechanical loading. This method was first used by Goldberg in 1967, who demonstrated a rapid hypertrophic response even after only 24 hours of synergistic ablation (80). Indeed, these models together with electrical stimulation in rats to mimic high-frequency muscle contractions, in a programme of stimulation designed to mimic resistance exercise in humans, were fundamental in the discoveries that the mammalian target of rapamycin (mTOR) signalling pathway was fundamental in mechanical load-induced muscle growth. Indeed, in the late 1990s, work by Baar and Esser identified that a major intracellular signalling molecule controlling protein synthesis, mTOR, was thought to be load- or 'mechano'-sensitive (via an assessment of mTORs downstream translation initiator, p70s6K). This was observed in rodents following electrical stimulation-evoked contractions (100 Hz) of the triceps surae muscles. Shortly after, Sue Bodine's group observed that load-induced muscle growth following synergistic ablation could be prevented using a chemical mTOR inhibitor, rapamycin, in rodents (81). Eight years later, it was confirmed in human skeletal muscle that mTOR was an important signalling molecule activated following exposure to physiological load via resistance exercise, where 12 mg of the mTOR inhibitor, rapamycin, was administered to subjects prior to performing 11 sets of 10 repetitions leg extension at 70% of the participants 1 repetition maximum (1RM). In this study, the rapamycin group did not show the increase in muscle protein synthesis normally seen following resistance exercise (82), consolidating mTORs role in resistance exercise-induced muscle growth in humans. The mTOR pathway is now one of the most studied pathways in resistance exercise-induced muscle growth. The mTOR pathway, its upstream and downstream regulators as well as other interacting 'cross-talk' pathways important in adaptation to resistance exercise are discussed in detail in Chapter 7 and 8.

MICROARRAY AND RNA-SEQUENCING ANALYSIS FOR GENOME-WIDE GENE EXPRESSION ANALYSIS

The technological and methodological advancements in the assessment of RNA and DNA in the early 2000s meant that Frank Booth's lab (see Box 1.1) in collaboration with Eric Hoffman were the first to use more extensive genome-wide gene transcription/expression techniques, also termed the 'transcriptome', via the use of microarray technology (microarray methods are discussed in Chapter 2). At that time, 'genome-wide' for this technology meant the coverage of around 3,000–4,000 gene transcripts, as well as around 3,000 expressed sequence tags (ESTs), as the method just predated the full publication of the human and mouse genome sequences, and the technology was in its relatively early stages. In this study, they compared the gene expression profiles of approximately 3,000 genes and 3,000 ESTs in both the 'faster' twitch vastus lateralis and the 'mixed-slow' fibres of the soleus muscles (83). They demonstrated that there were 59 differentially regulated genes (59 significantly increased or decreased in their expression level) between the two muscle fibre types. A year later in 2002, Esser's group undertook the first work investigating the transcriptome following muscle contraction. Electrical stimulation protocols in rodents were conducted in regimes that mimicked resistance exercise [as used in the study above (84, 85)], with corresponding gene microarrays being performed. These arrays (Affymetrix Rat U34A GeneChips) included around 7,000 genes and 1,000 ESTs. Importantly, a small subset of 18 genes were identified that were expressed 1 hour after exercise and around 70 genes demonstrated changes after 6 hours of exercise (86). Also published in the same year, a study that performed 9 weeks of resistance exercise in males, females and aged individuals demonstrated that almost 100 genes (from 4,000 profiled in total) were significantly differentially expressed in skeletal muscle across all groups in response to resistance exercise training. Interestingly, there were a large number of genes that demonstrated stark differences in males versus females as well as between young adult and aged individuals (87). By this time, and with gene array technology rapidly advancing, further research by Chen et al. (2003) profiled over 12,000 gene transcripts following an acute bout of resistance exercise in male humans, albeit in only three individuals (88). Each individual performed 300 concentric contractions with one leg and 300 eccentric contractions with the other leg. In this study, 50% of the gene expression changes identified were also seen in their earlier rodent studies following electrical stimulation, discussed above (86), partly validating their earlier animal data in humans. In addition, they identified 6 gene transcripts that were related specifically to eccentric muscle contractions. Due to the fully completed human genome in 2003 (89, 90), and with the latest knowledge of the 26,588 protein-encoding gene transcripts within the human genome, it was soon possible to undertake gene arrays for more than 23,000 gene transcripts. A study in 2007 performed Illumina microarrays, and out of 23,000 genes, nearly 600 genes were differentially turned on/off in the basal state in aged versus young adult skeletal muscle. Interestingly, after six months of resistance exercise in young versus aged adults (men and women), resistance exercise enabled the return of gene expression (for the majority of these genes) back towards levels observed in the young adult group (91). In terms of endurance exercise, in 2007, the first rat transcriptome was profiled following an acute bout of aerobic exercise and suggested that 52 genes were differentially expressed 1 hour post-exercise (92). In 2011, Timmons and Bouchard, using gene microarrays, showed that VO_{2max} responses to chronic endurance exercise (training) could be predicted by a 29 'gene expression signature' or 'training-responsive transcriptome' using baseline/pre-exercise skeletal muscle biopsies. With current microarrays, it is now possible to investigate close to 50,000 gene transcripts (i.e. known protein coding gene transcripts but also their variants). In 2012, using these powerful arrays, an investigation found that 661 genes were affected by both acute and chronic (12 weeks) resistance exercise in young and aged adult women, that was also associated with gains in muscle size and strength. Interestingly, the most pronounced changes in gene expression occurred in fast-twitch (MHC IIa) muscle fibres (93). Most recently, the development of RNA-sequencing (RNA-seq) as a technology is now beginning to surpass the use of microarrays, as it does not require the knowledge of the genome sequence that is constantly updated and reduces some of the issues with 'cross hybridising probes', as well as differences in data normalization in microarray technology. Furthermore, it can identify novel gene transcripts not previously identified. Having said this, microarrays are now good value for money, highly validated and still provide rich and useful data sets. The first mammalian RNA-seq study was conducted in 2008 (94). However, in human exercise studies, comprehensive RNA-seq was not described until 2015 by Hjorth et al. who described that 45 minutes of cycling increased the expression of 550 genes and 12 weeks of training in sedentary men increased 239 genes, where there was a trend towards extracellular matrix genes being increased in larger numbers (95). Lindholm's group in 2016 described RNA-seq data following endurance training, detraining and retraining in humans (96). They identified that there were 3,404 differentially expressed gene isoforms, mainly associated with oxidative ATP production and their study also identified 34 novel transcripts, that may not have been possible with microarrays, all with protein-coding potential (96). In 2017, an RNA-seq study was able to identify 161 and 99 gene transcripts that coded for secretory proteins (dubbed 'myokines' when secreted from muscle) after both acute exercise and chronic training, respectively; in addition, this identified 17 previously uncharacterized 'myokines' in skeletal muscle tissue after exercise (97). In 2018, a study also used RNA-seq to investigate gene expression in skeletal muscle after both endurance and resistance exercise (98), and in 2020 a comprehensive study undertook RNA-seq of muscle in trained (both resistance and endurance exercises) compared with untrained aged-matched males and females. This study demonstrated that exercise training throughout life was important for altering genes associated with reducing the likelihood of metabolic disease (99). One of the emerging areas within RNA-seq is single-cell RNA sequencing (scRNA-seq). While skeletal muscle tissue is predominantly made up of muscle cells, a biopsy can contain several different cell types such as (but not limited to) satellite cells, fibroblasts, endothelial cells, immune cells, neural/glial cells and fibroadipogenic progenitors (FAPs). To address this, there has been the development of scRNA-seq to identify gene expression of the cells that make up muscle tissue (100, 101); however, to our knowledge, this analysis has not yet been performed after exercise. However, these are likely to be performed very soon and will start to provide important insights into the gene regulatory networks that are integrated across the various cell types that make up skeletal muscle tissue. Of course, one of the hurdles is to be able to 'sort' and isolate these different cell populations from a muscle tissue biopsy. Luckily, these technologies using FACS (fluorescent activated cell sorting) technology or antibodies attached to magnetic beads have been developed already even for skeletal muscle; therefore, the combination of sorting cells from biopsies, unique cellular profiles identified and stored in databases, and single cell genome-wide/'OMIC' sequencing analysis will start to lead the way in molecular exercise physiology.

EMERGENCE OF EPIGENETICS IN MOLECULAR EXERCISE PHYSIOLOGY

We now know that the turning on or off of genes which is termed gene expression can be controlled 'epi'-genetically via modifications to DNA itself by methylation or by modifications to proteins (e.g. histones) that encase, wrap and surround DNA. These modifications involve adding or removing small chemical groups such as acetyl or methyl groups. In turn, these modifications allow altered chromatin access for transcription factor binding and therefore promotion or suppression of gene expression. For example, increased or reduced methylation to DNA itself is associated with reduced or increased gene expression, respectively. This is because the presence of methylation on a particular gene can impair the binding of transcription factors required to turn the genes on and thereby supress gene expression, or alternatively it can also cause DNA to be more compacted and inaccessible. This has been shown to be the case especially if the methylation is present in important gene regulatory regions called promoters or enhancers. The opposite occurs if methylation is reduced, with improved transcription factor binding and accessibility to DNA promoting a gene turning on environment. DNA methylation and the regulation of gene expression following exercise have only really been studied since 2009 onwards, as recently reviewed (102–105). Indeed, the first studies to investigate DNA methylation and its control of gene expression in skeletal muscle following exercise were by Juleen Zierath's group who showed reduced DNA methylation of PGC-1α, mitochondrial transcription factor A (TFAM) and pyruvate dehydrogenase lipoamide kinase isozyme 4 (PDK4), all associated with mitochondrial biogenesis, immediately post an acute bout of high intensity aerobic exercise. The modifications on these genes correlated with increases in their corresponding gene expression (106). Importantly, chronic exercise consisting of six months supervised aerobic exercise also altered the methylation of several genes associated with metabolism (107). The studies here were the first in molecular exercise physiology to look at epigenetic modifications, all conducted at the individual gene level. One of the first genome-wide 'OMIC' DNA methylation (or 'methylome') studies to demonstrate differential methylation in skeletal muscle after resistance exercise was conducted in 2018 by Robert Seaborne in the laboratory of Professor Adam Sharples that demonstrated resistance exercise evoked a genome-wide reduction in methylation, also known as a 'hypomethylated' genome signature. From a gene expression point of view, reduced/'hypo'methylated DNA, especially in important gene regulatory regions as described above, is associated with a gene turning on profile. The group also demonstrated that this epigenetic signature could be retained on some genes during detraining when exercise had completely ceased, or further hypomethylated and gene turned on during retraining if resistance exercise had been undertaken in the past (108, 109). Overall, such data suggest that these genes possessed an epigenetic memory of earlier exercise. A so-called 'epi-memory' (102) and its role in exercise adaptation are therefore of considerable interest to molecular exercise physiologists with the potential to enable an understanding for the role of previous exercise encounters, both acute and chronic, on the ability of skeletal muscle to respond to future exercise. This could, for example, enable better periodization of exercise bouts or help recovery from injury. Epigenetic muscle memory is discussed in detail in Chapter 6. Subsequently, it was later demonstrated that the hypomethylated profile was associated with a larger gene turn on profile of genes in specific pathways, such as actin cytoskeletal, extracellular matrix and other growth-related pathways in an integrative methylome/transcriptome study (110). The first genome-wide methylation studies after sprint interval exercise in humans were conducted in 2021 by Masaar et al., demonstrating that exercise evoking higher physiological and metabolic stress (change of direction running exercise versus straight line running) also promoted a hypomethylated signature; yet instead of this being observed in growth-related pathways like after resistance exercise, this occurred in metabolic pathways such as MAPK, AMPK and insulin pathways (111).

Other epigenetic modifications to DNA and chromatin such as histone modifications are also fundamental for allowing access for the transcriptional machinery to control gene expression. One example of a histone modification is acetylation. Exercise has been demonstrated to be associated with lysine acetylation on several different residues of histone proteins in human skeletal muscle, with acetylation generally associated with a compacting effect on the chromatin and therefore inaccessible DNA and reduced gene expression. Indeed, one of the first studies to link lysine acetylation of histone 3 of the Glucose transporter type 4 (GLUT4 – an important insulin responsive glucose transporter, especially in muscle and fat) gene promoter indicated that histones could play a key role in exercise-induced gene expression responses (112). Histone deacetylases (HDACs) and acetyltransferases (HATs) are enzymes that add or remove acetylate groups to histone proteins. Therefore, since these initial studies, HDACs have been established to play an important role in the gene expression response to exercise, where after 60 minutes of cycling, class II HDACs are phosphorylated and move to the nucleus which is associated with acetylation and an increase in gene expression associated with exercise adaptation (113). Also, lactate produced during exercise is an endogenous HDAC inhibitor targeting the histone 4 complex, thereby acetylating histone marks, and thus these modifications are associated with increased gene expression (114). Work thus far has only been conducted on a hand full of HATs for their role in exercise adaptation and therefore only a small role has been demonstrated for these enzymes in exercise (115–117). There are also several families of HATs that are completely unstudied in exercise adaptation and require future investigation. It is also worth noting that histones can be methylated; however, how this affects gene expression depends on which histone and sites are methylated. Generally speaking, methylation of histones such as lysine 4 on histone H3 (H3K4me3) is associated with increased gene expression as is acetylation of numerous lysine residues on histones H3 and H4 (acetyl H3 and acetyl H4), whereas trimethylation of lysines 9 and 27 on histone H3 (H3K9me3 and H3K27me3) and lysine 20 on histone H4 (H4K20me3) is involved in reducing gene expression. For example, after acute resistance exercise, it has been shown that H3 is methylated on both lysine 4 and lysine 27 (118). Only very recent studies have used current genome-wide 'OMIC' technology to assess histone and chromatin accessibility (ChIP-seq and ATAQ-seq, explained in Methods, Chapter 2) to investigate changes in skeletal muscle following exercise. The most notable study using genome-wide chromatin immunoprecipitation (ChIP-seq) demonstrated that aerobic exercise training led to reduce acetylation of histones close to important regulatory regions of genes called enhancers (119). This opens up the possibility that epigenetic modifications to histones are an important regulatory step in exercise responses and adaption. However, as molecular exercise physiologists, the impression is that we have only just started to scratch the surface in identifying these types of epigenetic modifications following exercise.

Finally, additional epigenetic modifications include post-transcriptional modifications to mRNA that can lead to altered protein structure and function. This can occur via small RNA species known as siRNA (small interfering RNA) and miRNAs (micro RNA) (120). Investigations have been conducted in response to both aerobic exercise and resistance exercise and miRNA profiles (121, 122) and miRNA expression can differ between high and low responders to exercise adaptation (123). Work by Timmons' group was really the first, most comprehensive study in this area, as genetic (genomic), gene expression (transcriptomic) and miRNA ('miRomic') profiles were undertaken within the same study. Indeed, they found that a gene network associated with the transcription factors Runx1, Pax3 and Sox9 were associated with adaptation to endurance exercise and that the regulatory miRNA's post-transcriptionally regulated these transcription factors (124). There are still only a few studies that investigate the role miRNAs have on transcribed mRNA sequences, and therefore the protein structure and function of the corresponding translated proteins, in response to exercise.

In the future, it is perhaps the integration of genomic, epigenomic (methylome, histone/chromatin modifications, miRomic) and their relation to genome-wide transcriptome profiles (discussed in 'Microarray and RNA sequencing' section above) that are likely to provide future advances in molecular exercise physiology. Importantly, however, it is of course the resulting protein that has been transcribed by the gene and 'activity' of a protein that formulate the cell signalling response to an exercise stimulus. Therefore, proteomics is also coming to the fore in molecular exercise physiology, as discussed directly below.

IMPORTANCE OF PROTEIN SIGNALLING AND PROTEOMICS IN MOLECULAR EXERCISE PHYSIOLOGY

Following exercise, we create exercise-induced 'signals', these signals come in the form of (but not limited to) changes in energy levels/glycogen, calcium fluxes caused by contraction, hypoxia, mechanical tension/load and extracellular 'signals' such as adrenaline or growth hormone. These types of signals are sensitively detected by molecular 'sensor' proteins within skeletal muscle. For example, AMPK is a molecular sensor of energy fluxes and changes in AMP/ADP. Calmodulin is a molecular sensor of calcium, HIF1-alpha a sensor of oxygen, mTORC a sensor of tension and amino acid availability, and the beta-adrenergic receptor and growth hormone receptor, sensors for adrenaline and growth hormone, respectively. The sensing of these molecular signals leads to 'signal transduction' or 'signalling cascades' of proteins that are inside the cell (intracellular) and typically activated/inactivated by phosphorylation/dephosphorylation. Ultimately, this increase in protein 'activity' leads to the activation of transcription factors (proteins that bind to DNA on gene regions important in turning specific genes on or off) and altered gene expression/transcription. Finally, after the gene is turned on, its messenger molecule (mRNA) can translocate to the ribosomes (the protein factories of the cell) and make new proteins (termed translation/protein synthesis). This can help replace those proteins that are damaged/lost due to exercise or help accrue new proteins for adaptation at the cell and tissue levels, such as those required for muscle growth or mitochondrial biogenesis. In 1992, the discovery of protein phosphorylation by Fischer and Krebs culminated in the award of the Nobel Prize in chemistry. With this discovery, earlier work in gene expression, and the development of Western blotting using phosphorylation specific antibodies, enabled the role of protein phosphorylation after exercise to be determined (125). Western blotting is still routinely used in molecular exercise physiology to measure protein activity (namely phosphorylation) and total abundance. However, due to advances in molecular biology, we also know that proteins within signal transduction networks are not only phosphorylated/dephosphorylated but also acetylated, deacetylated, sumoylayted, ubiquitinated, glycosylated, methylated and demethylated (126). These other processes are currently understudied in relation to the abundance of studies that investigate phosphorylation and its role in controlling protein activity and cell signalling cascades following exercise. With small pockets of the community investigating acetylation/deacetylation such as deacetylation of PGC-1 alpha after exercise (115), there is also the emergence of wider genome-wide or 'proteomic' approaches being taken using various cutting-edge methods (proteomics methods are covered in Chapter 2). Indeed, the first proteomic studies reporting training-induced adaptations in rat (127) and human skeletal muscle (128) were only published 7–8 years ago. In 2009, approximately 2,000 proteins from biopsy samples of human skeletal muscle were identified using mass spectrometry. The majority of those identified were abundant myofibrillar proteins, metabolic enzymes and kinases responsible for phosphorylation (129). This is because, unlike RNA, it is difficult to isolate all proteins as they cannot be amplified like cDNA (product of mRNA following reverse transcription) via PCR, and the proteome is even more diverse than the transcriptome, due to post-translational modifications. This is particularly challenging when the identification of proteins following mass spectrometry is based on assessing the unique 'masses' of a skeletal muscle sample in comparison to masses that are predicted within existing gene and protein databases. However, proteomics is becoming an extremely powerful tool, and doesn't require or rely on the commercial availability of suitable antibodies. Furthermore, proteins never before identified to play a role in exercise adaptation (e.g. PDIA3 – Protein Disulfide Isomerase Family A Member 3 (130)) are beginning to be discovered as regulators of aerobic capacity due to the latest 'mining'/profiling advancements within proteomics as opposed to undertaking Western blotting for known kinases. In more recent years, proteomic studies have provided unique insight into complex networks of post-translational modifications e.g. phosphorylation (131, 132), ubiquitination (133), fibre-type specific adaptation (134) and the contributions of synthesis and degradation to changes in the abundance of individual proteins (135) in exercised skeletal muscle.

THE FUTURE – INTEGRATIVE OMIC INVESTIGATION

Overall, in the future, more extensive integrative 'OMIC' investigations using current epigenomic, transcriptomic (discussed above), proteomic analysis at the tissue and single cell levels, in combination with genetic profiling using the latest technological sequencing advancements, are required in exercising knock-out, overexpression and compensatory hypertrophy rodent models, and importantly in human exercise intervention studies. This will enable the field to delve into the deepest regulatory networks and take molecular exercise physiology into the next generation of research to uncover the mechanistic underpinnings of exercise adaptation.

REFERENCES

1. Booth FW. *J Appl Physiol.* (1985). 1988. 65(4):1461–71.
2. Baldwin KM. *J Appl Physiol.* (1985). 2000. 88(1):332–6.
3. Bergström J. *Scand J Clin Lab Invest.* 1962. 14(suppl 68):1–110.
4. Bergström J, et al. *Nature.* 1966. 210(5033):309–10.
5. Ahlborg B, et al. *Acta Physiologica Scandinavica.* 1967. 70(2):129–42.
6. Hermansen L, et al. *Acta Physiologica Scandinavica.* 1967. 71(2–3):129–39.
7. Bergström J, et al. *Acta Physiologica Scandinavica.* 1967. 71(2–3):140–50.
8. Costill DL, et al. *J Appl Physiol.* 1971. 31(3):353–6.
9. Costill DL, et al. *J Appl Physiol.* 1971. 31(6):834–8.
10. Gollnick PD, et al. *Am J Physiol.* 1961. 201:694–6.
11. Hearn GR, et al. *Internationale Zeitschrift fur angewandte Physiologie, einschliesslich Arbeitsphysiologie.* 1961. 19:23–6.
12. Holloszy JO. *J Biol Chem.* 1967. 242(9):2278–82.
13. Baldwin KM, et al. *Pflugers Arch.* 1975. 354(3):203–12.
14. Holloszy JO, et al. *Annu Rev Physiol.* 1976. 38:273–91.
15. Booth FW, et al. *J Biol Chem.* 1977. 252(2):416–9.
16. Tucker KR, et al. *J Appl Physiol Respir Environ Exerc Physiol.* 1981. 51(1):73–7.
17. Morrison PR, et al. *Am J Physiol.* 1989. 257(5 Pt 1):C936–9.
18. Krieger DA, et al. *J Appl Physiol Respir Environ Exerc Physiol.* 1980. 48(1):23–8.
19. Booth FW. *Pflugers Arch.* 1978. 373(2):175–8.
20. Watson PA, et al. *Am J Physiol.* 1984. 247(1 Pt 1):C39–44.
21. Booth FW, et al. *Fed Proc.* 1985. 44(7):2293–300.
22. Booth FW, et al. Adv Myochem. 1987. 1:205–16.
23. Morrison PR, et al. *Biochem J.* 1987. 241(1):257–63.
24. Booth FW, et al. *J Appl Physiol Respir Environ Exerc Physiol..* 1979. 47(5):974–7.
25. Seider MJ, et al. *Biochem J.* 1980. 188(1):247–54.
26. Wong TS, et al. *J Appl Physiol.* (1985). 1990. 69(5):1718–24.
27. Wong TS, et al. *J Appl Physiol.* (1985). 1990. 69(5):1709–17.
28. Babij P, et al. *Sports Med.* 1988. 5(3):137–43.
29. Booth FW, et al. *Physiol Rev.* 1991. 71(2):541–85.
30. Booth FW. *Exerc Sport Sci Rev.* 1989. 17:1–27.
31. Saiki RK, et al. *Science.* 1985. 230(4732):1350–4.
32. Mullis KB, et al. *Methods Enzymol.* 1987. 155:335–50.
33. Wolff JA, et al. *Science.* 1990. 247(4949 Pt 1):1465–8.
34. Thomason DB, et al. *Am J Physiol.* 1990. 258(3 Pt 1):C578–81.
35. Bouchard C, et al. *Ann Hum Biol.* 1984. 11(4):303–9.

36. Perusse L, et al. *Ann Hum Biol.* 1987. 14(5):425–34.
37. Bouchard C, et al. *Med Sci Sports Exerc.* 1986. 18(6):639–46.
38. Bouchard C, et al. *Med Sci Sports Exerc.* 1989. 21(1):71–7.
39. Chagnon YC, et al. *J Sports Sci.* 1984. 2(2):121–9.
40. Dionne FT, et al. *Med Sci Sports Exerc.* 1991. 23(2):177–85.
41. Deriaz O, et al. *J Clin Invest.* 1994. 93(2):838–43.
42. Bouchard C, et al. *Hum Mol Genet.* 1997. 6(11):1887–9.
43. Bouchard C, et al. *Genetics of Fitness and Physical Performance.* Champaign, IL: Human Kinetics, 1997.
44. Bouchard C, et al. *Med Sci Sports Exerc.* 1998. 30(2):252–8.
45. Bouchard C, et al. *J Appl Physiol.* 1999. 87(3):1003–8.
46. Montgomery H, et al. *Nature.* 1998. 393(6682):221–2.
47. Williams AG, et al. *Nature.* 2000. 403(6770):614.
48. Teran-Garcia M, et al. *Am J Physiol Endocrinol Metab.* 2005. 288(6):E1168–78.
49. Teran-Garcia M, et al. *Diabetologia* 2007. 50(9):1858–66.
50. Klein RJ, et al. *Science.* 2005. 308(5720):385–9.
51. De Moor MH, et al. *Med Sci Sports Exerc.* 2009. 41(10):1887–95.
52. Timmons JA, et al. *J Appl Physiol.* (1985). 2010. 108(6):1487–96.
53. Bouchard C, et al. *J Appl Physiol.* (1985). 2011. 110(5):1160–70.
54. Robbins JM, et al. *JAMA Cardiol.* 2019. 4(7):636–43.
55. Robbins JM, et al. *Nat Metab.* 2021. 3, 786–797.
56. Ghosh S, et al. *J Appl Physiol.* 2019. 126(5):1292–314.
57. Wackerhage H, et al. *J Sports Sci.* 2009. 27(11):1109–16.
58. Webborn N, et al. *Br J Sports Med.* 2015. 49(23):1486–91.
59. Robertson E, et al. *Nature.* 1986. 323(6087):445–8.
60. Evans MJ, et al. *Nature.* 1981. 292(5819):154–6.
61. Martin GR. *Proc Natl Acad Sci U S A.* 1981. 78(12):7634–8.
62. Thomas KR, et al. *Cell.* 1987. 51(3):503–12.
63. Doetschman T, et al. *Nature.* 1987. 330(6148):576–8.
64. Virbasius JV, et al. *Proc Natl Acad Sci U S A.* 1994. 91(4):1309–13.
65. Scarpulla RC. *J Bioenerg Biomembr.* 1997. 29(2):109–19.
66. Puigserver P, et al. *Cell.* 1998. 92(6):829–39.
67. Wu Z, et al. *Cell.* 1999. 98(1):115–24.
68. Lin J, et al. *Nature.* 2002. 418(6899):797–801.
69. Baar K, et al. *FASEB J.* 2002. 16(14):1879–86.
70. Knutti D, et al. *Proc Natl Acad Sci U S A.* 2001. 98(17):9713–8.
71. Suwa M, et al. *J Appl Physiol.* (1985). 2003. 95(3):960–8.
72. Pilegaard H, et al. *J Physiol.* 2003. 546(3):851–8.
73. Handschin C, et al. *J Biol Chem.* 2007. 282(41):30014-21.
74. Islam H, et al. *Appl Physiol Nutr Metab.* 2020. 45(1):11–23.
75. Chin ER, et al. *Genes Dev.* 1998. 12(16):2499–509.
76. Naya FJ, et al. *J Biol Chem.* 2000. 275(7):4545–8.
77. Murgia M, et al. *Nat Cell Biol.* 2000. 2(3):142–7.
78. McPherron AC, et al. *Proc Natl Acad Sci U S A.* 1997. 94(23):12457–61.
79. McPherron AC, et al. *Nature.* 1997. 387(6628):83–90.
80. Goldberg AL. *Am J Physiol.* 1967. 213(5):1193–8.
81. Bodine SC, et al. *Nat Cell Biol.* 2001. 3(11715023):1014–9.
82. Drummond MJ, et al. *J Physiol.* 2009. 587(19188252):1535–46.
83. Campbell WG, et al. *Am J Physiol Cell Physiol.* 2001. 280(4):C763–8.
84. Baar K, et al. *Am J Physiol.* 1999. 276(1 Pt 1):C120–7.
85. MacKenzie MG, et al. *PLoS ONE.* 2013. 8(7):e68743.
86. Chen YW, et al. *J Physiol.* 2002. 545(Pt 1):27–41.
87. Roth SM, et al. *Physiol Genomics.* 2002. 10(3):181–90.

88. Chen YW, et al. *J Appl Physiol*. (1985). 2003. 95(6):2485–94.
89. Venter JC, et al. *Science*. 2001. 291(5507):1304–51.
90. Human Genome Sequencing C. *Nature*. 2004. 431(7011):931–45.
91. Melov S, et al. *PLoS ONE*. 2007. 2(5):e465.
92. McKenzie MJ, et al. *Med Sci Sports Exerc*. 2007. 39(9):1515–21.
93. Raue U, et al. *J Appl Physiol*. (1985). 2012. 112(10):1625–36.
94. Mortazavi A, et al. *Nat Methods*. 2008. 5(7):621–8.
95. Hjorth M, et al. *Physiol Rep*. 2015. 3(8):e12473.
96. Lindholm ME, et al. *PLoS Genet*. 2016. 12(9):e1006294.
97. Pourteymour S, et al. *Mol Metab*. 2017. 6(4):352–65.
98. Dickinson JM, et al. *J Appl Physiol*. 2018. 124(6):1529–40.
99. Chapman MA, et al. *Cell Rep*. 2020. 31(12):107808.
100. Rubenstein AB, et al. *Sci Rep*. 2020. 10(1):229.
101. Petrany MJ, et al. *Nat Commun*. 2020. 11(1):6374.
102. Sharples AP, et al. *Aging Cell*. 2016. 15(4):603–16.
103. Sharples AP, et al. Chapter Ten – Exercise and DNA Methylation in Skeletal Muscle. In: Barh D, Ahmetov II, editors. *Sports, Exercise, and Nutritional Genomics*. Academic Press; 2019. pp. 211–29.
104. Seaborne RA, et al. *Exerc Sport Sci Rev*. 2020. 48(4):188–200.
105. Widmann M, et al. *Sports Med*. 2019. 49(4):509–23.
106. Barres R, et al. *Cell Metab*. 2012. 15(3):405–11.
107. Nitert MD, et al. *Diabetes*. 2012. 61(12):3322–32.
108. Seaborne RA, et al. *Sci Rep Nat*. 2018. 8(1):1898.
109. Seaborne RA, et al. *Sci Data (Nat)*. 2018. 5:180213.
110. Turner DC, et al. *Sci Rep (Nat)*. 2019. 9(1):4251.
111. Maasar MF, et al. *Front Physiol*. 2021. 12:619447.
112. Smith JA, et al. *Am J Physiol Endocrinol Metab*. 2008. 295(3):E698–704.
113. McGee SL, et al. *J Physiol*. 2009. 587(Pt 24):5951–8.
114. Latham T, et al. *Nucleic Acids Res*. 2012. 40(11):4794–803.
115. Philp A, et al. *J Biol Chem*. 2011. 286(35):30561-70.
116. Dent JR, et al. *Mol Metab*. 2017. 6(12):1574–84.
117. LaBarge SA, et al. *FASEB J*. 2016. 30(4):1623–33.
118. Lim C, et al. *PLoS ONE*. 2020. 15(4):e0231321.
119. Ramachandran K, et al. *PLOS Biol*. 2019. 17(10):e3000467.
120. Valencia-Sanchez MA, et al. *Genes Dev*. 2006. 20(5):515–24.
121. Safdar A, et al. *PLoS ONE*. 2009. 4(5):e5610.
122. McCarthy JJ, et al. *J Appl Physiol*. (1985). 2007. 102(1):306–13.
123. Davidsen PK, et al. *J Appl Physiol*. (1985). 2011. 110(2):309–17.
124. Keller P, et al. *J Appl Physiol*. (1985). 2011. 110(1):46–59.
125. Yamaguchi M, et al. *Jpn J Physiol*. 1985. 35(1):21–32.
126. Consortium EP. *Nature*. 2012. 489(7414):57–74.
127. Burniston JG. *Biochim Biophys Acta*. 2008. 1784(7–8):1077–86.
128. Holloway KV, et al. *Proteomics* 2009. 9(22):5155–74.
129. Parker KC, et al. *J Proteome Res*. 2009. 8(7):3265–77.
130. Burniston JG, et al. *Proteomics* 2014. 14(20):2339–44.
131. Potts GK, et al. *J Physiol*. 2017. 595(15):5209–26.
132. Steinert ND, et al. *Cell Rep*. 2021. 34(9):108796.
133. Parker BL, et al. *FASEB J*. 2020. 34(4):5906–16.
134. Deshmukh AS, et al. *Nat Commun*. 2021. 12(1):304.
135. Hesketh SJ, et al. *FASEB J*. 2020. 34(8):10398–417.

CHAPTER TWO

Methods in molecular exercise physiology

Adam P. Sharples, Daniel C. Turner, Stephen Roth, Robert A. Seaborne, Brendan Egan, Mark Viggars, Jonathan C. Jarvis, Daniel J. Owens, Jatin G. Burniston, Piotr P. Gorski, Claire E. Stewart

CHAPTER DISCLAIMER

This chapter is written to support the theory and research discussed throughout this book. Therefore, feel free to dip in and out of this chapter as you come across these methods in later chapters. If molecular exercise physiology is a relatively new topic to you, and you have no experience of the methods used, we also recommend that before reading this chapter, you read **Chapter 3** – Genetics and exercise: an introduction (especially before reading methods to assess DNA and RNA), as well as **Chapter 7** – Signal transduction and exercise (before reading the methods to assess protein activity and abundance).

OVERARCHING LEARNING OBJECTIVES

At the end of this chapter, you should be able to:

1. Understand the potential resource that blood and in particular skeletal muscle provide for studies in molecular exercise physiology.
2. Explain the broad differences between DNA, RNA and protein extraction/isolation methods.
3. Understand the value of the data derived from the molecular and cellular biology techniques used in exercise physiology.
4. Discuss the overall validity of the methods used, in relation to cell and molecular responses and adaptation to exercise.

INTRODUCTION

When investigating the physiological responses to exercise, it is typical of exercise physiologists to harvest blood samples across numerous specific time points after an exercise intervention. The resultant samples provide important information relating to biochemical and endocrine adaptation, which may drive change or be driven by change at the level of peripheral tissues, such as skeletal muscle. Blood

DOI: 10.4324/9781315110752-2

sampling is of particular importance and provides novel data pertaining not only to peptide, hormone and cytokine changes, but also to cell-free ribonucleic acid (cfRNA), miRNA, DNA and variations in exosomes (extracellular vesicles that transport cellular components between cells). As such, serum biomarkers may be predictive of, or driven by the production within skeletal muscle tissue itself. The white blood cell component of whole blood can also be used to harvest DNA for genetic studies in molecular exercise physiology (see 'DNA isolation and genetic analysis' section below). Given that such genetic studies are investigating heritable DNA sequences that are shared across all cell types and are associated with sport/exercise and physiological traits, it is not essential that the DNA is isolated directly from the tissue of interest. However, given the peripheral nature of skeletal muscle, researchers in the field have a unique opportunity to harvest percutaneous (through the skin) muscle biopsies and thereby investigate the impact of exercise directly in this tissue by analysing DNA (genetics and epigenetics), RNA (gene expression), protein levels (e.g. cell signalling or total protein abundance) as well as isolating muscle-derived cells and performing histological analyses. Taken together, blood and muscle biopsy sampling provide a systems-based approach to the investigation of the molecular and adaptative response to exercise.

The capacity to obtain skeletal muscle biopsies enables molecular exercise physiologists to investigate the impact of exercise, nutrition, ageing and disease from a molecular to a cellular and structural level. The planning of the biopsy is key and determining the right time to harvest the muscle biopsy is very important. With the advent of adult human skeletal muscle stem cell isolation (1), the inclusion of such in-vitro (in-a-dish) techniques facilitates the testing of time-course questions that would otherwise be difficult to address via the collection of multiple biopsies over extensive time points due to ethical constraints. Therefore, the present chapter will first cover the methods required for genetic analysis of DNA (inherited DNA sequences) that do not require muscle biopsies. The chapter will then provide an oversight of the journey of the muscle biopsy from the harvesting of tissue itself, to the processing of the biopsy for analyses of DNA, RNA, protein, as well as the isolation of cells and histological analyses of both cells and tissue.

It is worth noting that although there is specific order to this chapter in terms of biopsy taking and the different analyses available to a molecular exercise physiologist, the chapter is also written to enable the reader to dip in and out of the specific or individual methods they wish to learn. Given the nature of the chapter being about methods, and that there are now many techniques that are used by molecular exercise physiologists, the chapter is quite lengthy and not everyone likes to learn about methodology all in one go (we understand that!). Therefore, if the content in this chapter becomes taxing, we recommend reading **Chapters 3 and 7** prior to tackling this methods section, as this will give you additional fundamental knowledge to tackle these methods sections. Alternatively, you may wish to read the specific methods sections as you come across the research that uses them in the later chapters. Finally, the methods described and explained here aim to provide undergraduate and postgraduate students undertaking research projects within molecular exercise physiology with the fundamental theory required to perform these methods.

DNA ISOLATION AND GENETIC ANALYSIS

The basics of DNA isolation and genetic analysis

Genetic analysis does not require a skeletal muscle biopsy. Instead, white blood cells or saliva and/or a cheek swab are used. As we consider the analysis of DNA sequences and the relationship of these sequences with sport- and exercise-related traits, we must address the very basic elements of where DNA comes from and how it is analysed. This section will therefore address the genetic analysis of DNA within a molecular exercise physiology context, including the processes of DNA isolation from blood/saliva/buccal cells, polymerase chain reaction (PCR)/DNA amplification, and analysis of DNA sequence variations using various techniques, including DNA sequencing and microarrays.

DNA isolation from blood or buccal swabs

While **genomic (gDNA)** and **mitochondrial DNA** (mtDNA) are present in nearly all the body's cells (an important exception being red blood cells), DNA collection for the purposes of research is typically limited to buccal epithelial cells (cheek cells) or leukocytes (white blood cells). For studies in which blood drawing is a typical procedure, white cells (or the 'buffy coat') from a typical non-coagulated blood sample may be used for DNA isolation. If blood sampling is not required for a research study, then collection of buccal cells from the inside of the cheek using a cotton-type swab is a less invasive way to collect DNA. These cells are then processed to burst ('lyse') the cells and release the DNA from the cell nuclei, most often using commercially available DNA isolation kits (e.g. DNeasy Blood and Tissue Kit, Qiagen, Manchester, UK), which contain the chemicals needed for cell lysis, protein denaturation (to remove histone and other proteins from the DNA) and purify the remaining DNA. DNA isolation typically requires a couple of hours using the reagents provided in the kits across multiple steps involving centrifugation at high speeds, but many samples can be processed at the same time for efficiency. Once DNA is collected and purified, it can be stored in a sealed tube in a typical refrigerator or freezer for many years; it is one of the most stable biological molecules known. This DNA can then be assessed across a variety of DNA analysis methods typically employed in the field of molecular exercise physiology.

Polymerase chain reaction (PCR) for DNA amplification and genetic analysis

While DNA is relatively easy to collect and isolate, the samples obtained from research volunteers are precious. Planning experiments properly, gaining relevant ethical approvals and recruiting participants are often time-consuming and complex processes. As such, using as little amounts of the DNA as possible is paramount to ensure adequate quantities for future research questions arising from the initial studies (that may perhaps still be subject to new or amended ethical approvals). Moreover, some researchers are interested in only a very small region or regions of the entire human genome for a particular analysis (e.g. studying a certain number of single nucleotide polymorphisms or **SNPs**), with the rest of the genome being unneeded. Both of these issues are addressed by the PCR DNA amplification technique (2), which targets a small DNA region of interest and makes many copies of it. In this way, researchers can take a very small amount of DNA for each research volunteer, identify the region or fragment of DNA they wish to study, then make ample quantities of that small DNA region, which will serve for all later 'downstream' analytical methods.

In brief, a PCR reaction requires a so-called **mastermix**, which comprises of **forward and reverse primers** (≈20 bp short single DNA strands) that bind to the start and end of the DNA fragment selected for amplification, **Taq polymerase**, which synthesises a DNA double strand from a single strand, starting at the point where the primer binds to a DNA single strand, **dNTPs or nucleotides** (i.e. the dTTPs, dATPs, dGTPs, dCTPs that are used to synthesise new DNA), and a chemical **buffer** which includes Mg^{2+} as a co-factor that keeps the pH stable.

The mastermix and the gDNA of each individual being studied are then combined in thin-walled PCR tubes or plates and inserted into a **thermal cycler**, which allows for rapid heating and cooling of the mixtures in the tubes. The 'cycles' of the thermal cycler are repeated cycles of **denaturation**, which causes the DNA double strand to separate; **annealing**, in which the sample is cooled, and the PCR primers bind to the gDNA region of interest; and **extension**, in which the sample is warmed again to a temperature at which Taq polymerase extends the DNA double strand at the point that the primers have annealed to the gDNA. The PCR cycles are repeated until a sufficient amount of gDNA is made, which is commonly over >25 cycles (see **Figure 2.1**).

Analysis of DNA sequence variation

PCR products, which are double-stranded DNA (dsDNA), are invisible to the naked eye and thus require specific methods to enable visualisation. There are a variety of ways of visualising PCR fragments,

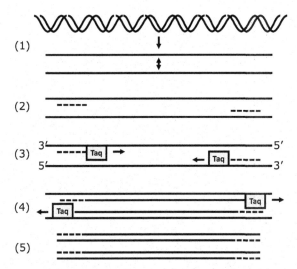

Figure 2.1 Polymerase chain reaction. (1) Denaturing or melting of a double-stranded DNA (dsDNA) into single strands. (2) Binding of primers (dotted lines) to single DNA strands. (3) Taq polymerase recognises the bound primer sites and extends or synthesises the double strand. (4) Taq polymerase continues to synthesise DNA with the addition of dNTPs until the end of the single strand is reached or the temperature is changed. (5) Two dsDNA molecules are synthesised after the PCR cycle is complete. The number of dsDNA molecules doubles after each cycle. By 25–35 cycles, there are millions of copies of the PCR product.

including **DNA fragment separation and visualisation**, in which DNA fragments are separated by size on an **agarose gel** followed by **ethidium bromide** binding and **UV light visualisation**; **restriction enzyme digestion and visualisation**, in which PCR products are cut with a **restriction enzyme** (that recognise a specific sequence) and then separated on an agarose gel followed by ethidium bromide-UV light visualisation; **fluorescent labelling of alleles**, in which fluorophores (fluorescent primers or hybridisation probes) are added to the PCR reaction and the thermal cycler is designed to analyse the colour directly at the end of the PCR reaction, with different DNA variants producing different colours that correspond to a given genotype; and **DNA sequencing**, in which the PCR fragments are directly sequenced using a **DNA sequencer** that reveals all of the bases within the whole DNA/PCR product.

DNA separation and visualisation can be used if the size of the PCR product varies between genotypes. For example, in the angiotensin converting enzyme (ACE) gene (where an insertion/deletion or I/D polymorphism has been identified as a putative endurance exercise-related polymorphism, covered in **Chapter 5**), there can be either a 287-bp-long insertion (I) or this fragment can be missing (the so-called deletion or D allele) (3). PCR primers can be designed to amplify this DNA region, producing PCR fragments of differing sizes depending on the presence or absence of the insertion. Using agarose gel separation, the insertion allele will appear longer than the deletion allele, and genotypes (i.e. II, ID and DD) can be distinguished on the gel using ethidium bromide and UV light visualisation. Similarly, restriction enzymes can be used to cut DNA at the so-called restriction sites which are short, specific DNA sequences. These enzymes can be used to cut a PCR product into fragments if a certain allele is present, allowing the different-sized fragments to be visualised on an agarose gel and genotypes visually determined. For example, PCR followed by *Dde*I restriction enzyme digestion is used to identify the ACTN3 R577X genotype (**Figure 2.2**), a genotype which is associated with speed and power performance (4), and discussed in detail in **Chapter 4**. Such a genotype is known as a **restriction fragment length polymorphism** or **RFLP**. These agarose gel techniques are rapidly giving way to the

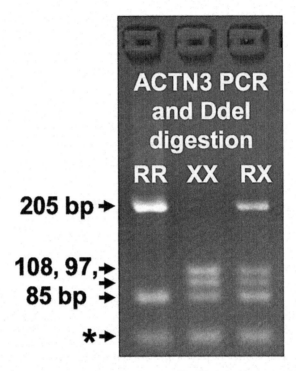

Figure 2.2 Example of a result of an *ACTN3* R577X PCR genotyping assay, a genotype associated with speed and power performance (4). The 290 bp PCR product, which has been amplified by PCR, has been cut with DdeI either once into 205 bp and 85 bp fragments (ACTN3 577 RR) or twice into 108 bp, 97 bp and 85 bp fragments (ACTN3 577 XX). '*' refers to primer-dimers which are an artefact.

more modern fluorescent techniques, which eliminate the need for the agarose gel step and simplify the entire procedure, though at a higher cost in thermal cycling equipment and supplies.

Increasingly common are direct **DNA sequencing** approaches, in which either a DNA fragment or the entire genome of an individual is completely sequenced, base by base. DNA sequencing relies on a similar fluorescent technique as mentioned above, except that each DNA base (A, T, G, C) has a unique colour and the DNA fragment (or entire genome) is amplified by a PCR technique that results in many fragments of varying sizes. These different-sized fragments are then ordered by size using electrophoresis, after which computational analysis determines the base sequence. For the entire genome, this process is done in parallel across thousands of fragments, which are then spliced together using a computer analysis of the resulting sequences. The latter process, in which the entire genome is analysed simultaneously in large fragments, is known as **next generation sequencing (NGS)** (5) and the cost has dropped to such an extent that companies are now advertising the direct sequencing of a person's entire genome for costs (~$1,000) that were considered unthinkable only ten years ago. NGS is covered in more detail in this chapter below.

As discussed in **Chapter 1**, NGS is now becoming much more common, yet thus far, is not as widely used in exercise genetics due to the cost and the requirement of large sample sizes. Indeed, NGS is perhaps being more frequently used for the analysis of RNA/gene expression analysis e.g. RNA-sequencing (RNA-seq) where lower sample number is feasible, as discussed below in this chapter. Therefore, in exercise genetics, researchers have been typically focused on a smaller number of specific alleles of interest, perhaps numbering in the thousands. Researchers also often rely on **microarray** technology to determine a large number of alleles in a single individual at one time. This is often seen in **genome-wide association studies** (GWAS) in which researchers use SNP chips (microarrays) that measure hundreds of thousands of SNPs all over the human genome and look for associations with the trait of interest (6). Whether using a targeted microarray of many alleles specifically chosen by researchers or using commercially available chips with thousands of alleles across the genome to allow more unbiased exploratory analyses, the microarray technique allows for a small amount of gDNA that provides a considerable amount of information in a single assay. The greater task with microarray analysis (or sequencing) is the bioinformatics and statistical procedures required later to properly analyse the data. DNA microarrays are also covered in more detail below in sections for DNA methylation/methylome analysis and RNA/gene expression/transcriptome analysis.

BIOPSY AND SAMPLE PROCESSING FOR DOWNSTREAM ANALYSIS (DNA, RNA, PROTEIN ISOLATION, IMMUNOHISTOLOGY AND CELL ISOLATION)

Introduction to biopsy sampling and processing

The development of the trocar biopsy needle was published by Duchene in 1868 (7). Almost 100 years later, Bergström reported the use of a modified needle for the analyses of skeletal muscle biopsies (8, 9). Fifteen years later, Evans et al. introduced the application of suction to the Bergström needle in order to increase biopsy yield (10). In parallel with the development of the Bergström needle, researchers in Scandinavia were reporting the use of the conchotome for obtaining skeletal muscle tissue (11, 12). The latter provides even greater yield and retains the structural integrity of the sample harvested, which can sometimes be lost due to suction, thereby impacting negatively on some types of histological analyses. The chosen biopsy technique must therefore depend on the research question being addressed and the resultant sample size needed. For example, if only DNA and RNA analyses are to be performed, then needle core biopsy (using commercially available, disposable needle core guns) will provide a sufficient yield of tissue, whereas if histological analyses and cell isolations are required, then conchotome biopsies with larger yields may be the preferred approach (**Figure 2.3**).

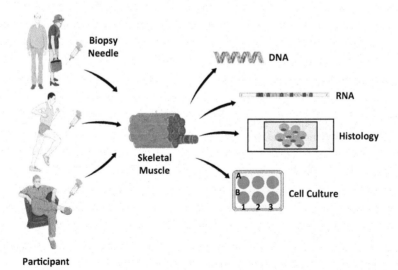

Figure 2.3 The potential of a skeletal muscle biopsy for further cellular and molecular analyses. Created using smart.servier.com. No permissions required. Attribution 3.0 Unported (CC BY 3.0) licence, https://creativecommons.org/licenses/by/3.0/.

Biopsy procedure

Following ethical approval and informed consent, strict aseptic technique and confirmation that the participant is not allergic to the local anaesthetic being applied, an appropriately trained scientist or medical doctor is able to perform the biopsy procedure. In brief, the biopsy site (that is appropriately selected according to the research question such as the assessment of glycogen utilisation of the quadriceps muscle after cycling or protein synthesis after knee extensor exercise) is shaved, sterilised and anaesthetised, prior to a small incision being made through the skin and fascia with a sterile scalpel. The needle/conchotome of choice is inserted through the incision site and a small muscle biopsy (from ~20 mg to 350 mg of tissue depending on the method used) is retrieved. Pressure is applied to the wound site to stop any bleeding, prior to cleaning with surgical scrub, closure with steri-strips and dressing with waterproof dressings. For a more detailed step-by-step protocol for muscle biopsy procedures, see methodological chapter by Turner et al. (13). The procedure from start to finish takes approximately 30 minutes, the majority of which is required for the preparation of the biopsy being harvested and the closure of the wound post-harvest. The sampling itself takes seconds to minutes, depending on whether a single or multiple biopsies (known as 'passes') are harvested. If more than one biopsy is taken from the same site within a short time span (e.g. immediately, 3 hours and 24 hours after exercise), careful consideration regarding the potential effects of the initial biopsy on subsequent biopsy samples is important. Obtaining muscle tissue from alternate legs (if both legs underwent the same intervention, yet perhaps with a confounding variable of leg dominance) or making new incisions in the same limb ~3–5 cm apart are potential strategies for harvesting the required yield of tissue for different timepoints. Of further importance is to carefully consider the impact of the biopsy procedure on the experimental variables of interest, together with the paramount concern for the health and safety of the human participants.

Once isolated, the muscle biopsy should immediately be transferred to sterile containers (e.g. nuclease-free Eppendorf tubes or specimen pots) on ice, under sterile conditions and prepared for storage or extraction procedures (e.g. DNA, RNA, protein, histology; see **Figure 2.3**). If the sample is to be used for multiple analyses, then this should be considered in advance to enable appropriate preparation made to ensure the stability and validity of all samples harvested. Furthermore, it might be necessary to have more than one person processing the biopsy as it is harvested to ensure the integrity of not only the sample, but also the data obtained in the later analyses.

Following removal of the biopsy, the sample is placed into a sterile petri dish on ice and carefully cleaned and dissected to remove any fat or fibrous tissue using a sterile scalpel prior to storage and processing. If cells are to be isolated in addition to other analyses, including histology, then it may be necessary to obtain multiple biopsy 'passes' from the same site to facilitate appropriate processing for each methodology.

DNA isolation from a muscle Biopsy

In instances where DNA is isolated from a muscle biopsy, samples are immediately snap frozen in liquid nitrogen prior to storage at −80°C for later analyses. Samples can also be processed and stored frozen in special commercially available solutions that help maintain the integrity of the nucleic acids. If only conducting DNA analysis, this may not be necessary as it is fairly stable. However, if also isolating RNA (described below) from the same sample, this might be an important consideration as RNA is more likely to degrade with storage and repeated freeze-thaw cycles. DNA can then be extracted from the stored frozen samples by first placing the tissue in lysis buffers from commercially available column-based kits (e.g. DNeasy Blood and Tissue Kit, Qiagen, Manchester, UK) and homogenised using mechanical means such as handheld homogenisers (e.g. TissueRuptor, Qiagen, Manchester, UK) allowing single sample preparation, or homogenising machines (e.g. MagNA Lyser, Roche, Germany), allowing multiple samples to be homogenised at one time. This technique uses small sterile ceramic beads contained within the sample tube with the lysis buffers, which are then mechanically 'shaken' to efficiently break down the tissue and lyse the cells. The DNA is then isolated in the same way as described above for white blood or buccal samples in a process of protein denaturation and DNA purification. Samples are then analysed for yield and quality. Such measurements are commonly performed using **spectrophotometry** which relies on the absorption of UV light at specific wavelengths within the sample. Small volumes (~1 μl) of DNA sample are pipetted onto a spectrophotometer (e.g. Nanodrop, ThermoFisher Scientific) probe which measures the amount of UV light absorbed at 260 nM (the optimal wavelength for DNA) and are quantified by a photodetector to determine the concentration of DNA via the **Beer–Lambert law**. The purity or 'quality' of DNA is indirectly quantified from the absorbance ratio at 260 nM (optimal for nucleic acid absorption) to 280 nM (the optimal wavelength for protein absorption) and 230 nM (which detects other potential chemical impurities because of inadequate extraction/isolation procedures). An A_{260}/A_{280} ratio of 1.8–2.0 is optimal for an indication of 'pure' DNA. One of the issues with **spectrophotometry** for DNA quantification is that RNA also absorbs light at a similar wavelength. Therefore, this technique does not detect RNA contamination within a DNA sample. Samples are often therefore treated with RNase enzymes to degrade any RNA contaminants within the DNA sample. Commercially available column-based kits often include RNase enzymes within the extraction protocol. Another more direct assessment of DNA is via **fluorometry** that relies on the use of fluorescent dyes that intercalate with DNA only (not RNA) and can more accurately measure DNA quantities compared with UV spectrophotometry, especially when total DNA yields are low. However, this method is more expensive due to the requirement of DNA standards and fluorescent reagents. Moreover, this method is also unable to detect potential protein or chemical contaminants within the sample that UV spectrophotometry can indirectly determine using the appropriate absorbance readings described above.

RNA isolation from a muscle biopsy

When wishing to isolate RNA, samples stored at −80°C (or in buffers that maintain RNA integrity) are immersed in Tri-Reagent (also termed TRIzol) or in the lysis buffers provided in commercially available kits. The Tri-Reagent method is popular, as it is relatively inexpensive (compared with commercially available kits). Tri-Reagent is a monophasic solution containing two chemicals (**phenol and guanidine isothiocyanate**) which maintains the 'quality' of the RNA by inhibiting the activity of RNA-specific degrading enzymes, RNases (enzymes that degrade RNA), while disrupting and dissolving cellular components during the lysis and homogenisation steps. The addition of the third chemical, **chloroform,** and centrifugation allow the RNA to become soluble in the upper 'aqueous phase' in a process called phase separation. Careful removal of this aqueous phase followed by the addition of **isopropanol** precipitates the RNA out of the solution, and after centrifugation, the resultant RNA pellet is washed with 75% ethanol to remove any remaining chemicals from the previous steps. RNA pellets are resuspended in specific RNA storage solutions or TE (Tris-EDTA) buffers to help maintain the integrity of the much less stable RNA. Finally, the quantity and quality of the RNA are then analysed in the same way as DNA described, although the optimal A_{260}/A_{280} nM ratio for RNA being 2.0–2.2 (compared with 1.8–2 for DNA). Given DNA absorbs light at a similar wavelength to RNA, the UV spectrophotometry (e.g. Nanodrop, ThermoFisher Scientific) method does not identify DNA contamination within the RNA sample. Therefore, samples isolated by the Tri-Reagent method are often treated with DNase enzymes to degrade any DNA contaminants within the RNA sample. Often, commercially available column-based kits for RNA isolation contain DNase enzymes within the extraction protocol and therefore do not need to be repeated when quantifying RNA quantity and quality. Finally, as with DNA described above, fluorometry can also be used for RNA quantification using dyes that only bind to RNA to give an accurate concentration of the RNA material in a given sample.

Protein isolation from a muscle biopsy

For protein isolation, samples stored at −80°C are first pulverised in liquid nitrogen (LN_2) prior to homogenisation on ice, using handheld or machine homogenisers (described above) in relevant lysis buffer (depending on the analyses to be performed e.g. Western blotting and proteomics as described later in this chapter), typically contain protease and phosphatase inhibitors, that prevent proteins from degrading and halt any continued phosphorylation. Preparation on ice is also important (in combination with phosphatase inhibitors), particularly if phospho-protein analyses are to be performed. Following homogenisation, samples are centrifuged at, e.g., 1,000 g at 4°C for 5 minutes. Soluble proteins are contained within the supernatant and non-soluble (e.g. myofibrillar proteins) in the ensuing pellet. The pellet can be resuspended in relevant assay buffer and protein concentrations determined by specific protein assays such as the bicinchoninic acid (BCA) or Bradford protein assay, prior to aliquoting for long-term storage or for further analyses. As an example, the BCA protein assay detects Cu^{+1}, and is created when Cu^{+2} is reduced by protein in an alkaline environment (14). When the reagent supplied in the BCA assay is mixed with the protein sample, two molecules of BCA join with one cuprous (Cu^{+1}) molecule and the reaction becomes purple-coloured. This purple colour is due to the macromolecular structure of protein, the number of protein bonds and the presence of the four amino acids: Cysteine, Cystine, Tryptophan and Tyrosine (15). The darker the colour, the more protein present. The absorbance of this complex is then measured using a spectrometer between 540 and 590 nM which is linear with increasing protein concentrations.

Histology and sample preparation from a muscle biopsy

Although many histological techniques do not require samples to be stored on ice, if any phospho-protein analyses are to be performed, then it is suggested that samples are maintained on ice. Samples are flash frozen in isopentane, and pre-cooled in LN_2, with rapid agitation, to prevent ice crystal formation. Samples can be mounted onto cork blocks (with muscle fibres perpendicular to the blocks) for sectioning via a cryostat or mounted in paraffin for microtome sectioning. Both methods allow fine, sequential sections of ~5–10 μm to be harvested. Samples can be mounted on glass slides, fixed and subjected to colorimetric (e.g. haemotoxylin and eosin/H&E staining for visualising nuclei and cytoplasmic components, respectively), histochemical (e.g. ATPase stain allowing enzymatic conversion of substrate to colorimetric output) or immunohistochemical (using relevant primary and secondary antibodies) analyses using relevant microscopy techniques. For more details, see immunohistology/immunocytology section below.

Cell isolation from a muscle tissue biopsy

Following muscle biopsy procedures, muscle tissue is first transported to a Class II tissue culture hood in pre-cooled (4°C) cell culture medium. Samples are then washed several times in sterile phosphate buffered saline (PBS) and any connective tissue is removed prior to mincing with forceps, scalpel and/or scissors in trypsin-EDTA. The trypsin enzymatically digests proteins, whereas EDTA removes calcium and magnesium from cell surfaces that then allows the trypsin to digest cell to cell adhesion molecules and disassociate cells from the tissue. The resulting minced/digested tissue 'slurry' is incubated and mixed at 37°C for 10 minutes, and then the trypsin is neutralised using pre-heated (37°C) culture serum. The remaining tissue is subjected to two further rounds of digestion, mincing and neutralising. The solution is then centrifuged (e.g. 340 g for 5 minutes at room temperature), and the resultant pellet is resuspended in fresh pre-heated (37°C) growth medium and cells are transferred to pre-gelatinised (to help cells adhere to the surface) cell culture flasks and cultured at 37°C in 5% CO_2 with regular media top-ups, after being left undisturbed for ~1 week, until ~80% confluency is achieved (for further details, see below Section 'Monolayer cell culture'). Cells can be expanded for freezing or used for immediate experimentation such as examining cell growth, migration, fusion, differentiation and/or cell death in the absence or presence of relevant environmental treatments and the cells/myotubes lysed for output measures, including DNA, mRNA and protein analyses or 'fixed' for histochemical analyses.

EPIGENETIC ANALYSIS OF DNA

LEARNING OBJECTIVES

At the end of this section, you should be able to:

1. Describe the different methodologies required to investigate DNA methylation.
2. Explain the key differences between these methods.

3. Discuss and critique the different methods for addressing research-specific questions.
4. Describe the methods required to analyse histone and chromatin modifications.
5. Explain the advantages and disadvantages of using ChIP-sequencing for histone modification analysis.
6. Explain the similarities and key differences between CUT&RUN and CUT&Tag protocols when investigating histone modifications.

DNA METHYLATION

Introduction to DNA methylation

Exercise can 'epigenetically' modify our inherited genetic code and lead to altered molecular responses involved in exercise adaptation, something that will be covered in much more detail in **Chapter 6**. While several known epigenetic modifications of DNA exist, the most coveted of these is DNA methylation. Changes in DNA methylation involve the addition or removal of methyl molecule from the 5th carbon position of a cytosine residue. The purpose of this methods section is to therefore provide an overview of the most prominent methodologies (single DNA base pair resolution to genome-wide level analyses) that are currently used to study DNA methylation and the applicability of these methods in the field of molecular exercise physiology. The choice of method employed is dependent on the specific research question, sample size and cost-effectiveness.

Bisulphite treatment of genomic DNA

Following isolation of DNA from the tissue or cells of interest, a fundamental process for DNA methylation analysis is the treatment of DNA with sodium bisulphite. This treatment converts unmethylated cytosine nucleotides to uracil. Following PCR amplification (discussed above), the uracil nucleotide is then converted into thymine. Cytosines that are methylated, however, are protected during this process, and therefore do not undergo any changes (see **Figure 6.1, Chapter 6**). This process therefore enables methylated and unmethylated cytosine residues to be easily distinguished during subsequent 'downstream' analysis, described below, and shown in **Figure 2.4**.

Genome-wide methylation sequencing

With the rapid development and declining costs of DNA sequencing, there has been a surge in methodologies that utilise the power of sequencing. None more so than **W**hole **G**enome **B**isulphite **S**equencing (**WGBS**). What distinguishes WGBS from regular DNA sequencing is that the DNA is first treated with sodium bisulphite, described above. Depending on when this treatment is performed, protocols will differ slightly, but generally exist in a pattern similar to the following:

1. DNA is first fragmented by sonication – the DNA is cut up using ultrasonic sound energy, to create smaller DNA molecules, usually around 200–500 bp in length.
2. Fragmented DNA is then base-end repaired – the 5' and 3' ends of the DNA fragments are removed of any unwanted overhanging bases and a single 'A' is tagged on the end.
3. Addition of adapters – special adapters are then joined ('ligated') to the ends of the repaired DNA fragments. These adapters are a long nucleotide sequence that enables the DNA fragment to be PCR-amplified, together with allowing the incorporation of unique index/barcode sequences and the fragments to adhere to the sequencing flowcell (all of these are described below).
4. Adapter-ligated DNA is cleaned up and sequencing indexes/barcodes are added, and the DNA is amplified via PCR. The adapter-ligated DNA is then mixed with sequence-specific 'indexes'. Indexes are usually a 6- or 8-bp-long nucleotide sequences that are attached to the DNA fragment, in a sample-specific manner. These indexes enable researchers to mix multiple biological samples together (e.g. baseline control and post-exercise conditions, reducing sequencing costs), while still

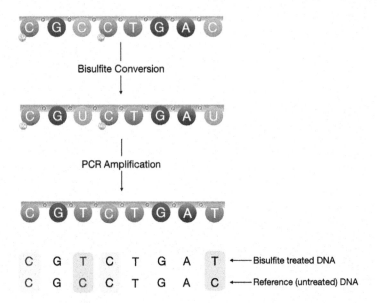

Figure 2.4 Simplified schematic representation of bisulphite conversion workflow. Non-methylated cytosine nucleotides (light green) are converted to uracil following bisulphite treatment which later become thymine (blue) upon amplification via PCR. Conversely, methylated cytosine residues (dark green) remain unchanged upon bisulphite treatment, making it possible to distinguish methylated and unmethylated cytosines during subsequent analysis.

being able to distinguish each sample by their unique sequencing index. The DNA sample is then amplified via PCR to allow these indexes to be incorporated and to increase the yield of the sample.

5. The resulting DNA samples are commonly referred to as DNA libraries and are then ready to be sequenced and undergo various quality control measures before being analysed for biological meaning.

Covering ~28 million CpG sites in the human genome, WGBS is the most comprehensive DNA methylation methodology currently available (16). However, a lot of sequencing data is required for accurate and reliable results which may impose complications (such as cost) if one wishes to investigate numerous samples, from several conditions, across multiple time points. In a bid to make sequencing more widely accessible, other methodologies have been developed.

Reduced representative sequencing

Reduced **R**epresentative **B**isulphite **S**equencing **(RRBS)** offers one of the most cost-effective methods for investigating genome-wide methylation levels (17, 18). RRBS protocols generally follow that of WGBS, but with one crucial difference. Unlike WGBS, DNA used during RRBS is not randomly fragmented. It is instead 'cut' using restriction enzymes that recognise specific DNA sequences. For example, the restriction enzyme, *MspI*, the most commonly utilised enzyme for RRBS experiments, recognises and cuts the DNA sequence CCGG (see **Figure 2.5**).

RRBS has proven very effective both from a biological and financial perspective. It is estimated that this method only sequences ~1% of the genome but covers ~12% of all CpG sites and >80% of all known methylation sites within gene promoter regions (18, 19). It is important to consider, however, that due to

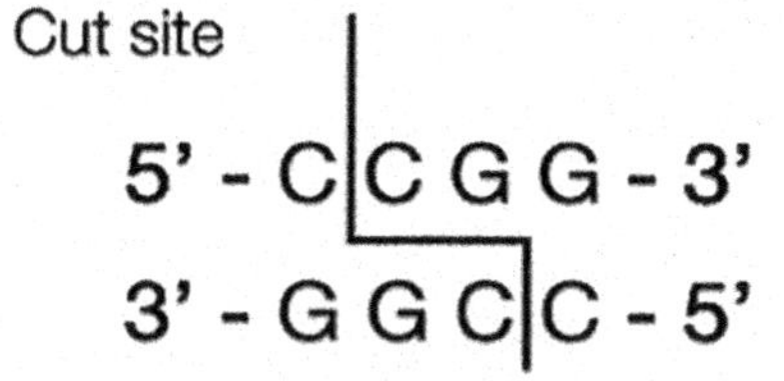

Cut site

$$5' - C | C G G - 3'$$
$$3' - G G C | C - 5'$$

Figure 2.5 This cut in the DNA creates fragments that have a cytosine residue, and a CpG site, at both ends. As CpG islands are heavily enriched in promoter and enhancer regions of genes, this enzyme will create small DNA fragments that are derived from these interesting regulatory regions. By filtering out the larger fragments, it is possible to create a DNA pool that is heavily enriched for fragments derived from these interesting areas of the genome (i.e. CpG sites within promoter/enhancer and regions).

the nature of this protocol, it is not applicable for experiments that wish to explore areas outside of these regions (e.g. intergenic regions).

Sequencing and computational workflow

Once created, DNA sequencing libraries usually undergo a sequencing process known as 'sequencing by synthesis'. First, the DNA library sample is transferred to a flowcell which is placed in the sequencing platform. The single-stranded DNA (ssDNA) fragments then adhere to the flowcell via hybridisation of the adapter regions (as described directly above). The attached DNA strands are then amplified to create 'clusters' of the exact same ssDNA molecules – these represent the template strand. Sequencing by synthesis is then performed by the creation of a copy strand of DNA, where fluorescently labelled nucleotides of different colours are added in a stepwise manner, each giving off a different coloured fluorescence signal depending on the nucleotide (e.g. the four nucleotides have four different colours). Depending on the length of the DNA libraries, this process is repeated until the entire DNA length has been sequenced. At the end of the sequencing process, the original DNA fragment will now be computationally identified in a series of nucleotide sequences. These computational sequences are commonly referred to as sequencing reads.

DNA sequencing reads are then separated by the unique index used during library preparation, to ascertain which data (or DNA reads) is derived from which experimental sample. Once collated, these DNA reads are then ready to undergo computational pipelines, which will commonly follow processes outlined below:

1. Initial quality control of the sequencing reads to ensure that the sequencing process was performed adequately.
2. Trimming of sequencing reads to remove misleading, biased or poor-quality parts of each DNA sequencing read, so that only a region of high-quality reliable data is left on each sequencing read.
3. Alignment to the genome of interest (e.g. humans/*homo sapiens*, mouse/*mus musculus*).
4. Followed by experiment-specific analyses. This may include the identification of differentially methylated regions (DMRs) and differentially methylated CpG sites (DMCpGs), also known as differentially methylation positions (DMPs) or loci (DML), between two different conditions and/or across different time points.

There are multiple factors that dictate the requirements for a sequencing experiment. These factors, such as the number of experimental samples, the sample source (e.g. origin of tissue and organism) and the type of analyses to be performed, need to be carefully considered during the planning stages of an experiment as they will dictate the reliability and validity of the dataset and subsequent analyses (20, 21). Both WGBS and RRBS methodologies rely on quality sequencing libraries and computational expertise in order to analyse the data. Use of these methods, especially WGBS, has the capacity to detect genomic regions that are currently unknown or less characterised which is more important if one wishes to investigate undiscovered DNA sites and genes.

DNA methylation microarrays

In contrast to WGBS and RRBS methods, microarrays rely entirely on using known sequences of DNA. Methylation-based microarray chips work by having two versions (one methylated and one unmethylated) for every CpG target site on the chip, in a complimentary fashion to the input DNA. Bisulphite-converted DNA from an experimental sample (e.g. a pre-exercise muscle biopsy) will be complimentary either to the methylated or the unmethylated strand that is on the microarray chip, giving off a specific fluorescence signal depending on which one it binds to. This process happens with all bisulphite-converted DNA fragments in the experimental sample. A CpG site-specific beta (β) methylation value (a number between 0 and 1) is then calculated depending on the number of fragments that are methylated or unmethylated. It is via this process that the methylation profile of hundreds of thousands of individual CpG sites can be simultaneously characterised.

In contrast to the sequencing workflows, the output of methylation microarray platforms is already quantified (e.g. each sample, for each CpG site, has a quantified β-methylation value). This makes data analysis more computationally accessible and easier to perform.

Microarray technology is more user-friendly, time-efficient and less computationally demanding than WGBS analysis. Furthermore, technology of this kind continues to improve, with recent methylation EPIC arrays (Illumina, USA) now able to analyse >850,000 CpG sites of the human genome (22). This includes excellent and dense coverage of gene regions, including CpG islands, RefSeq genes, ENCODE transcription factor binding sites and FANTOM5 enhancers. Furthermore, the relatively new array technology covers 90% of the previous models sites (HumanMethylation 450K BeadChip), but with an added 350,000 CpG sites of regions identified as potential enhancers by FANTOM5 and ENCODE. Collectively, the array provides one of the most comprehensive but discernible explorations of known CpG sites in the human genome. The major drawback of methylation arrays, however, is the reliance of known sequences to create the array chip, which means that they are designed to be species-specific. For example, while the human 850K EPIC array does contain 'probes' for sequences that are also present in the rat and mouse genome, the number of CpGs that can be reliably investigated in rat or mouse samples using the human arrays is significantly reduced (from >850,000 to 20–30,000 CpGs). However, different arrays for different species have started to emerge to circumvent this issue. Moreover, arrays are unable to identify unknown or less characterised regions of DNA (e.g. intergenic regions, ribosomal or mtDNA). A deeper comparison of array and sequencing-based methodologies for DNA methylation analysis can be found elsewhere (23).

Loci-specific methylation sequencing

For targeted region-specific DNA methylation analysis, there are several 'sequencing by synthesis' methods. The majority of which require bisulphite treated DNA, amplicon-specific forward and reverse primers, PCR to amplify the region of interest and electrochemical or light-detective methods to provide a 'read out' for the data. These methods can be viewed, in part, as a minimised version of sequencing technologies mentioned above. For example, loci-specific **pyrosequencing** utilises bisulphite-converted DNA and special biotin-labelled primer sets to amplify a DNA region of interest (typically no greater than 350 bp). Through base-by-base addition of nucleotides, a resultant light signal (that is specifically proportional to the number of nucleotides added to the DNA template) enables identification of methylation levels of a specific genomic region (24). More loci-specific methodologies, such as Bis-PCR sequencing (25) and targeted bisulphite sequencing, rely on amplifying (via PCR) the region of interest, creating sequencing libraries from these amplified regions before performing NGS.

ANALYSIS OF HISTONE MODIFICATIONS AND CHROMATIN ACCESSIBILITY

Introduction to histone and chromatin analysis

Epigenetic modifications can also occur on the proteins that tightly bind DNA into its compact structure. These proteins, known as histones, have amino acid 'tails' that protrude from them which are susceptible to modifications such as methylation, ubiquitination and acetylation to name a few. Depending on the type of modification, the specific histone protein and the position along the tail, these alterations can lead to changes in the function of DNA. Furthermore, a higher level of regulation also occurs to our chromatin, the structure that encompasses the DNA-bound histone complex. For example, some histone modifications can attract other molecules that either tighten ('heterochromatin') or loosen ('euchromatin') the compact nature of the surrounding chromatin, affecting the accessibility of other molecules to enter and regulate certain parts of the DNA. More information regarding histone modifications, chromatin accessibility and the link between these two regulatory mechanisms and exercise can be found later in this book (**Chapter 6**) as well as in a series of more generalised (26, 27) and exercise-specific review articles (28). The purpose of this section, however, is to provide an overview of current methodologies that can be utilised to analyse both histone modifications and the accessibility of the chromatin.

Histone modifications

The most common methodology for analysing histone modifications is **Ch**romatin **I**mmuno**P**recipitation or **ChIP**. This method utilises a fixative (e.g. formaldehyde) to 'lock' the DNA into its current state, thereby freezing the interactions that occur between molecules and histones. Using a ChIP-specific cell lysis buffer to lyse and extract DNA that is still bound to the histones, the sample is then randomly fragmented using sonication, and an antibody is used to bind to specific modifications of choice (e.g. tri-methylation of the 4th lysine residue on histone 3 or H3K4me3). Those histones that are bound by the antibody are pulled out ('immunoprecipitated'), the nearby DNA is then purified and the researcher is left with a pool of DNA fragments that were originally located near the histone mark of interest. Three methods are then utilised to find out exactly where in the genome these DNA fragments originally came from.

For a genome-wide approach, ChIP-on-chip was the first widespread method in which DNA fragments are analysed via a microarray analysis. However, with rapid development of NGS, ChIP-sequencing has become the more common methodology employed. In this process, DNA fragments are purified from the ChIP protocol detailed above and used for generating DNA sequencing libraries. Nonetheless, both approaches provide information regarding the binding location and amount of binding of the histone modification of choice (e.g. the histone mark used for the antibody immunoprecipitation step) (29). Each of these protocols however is costly, and produces a large amount of data and requires advanced expertise in bioinformatics and data analysis (30). To overcome these limitations, researchers are now able to analyse the purified DNA fragments via quantitative PCR (qPCR), in a far more targeted approach. This protocol relies heavily on the researchers knowing exactly the region in the genome they wish to target, and the ability to design PCR primers to investigate this area. ChIP protocols require large quantities of starting material, are relatively labour-intensive and, for sequencing protocols, require large amounts of sequencing data in order to reliably determine meaningful differences across datasets.

CUT&RUN

To overcome issues surrounding background noise, together with the amounts of sequencing data and starting material required for ChIP-sequencing, two new methods have been developed. The first is **C**leavage **U**nder **T**argets and **R**elease **U**nder **N**uclease (or **CUT&RUN**) which was first established in 2017 as a more efficient and cost-effective approach for analysing molecule binding sites (31). This approach uses a similar antibody immunoprecipitation method as suggested previously, but then exposes the DNA to an enzyme that can recognise the antibody and cut the DNA around it. The resultant DNA fragments are then purified and can be used for making DNA sequencing libraries through standardised protocols.

Importantly, the enzyme cutting process occurs in a highly specific and effective manner and without the need for prior fixation of the tissue or cell sample, which collectively, greatly reduces both background noise and the potential discovery of false positives (32). Indeed, it is suggested that reliable data can be obtained from as little as 100–1,000 cells (31).

CUT&Tag

Since its inception, the CUT&RUN protocol has been further modified to dramatically reduce the time required to generate sequencing-worthy DNA fragments as well as having the ability to be performed at single-cell resolution. The **C**leavage **U**nder **T**argets and **T**agmentation protocol (**CUT&Tag**) utilises a similar approach to CUT&RUN but the enzyme responsible for cutting the DNA fragments is preloaded with DNA sequencing adapters that are important for subsequent amplification and sequencing of DNA fragments (33). Once the DNA fragments are purified, they are then amplified via PCR ready to be sequenced.

Both CUT&RUN and CUT&Tag benefit from a far superior noise to signal ratio and a lower false discovery rate (31), which makes them more advantageous over traditional ChIP-sequencing protocols. Furthermore, the slightly modified chemistry and enzyme used for the CUT&Tag assay makes it highly applicable for single-cell analysis, whereas traditional ChIP and CUT&RUN protocols are not. However, despite undergoing continued development and improvement, the methods for these two protocols are still relatively new. Therefore, the reliability and reproducibility are important factors worth considering when planning to use these methods for experiments.

Chromatin accessibility assay

The ability for molecules and enzymes to access specific DNA regions is crucial for DNA function. For example, the ability for polymerases to access parts of our DNA is critical for increased gene transcription and therefore, greater expression of messenger RNA (mRNA), explained in more detail in **Chapter 3 and 6**. The tightening (heterochromatin) and loosening (euchromatin) of the chromatin structure are closely linked with the epigenetic state of certain histone modifications, and thus, accessibility assays are commonly utilised in conjunction with histone modification assays to understand more about the genomic landscape of the sample. **A**ssay for **T**ransposase-**A**ccessible **C**hromatin (**ATAC**) sequencing is the most common method for ascertaining accessible regions of DNA. This method employs a similar enzyme to the CUT&Tag protocol described above, whereby a special enzyme that is preloaded with sequencing adapters can enter, bind and cut dsDNA. Importantly, this enzyme is only able to perform this function in accessible regions of the genome. Thus, following DNA purification in this process, the researcher is left with DNA fragments that are enriched from regions that are highly accessible (34). After these fragments have been sequenced, the data is aligned to the genome of interest with the fragments that cover a certain area correlating with the accessibility of that genomic region. This method is suggested to be far more efficient, sensitive and cost-effective than alternative methods (35, 36).

GENE EXPRESSION ANALYSIS OF RNA

LEARNING OBJECTIVES

At the end of this section, you should be able to:

1. Describe the process of gene transcription/mRNA expression.
2. Discuss the application of reverse transcription (rt) real-time (RT) quantitative polymerase chain reaction qPCR or 'rt-RT-qPCR' for assessing gene/mRNA expression.

3. Explain the overarching laboratory and analytical methods for assessing 'targeted' gene/mRNA expression.
4. Describe the principles of genome-wide analysis of gene expression or 'transcriptomics'.
5. Explain the overarching methods of gene expression analysis for both microarrays and RNA sequencing (RNA-seq) for use in molecular exercise physiology.
6. Explain the key differences between microarrays and RNA-seq methods for gene expression analysis.

GENE TRANSCRIPTION OR EXPRESSION

Introduction to gene expression

Inherited genetic variations or epigenetic modifications, as a result of exercise, are able to effect whether a gene is turned on or off. The process of turning a gene on or off is termed **gene transcription or expression** (the full theory for gene transcription is described in detail in **Chapter 3**). In order to measure gene expression, **RNA** must first be isolated. As with DNA, researchers are able to directly extract RNA from skeletal muscle tissue to study gene expression following exercise. RNA is very similar to DNA; however, RNA is single-stranded, whereas DNA is double-stranded (i.e. the double helix structure); the sugar backbone in RNA is ribose compared with deoxyribose within DNA (hence why RNA is much less stable than DNA) and there is a replacement of **thymine (T)** in DNA with **uracil (U)** in RNA (see **Chapter 3, Figure 3.7**). Also known as **mRNA**, the RNA copies and delivers the DNA 'message' to the protein-making machinery of the cell (i.e. the ribosome) to initiate the production of protein. Therefore, as suggested above, the making or synthesis of RNA from DNA is termed **transcription**, whereas the synthesis of protein from transcribed mRNA is termed **translation**.

Gene transcription is first initiated by the protein, **RNA polymerase**, that binds to a transcription factor complex located within the '**promoter**' region of the gene on the DNA. This process begins at the transcription initiation site where RNA polymerase begins transcribing the corresponding single-stranded RNA (ssRNA) encoded by the targeted dsDNA sequence in the 5' to 3' (or left to right) direction. The resultant **pre-mRNA** molecule then undergoes several post-transcriptional modifications to remove or '**splice**' the non-coding regions (i.e. **introns**) of the mRNA. Therefore, the final **mature mRNA** product only contains the coding regions of the gene (i.e. **exons**) ready to be subsequently translated into protein by the ribosomes in the cytoplasm (**Chapter 3, Figures 3.8 and 3.9**). Many copies of RNA can then be produced from one gene. It is often the amount of mature mRNA that researchers will want to measure due to its protein coding potential. To analyse whether an exercise intervention has increased or decreased gene transcription levels of individual genes, RNA must be first isolated, quantified and checked for its 'quality' as described earlier in this chapter. Finally, RNA can then be analysed via a method called reverse transcription real-time quantitative polymerase chain reaction (**rt-RT-qPCR**).

Importantly, as mentioned above, RNA is much less stable than DNA and therefore presents some key methodological challenges. Most importantly, RNA degrading enzymes known as **ribonuclease (RNase)** can break down the RNA molecule into smaller fragments that are unusable for experiments. It is therefore recommended to always ensure good laboratory practice (i.e. always wearing clean gloves, a lab coat and cleaning all surfaces and equipment with RNase-inhibiting solution) and always use RNase-free plasticware and diethylpyrocarbonate (DEPC)-treated water when making relevant buffers and reagents.

Reverse transcription, real-time, quantitative polymerase chain reaction (rt-RT-qPCR) for measuring gene expression

After RNA has been isolated and quantified, the ssRNA is first 'reverse transcribed' by treating with the enzyme, **reverse transcriptase,** which converts the ssRNA into a dsDNA molecule known as **complimentary DNA (cDNA).** This is important because cDNA can then undergo **PCR** amplification (as described earlier in this Chapter and visualised in **Figure 2.1**) using a series of heating and cooling steps to amplify a gene of interest (also known as a 'target gene').

cDNA is placed into a thermal cycler for an initial **denaturation** step (at around 95°C) which occurs to separate the cDNA into single strands. At a lower temperature (around 55°C), a 'mastermix' containing **forward and reverse primers** (≈20 bp short single DNA strands) bind or **anneal** to complimentary matches on the ssDNA (forward primer at the start of one of the single strands and the reverse primer at the end of the other single strand). Then at a slightly higher temperature usually (around 72°C), **Taq polymerase** then binds to the primer sequences and **extends** the single strands into double strands using a chemical buffer of **dNTPs or nucleotides** (i.e. dTTPs, dATPs, dGTPs, dCTPs) which includes Mg^{2+} as a co-factor that keeps the pH stable. This completes the first cycle. You can then visualise if there was 1 cDNA strand to start with; after the process of denaturation, annealing and extension, there would be a doubling of the original cDNA molecule. Samples then undergo repeated cycles of; **denaturation**, which causes the dsDNA to separate into two single strands again; **annealing**, where the temperature is reduced once more to allow PCR primers to bind to the gene region of interest; and **extension**, in which the temperature is increased to allow extension and attachment of dNTPs via the enzyme, Taq polymerase, to create a newly synthesised dsDNA molecule, or cDNA molecule (see **Figure 2.1**). After each PCR cycle, the abundance of new copies of the region of interest (termed a gene 'product') is quantified in 'real-time' by the incorporation of fluorescent molecules that bind to the cDNA during each PCR cycle, where the amount of gene product is directly proportional to the amount of fluorescence signal. This differs from conventional PCR methods that use gel electrophoresis methods (see below) to quantify the amounts of a gene product present within each sample following PCR amplification. Hence, the real-time qPCR technique adopts its name due to the ability to detect differences in gene expression after each cycle in 'real-time' rather than after the entire PCR amplification procedure has completed.

rt-RT-qPCR method

SYBR Green dye provides the simplest and most commonly used fluorescent molecule for detecting and quantifying PCR products in real time. SYBR Green binds dsDNA, and upon excitation emits light. Thus, as a PCR product of cDNAs representing the gene of interest accumulates after each cycle, so does the fluorescence signal. The amount of bound SYBR Green fluorescence in each sample is calculated by counting the number of PCR cycles required to reach a defined level of fluorescence, termed as the fluorescence threshold (**Figure 2.6**). Therefore, the fewer PCR cycles required to reach a defined fluorescence threshold suggest larger quantities of the gene product present within each sample and therefore greater expression of the target gene. In other words, the more a target gene is expressed in a sample, the lesser the number of amplification cycles are required to accumulate enough gene products to give a fluorescent signal against the defined fluorescent threshold level. This relationship can be defined mathematically: the number of cycles required to reach the fluorescent threshold (i.e. the **cycle threshold** or C_T **value**) is inversely proportional to the log of the initial template number. Therefore, C_T values can be used to calculate initial template quantities and thus gene expression. To do this, the fluorescent threshold where samples and reference genes will be compared needs to be defined. To achieve this, the fluorescence is plotted on a graph against cycle number, and the threshold line is positioned where there is the emergence of an exponential rise in fluorescence above background levels (see **Figure 2.6**). The C_T for an individual sample is defined as the cycle number at which the sample's fluorescence trace crosses the threshold line. The C_T values obtained across experimental samples (such as post-exercise) are compared to the C_T of reference (or 'housekeeping') genes as well as calibrator (or control) samples (such as baseline/pre-exercise) to calculate relative changes in the mRNA expression (equation described below). Reference genes are 'stable' genes whose expression is unaffected by experimental factors. These are typically genes that have cellular 'housekeeping' properties such as those involved in cell survival and are consistently expressed, despite any changes caused by exercise.

Melting curve analysis is also performed to distinguish a gene products 'melting' temperature (**Tm**). This so-called 'melt curve analysis' can be used to verify that only one intended gene target/gene of interest was amplified, indicating PCR product homogeneity (or specificity) and that the sample was not contaminated. To generate the melting curve, fluorescence of the sample product is measured as the temperature is gradually increased. A sharp increase in fluorescence is observed as the product is

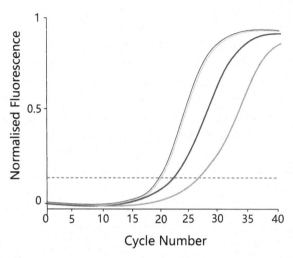

Figure 2.6 Schematic representation of the fluorescence readout from a typical rt-RT-qPCR experiment. The four curves correspond to two different genes across two samples from separate experimental conditions. For example, red and yellow curves represent a reference / housekeeping gene (i.e. RPIIß) and green and blue curves represent the gene of interest (i.e. IGF-I). Red and blue lines represent the same post-exercise condition, whereas yellow and green represent the pre-exercise condition from the same individual. The horizontal dotted line represents the defined **cycle threshold (C_T)** line where there is an exponential rise in fluorescence against background levels. Notice the similar rise in fluorescence for both the pre- (yellow) and post-exercise (red) curves for the reference/ housekeeping gene and this therefore provides similar C_T values ($C_T = 20.59 \pm 0.14$, mean $\pm$ SD), suggesting that the expression levels have remained stable between conditions in this reference gene as expected. However, the C_T value for the gene of interest in the post-exercise condition (blue, $C_T = 22.18$) is much lower than that in the pre-exercise condition (green, $C_T = 26.86$), suggesting on first inspection that there is an increase in gene expression of IGF-I (accumulation of fluorescence earlier) in the post-exercise condition. From these curves, a fold increase in IGF-I between pre- and post-conditions can be calculated, which is detailed in the equation explained below.

amplified followed by a sharp decrease in fluorescence that occurs as the product denatures, marking its 'melting' temperature (**Figure 2.7**). If there are multiple peaks, this may suggest contamination or unspecific amplification, or that primers have been unable to anneal to the target gene (**Figure 2.7**). Multiple smaller peaks at lower temperatures can also suggest primer-dimer issues where forward and reverse primers with complimentary sequences may have annealed to each other (or themselves) rather than the gene of interest. If this occurs, these samples cannot be used in any analyses, and primers may need to be redesigned until the melt curve analysis demonstrates primer 'specificity' (amplification of only one product in the melt curve analysis) following RT-qPCR. Products can also be sequenced to confirm that the gene amplified is indeed the gene of interest. Alternatively, the size of the PCR product can be confirmed by running the products on an agarose gel followed by ethidium bromide binding and UV light visualisation against known DNA molecular weight markers (see gel electrophoresis section below).

Relative gene expression analysis

The cycle threshold or C_T values (defined as the cycle number at which the samples fluorescence trace crosses the threshold line, **Figure 2.7**) for a target gene of interest in the experimental condition (e.g. post-exercise), a reference (or housekeeping) gene and a calibrator/control condition

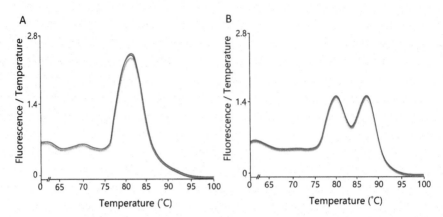

Figure 2.7 Schematic representation of successful **(A)** or unsuccessful **(B)** melt curve analysis following rt-RT-qPCR. **(A)** Examples of single 'peak' curves showing that only one product was amplified, together with no primer-dimer issues (that might show multiple peaks at low melt temperatures). **(B)** Example of two distinct peaked curves perhaps due to non-specific amplification. In this situation, the researcher should look to see if there were any reasons for a specific sample error and if they think this might have been the case, repeat the PCR assay again; then if the problem persists, redesign the primers and repeat once more.

(e.g. baseline/pre-exercise) are used to generate relative expression values. An expression value of 1 represents the calibrator/control sample (e.g. baseline/pre-exercise). Therefore, any changes from the value of 1 observed in the experimental condition (e.g. post-exercise) provide a relative expression value between pre- and post-exercise. For example, an expression value of 2 observed in the experimental condition would suggest a doubling or 100% increase in gene expression post-exercise relative to pre-exercise. To obtain relative expression values, raw C_T values derived from RT-qPCR experiments are input into specific equations such as the Delta Delta C_T ($\Delta\Delta C_T$) equation, otherwise known as the Livak Method (37).

Equation 1: $\Delta C_T = \text{Mean } C_{T\,(\text{Target gene, test})} - \text{Mean } C_{T\,(\text{Reference gene, test})}$.

Equation 2: $\Delta C_T = \text{Mean } C_{T\,(\text{Target gene, calibrator})} - \text{Mean } C_{T\,(\text{Reference gene, calibrator})}$.

Equation 3: $\Delta\Delta C_T = \Delta C_T$ of Equation 1 $- \Delta C_T$ of Equation 2.

Equation 4: $2^{-\Delta\Delta CT}$ (gives a normalised expression ratio).

Where:

Mean $C_{T\,(\text{Target gene, test})}$ = The average C_T value of the target gene (e.g. IGF-I) in the experimental 'test' condition (e.g. post-exercise).

Mean $C_{T\,(\text{Reference gene, test})}$ = The average C_T value of the reference/housekeeping gene (e.g. RPIIβ) in the experimental 'test' condition (e.g. post-exercise).

Mean $C_{T\,(\text{Target gene, calibrator})}$ = The average C_T value of the target gene (e.g. IGF-I) in the calibrator condition (e.g. pre-exercise).

Mean $C_{T\,(\text{Reference gene, calibrator})}$ = The average C_T value of the reference/housekeeping gene (e.g. RPIIβ) in the calibrator condition (e.g. pre-exercise).

FOR EXAMPLE:

Equation 1: $\Delta C_T = 22.18_{(\text{IGF-I, post-exercise})} - 20.69_{(\text{RPII}\Delta,\ \text{post-exercise})}$.

Equation 2: $\Delta C_T = 26.86_{(\text{IGF-I, pre-exercise})} - 20.49_{(\text{RPII}\Delta,\ \text{pre-exercise})}$.

Equation 3: $\Delta\Delta C_T = \Delta C_T$ of Equation 1 (1.49) $- \Delta C_T$ of Equation 2 (6.37).

Equation 4: $2^{-(4.88)}$ (gives a normalised expression ratio of 29.45).

Therefore, in this example (which is hypothetical data and corresponds to no true analysis), the expression ratio of the target gene (i.e. IGF-I) post-exercise compared with pre-exercise is approximately 29:1, suggesting that IGF-I is highly expressed after exercise (29 − normalised expression values higher often referred to as 29 'fold' higher).

It is important to note that this relative gene expression analysis method assumes that both the reference and target genes are amplified in each PCR cycle with efficiencies near 100% and within 5–10% of each other (where 100% efficiency refers to a perfect doubling of the template DNA during each cycle). Indeed, as long as the reference and target genes have similar and reasonably high efficiencies (i.e. both ~ 85%), this does not make a difference to the output as the relative differences would be the same. However, if efficiencies are excessively low (i.e. <70%), then this may indicate an issue with low quality or contaminated samples or primer issues. In this instance, experiments should perhaps be undertaken again to derive new samples until higher efficiencies that are consistent across samples are achieved. If efficiencies are still high across all samples but differ (>10%) between target and reference genes, this can be corrected for by inputting all sample efficiencies into a specific equation (38) and can be applied using the 'Relative Expression Software Tool' or 'REST' software. However, if this occurs frequently, then it might be an indication that primers for target genes and reference genes need to be redesigned according to more standardised specifications.

TRANSCRIPTOMICS

Introduction to transcriptomics

In the sections above, we focused on the isolation of RNA directly from skeletal muscle tissue to enable downstream analysis of mRNA expression. Specifically, the application of reverse transcription real-time quantitative polymerase chain reaction (rt-RT-qPCR) for assessing gene expression levels of a 'target' gene of interest. Targeted gene expression analysis is, and has been, very important in the field of molecular exercise physiology when determining changes in mRNA levels after exercise. While rt-RT-qPCR may be considered complex to the novice user, this method is well-recognised and a commonly used technique in the field of molecular biology. Indeed, it is now commonplace for laboratories to have their own thermal cycler machines. While RT-qPCR is relatively expensive (for PCR reagents and equipment) compared to traditional non-invasive physiology techniques, the cost is still affordable within moderately funded studies. A limitation to this technique, however, is that scientists are restricted to the number of individual genes that can be analysed at once. Given there are approximately 27,000 protein coding genes in the human genome, scientists may want to discover new genes or pathways associated with exercise, rather than focusing on well-known important regulators at the targeted gene level. This might be particularly important when analysing precious muscle biopsies from unique populations (e.g. elite/ injured athletes, diseased/aged individuals or after complex exercise or nutritional interventions). With the advent of **genome-wide** analysis of transcription/gene expression (also known as '**transcriptomics**') using **microarrays** or **RNA-seq**, scientists are now able to simultaneously assess the expression levels of larger numbers of genes (or even all genes) using comparable amounts of RNA used to investigate several genes via rt-RT-qPCR. However, these techniques are much more expensive making it less affordable for researchers to analyse the desired number of samples that are typically required for exercise studies, more so, if numerous time points after an acute bout of exercise or across a chronic exercise training intervention are to be explored. In **Chapter 1**, the emergence and historical narrative for the use of microarrays and RNA-seq technologies within molecular exercise physiology were discussed. Moreover, the use of microarray technology for genotyping/SNP and DNA methylation arrays has been discussed above. In terms of high-throughput sequencing techniques, similar methods described above for sequencing DNA are also used for the sequencing of RNA. As with rt-RT-qPCR, RNA must first undergo reverse transcription to produce cDNA before proceeding with downstream analysis. It is also fundamental that RNA has high integrity (how well conserved the RNA is) for microarray or RNA-seq. Therefore, as well as the standard 'quality' checks of RNA using UV spectrophotometry as described above, additional analyses of the integrity of the RNA are required using gel electrophoresis and analysing the ratios of 28S to 18S ribosomal bands (typically using an Agilent Bioanalyzer). This provides an RNA Integrity Number (RIN) between 1 and 10, with 10 being the highest quality samples with the least degradation.

Microarrays for transcriptome analysis

The DNA microarray involves the use of small, microscopic slides, also known as a DNA/gene chip, that are robotically printed. Thousands of individual wells (or 'spots') are embedded within each chip, each containing laboratory-made oligonucleotides ('oligos' for short)' which are small fragments of DNA sequences (also called 'probes' or 'reporters') that correspond to a specific gene. These act in a similar way to 'primers' in rt-qRT-PCR targeted gene expression analysis above; yet in microarrays, there are oligo sequences that code for all the known genes in the genome, one gene in each well or spot. During microarray experiments, an individual mRNA sample is first converted to cDNA, like when using RT-qPCR (**Figure 2.8**). Synthesised cDNA samples that represent different conditions (e.g. pre- and post-exercise) are then fluorescently labelled before being mixed and loaded onto the microarray chip (**Figure 2.8**). Samples then bind, or **'hybridize'**, to the complimentary oligo sequence probe within each well. Any changes in gene expression can then be quantified via a laser that scans and detects the **fluorescent probe** that has been attached to the samples (see **Figure 2.8**). The colour of the fluorescence signal then indicates whether mRNA expression levels within a subset of samples have either increased, decreased or remained unchanged (or expressed similarly) when compared to control samples. For example, control samples (i.e. a pre-exercise at rest/baseline) are labelled with a fluorescent probe of one colour (e.g. green) whereas the experimental samples (i.e. post-exercise) are labelled with a fluorescent probe of a different colour (e.g. red). The samples are then typically, but not exclusively, mixed together and applied to the microarray chip. As the chip is flooded with the mixture of labelled samples, the cDNA from both the control and experimental groups can simultaneously bind to the probes within each well, providing a measurement of differential mRNA expression levels for all genes assessed. This is called 'two colour profiling' or 'two channel microarrays' and it is probably the most used method for detecting **differentially expressed genes (DEGS)** between the control and experimental samples. In this example, the well will fluoresce red if the expression level is greater (commonly referred to as **'up-regulated'**) in the experimental samples (i.e. post-exercise) compared with the control samples (i.e. pre-exercise), whereas a neutral colour, such as yellow, will indicate that expression levels are similar or unchanged between groups. Conversely, the well will fluoresce green if the gene expression is higher in the control (i.e. pre-exercise) group compared with the experimental group, suggesting that gene expression has reduced (commonly referred to as **'down-regulated'**) after exercise (see **Figure 2.9**). Following experimental procedures, the resultant data then undergoes quality assurance (QA) and control (QC) measures (to confirm that there are no issues with the fluorescent intensities that may suggest a potential issue with individual probes or the entire array), normalisation (that eliminates systematic experimental bias and technical variation while preserving biological variation) and differential gene expression analysis between control and experimental conditions. Researchers can also conduct Gene Ontology (GO terms) (http://geneontology.org/docs/ontology-documentation/) or KEGG pathway analysis that determines gene expression profiles of several/hundreds of genes within a specific gene ontology 'term' or defined pathway. For example, overarching GO term 'molecular function' can include terms such as

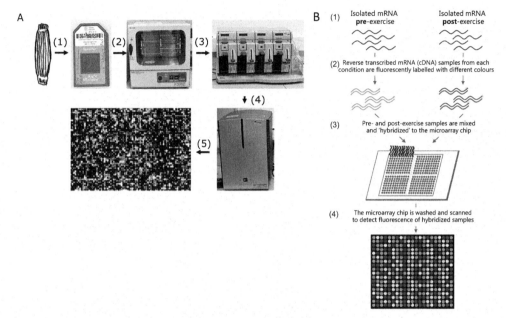

Figure 2.8 (A) Workflow of procedures for assessing gene expression from skeletal muscle via transcriptome microarray analysis. (1) RNA is first extracted from muscle tissue, and cDNA is synthesised and labelled with a fluorescent probe. (2) The fluorescently labelled cDNAs are then added to a microarray chip and hybridised to the probes on the chip in a hybridisation oven. (3) The samples are washed and (4) scanned by a microarray scanner. (5) A magnified output of a microarray in black and white. Each dot corresponds to a single probe and the intensity of the probe corresponds to the concentration of mRNA in the original sample. **(B)** Schematic representation of microarray experimental procedures. (1) RNA is first isolated from muscle tissue pre- and post-exercise and (2) cDNA is synthesised from the mRNA and labelled with fluorescence probes of different colours that represent individual conditions. In this case, pre-exercise samples are labelled green, and post-exercise samples are labelled with red. (3) All samples are then mixed and loaded onto the microarray chip to enable binding (or 'hybridization') of mRNA that is complimentary to the oligo probe within each well. The microarray chip is also washed several times to remove any unbound sample. (4) Finally, the microarray chip is scanned so that the colour and intensity of fluorescence in each well can be analysed for gene expression. In this example, the colours represent genes that have increased (green), decreased (red) or remained unchanged (yellow) after exercise. Note that the magnified output (4) represents only a small segment of 378 genes and that thousands of genes can be analysed simultaneously by microarray.

'protein kinase activity', and examples of KEGG pathways include the 'AMPK pathway' or 'mTORC pathway'. Statistical analysis can then suggest that, for example, the AMPK pathway is significantly enriched in post- versus pre-exercise and using software researchers can create gene ontology or pathway figures that visually represent (typically using different colours) which genes in those specific terms or pathways are either up- or down-regulated. Following DEG analysis, it is also common to further analyse gene expression of the most statistically signficant differentially expressed genes identified via rt-RT-qPCR in order to validate or confirm the array data. As alluded to previously, microarrays for gene expression profiling are still fairly expensive, given the specific equipment needed (e.g. laser scanning equipment) together with the microarray chips and reagents required for each sample as it is only possible to analyse a single sample per array. Furthermore, prior and up to date knowledge of all gene sequences are required and this technique will not identify any single nucleotide variants (or SNPs), alternate splicing or non-coding RNA (although most recent gene expression microarrays now profile over 50,000+ genes, that includes known protein coding genes as well as their many non-coding transcript variants). However, RNA-seq can negate some of these limitations and therefore extends the capacity of microarrays.

RNA-sequencing (RNA-seq)

RNA-sequencing, typically referred to as 'RNA-seq', is a thorough method for analysing all RNAs simultaneously. Unlike microarrays, RNA-seq does not require prior knowledge of the RNA sequence that is explored and, therefore, has the capacity to detect single nucleotide variants, alternate splicing, post-transcriptional modifications and silencing RNAs (also called micro-RNAs/miRNA), non-coding RNA (ncRNA), exon-intron boundaries and pre-mRNA via NGS technology. The present section will focus on the use of RNA-seq for quantifying gene expression, as this is the analysis that is becoming more frequently used in molecular exercise physiology. To undertake RNA-seq, sequencing 'libraries' are first created using the isolated RNA. As with RT-qPCR and microarrays, RNA is first isolated, and reverse transcribed to cDNA. The cDNA is then fragmented into small consistent sizes and sequencing 'adaptors' are added or 'ligated' to the ends of the cDNA fragments. These adaptors contain constant sequences that enable the sequencers to recognise where to start sequencing. NGS is then performed similar to that explained in the above sections. Prior to these crucial steps, the total RNA obtained from the muscle biopsy may be first treated to remove ribosomal RNA that makes up the majority (~90%) of extracted RNA. Therefore, if left untreated, sequencing data would be representative of predominantly rRNA rather than other RNAs of interest (e.g. pre-mRNA, mRNA, miRNA and ncRNA).

In terms of RNA-seq analysis, raw sequences are first aligned to a species-specific reference genomes and the gene transcripts are 'assembled' using this genome reference. The expression level of each gene can then be estimated by counting the number of 'reads' measured that align to each exon or full-length gene transcript. As with microarrays, normalisation of RNA-seq data is required to measure gene expression, with the read counts normalised to account for variability in library fragment size and read depths. For example, a specific transcript read count is normalised by the gene length and the total number of mapped reads in the sample to provide relevant units of measurement. In the case of single-end sequencing (the sequencer reads a fragment from only one end to the other), this results in 'Reads Per Kilobase of transcript per Million mapped reads (**RPKM**)'. For paired end-reads (where the sequencer reads the transcript at one read, finishes this direction at the specified read length and then starts another round of reading from the opposite end of the fragment, enabling more accurate read alignment), the reads are normalised using 'paired Fragments Per Kilobase of transcript per Million mapped reads (**FPKM**)'. Differential gene expression (DEG) analysis can then be performed to compare expression levels between samples. Other advantages of RNA-seq is that the analysis can detect allele-specific expression (e.g. transcription of both maternal and paternal alleles and SNPs) as well as identify expression quantitative trait loci (eQTLs), which are the variations of specific loci that are associated with how the gene is expressed. This might inform the researcher of those individuals who possess a specific genetic variation that may predispose them to altered gene expression after exercise.

PROTEIN ANALYSIS

LEARNING OBJECTIVES

At the end of this section, you should be able to:

1. Describe the principles of the Western blot method.
2. Explain the key methodological features of the Western blot method.

3. Discuss the application of using the Western blot method for measuring post-translational modifications and protein abundance within the context of molecular exercise physiology.
4. Describe the principles of proteomics.
5. Explain the key features of proteomic methodology and analysis and discuss its use in molecular exercise physiology.
6. Describe the principles of immunohistochemistry and immunocytochemistry.
7. Understand the key methodological features of immunolabelling.
8. Discuss the application of using immunolabelling methods for measuring skeletal muscle phenotype, protein localisation, co-localisation/occurrence and protein translocation.

Introduction to protein analysis

Once a gene is transcribed to mRNA, the mRNA transcript is translocated (or moved) out of the nucleus into the cytoplasm where it can be **translated** into protein by the **ribosome**. Many of the important effects of exercise that occur at the protein level are mediated by changes in the location, activity and total abundance of protein, which is often referred to as protein 'expression'. Furthermore, not all gene expression changes are 'functional' and therefore altered gene transcription may not always lead to a change in protein translation. Therefore, the measurement of protein levels and modifications (e.g. **phosphorylation** and **acetylation**) are arguably the most important methods of molecular exercise physiology to inform key exercise responses such as the activation of signal transduction pathways that can, in turn, regulate gene transcription (see **Chapter 7**), protein synthesis/degradation (see **Chapters 7 and 8**) or the adaptive changes in metabolic, mitochondrial and myofibrillar/sarcomeric proteins associated with enhanced function of skeletal muscle (see **Chapters 8 and 9**). For these reasons, molecular exercise physiologists often measure the extent of protein modifications such as phosphorylation and acetylation (introduced in **Chapter 1** and discussed in more detail in **Chapter 7**) that determine protein activity, as well as the abundance/concentrations of individual proteins with known functions in cells such as the regulation of transcription, or the metabolism of carbohydrate and fat in the resynthesis of ATP. While in recent years there is increasing interest in, and availability of, methods for large-scale and non-targeted analysis of protein (see Proteomics section below), a staple of most molecular physiology labs for the past 40 years has been **Western blotting** for the detection of modifications to and abundances of protein targets. This method was first developed across multiple labs between the late 1970s and early 1980s (39, 40) and its utility, low cost and accessibility of a catalogue of antibodies have since remained an attractive analytical method across the life sciences.

WESTERN BLOTTING

Introduction to the Western blot method

The measurement of proteins using Western blotting involves four essential steps:

1. Protein extraction and denaturation.
2. Separation of proteins by gel electrophoresis.
3. Electrotransfer of proteins from the gel to specific membranes (i.e. nitrocellulose or polyvinylidene difluoride, PVDF).
4. Detection and visualisation of the protein of interest using primary and secondary antibodies.

Because of the use of antibodies in the final step, the name 'immunoblotting' is often used interchangeably, and is synonymous with Western blotting. The name 'Western' blotting coined by W Neal Burnette (41) is a pun on 'Southern' blot, which is a similar method for DNA developed by Edwin Southern in 1973 that was later published in 1975 (42).

Protein extraction and denaturation

The extraction of protein from tissue samples was briefly introduced earlier in this chapter. In the case of Western blotting of skeletal muscle samples, the ordered structure of myofibrils and connective tissue means that skeletal muscle is relatively difficult to prepare for protein assays compared to cell culture lysates. Grinding the muscle samples under liquid nitrogen using a mortar and pestle (or similar approach), and then processing the debris/powder in a relevant homogenisation/lysis buffer with an electric tissue homogeniser have the effect of shearing the cells and organelles and breaking up the connective tissue. A typical homogenisation buffer for extracting soluble proteins is a pH-buffered solution (pH 7–9) containing non-ionic detergents such as Triton x-100 to disrupt membranes and protein interactions, reducing agents such as dithiothreitol (DTT) to break disulphide bonds, and other agents such as EDTA and EGTA to inhibit protein-degrading proteases. When post-translational modifications are of interest, other agents such as phosphatase inhibitors (e.g. sodium orthovanadate and sodium fluoride) or deacetylase inhibitors (e.g. trichostatin A and nicotinamide) are included to avoid dephosphorylation and acetylation, respectively, during protein extraction. Inhibitor 'cocktails' that can be added to standard homogenisation buffers are now commercially available and are formulated specifically for preserving the various post-translational modifications of interest.

After homogenisation, the homogenate is centrifuged and the supernatant is collected, which in the crudest approach contains the soluble proteins, whereas myofibrillar proteins remain in the pellet. Specific cellular fractions (e.g. myofibrils, sarcoplasm, mitochondria and nuclei) can be isolated as part of the homogenisation process and may be of interest depending on the protein or process of interest, but such approaches require additional buffer and centrifugation considerations (43). The protein concentration of the supernatant is measured using a protein assay such as the Bradford or BCA assay. This assay is an important step for measuring the yield of protein from the homogenisation process, but also important for standardising the protein concentrations during the loading of samples in later steps.

Next, proteins are denatured and negatively charged in preparation for electrophoresis by mixing the sample with Laemmli sample buffer (44) which contains a strong negatively charged (anionic) detergent named sodium dodecyl sulfate (SDS). The samples are heat-denatured by incubation in this buffer, usually at ~95 C for 3–5 minutes. Under these conditions, proteins unfold completely and the SDS binds with the proteins and prevents them from refolding when the sample is cooled.

Separation of proteins by gel electrophoresis

A method known as SDS-polyacrylamide gel electrophoresis (SDS-PAGE) is then used to separate the denatured and negatively charged proteins within each sample by passing the samples through a gel constructed from acrylamide polymers, which cross-link to form a molecular sieve. Samples containing the standardised protein concentrations are loaded into wells at the top of the gel and an electric current is passed through the gel. This causes the negatively charged proteins to migrate towards the positive electrode at the bottom of the gel. Small proteins pass through the gel faster than large proteins and thus the procedure effectively orders proteins according to their size, i.e. their molecular weight in kDa (**Figure 2.9**).

Electrotransfer of proteins from gel to membrane

After protein separation by gel electrophoresis, the proteins are embedded within the gel and are not accessible to antibodies. However, it is the targeting of a specific protein with a relevant antibody that is the essential aspect of analysis in Western blotting. To achieve this, the gel is sandwiched against a sheet of nitrocellulose or PVDF membrane that is designed to capture proteins and an electric current is used to elute the proteins from the gel onto the surface of the membrane. This process makes the proteins

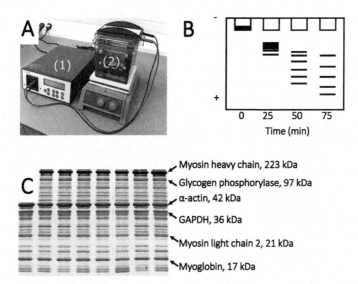

Figure 2.9 Protein electrophoresis equipment and results. **(A)** SDS-polyacrylamide gel electrophoresis (SDS-PAGE) set-up, wherein a power pack (1) is connected to an electrophoresis tank (2) with protein-loaded gels included. **(B)** Schematic illustration of how proteins, which are denatured and negatively charged as a result of SDS treatment, migrate over time towards the positive electrode at the bottom of the gel. This allows the separation of denatured protein based on their molecular weight, given in kDa. **(C)** Result of an electrophoresis experiment where skeletal muscle proteins have been separated as described. For Western blotting, these proteins are then transferred onto a membrane and incubated with antibodies against a specific protein which is then visualised.

accessible to antibodies as the proteins are now, in effect, fixed on the surface of the membrane. After proteins are transferred from the gel to the membrane, the membrane must be 'blocked' by incubating it with a generic protein mixture such as non-fat dried milk or bovine serum albumin (BSA). The abundance of milk or serum proteins ensures that antibodies that are subsequently incubated with the membrane bind selectively to their target proteins rather than non-specifically to random protein-binding sites on the membrane.

Detection and visualisation of the protein of interest

For the subsequent quantification of a protein of interest, the membrane is typically incubated with the **primary** and **secondary antibodies** separately. The primary antibody binds the protein of interest, whereas the secondary antibody binds and detects the primary antibody. The secondary antibody is conjugated to a molecule or enzyme that enables it to be visualised. For example, the enzyme horseradish peroxidase (HRP) is commonly used to catalyse a light-emitting reaction which is known as enhanced chemiluminescence (ECL). The light emitted from this process is too weak to be seen by the naked eye, but it can be captured either on photographic film or by a camera system that captures light from the membrane over several minutes to create a digital image **(Figure 2.10)**.

One of the main purposes of Western blotting in molecular exercise physiology is the detection of post-translational modifications of proteins in response to acute exercise as a means of understanding

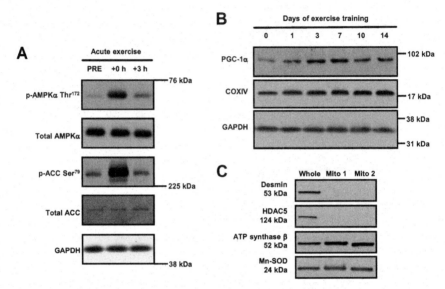

Figure 2.10 Original Western blot results. **(A)** Changes in phosphorylation status of AMPK at Thr[172], and its downstream target acetyl CoA carboxylase (ACC) at Ser[79], in human skeletal muscle in response to ~35 minutes of cycle ergometer exercise at ~80% VO_{2peak}. Increases in band intensity immediately after exercise (+0 h) for p-AMPK and p-ACC are indicative of increased phosphorylation of the respective protein targets at the named phospho-sites. The return to pre-exercise values after 3 hours of recovery (+3 h) demonstrates the transient nature of the exercise-induced changes in signal transduction and metabolic regulation. Total protein abundance of AMPK and ACC are unchanged at any time point, as is the abundance of the loading control/reference protein GAPDH. **(B)** Changes in the total protein abundance of the transcriptional co-activator PGC-1α (increases of ~50–90%) and the mitochondrial protein cytochrome c oxidase subunit 4 (COXIV) (increases of ~20%) in human skeletal muscle in response to 14 consecutive days of aerobic exercise training for 60 minutes daily at ~80% VO_{2peak}. Total protein abundance of the loading control, GAPDH, is unchanged at any time point. **(C)** Protein abundance in rat skeletal muscle of the myofibrillar protein desmin, the nuclear histone deacetylase HDAC5, and the mitochondrial proteins ATP synthase β and manganese superoxide dismutase (Mn-SOD) in either whole muscle lysates (Whole) or mitochondrial isolations produced via differential centrifugation (Mito 1) or a commercially available mitochondrial isolation kit (Mito 2). The absence of myofibrillar and nuclear proteins in the lanes from the mitochondrial isolations is used as an indication of purity and/or successful isolation.

intracellular signal transduction pathways that are altered after exercise. This requires antibodies that detect a particular type of modification at a specific residue in the protein of interest, such as the phosphorylation of the threonine 172 site (Thr[172]) of AMPK (Figure 2.10), or the lysine acetylation status of a whole muscle lysate. Numerous commercial vendors have extensive catalogues of antibodies, which encompass a broad spectrum of proteins and their different modifications, thereby allowing researchers to interrogate both total and modified proteins as needed.

Although care is taken to load equivalent amounts of sample in the gel, it is necessary to normalise the resultant data by expressing the quantity of the protein of interest relative to either the total amount of protein in the gel, or the total amount of protein on the transfer membrane, or a reference/'housekeeping' protein that should not change or differ between samples. For example, a high abundance protein such as the glycolytic enzyme glyceraldehyde 3-phosphate dehydrogenase (GAPDH) (Figure 2.10) or an actin is often used as a reference protein, although the chosen reference protein should depend on the exercise intervention being studied as, for example, actin can change with resistance training.

Because of this approach to express the abundance of a protein relative to total protein or a reference protein, Western blotting is therefore considered to be a semi-quantitative assay, and can be subject to high coefficients of variation (CV). For example, CVs of ~40% may be seen if normalised to a high abundance reference protein, but this can be reduced to ~10% if normalised to total protein in the gel (45). Alternatively, with appropriate loading of a protein of several known concentrations over a linear range, it is possible to more accurately and directly quantify the abundance of a target protein (46).

Further considerations for Western blotting

While the overview of western blotting provided is necessarily brief for this chapter, there are sometimes several sub-steps for each of the aforementioned steps described above, which require technical considerations and optimisation in order to ensure adequate sensitivity and to obtain valid and reliable results. Such detail is beyond the scope of this chapter but interested readers are directed elsewhere (47, 48), including an excellent description of issues related to skeletal muscle and molecular exercise physiology (43).

PROTEOMICS

Introduction to proteomics

Proteomics is a method for the large-scale and non-targeted analysis of protein abundances, turnover rates or modification states that offers insight into muscle responses to exercise and a data-driven approach to developing new hypotheses (49). Proteins exhibit a broad array of different physical and chemical properties, which makes the extraction and analysis of proteins challenging compared to other molecular classes (e.g. RNA and DNA) that share more similar physio-chemical properties. Skeletal muscle is a challenging tissue for proteomics because muscle contains a small number of highly abundant myofibrillar proteins and metabolic enzymes, which dominate a large proportion of the analytical landscape, but this characteristic can also be seen as an opportunity. The high abundance of proteins that underpin the functional properties of muscle, and are associated with exercise performance, means that muscle is essentially a 'self-enriching' substrate for phenotyping studies. Relatively simple proteomic methods can be applied to phenotype muscle in a more efficient manner than longer established techniques such as myosin heavy chain profiling or enzyme activity studies (50). Proteomic studies are also a pragmatic means to discovering new information about muscle responses to exercise, particularly in humans where the small size of biopsy samples limits the number of targeted protein assays (such as Western blotting, described above) that can be performed in any one experiment. In recent years, proteomic studies have provided unique insights into the complex networks of post-translational modifications such as phosphorylation (51, 52) and ubiquitination (53, 54), fibre-type specific adaptation (55) and the synthesis and degradation rates of individual proteins (56) in exercised muscle. Each of these areas has necessitated the development of specialist techniques; yet generally, all proteomic studies share 3 core components of **sample separation**, **mass analysis** and **data processing**.

Separation techniques for proteomic analysis

The first proteomic studies of rodent (57) and human (58) muscle responses to exercise training used 2D gel electrophoresis to separate proteins into '**proteoforms**' based on their isoelectric point and molecular weight. Proteoforms consist of different combinatorial patterns of splice variations and post-translational states of each protein and are the true functional entities of the proteome. However, 2D electrophoresis has limited power to resolve all proteins; for example, most studies in muscle report a few hundred proteins, which each might be observed as several different proteoforms (59). Most contemporary studies now use enzymatic digestion strategies that cleave proteins into peptides to offer more comprehensive coverage of the muscle proteome. Peptides are more soluble and, therefore, easier to handle than proteins and still contain enough information to infer the abundance, turnover rate or modification status of the parent proteins. Complex mixtures of peptides from muscle digests can be separated based on their relative hydrophobicity using **reverse-phase liquid chromatography**. This provides a reproducible means of delivering peptides to the **mass spectrometer** (see section directly below) over a defined period of time (60). Numerous different peptides may have a similar level of retention on the column and therefore elute into the mass spectrometer at the same time. Therefore, it is important to match the liquid chromatography separation with the speed and resolution of the mass spectrometer. If too many peptides are delivered to the mass spectrometer at any one point in time, information will be lost. Often, the very large numbers of peptides generated by digestion of whole tissues such as muscle necessitate prior fractionation of the sample in order to achieve the deepest levels of proteome coverage (55). Moreover, only a small proportion of the total number of copies of a protein carry post-translational modifications. Therefore, enrichment techniques are needed to study residue-specific covalent modifications, including phosphorylation (61), acetylation (62) and ubiquitination (53). Using these techniques, the connection between the complex combinatorial patterns of different modifications at different residues within each protein is lost, so these data do not provide insight at the proteoform level. Modern-day approaches to proteoform analysis include partial digestion strategies and sophisticated mass spectrometry of intact proteins but typically need to be targeted to individual proteins (63), so there is still a place for 2D gel electrophoresis, particularly in skeletal muscle where the majority of the myofibrillar proteome consists of closely related protein isoforms that exist as different splice variants or post-translational states, and have different responses to exercise stimuli (56).

Mass spectrometry

Mass spectrometers that are used for proteomics consist of a 'source' that generates charged peptide ions and a system of filters that are used to transfer the ionised peptides into a mass analyser. Electrospray ionisation (ESI) is the most commonly used ionisation source and also provides an interface between liquid chromatography and mass spectrometry. As peptides elute from the liquid chromatography column, they are sprayed through a fine needle near the entrance of the mass spectrometer. The liquid from the chromatography system is evaporated and the peptides enter the 'gas phase' and become protonated (i.e. 'charged'). Positively charged peptides are drawn into the mass spectrometer and are then directed by ion filters that can either permit or exclude different peptides from travelling to the mass analyser. There are several different configurations of mass analysers used in proteomic studies, but they all rely on a general process known as **tandem mass spectrometry**, which simply means 2 levels of mass analysis are performed. The first level, 'MS' or 'MS1', measures the mass of intact peptides and the second level, 'MS/MS' or 'MS2', measures the mass of the fragment ions that are produced when the peptides are broken into smaller pieces through a process known as collision-induced dissociation (CID). The key performance parameters of mass spectrometers are speed, resolution and mass accuracy. Most instruments continuously run through alternating cycles of MS1 and MS2 data collection, and each cycle needs to be completed in less than 1 second to achieve good quality proteomics data. Numerous different peptides are delivered to the mass spectrometer at any particular moment, so the mass analyser must have sufficient power to resolve (i.e. distinguish between) ions of very similar masses. Good mass accuracy is also essential for the unambiguous identification of peptides and post-translationally modified residues, and modern-day instruments routinely achieve a level of mass accuracy of close to 1 parts per million (ppm). It is also important to appreciate that the key performance parameters of mass spectrometers are interrelated. For example, most mass spectrometers achieve better levels of mass resolution at slower rates of data acquisition. Therefore, it is important to optimise instrument settings to suit the aims of different types of experiments.

Data processing

Mass spectrometry is used to identify proteins by comparing the accurately measured masses of their peptides and peptide fragments against protein and gene databases. Each protein has a unique linear sequence of amino acids that is created during ribosomal translation based on instructions from mRNA. Likewise, each amino acid has a characteristic elemental composition (e.g. carbon, hydrogen and nitrogen) and therefore a predictable mass. The preparation of samples for mass spectrometry also involves the digestion of proteins into peptides using proteases that have a specific cleavage specificity, which imparts fixed information onto the samples. For example, peptides generated by trypsin digestion have either a lysine (K) or arginine (R) residue at their C terminus. When measured at a sufficient level of accuracy, the peptide mass along with the masses of the fragments from that peptide are sufficient to unambiguously identify the peptide and therefore the parent protein (**Figure 2.11**). The intensity of peptide and fragment ion peaks can also be used to infer the relative abundance of proteins amongst different samples, and if deuterium oxide (D_2O or 'heavy water') was provided during the experiment (64), the peptide mass spectra can be used to calculate the synthesis rate of each protein.

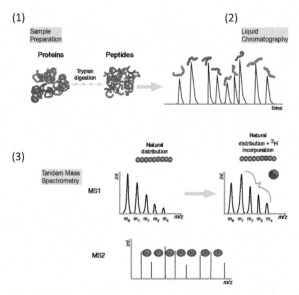

Figure 2.11 Proteomic profiling of deuterium-labelled muscle. (1) Proteins extracted from skeletal muscle samples taken prior to and after a period of ^2H_2O labelling *in-vivo* are digested into peptides of ~6–20 amino acids (circles). (2) The peptide mixture is separated in time by reverse-phase liquid chromatography. (3) Peptides elute based on their relative hydrophobicity and are delivered to the mass spectrometer. Tandem mass spectrometry (MS/MS) involves 2 levels and mass analysis: Level 1 (MS1) records the mass spectra of the intact peptide and level 2 (MS2) records the mass spectra of each peptide after it has undergone fragmentation. MS1 data are used to analyse both the relative abundance (based on the intensity of all mass isotopomers) and deuterium (^2H) incorporation (based on the relative distribution of mass isotopomers) of each peptide. MS2 data provide information on the peptide sequence (i.e. 'AFAHWGR') and are used to identify peptides by comparing the pattern of fragmentation against protein databases. ^2H-labelled amino acids incorporated into protein during synthesis *in-vivo* cause a shift in the distribution of MS1 peptide mass isotopomers. Peptides that contain ^2H-labelled amino acids can only contribute to the abundance of m_1, m_2, m_3... mass isotopomers and the relative abundance of the m_0 mass isotopomer declines as a function of deuterium incorporation (protein synthesis). Reprinted from Burniston, J.G et al., 2019 (64) by permission from Springer Nature: Springer. Omics Approaches to Understanding Muscle Biology. Methods in Physiology by Burniston, J., Chen, YW. (eds). Copyright 2019.

IMMUNOHISTOCHEMISTRY AND IMMUNOCYTOCHEMISTRY

Introduction to immunolabelling

As discussed previously, both acute exercise stimuli and long-term exercise training can modulate changes in protein activity (phosphorylation), protein abundance and the location of a protein within skeletal muscle cells or fibres. Acutely, changes in activity, localisation or the co-localisation of two proteins may be indicative of ongoing signalling cascades, or protein interactions pertinent to skeletal muscle responses to exercise or nutritional intake (65, 66). Similarly, to Western blot, immunohistochemistry takes advantage of the binding of monoclonal or polyclonal antibodies that are complementary to the antigens on a target protein of interest. Monoclonal antibodies will only bind to one epitope on the antigen molecule on the protein of interest, whereas polyclonal antibodies will bind to multiple epitopes on the target antigen. Once bound, a secondary antibody conjugated to an enzymatic or fluorescent molecule can be added which will bind to the primary antibody. As immunohistochemistry is reliant on tissue being preserved as it where in-situ, using chemical fixation or snap-freezing rather than being homogenised, researchers are able to visualise the protein itself using miscroscopy.

While it is discouraged to use immunohistochemistry to accurately quantify a proteins abundance when compared to other available methods (Western blotting or, mass spectrometry etc.), immunohistochemistry has a unique advantage in that it allows the molecular exercise physiologist to study the location of a protein (e.g. nuclear, perinuclear, cytosolic and sarcomeric) and whether this changes in response to an acute insult or with adaptation followed chronic training. The following section will cover both immunohistochemistry (performed on frozen/chemically fixed ultrathin tissue sections from a human tissue biopsy/rodent muscle tissue) and immunocytochemistry (performed on fixed cells cultured in-vitro) and will therefore be referred to as **immunolabelling**.

Overview of the immunolabelling methods

The detection and visualisation of proteins with immunolabelling involves, broadly speaking, three key steps:

1. Tissue preparation (sectioning of muscle tissue sample or fixation of cells in-vitro).
2. Immunolabelling of proteins of interest using antibodies.
3. Detection and visualisation of the protein of interest using fluorescent microscopy.

Tissue preparation

Fresh human biopsies are most commonly prepared for transverse cross-sectioning rather than being aligned for longitudinal sectioning. For transverse cross-sectioning, muscle fibres are typically aligned perpendicular to their longitudinal axis and attached/embedded in optimal cutting temperature (OCT) compounds, such as Tissue-Tek (Sakura Finetek Europe, Zoeterwoude, The Netherlands) on a cork disc. The biopsy sample is then snap-frozen, agitated (to prevent ice crystal formation) in isopentane (2-methylbutane), pre-cooled in liquid nitrogen (LN_2) and stored at $-80°C$. This can also be performed on whole rodent muscle, typically by cutting out the mid-belly of the muscle and completely embedding it in OCT/Tissue-Tek while placed on cork. Freezing muscle tissue in this manner allows tissue to be preserved quickly without affecting its morphology, while also preventing the actions of proteases that may degrade proteins of interest. Alternatively, to better preserve muscle length and structure (67), whole animal limbs can be fixed in formalin (usually 10% formaldehyde in water) to quickly penetrate and fix tissues at the desired length. The aldehyde in this process acts to cross-link all proteins and lipids, therefore preserving cell structures and membranes. However, this process may modify epitopes, subsequently affecting immunoreactivity with specific target antibodies, especially when using monoclonal antibodies. While antigen retrieval is possible through heating or enzymatic procedures, it can be a timely and difficult process to optimise and can also generate non-specific binding that can then cause autofluorescence when visualising the proteins of interest using a fluorescent microscope (described below).

Frozen tissue can then be prepared into ultrathin cryosections, typically between 5 and 12 μm (0.005 and 0.012 mm) using a cryostat machine which contains a freezing chamber set between −20°C and −24°C, where the samples are first allowed to equilibrate for 30 minutes before attempting to cut them into sections. Ultrathin sections are then made by bringing the sample closer to a blade. The ultrathin slice is then collected under a 'roll-plate' and can be collected onto positively charged, pre-warmed slides (e.g. Superfrost Plus microscope slides, Fisher Scientific, Pittsburgh, PA), by carefully lowering the glass slide above the sections without making contact with the metal roll-plate. The resultant cryosections are negatively charged so should adhere to the glass slide without actual contact. Sections on glass slides are then stored at −80°C to preserve the antigen integrity on the target proteins of interest. Cells grown in-vitro (see below) can also be fixed in time for immunocytochemistry using formalin fixation, or alternatively methanol/acetone fixation by immersing cells for a short period. This alcoholic method dehydrates cells instantly but may damage some internal structures, so fixation protocols should be optimised.

Immunolabelling of proteins of interest using antibodies

Similar to Western blotting described above, pre-prepared tissue sections must be 'blocked' to prevent non-specific binding of primary or secondary antibodies. While Western blotting traditionally relies on the chemiluminescence of HRP for visualisation, immunohistochemistry/immunocytochemistry involves immunolabelling tissue sections/cells with secondary antibodies that are conjugated to fluorescent molecules/dyes (i.e. fluorophores) that can be detected at specific wavelengths using **widefield fluorescence** or **confocal** microscopy. Tissue sections or cells are then often co-stained with nucleic acid markers such as 4′,6-diamidino-2-phenylindole (DAPI), which binds to adenine–thymine-rich regions in DNA and can be excited at a wavelength of 405 nM. Molecular exercise physiologists often employ immunohistochemistry, specifically to label and phenotype changes in myosin heavy chain isoform content (I, IIA, IIX, IIB) between muscle groups, endurance and resistance-trained individuals and across the lifespan (68, 69). These antibodies are often used in combination with antibodies against proteins that demarcate the myofibre membrane such as dystrophin and laminin, in order to analyse fibre type and size using semi-automated or automated computer analyses programmes (70, 71). The same process occurs for immunocytochemistry but it is often combined with fluorescently conjugated phalloidin, a bicyclic peptide produced by the death cap mushroom (*Amanita haloids*) that labels all actin filaments to reveal myoblast cytoskeleton organisation and muscle sarcomeres in myotubes (myoblasts and myotubes are described in section 'Cell culture models of exercise' in the below section).

Detection and visualisation of the protein of interest through fluorescent microscopy

Once labelled with antibodies/dyes, tissue sections/cells must be covered with an antifade mounting medium to protect fluorescent dyes from fading when exposed to light (termed photobleaching), followed by covering the sample with a glass cover slip. This flattens the tissue section, preventing the tissue section from drying out and protecting the objective lens of the microscope. Tissue sections or cells, and more specifically the fluorescent molecules bound to the antibody complex of the proteins of interest, can be excited using specific wavelengths of light generated by a laser or light/filter system and the resultant emission captured with a monochromatic camera. This is generally performed on a confocal microscope which provides greater control of depth, excitation/emission spectra and reduction of background information when compared with traditional widefield fluorescence microscopy. However, confocal microscopes can be very expensive and most fluorescent microscopes will perform most standard procedures to an acceptable level.

Further considerations for immunolabelling

Again, for further information on the technical considerations of primary and antibody selection, tissue sectioning, antibody incubation, fluorophore selection, imaging and image analysis for immunolabelling, we direct readers to the following referenced resources (68, 70–74).

CELL CULTURE MODELS OF EXERCISE

LEARNING OBJECTIVES

At the end of this section, you should be able to:

1. Describe what satellite cells, myoblasts and myotubes are.
2. Describe the application of cell culture methods in molecular exercise physiology.
3. Describe the advantages and limitations of cell culture models.
4. Describe the use of mechanical stretch and electrical stimulation techniques to investigate the molecular responses to exercise in-vitro.
5. Discuss, compare and contrast the distinct cell culture methods.

Introduction to satellite cells

During muscle tissue development in gestation, myogenic progenitor cells derived from embryonic stem cells migrate and differentiate into committed myoblasts (muscle cells) that then fuse together to form primary multinucleated myotubes (primary muscle fibres) attached via tendons to bones. These primary fibres grow in length and size during gestation. However, the number of fibres is thought to be set in gestation, with any further growth occurring because of increased fibre cross-sectional area (size), rather than fibre number. Also, adult skeletal muscle fibres are multinucleated and thus terminally differentiated, meaning skeletal muscle fibres with many nuclei are incapable of replication or fibre division (known as hyperplasia) say compared to single cells with one nucleus that can divide (i.e. mitosis). Indeed, under normal physiological stress such as exercise, hyperplasia has not been observed in adult human muscle. Only when non-physiological stress such as synergist ablation (surgical removal of the agonist muscles to supra-physiologically load the antagonist muscles) is performed, hyperplasia is observed in mammalian (typically rodent) models. However, despite no evidence of hyperplasia in adult human skeletal muscle under normal physiological conditions, skeletal muscle is well documented to respond to environmental cues such as repeated exercise with an increase in size (hypertrophy) or to periods of disuse/physical inactivity with muscle loss (atrophy), which is perhaps in part as a consequence of the existing fibres increasing or decreasing in size, but also partly determined by the number of nuclei in a given area of the muscle fibre (myonuclear domain) discussed in **Chapter 13** – 'Satellite Cells and Exercise'. Importantly, skeletal muscle also has the ability to regenerate and repair after injury and/or microtrauma (e.g. from eccentric damage following exercise). This begs the question of how sustained growth, repair and regeneration after exercise are achieved in skeletal muscle tissue without the ability of fibre replication.

Satellite cells and in-vitro models

It was in 1961 that Alexander Mauro and Bernard Katz independently identified single-celled population underneath the basal lamina of mature muscle fibres in the frog (75, 76). Mauro called these 'satellite' cells due to their peripheral/'satellite' location. He hypothesised that these were quiescent (dormant) single-celled myoblasts that had not yet fused into muscle fibres. Indeed, the concept of muscle cell fusion was proposed as early as the mid-1850s to 1860s after experiments into muscle degeneration and regeneration were first documented by German scientists, who went on to provide the first evidence of in-vitro myotube formation (fusion of single muscle cells/myoblasts into primary multinucleated fibres) before the end of the second decade of the 20th century. Even the concept of myonuclei (nuclei of myotubes/fibres created from the nuclei of single myoblasts that have fused together) being terminally differentiated/incapable of replication was hypothesised by these early pioneers (for a historical insight into satellite cells, see (77)). Since the formal identification of the satellite cell, in 1961, it has since been substantiated that these resident adult muscle cells, which remain as single cells underneath the basal lamina, are indeed able to activate, proliferate, differentiate and fuse to repair skeletal muscle fibres and contribute myonuclei to maintain increases in muscle size. The role of satellite cells in these processes and with exercise is discussed in detail in **Chapter 13** – 'Satellite Cells and Exercise'.

As such, these muscle cells 'in a dish' (in-vitro) can be used to model: repair, regeneration, hypertrophy and atrophy in the field of molecular exercise physiology. Indeed, commercially available muscle cell lines predominantly from mice or rats (frequently used include the L6s, Sol8s, C2 and C2C12s), as well as primary muscle cell cultures derived from muscle biopsies have now been extensively studied within the field. There are various advantages, as well as limitations, for the use of these muscle cells and various experimental models have been used to mimic exercise to investigate the molecular responses of cultured muscle cells and myotubes to these stimuli. While some may suggest that cell models are perhaps reductionist in terms of the cells not then being situated within the niche in which they were derived in the body, muscle cells have been shown by some to retain many characteristics from their in-vivo niche. For example, cells isolated from physically active/exercised, or diseased individuals such as those suffering with obesity, type-II diabetes and cancer cachexia, as well as those from older donors, remember aspects of the environment they resided in and exhibit cellular and molecular processes and phenotypes in-vitro that are characteristic of the environment they experienced in-vivo (for reviews, see (78, 79)). Therefore, these cells can provide the opportunity to investigate both the internal/intrinsic influencers of muscle cell responses to different in-vivo environments and the external influence of physiological stress such as exercise. For example, using in-vitro damage protocols, mechanical loading/ stretching and electrical stimulation can be performed to mimic the environmental stress of exercise on skeletal muscle. Furthermore, cells are easily dosed in culture with specific growth factors, steroids, nutrients, pharmacological agents and can also be genetically engineered (performing gene silencing/ knockdown or overexpression) to mechanistically ratify the role of specific genes in the molecular response of muscle to various stimuli. Cell studies also allow more extensive time-courses to be investigated that might not be possible due to ethical constraints because of the sheer number of repeated biopsies that would be required. Cell models can also be combined with in-vivo biopsy studies in order to evaluate both the in-vivo and in-vitro responses to exercise, injury, nutrition, age and disease. Disadvantages of in-vitro models include the requirement for expensive facilities, equipment and reagents. For example, sterile culture environments require specialised lamina flow/filtered workspaces (called culture hoods), as well as CO_2 incubators, light and/or fluorescent microscopes, liquid nitrogen vessels and a continued commitment to running costs, servicing and regular reagents/consumables to maintain or store cells and undertake relevant analysis. Despite an advantage of repeated sampling as suggested above, there is also still a finite amount of time the cells can be cultured without becoming non-viable. For example, mature myotubes can spontaneously contract and detach from cell culture dishes. Long-term culture can also make it more likely for the cells to become overpopulated or infected.

Before outlining some of the most relevant in-vitro models to mimic exercise stimuli, it is also important to describe the two methodological schools of thought for the isolation of muscle-derived cells, as this can alter the interpretation of the results from the in-vitro model used. From the biopsy slurry, some researchers use antibodies that are specific to muscle cell surface proteins and use magnetic sorting or Fluorescent Activated Cell Sorting (FACS) to 'pull out' a purer population of only muscle cells. Allowing very accurately the study of the molecular responses in just muscle cells only (e.g. via downstream analysis of DNA, RNA and protein produced from muscle cells). However, some take the view that in this scenario the cellular milieu is no longer retained. The second method omits this purification step, retaining all different types of cells from the biopsy which might then be more representative of the cellular milieu. However, the resulting molecular responses are then reflective of a mixture of predominantly myoblast and fibroblasts as well as (a much lower proportion) other resident cell types isolated from biopsies such as endothelial cells, immune cells, neural/glial cells and fibroadipogenic progenitors (FAPs). It is important to note that a similar issue of cell heterogeneity is the same for a muscle tissue biopsy homogenate, where isolations of DNA, RNA and protein are representative of all the different cell types found in a whole muscle biopsy (even if any visible matrix or fat is removed from the isolation procedure). With advancements in purifications of different cell types and nuclei specific or single-cell OMIC technologies (introduced in **Chapter 1**), these issues around cell or tissue sample heterogeneity may become a thing of the past. While these advancments are currently in progress, most of our knowledge to date is derived from homogenous muscle-derived cells and tissue. What is a slight advantage in-vitro is that it is quite easy to characterise the proportion of myogenic to fibrogenic cells by routine immuno-staining (this is also possible in a tissue sample but it is not as straight forward). However, what is clear is that this characterisation of the main cell populations is something that is crucial to every in-vitro study so that any results obtained from later analyses can account for differences (if any) in the cell population proportions between experimental conditions. Most of the time if cells are isolated from a group of young healthy individuals (perhaps matched for their physical activity status) by the same researcher performing the procedure, then myogenic/fibrogenic ratios can be quite consistent. However, the ratio of myogenic to fibrogenic cells can alter if cells are isolated from older or diseased individuals. Therefore, comparisons between young healthy versus old or diseased populations should be done with this caveat in mind. In the next section, we will now briefly outline the methods for the use of these cells in culture and the experimental means to mimic exercise in these in-vitro models.

Monolayer cell culture

Following the cell isolation steps described earlier in this chapter, cells can be grown in plastic tissue culture flasks or dishes, commonly coated with specific cell-adhesive proteins (typically gelatin/laminin or collagen) to enhance cell attachment with the addition of 'growth media', a solution containing nutrients, serum and antibiotics to support cell proliferation/mitosis and promote cell survival once isolated from the body. Whenever cells are cultured in flat-bottomed plastic dishes/flasks, the method is typically referred to as 'monolayer', which essentially means a single layer of cells/myoblasts (or myotubes in the case of fused myoblasts). It is important that the myoblasts are grown in optimal conditions within an incubator (i.e. 37°C, humidity control & 5% CO_2) which allows cells to proliferate (undergo mitosis/cell division) and increase in number until cells are confluent and cover around 80–90% of the surface area of the cell culture plates or wells (**Figure 2.12**). This ensures the cells do not become overcrowded or 'contact inhibited' whereby they have little space to divide or differentiate and in the case of myoblasts, fuse into myotubes.

Once cells are confluent, the cells are treated with proteolytic enzymes like trypsin/EDTA that detaches the cells from the surface of the cell culture flasks or dishes. The cells can then be frozen and stored in liquid nitrogen for future experiments or split onto new flasks to increase cell number (depending on the number of cells needed for future experiments) or put into smaller well plates, e.g., 6 or 12 well plates to enable assays for investigating muscle cell growth, differentiation, fusion, myotube formation/hypertrophy across various conditions. Muscle cells, especially if wanting to investigate myotubes, are typically cultured on 6 or 12 well plates (rather than 24/36/96 well plates) as the cells need space to fuse and mature into multinucleated myotubes (**Figure 2.12**). In order to investigate differentiation and myotube formation (and to use myotubes that can then be treated with exercise mimicking stimuli, or, for example; endogenous agents, nutrients, pharmacological agents or gene silencing/overexpression reagents), cells are typically grown to confluence in 6 well plates by seeding 80–150,000 cells per ml in 2 ml of growth media (total of 160–300,000 cells) and left for 24 hours to attach and undergo an initial round of proliferation. The cells are then washed twice using 2 ml's of and then, to initiate differentiation and cell fusion into myotubes, the amount of serum in the media solution is dropped

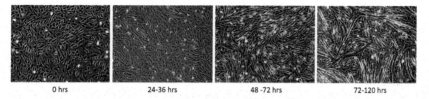

| 0 hrs | 24-36 hrs | 48 -72 hrs | 72-120 hrs |

Figure 2.12 Representative images of myoblasts (100% myogenic C2C12 cell line) in monolayer culture post-transfer into low serum differentiation media /DM. Images left to right are of cells at 0 hours (30 minutes post-transfer into DM) and then approximately between the times 24–48 hours, 48–72 hours and 72–120 hours. Cells are approximately 80–90% confluent at 0 hours, cells continue to proliferate up to 24–36 hours, and then begin to fuse and form myotubes at 48–72 hours; multiple multinucleated myotubes are then clearly visible by 72–120 hours.

from 20% serum content in the 'growth media' to 2% serum content, a media commonly known as 'differentiation media'. Dropping the serum content in the media removes the high concentrations of proteins (typically mitotic growth factors) that are contained within the serum and are involved in cell proliferation/division (examples include, not exclusively, mitotic growth factors such as Hepatocyte growth factor, Transforming growth factor Beta, Fibroblast growth factor). Removal of the serum also means that the cells start to upregulate and secrete their own growth factors, for example, Insulin like-growth factors I (IGF-I) and IGF-II that are fundamental for the differentiation of skeletal muscle cells. Commercially purchased mouse C2C12 cells increase their own IGF-I and IGF-II levels very efficiently, while primary-derived human muscle cells sometimes require additional IGF-I to be added to the media to help initiate this process.

If interested in investigating early proliferation of myoblasts in the presence of nutrients, growth factors, steroids, stretch, electrical stimulation, etc., this can still be performed upon transfer to low serum media. This is because for the following 12–36 hours, before the cells fuse into myotubes, they are still cycling in S/G2 phase of the cell cycle (undergoing mitosis/division). Indeed, upon damage of muscle tissue in-vivo, satellite cells activate, migrate to the site of damage/injury, proliferate for short period and then fuse to the existing fibres. Therefore, by dropping the level of serum in cell culture, an in-vitro model of this initial proliferation and then fusion process can exist. Then after approximately 36–72 hours post-transfer to lower serum media, the cells exit the cell cycle (in the G1 phase) to differentiate and begin to upregulate important Myogenic Regulatory Factors (MRFs) such as MyoD, Myogenin and Myomaker which are involved in myoblast fusion in order to form multinucleated myotubes, with this continuing for approximately another 4–5 days (**Figure 2.12**). Here, if cells were dosed or stretched/loaded (see below), then morphological indices of myonuclear accretion (myotube number, nuclei per myotube) and myotube size can be measured. At day 7–10, most myotubes are established and they begin to express a mature myotube phenotype (striated appearance, upregulation of adult myosin heavy chains and down-regulation of neonatal or embryonic myosin heavy chains). Myotubes can also spontaneously contract, and if fortunate, this myotube 'twitching' can be observed under the light microscope. If the research interest relates to exercise-type stimuli on mature myotubes (which might better represent a model of muscle in-vivo, compared with proliferating/fusing cells that might mimic more the regenerative process of satellite cells to damage/injury), then the myotubes can be utilised in these types of exercise mimicking experiments detailed below.

Models of exercise in-vitro using mechanical loading

Very early work into the mechanical loading of muscle in-vivo demonstrated that loading (stretching) of the plantaris and soleus muscles via synergistic ablation (in this case the removal of the gastrocnemius tendon) evoked rapid hypertrophy (80). Mechanical loading has now been extensively used, both in-vivo and in-vitro, to investigate the molecular response and adaptation of muscle tissue and cells to eccentric exercise like stimuli or muscle damage protocols. Mechanical loading of monolayer cells was first conducted in embryonic chick myoblasts that were multiaxially (different directions) stretched (or 'loaded') for 18 hours on silicone membranes, with the researchers identifying alterations in amino acid uptake, protein synthesis and myotube size (81, 82). Since then, there have been a substantive number of studies mechanically loading myoblasts and myotubes and this is reviewed elsewhere (83). The most well-characterised in-vitro system to perform these types of stretch-loading protocols in monolayer cultures is the Flexcell® FX-5000™ Tension system that uses vacuum pressure to deform flexible silicon membranes on which the cells are cultured. Typically, cells are differentiated into myotubes in low serum (2%) differentiation media for 7–10 days on fibronectin-coated flexible-bottomed culture plates (25 mm BioFlex®, Dunn Labortecknik, Germany), which are then connected to the Flexcell® system. The Flexcell® system then allows either uniaxial or equibiaxial loading of the resultant myotubes. High frequency intermittent loading or low frequency continuous loading regimes then can be used to mimic resistance or endurance stimuli, respectively. Loading of monolayer cultures has been shown to evoke some of the molecular responses observed after exercise in-vivo (reviewed in (83)).

One of the issues with loading monolayer cultures is that cells typically differentiate into myotubes while not under tension; therefore, myotube formation is not aligned, in contrast to muscle tissue that would possess aligned fibres in a parallel muscle in-vivo. Furthermore, monolayer culture is a single sheet of myoblasts or myotubes, which is perhaps not representative of the three-dimensional (3D) structure of myofibres and the surrounding extracellular matrix (ECM) seen in-vivo. Lack of directionality in monolayer cultures also makes it difficult to measure functional responses such as force production. One way to get around this is to use 3D cell culture van techniques. The advent of bioengineering or 'tissue engineering' of skeletal muscle to create 3D muscle cultures in-vitro was introduced over 30 years ago, whereby primary myotubes were cultured within a collagen matrix, that could be subsequently stretched using a computer-controlled system (Vandenburgh, 1988; Vandenburgh et al., 1988). Since then, many laboratories have experimented with different biomaterials as relevant matrices such as collagen, fibrin and laminin (along with many others, either naturally occurring or synthetic biomaterials) in an attempt to produce a more representative 3D 'muscle in a dish'. Culturing cells on a representative ECM under tension can result in aligned myotubes between the points of tension and an improvement in myotube maturation due to enhanced cell attachment and longer time in culture. There are many other considerations when culturing 3D muscle constructs, including increased cell numbers required versus monolayer culture (which can become a problem when using primary-derived human cells that are often limited in number due to small biopsies), together with additional and specialised media components/ supplements compared with the standard components in monolayer cultures (Khodabukus & Baar, 2016). As described above, the cells acquired from muscle biopsy contain a heterogenous cell population (mixture of myoblasts, fibroblasts and other resident muscle-derived cells). Therefore, the percentage of muscle-specific cells within the sample can alter differentiation capacity, maturation and survival of the 3D muscle constructs (84) and are important considerations when engineering 3D muscle in-vitro. The fabrication of 3D muscle can also be a lot more complex, labour-intensive and expensive than monolayer culture. To provide an example, we will briefly describe a bioengineered muscle model that uses a 'self-assembling' 3D culture technique (**Figure 2.13**).

Following myoblasts being grown to confluence in a cell culture flask, cells are seeded in high serum (20%) media onto a fibrin scaffold within 6 well plates that have been pre-coated in a special type of silicone. Within the well, there are a fabricated and aligned series of pins that attach pieces of suture material (mimicking two tendons at each end of the muscle) at standardised distances. When the cells are added to the wells and adhere to the fibrin matrix, the cells themselves produce force on attachment, and the fibrin begins to lift and roll in line with the tension caused by the pins and sutures (**Figure 2.13**). This is supported using a specialised type of silicone that is non-adhesive which allows the fibrin matrix containing the adherant cells to come away from the bottom of the culture dishes (**Figure 2.13A**). Following 2–3 days in growth media, the cultures are switched to low serum (2%) differentiation media and allowed to differentiate/fuse between the pins and suture for 2–3 days (**Figure 2.13B**). After this

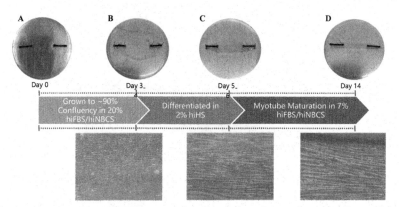

Figure 2.13 Schematic representation of methods for bioengineering 3D fibrin skeletal muscle.
Macro- and microscopic images of fibrin bioengineered muscle at **(A)** 0 days, **(B)** 3–4
days grown to confluency in media containing high serum (20% heat inactivated Fetal
Bovine Serum (hiFBS)/heat inactivated Newborn Calf Serum (hiNBCS), **(C)** 5–6 days
differentiated in low serum media (containing 2% heat inactivated horse serum (hiHS))
for 48 hours and **(D)** myotubes matured up to 14 days in media containing 7% serum
(3.5 hiFBS/3.5% hiNBCS) (10 × magnification, scale bar = 50 μm). The figure is taken
from Turner *et al.* (93), as a full open access article where permissions are not required
provided the work is properly cited. *The Journal of Cellular Physiology* (John Wiley and
Sons 2021).

point, myotubes form in a highly aligned arrangement between the pins and sutures in a 3D muscle
bundle otherwise known as a 'myoid' (**Figure 2.13C**). After this, the myotubes are then maintained in
media with 7% serum for approximately 7–8 more days. This is when myotubes are mature, highly
aligned cylindrical structures (**Figures 2.13D** and **Figure 2.14C**). Fibrin-muscle-constructed diameters
are then measured with, for example, digital Vernier callipers. This is due to the dimensions of
construct/'myobundle' width of <4 mm at the narrowest point that has been associated with the most
mature myotube phenotypes (85–90). Before any loading studies can commence on the 3D muscle
constructs, myotube maturation can be further confirmed by immunocytochemical analysis of adult
myosin heavy chains, as well as number and size of myotubes, number of nuclei per myotube and other
morphological measures by staining with phalloidin (actin) or desmin (intermediatory protein filament
in muscle) and a nuclear stain such as DAPI (described above in immunolabelling). qRT-PCR is also
usually performed to determine the embryonic, neonatal and adult myosin heavy chain gene expression
levels, to confirm that embryonic and neonatal expression is low and adult myosin heavy chain is high,
suggesting a mature myofibre phenotype in the resulting myotubes. Mature constructs can also be found
to spontaneously twitch under the light microscope. Finally, electrical stimulation in these fibrin myoids
has been used to measure force-frequency and length-tension relationships, to confirm similar profiles to
muscle tissue, by measuring peak twitch force and tetanic force (91, 92).

Following the fabrication of bioengineered skeletal muscle constructs, a **bioreactor** system can then be
used to apply a mechanical loading stimulus (13, 93) to mimic eccentric loading. Early bioreactors could
apply uniaxial stretch to single bioengineered myoids using computerised step motors (94, 95), with
similar systems also allowing higher throughput (i.e. in 6 well culture dishes (96)). Most recently,
uniaxial stretch of up to 15 bioengineered myoids muscles has been achieved using the TC-3 tension
bioreactor system (EBERS Medical Technology, Spain). This system allows loading to manipulate the
amount of stretch (compared to resting length), frequency of loading, repetitions, sets and rest times
across several 3D muscle cultures simultaneously allowing for high-throughput and highly controlled

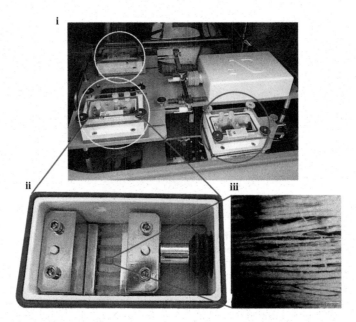

Figure 2.14 Diagrammatic representation of mechanical loading in 3D bioengineered fibrin skeletal muscle. **(i)** The TC-3 bioreactor system (EBERS Medical Technology, Spain) used to mechanically load 3D bioengineered skeletal muscle consists of 3 chambers that can house approximately 5 bioengineered muscles per chamber (total approx. 15 muscles in one experiment). In the image, the 'loaded' muscles are highlighted with yellow circles and the 'non-loaded' controls are highlighted in the blue circle. **(ii)** 5 × fibrin bioengineered skeletal muscle constructs clamped within a single bioreactor chamber. **(iii)** Microscopic image taken of a single bioengineered muscle construct immuno-stained for f-actin (phalloidin-FITC, green) and myonuclei (DAPI, blue) and imaged using confocal microscopy. The figure is taken from Turner *et al.* (93), as an open access article where permissions are not required provided the work is properly cited, *The Journal of Cellular Physiology* (John Wiley and Sons, 2021). Figure Biii (immuno-image) is originally taken from Seaborne et al. (97) Title: UBR5 is a novel E3 ubiquitin ligase involved in skeletal muscle hypertrophy and recovery from atrophy. with permission (Copyright-2019). *The Journal of Physiology* (Copyright-2019 The Physiological Society).

loading regimes within sterile cell culture incubators (**Figure 2.14**) (93). The mechanical loading using this model has been used to demonstrate a similar transcriptional response to human and rodent muscle tissue after exercise *in-vivo* (93). Briefly, to achieve this, matured bioengineered 3D muscle (described above) can be transferred to a TC-3 tension bioreactor system (EBERS Medical Technology, Spain; see **Figure 2.14i**) where constructs are submerged in buffer within individual chambers (**Figure 2.14ii**) that are then attached to the loading arm of the bioreactor (**Figure 2.14iii**) which is housed in a sterile cell culture incubator. Non-loaded 3D muscles, i.e. inserted into bioreactor but not stretched, can also be used as relevant controls within the study design.

Muscles can be loaded with regimes that either mimic endurance or resistance-type loading by the degree of stretch being moderate and continuous (e.g. 5% stretch for 1–3 hours for endurance) or a more severe degree of stretch with an intermittent pattern (e.g. 10% stretch for 4 sets × 10 repetitions with rest between sets, repeated 4–5 times for resistance). Regimes can also mimic other *in-vivo* conditions/models such as bone growth during development (slow elongation – called continuous ramp loading) or synergistic ablation (called static stretch, where the muscle is stretched quickly, e.g. 15%+ its resting length and held at this length for a prolonged period of time). Detailed examples of mechanically loading 3D muscle including the frequencies of loading can be found here (13, 93).

From an exercise modelling perspective, a current major limitation of these in-vitro cultures is the difficulty of performing repeated exercise or chronic training in-vitro. This is because there are issues with maintaining sterile cultures for long periods of time to enable repeated loading regimes, as well as keeping the cells and myotubes viable and the matrix in a suitable condition to sustain the 3D muscles for extended periods. For example, commonly used C2C12s make quick and large myotubes in 3D cultures as these cells are 100% myogenic (all myoblasts). However, they produce higher levels of proteins that can degrade the matrix (e.g. matrix metalloproteinases or MMPs); therefore once myotubes are mature, the window for experimental testing is narrow as the matrix will eventually degrade and the construct may snap. Primary human cells that keep their fibroblast population are not often fast at making large myotubes (and therefore making the 3D muscles in the first instance becomes more challenging). However, the fibroblasts being present in the cell population can help to produce and lay down matrix proteins to maintain matrix structure for longer, and therefore human 3D muscle constructs can be viable for extended periods (84). Furthermore, human muscle cells do not as aggressively produce high levels of matrix degrading factors such as MMPs which can also help sustain the 3D cultures for longer. As mentioned, human 3D muscles can be extremely difficult to create due to different growth/fusion rates of the cells depending on the donor characteristics. Overall, however, the future is bright, as the optimisation of cell proportions, reagents, matrices and bioengineered model systems are being rapidly refined. In the future, these in-vitro models may offer the potential to investigate chronic exercise mimicking stimuli.

Models of exercise in-vitro using electrical stimulation

One of the limitations of mechanical loading to mimic exercise in-vitro is the lack of active contraction. Therefore, electrical stimulation regimes have been used to model concentric contractions in-vitro. Indeed, early studies using electrical stimulation in monolayer cultures have demonstrated that continuous low frequency stimulation can increase AMPK, enhance glucose metabolism and fatty acid oxidation, and high frequency intermittent stimulation can increase protein synthesis, reviewed in (83, 98) in monolayer as well as in bioengineered 3D muscles (83).

A popular method for undertaking electrical stimulation in monolayer is the use of the IonOptix C-Pace EM® (IonOptix, Dublin, Ireland) system. This passes an electrical current through the cell culture media using carbon electrodes that are inserted into the cell culture plate. Using this system, you can alter voltage, frequencies and pulse durations. For example, cells would be grown to 80–90% confluency in growth media and then evoked to differentiate in differentiation media for 7–10 days. Resulting myotubes could then be stimulated using endurance-type stimuli, for example, continuous longer pulses at a lower frequency (13 V, 2 ms pulse duration, 2 Hz continuously for 3 hours) (99), where this stimulus has been demonstrated to increase oxidative capacity of the myotubes. Alternatively, a resistance exercise mimicking regime may consist of higher frequency short pulses for shorter or intermittent durations to elicit more forceful contractions (15 V, 0.4 ms pulse duration with 4 s rest between each contraction at 100 Hz for 30 minutes), that have been shown to increase protein synthetic cell signalling that has been also been observed after resistance training in human muscle tissue in-vivo (100). The electrical stimulation of 3D cultures uses similar techniques; for example, the electrical stimulation of fibrin myoid bioengineered muscle constructs has been also been undertaken in 6 well plates, in a similar fashion to monolayer cultures (101). Careful consideration when using 3D muscle cultures for electrical stimulation studies is required to ensure that it is sufficient to evoke force-frequency and length-tension relationships that are akin to muscle in-vivo (91, 92), as there is more resistance through a 3D construct than a monolayer culture. However, as alluded to above, the advantage of 3D culture is being able to measure force more easily following electrical stimulation in aligned myotubes compared to monolayer that have swirling myotubes producing multidirectional force patterns. Other more recent models of electrical stimulation use 'hydrogel' systems (**Figure 2.15**) where cells and matrix (e.g. fibrin or collagen) are polymerised together in a pre-established 3D cast or frame (102). These systems have also been

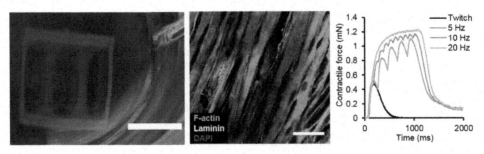

Figure 2.15 Left – 3D fibrin matrix hydrogel system with human derived skeletal muscle cells; **middle** – immuno-cytology image of the mature myotubes in the 3D muscle construct demonstrating aligned myotubes with actin filaments, surrounding a laminin ECM and myonuclei (DAPI stained); **right** – contractile force produced after various electrical stimulation twitches of varying frequencies. All images are taken from Madden L, et al. 2021 (102) as an open access (Attribution 4.0 International -CC BY 4.0 https://creative commons.org/licenses/by/4.0/) article in *Elife* where permissions are not required provided the work is properly cited.

subject to electrical stimulation and demonstrated relevant twitch and tetanic contractions (**Figure 2.15**), as well as functional and mature acetylcholine receptors and even hypertrophy following administration of anabolic agents.

In summary, cells can be isolated from biopsies and used in culture systems to mimic exercise stimuli. These can serve as useful tools to investigate mechanisms of skeletal muscle adaptation in highly controlled laboratory conditions. The research question, the type of in-vitro model and whether the results are relevant to human exercise in-vivo should be considered when using these types of cell models in exercise studies. In-vitro systems that can simultaneously electrically stimulate and mechanically load to represent both the stimuli of muscle contraction and lengthening's are likely to continue in their development in the near future (13).

References

1. Crown A, et al. *J Endocrinol*. 2000. 167(3):403–15.
2. Mullis KB. *Sci Am*. 1990. 262(4):56–61, 4–5.
3. Rigat B, et al. *J Clin Invest*. 1990. 86(4):1343–6.
4. Yang N, et al. *Am J Hum Genet*. 2003. 73(3):627–31.
5. Alekseyev YO, et al. *Acad Pathol*. 2018. 5:2374289518766521.
6. Loos RJF. *Nat Commun*. 2020. 11(1):5900.
7. Duchenne GB. *De la paralysie musculaire pseudo-hypertrophique ou paralysie myo-sclérosique: P. Asselin*; 1868.
8. Bergström J. *Scand J Clin Lab Invest*. 1962. 14(suppl 68):1–110.
9. Bergström J. *Scand J C Lab Invest*. 1975. 35(7):609–16.
10. Evans W, et al. *Med Sci Sports Exerc*. 1982. 14(1):101–2.
11. Henriksson KG. *Acta Neurol Scand*. 1979. 59(6):317–23.
12. Dietrichson P, et al. *J Neurol Neurosurg Psychiatry*. 1987. 50(11):1461–7.
13. Turner DC, et al. *Methods in Molecular Biology*. Clifton, NJ: SpringerNature; 2019. 1889. pp. 55–79.
14. Smith PK, et al. *Anal Biochem*. 1985. 150(1):76–85.
15. Wiechelman KJ, et al. *Anal Biochem*. 1988. 175(1):231–7.
16. Lister R, et al. *Nature*. 2009. 462(7271):315–22.
17. Meissner A, et al. *Nucleic Acids Res*. 2005. 33(18):5868–77.

18. Gu H, et al. *Nat Protoc.* 2011. 6(4):468–81.
19. Smith ZD, et al. *Methods.* 2009. 48(3):226–32.
20. Grehl C, et al. *Epigenomes.* 2018. 2(4):21.
21. Olova N, et al. *Genome Biol.* 2018. 19(1):33.
22. Seaborne RA, et al. *Sci Data (Nat).* 2018. 5:180213.
23. Stirzaker C, et al. *Trends Genet.* 2014. 30(2):75–84.
24. Tost J, et al. *Nat Protoc.* 2007. 2(9):2265–75.
25. Holland ML, et al. *Science.* 2016. 353(6298):495.
26. Zhou VW, et al. *Nat Rev Genet.* 2011. 12(1):7–18.
27. Bannister AJ, et al. *Cell Res.* 2011. 21(3):381–95.
28. Seaborne RA, et al. *Exerc Sport Sci Rev.* 2020. 48(4):188–200.
29. Kimura H. *J Hum Genet.* 2013. 58(7):439–45.
30. Park PJ. *Nat Rev Genet.* 2009. 10(10):669–80.
31. Skene PJ, et al. *eLife.* 2017. 6: e21856.
32. He C, et al. *eLife.* 2017. 6:e25000.
33. Kaya-Okur HS, et al. *Nat Commun.* 2019. 10(1):1930.
34. Buenrostro JD, et al. *Nat Methods.* 2013. 10(12):1213–8.
35. Buenrostro JD, et al. *Curr Protoc Mol Biol.* 2015. 109:21.9.1–.9.9.
36. Schep AN, et al. *Genome Res.* 2015. 25(11):1757–70.
37. Schmittgen TD, et al. *Nat Protoc.* 2008. 3(6):1101–8.
38. Pfaffl MW. *Nucleic Acids Res.* 2001. 29(9):e45.
39. Renart J, et al. *Proc Natl Acad Sci U S A.* 1979. 76(7):3116–20.
40. Towbin H, et al. *Proc Natl Acad Sci U S A.* 1979. 76(9):4350–4.
41. Burnette WN. *Anal Biochem.* 1981. 112(2):195–203.
42. Southern EM. *J Mol Biol.* 1975. 98(3):503–17.
43. Bass JJ, et al. *Scand J Med Sci Sports.* 2017. 27(1):4–25.
44. Laemmli UK. *Nature.* 1970. 227(5259):680–5.
45. Vigelsø A, et al. *J Appl Physiol.* (1985). 2015. 118(3):386–94.
46. Pillai-Kastoori L, et al. *Anal Biochem.* 2020. 593:113608.
47. Ghosh R, et al. *Expert Rev Proteomics.* 2014. 11(5):549–60.
48. Mishra M, et al. *Expert Rev Proteomics.* 2017. 14(11):1037–53.
49. Hesketh SJ, et al. *Expert Rev Proteomics.* 2020. 17(11–12):813–25.
50. Malik ZA, et al. *Proteomes.* 2013. 1(3):290–308.
51. Potts GK, et al. *J Physiol.* 2017. 595(15):5209–26.
52. Steinert ND, et al. *Cell Rep.* 2021. 34(9):108796.
53. Parker BL, et al. *FASEB J.* 2020. 34(4):5906–16.
54. Baehr LM, et al. *Function.* 2021. 2(4):zqab029.
55. Deshmukh AS, et al. *Nat Commun.* 2021. 12(1):304.
56. Hesketh SJ, et al. *FASEB J.* 2020. 34(8):10398–417.
57. Burniston JG. *Biochim Biophys Acta.* 2008. 1784(7–8):1077–86.
58. Holloway KV, et al. *Proteomics.* 2009. 9(22):5155–74.
59. Burniston JG, et al. *J Proteomics.* 2014. 106:230–45.
60. Burniston JG, et al. *Proteomics.* 2014. 14(20):2339–44.
61. Guo H, et al. *J Mol Cell Cardiol.* 2017. 111:61–8.
62. Overmyer KA, et al. *Cell Metab.* 2015. 21(3):468–78.
63. Gregorich ZR, et al. *J Proteome Res.* 2016. 15(8):2706–16.
64. Burniston JG. Investigating Muscle Protein Turnover on a Protein-by-Protein Basis Using Dynamic Proteome Profiling. In: Burniston JG, Chen YW, editors. *Omics Approaches to Understanding Muscle Biology.* New York, NY: Springer US; 2019. pp. 171–90.
65. Hodson N, et al. *Exerc Sport Sci Rev.* 2019. 47(1):46–53.
66. Song Z, et al. *Sci Rep.* 2017. 7(1):5028.
67. Willingham TB, et al. *Nat Commun.* 2020. 11(1):3722.

68. Murach KA, et al. *J Appl Physiol. (1985)*. 2019. 127(6):1632–9.
69. Schiaffino S, et al. *J Appl Physiol. (1985)*. 1994. 77(2):493–501.
70. Desgeorges T, et al. *Skelet Muscle*. 2019. 9(1):2.
71. Wen Y, et al. *J Appl Physiol. (1985)*. 2018. 124(1):40–51.
72. Feng X, et al. *J Vis Exp*. 2018. (134):e57212.
73. Kumar A, et al. *J Vis Exp*. 2015. (99):e52793.
74. Meng H, et al. *J Vis Exp*. 2014. (89):e51586.
75. Mauro A. *J Biophys Biochem Cytol*. 1961. 9:493–5.
76. Katz B. *Sci Am*. 1961. 205:209–20.
77. Scharner J, et al. *Skelet Muscle*. 2011. 1(1):28.
78. Sharples AP, et al. Epigenetics of Skeletal Muscle Aging. In: Vaiserman AM, editor. *Epigenetics of Aging and Longevity*. 4. Boston: Academic Press; 2018. pp. 389–416.
79. Sharples AP, et al. *Aging Cell*. 2016. 15(4):603–16.
80. Goldberg AL. *Am J Physiol*. 1967. 213(5):1193–8.
81. Vandenburgh H, et al. *Science*. 1979. 203(4377):265–8.
82. Vandenburgh H, et al. *J Biol Chem*. 1980. 255(12):5826–33.
83. Kasper AM, et al. *J Cell Physiol* 2018. 233(3):1985–98.
84. Martin NR, et al. *Biomaterials*. 2013. 34(23):5759–65.
85. Khodabukus A, et al. *Tissue Eng Part C Methods*. 2009. 15(3):501–11.
86. Khodabukus A, et al. *Tissue Eng Part C Methods*. 2012. 18(5):349–57.
87. Khodabukus A, et al. *J Cell Physiol*. 2015. 230(10):2489–97.
88. Khodabukus A, et al. *J Cell Physiol*. 2015. 230(6):1226–34.
89. Khodabukus A, et al. *J Cell Physiol*. 2015. 230(8):1750–7.
90. Khodabukus A, et al. *Tissue Eng Part A*. 2015. 21(5–6):1003–12.
91. Dennis RG, et al. *In Vitro Cell Dev Biol Anim*. 2000. 36(5):327–35.
92. Huang YC, et al. *J Appl Physiol. (1985)*. 2005. 98(2):706–13.
93. Turner DC, et al. *Journal of Cellular Physiology*. 2021. 236(9):6534–47.
94. Player DJ, et al. *Biotechnol Lett*. 2014. 36(5):1113–24.
95. Eastwood M, et al. *Cell Motil Cytoskeleton*. 1998. 40(1):13–21.
96. Powell CA, et al. *Am J Physiol Cell Physiol*. 2002. 283(5):C1557–65.
97. Seaborne RA, et al. *J Physiol*. 2019. 597(14):3727–49.
98. Nikolic N, et al. *Acta Physiol (Oxf)*. 2017. 220(3):310–31.
99. Tamura Y, et al. *Am J Physiol Cell Physiol*. 2020. 319(6):C1029-C44.
100. Valero-Breton M, et al. *Front Bioeng Biotechnol*. 2020. 8:565679.
101. Donnelly K, et al. *Tissue Eng Part C Methods*. 2010. 16(4):711–8.
102. Madden L, et al. *eLife*. 2015. 4:e04885.
103. Vandenburgh HH, et al. *In Vitro Cellular & Developmental Biology*. 1989. 25(7):607–16.

CHAPTER THREE

Genetics and exercise: an introduction

Claude Bouchard and Henning Wackerhage

This chapter in the first edition of the book had been written by Dr. Stephen Roth and Dr. Henning Wackerhage.

LEARNING OBJECTIVES

At the end of the chapter, you should be able to:

1. Describe the fundamentals of genetics.
2. Explain how genetic variation contributes to the variability of sport- and exercise-related traits.
3. Describe the structure and function of DNA and the major types of DNA sequence variants.
4. Describe the "central dogma" of molecular biology (DNA to RNA to protein).
5. Discuss genetic testing, gene doping and other concepts that are important for sports and society.

INTRODUCTION

Genetic inheritance is critical to success in sports. One example of the importance of genetics in sports are NBA basketball players. On average, NBA basketball players are ≈2 m tall, but some players are even taller. For example, Gheorghe Muresan and Manute Bol are amongst the tallest NBA basketball players with a height of 2 m 31 cm. In comparison, the average height of an American man is 1 m 76 cm. The average NBA player's body height corresponds to the 99.9th percentile of the US height distribution and Gheorge Muresan and Manute Bol are among the tallest people in the world. A man of average height has almost no chance of ever playing NBA basketball, no matter how good his basketball skills are. Since about 65–80% of human variability in height is explained by genetic differences (1), becoming a world-class basketball player is thus highly dependent on genetic inheritance. Basketballers are a striking example of the importance of genetics in sports, but genetics plays an important role in all sports. The importance of genetics for basketball performance is also a key example why the so-called 10,000 h rule, i.e. that 10,000 h of practice will result in expert performance, is wrong. Ten thousand hours of basketball practice will not change the height of an average-sized individual, showing that genetics and sports practice are both needed to become a high-level NBA basketball player (2).

According to the World Health Organization, genetics is the *study of heredity*. Genetics is concerned with the study of genes, genetic variation and heredity in organisms. We begin this chapter by discussing the field

DOI: 10.4324/9781315110752-3

of genetic epidemiology where researchers use twins, nuclear families and other relatives by descent or by adoption to evaluate whether sport- and exercise-related traits such as body height, VO_{2max}, strength, muscle mass or trainability are inherited. Because biological inheritance is encoded in the cellular deoxyribonucleic acid (DNA), we then describe the DNA molecule, the central dogma of molecular biology (DNA to RNA to protein) and the human genome. Next, we explain the link between inheritance and the human genome with an emphasis on variation in the DNA sequence. Finally, we discuss practical issues such as genetic testing, gene therapy and gene doping.

WHAT IS THE EVIDENCE THAT SPORT- AND EXERCISE-RELATED TRAITS ARE PARTLY INHERITED?

To answer this question, we need to estimate the so-called heritability, which informs us about how much the variation of a sport- and exercise-related trait in a given population is accounted for by genetics. For example, there is a large variation of human body height ranging from a tiny 57 cm to a giant 272 cm. The heritability of body height explains how much of the population variation in body height can be accounted for by genetics.

HOW DO YOU QUANTIFY THE CONTRIBUTION OF NATURE (GENETICS) VERSUS NURTURE (ENVIRONMENT) TO A TRAIT?

Humans have different eye, hair and skin colours. They can be small or tall, weak or strong, run slow or fast, adapt poorly or well to exercise training and have a low or high risk of developing diseases such as type 2 diabetes or cancer. Given this extensive variability, it is legitimate to ask questions such as: Why am I a good sprinter but a poor endurance athlete? Why do I respond so poorly to exercise training? Is it in my genes or is it because of my diet or exercise programme? The Victorian-era scientist, Francis Galton (1822–911), argued that both "*nature*" and "*nurture*" contribute to human variation. Galton used the term *nature* to describe biological inheritance (genetics) and *nurture* to describe lifestyle and environmental factors, which are not only the physical environment but also the social environment and lifestyle factors such as exercise training or diet. Genetics implies that there are DNA sequence variants influencing human variability for a trait of interest. In contrast, the environment includes not only the physical environment but also the social environment and lifestyle factors such as exercise training or diet.

From a biological or behavioural perspective, there are two types of traits or characteristics of an individual. Some traits are binary such as either having a disease like Duchenne Muscular Dystrophy or not. These types of disease often depend on one gene, which can be either normal or defective. Other traits become mature under the influence of a number of genes: the eight genes (and perhaps more) contributing to eye colour would be a good example.

In contrast, the so-called **quantitative traits**, such as body height or VO_{2max} trainability, tend to be normally distributed, and their variation in the population is typically influenced by many DNA variants, environmental factors and their interactions. The most thoroughly investigated quantitative trait is human body height. Studies involving hundreds of thousands of subjects suggest that body height is influenced by thousands of DNA variants (3). Each of the body height-influencing DNA sequence variants adds or subtracts a fraction of a millimetre to or from the individual's height. Studying such traits requires the analysis of DNA variants across the entire genome in many thousands of subjects. The inability to meet these requirements remains a limitation of sport and exercise genetics studies to this day (4).

The research field that aims to measure the contribution of genetics and environment (including lifestyle factors) to the variation of a quantitative trait is named **genetic epidemiology**. It seeks to quantify by how much a trait, such as body height, is dependent on genetics and by how much a trait is dependent on environmental factors, such as physical activity or diet, in a given population. Genetic epidemiology studies are typically observational, but when a trait is significantly influenced by genetics, it justifies large-scale projects aimed at discovering the exact DNA variants and molecular mechanisms underlying variability in a given trait. In humans, researchers have used nuclear families, monozygotic and dizygotic twins, and families of relatives by adoption to estimate by how much traits such as VO_{2max} (5) or skeletal muscle fibre-type distribution (6) are inherited in a given population.

THE VARIANCE COMPONENTS OF EXERCISE TRAITS

How do genetic epidemiologists estimate the fraction of variation of a quantitative trait that is explained by genetics? First, they must quantify the overall variance of a trait. Then, they break down the overall variance into several contributors to the overall variance. Thus, for a trait such as muscular strength, the total variance ($\mathbf{V_P}$) in a population is divided into the variance explained by genetics ($\mathbf{V_G}$), the variance due to environmental conditions, including exercise and diet ($\mathbf{V_E}$), and the variance that is due to interactions between genetics and environment ($\mathbf{V_{GE}}$). The variance that remains is explained by errors, especially measurement errors but also day-to-day variability ($\mathbf{V_{Er}}$). Expressed in the form of an equation, this looks as follows:

$$\mathbf{V_P = V_G + V_E + V_{GE} + V_{Er}}$$

Three illustrations will help you get a better grasp of the significance of the environment (E), genetic (G) and genetic by environment (GE) components of the variance of a trait in a population. Let's use VO_{2max} training response as our trait of interest. In **Figure 3.1**, **Panel A**, endurance training, an environmental factor, increases VO_{2max} in all subjects similarly, with no variation in baseline VO_{2max} or VO_{2max} response to endurance training. The variation in VO_{2max} is only attributable to endurance training (V_E) and error (V_{Er}) but there is no genetic (V_G) or genetic by environment component to the training response. In **Panel B**, the subjects differ at baseline depending on their genotype. This means that there is a G component.

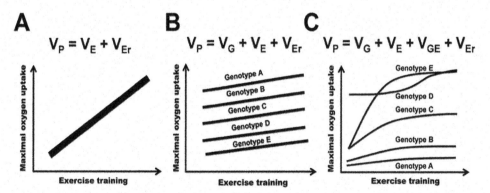

Figure 3.1 Three scenarios depicting the relationship between exercise training and VO_{2max}. **Panel A**: the variation in VO_{2max} is defined entirely by the level of exposure to exercise training. There is no genetic source of variability. **Panel B**: there are individual differences in VO_{2max} accounted for by the genotype at baseline. However, all genotypes respond equally well to exercise training. Thus, we have both G *and* E components of the variance but no GE component. **Panel C**: here, all components (G, E and GE) are contributing to the variability in VO_{2max} trainability. In these illustrations, the error component (Er) is assumed to be negligible. See text for more explanation.

However, they improve similarly with exercise training and have a similar VO_{2max} trainability. Thus, genetics does not affect how VO_{2max} responds to endurance training and therefore, there is no GE source of variance. Finally, in **Panel C**, we have a scenario where the different genotypes respond differently to endurance training. This means that there are E and G (main genetic effect) components but also a GE source of variance accounting for VO_{2max} trainability.

HOW IS HERITABILITY ESTIMATED?

Heritability is the fraction of the variation in a quantitative trait explained by variation due to genetic differences in a given population (i.e. V_G/V_P). To calculate heritability, human studies rely on various types of relatives by biological descent or by adoption. Commonly used study designs include twin studies, adoption studies and nuclear family studies.

The first question to be asked is whether there is familial aggregation for the exercise trait of interest. To address this question, a comparison of the variance between families compared to the variance within-family members is undertaken. Thus, a higher between-family than within-family variance suggests that members of a given family are more similar than individuals of different families. For instance, in the HERITAGE Family Study, there was 2.7 times more variance between families than within families for VO_{2max} in the sedentary state (7).

If familial aggregation is confirmed, the next step is to estimate the heritability level for the trait. For this purpose, studies rely on naturally occurring human relationships such as the two types of twins (identical/non-identical), nuclear family-based quantifications, comparisons of adopted offspring with their foster and biological parents or brothers and sisters, or finally, some other types of relatives. **Table 3.1** summarizes the expected sources of covariance in various pairs of relatives in terms of shared genetic fractions and common environment (cohabitation) levels.

Table 3.1 Expected genetic and shared environmental covariance for various pairs of human relatives

Types of relatives	Genetic variance shared by descent	Common environmental variance based on cohabitation
Spouse-spouse	0	1
Parent-child (living together)	1/2	1
Full siblings (living together)	1/2	1
Full siblings (living apart)	1/2	0
Half siblings (living together)	1/4	1
Half siblings (living apart)	1/4	0
Aunt/uncle-niece/nephew	1/4	0
First cousins (living apart)	1/8	0
DZ twins (living together)	1/2	1
MZ twins (living together)	1	1

DZ, dizygotic (non-identical); MZ, monozygotic (identical).

Equations have been developed to take advantage of shared variances, as defined in the table, to estimate the heritability of complex human traits.

Modified from Bouchard, Rankinen (8).

Based on these expected levels of covariance, heritability (**H²**) can be estimated. For instance, in twin studies, researchers recruit monozygous and dizygous twins, measure the trait of interest and then compare the correlation or within-pair variability between monozygous and dizygous twin sets. Given that there is no genetic variation between identical monozygous twin brothers or sisters as they develop from the same cell (zygote/fertilized egg), the within-pair variability is driven mainly by environmental factors and error. In contrast, since non-identical dizygotic brothers or sisters share only 50% of their genes as they develop from different cells (zygotes from separate fertilized eggs), the trait variation between members of dizygous twin pairs is explained by genetic variation, environmental variation, plus interactions between these factors and the error component. Although there are more accurate and sophisticated ways of computing the heritability coefficient, a quick approximation is obtained with the Falconer formula, where r is a correlation coefficient that describes how close the values of the phenotype are in twin pairs:

Heritability $= 2\left(r_{monozygous} - r_{dizygous}\right)$

As an example, let's use data of a twin study on the methylation level (i.e. the addition of CH_3/methyl groups) in a specific stretch of DNA (9). A higher correlation of 0.88 was found among pairs of monozygous twins when compared to a correlation of 0.48 in dizygous twins. Using the correlations from **Figure 3.2** and the Falconer formula yields the following heritability estimate:

Heritability $= 2(0.88 - 0.48)$ or 0.8

Thus, the variance of DNA methylation level explained by genetics reaches 0.8 or (0.8×100) 80%. This suggests that 80% of the individual differences in DNA methylation are potentially explained by DNA sequence variants. A word of caution: this approach generates heritability levels that are inflated compared to other methods. Complex models and related equations have been used to extract all the information from the covariance pattern and develop optimized estimates of heritability (10).

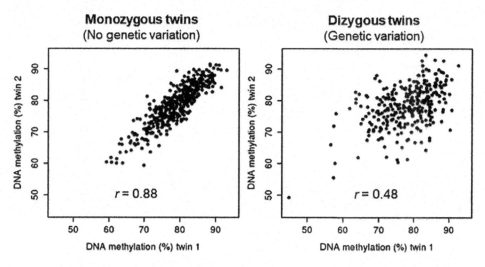

Figure 3.2 Correlation of the methylation level at DNA sites (in %) in genetically identical twins and dizygotic twins. Genetically identical twins are much more similar than the twin pair members sharing only 50% of their genes by descent for methylation level. Figure redrawn from Hannon E, et al. 2018 (9) as an open access (Attribution 4.0 International - CC BY 4.0, https://creativecommons.org/licenses/by/4.0/) article in *PLOS Genetics* where permissions to share, reuse or adapt are not required provided the work is properly cited.

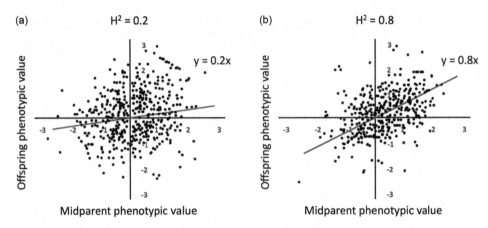

Figure 3.3 Schematic redrawn from Visscher, Hill et al., (11) demonstrating heritability can be estimated by comparing the regression level of a trait between both parents (midparent value) and thier offspring. **Panel A** shows that the phenotype of the parents (midparent is the mean of both parents) correlates poorly with the phenotype in the offspring (low slope of the regression line). In **Panel B**, there is a much closer relationship between the mean phenotype of parents and that of their offspring (steeper slope), suggesting that the variation of the trait under the scenario in **B** is more dependent on genetics than the trait shown in **A**. However, under both scenarios, there are wide inter-individual differences as shown by the scatter of individual values. Redrawn (schematic only and dots do not represent original values) from Visscher, Hill et al., Nature Reviews Genetics 9: 255–66, 2008, Springer Nature. (11).

Since heritability is a *population estimate* of the relative contribution of genetic differences to a given trait, it does not provide useful information for a given individual. For instance, consider the data depicted in **Figure 3.3.** Using family data, heritability can be estimated by comparing the regression level of a trait between both parents (midparent value) and their offspring (11). The figure illustrates the distribution of individual scores for two levels of heritability, namely 0.2 and 0.8. Note that even under the conditions of high (80%) heritability, there are very large inter-individual differences in the relation between the midparent value for the trait and the values of their offspring. This is an illustration that heritability is an average population value and is not applicable to a given individual.

A complementary strategy to determine whether there is a genetic component to an exercise-related trait is to rely on the comparison of the variance between inbred (genetically identical) strains of rodents to the variance within strains. In one experiment, the endurance performance of 11 inbred strains of completely untrained rats (6 animals per strain) was tested (12). The highest performing strain could run 2.5 times further in distance than the lowest performing strain. This resulted in an endurance performance heritability coefficient of 0.50 or 50%, even in animals that were never exercise-trained.

FROM GENETIC EPIDEMIOLOGY TO THE HUMAN GENOME

If genetic epidemiology studies show that a trait such as muscle strength or VO_{2max} trainability is significantly inherited within a population, then the next goal is to identify the variations in the DNA sequence that explain the heritability of the trait. Why? First, as scientists we aim to not only describe but also explain phenomena. Second, the discovery of causative DNA variants could illuminate the fundamental biology underlying a trait. Third, such advances could also have some practical application. For instance, we know from the HERITAGE study that there are large individual differences in VO_{2max} response to an endurance training programme, with a genetic component in the order of 50% (13).

Imagine that all DNA variants associated with VO_{2max} trainability have been discovered. People could be genotyped for these variants and the information used to inform individuals on the probability that they will respond well or poorly to endurance training. In other words, one could expect that genetic-based counselling could improve the efficacy of exercise prescription in the future.

However, even though we can now measure millions of common DNA variants on a DNA chip for $100 and every single adenine (A), cytosine (C), guanine (G) and thymine (T) nucleotide of the genome for about $1,000, identifying the specific DNA variants that explain individual differences in sport- and exercise-related traits has proved difficult, something we will discuss later in this chapter.

THE SEQUENCE OF DNA NUCLEOTIDES CONSTITUTES THE BLUEPRINT DEFINING THE DEVELOPMENT OF A LIVING MAMMALIAN ORGANISM

DNA is the molecule that encodes the blueprint of our organism, including its sporting potential. DNA is composed of four nucleotides with each including a phosphate, a sugar termed deoxyribose and a base which could be either a purine (i.e. adenine or guanine abbreviated as A and G) or a pyrimidine (i.e. cytosine or thymine abbreviated as C and T).

Until about the early 1950s, there was still a debate whether the genetic information was contained in the strings of nucleic acids or in the diversity of proteins. Then in 1944, Oswald T. Avery demonstrated with viruses that nucleic acids were the carriers of genetic information. However, it was still unknown how the genetic information was encoded in the DNA of an organism. In 1953, Francis Crick and James Watson resolved the issue in a 1-page paper in *Nature*. Based on X-ray of the DNA structure by Rosalind Franklin and Maurice Wilkins, and other data on nucleic acid structure, they proposed the now famous double helix structure of DNA (14). They concluded, correctly, that DNA consists of two strands that are coiled around each other to form a double helix, with the informative base on the inside (**Figure 3.4**). In each strand, the various bases (i.e. A, T, C and G) are connected by a strong phosphate bond. The two

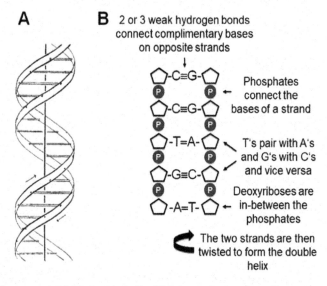

Figure 3.4 (A) Watson and Crick's double helix. **(B)** Schematic showing a linear representation of how nucleotides of the double helix are connected by phosphates within each strand and how weak 2–3 hydrogen bonds connect the two strands.

strands are also bound together because A binds to T and G binds to C on the other strand via two or three hydrogen bonds which are, however, weak when compared to the phosphate bonds.

WHAT ARE CHROMOSOMES?

The DNA of the human genome is divided into 46 chromosomes. Normally, the chromosomes are dispersed within a nucleus and individual chromosomes cannot be recognized under the microscope. However, during the so-called metaphase period of cell division, DNA condenses to form tightly packed chromosomes (15) that segregate into two daughter cells. The human genome is composed of 22 pairs of non-sex chromosomes, termed **autosomes**, plus two **sex chromosomes**, which are XX in females and XY in males. **Figure 3.5** depicts the chromosomes in a human cell.

It is important to understand the distinction between germ cells and somatic cells. Male and female **germ cells**, sperm and oocyte, respectively, are termed **gametes**. A mature germ cell has only 22 autosomes plus one sex chromosome (i.e. 23 chromosomes in total) in its nucleus instead of the full complement of 46 chromosomes seen in somatic cells (i.e. body cells or non-germ cells). A germ cell containing 23 chromosomes is said to be "**haploid**", whereas normal somatic cells contain a "**diploid**" set of chromosomes. When a haploid sperm and a haploid oocyte fuse, they form a diploid, fertilized oocyte, termed zygote, which is the first cell of the new organism. It has 2 × 22 autosomes plus either two X chromosomes (XX, a female) or an X and a Y (XY, a male) chromosome.

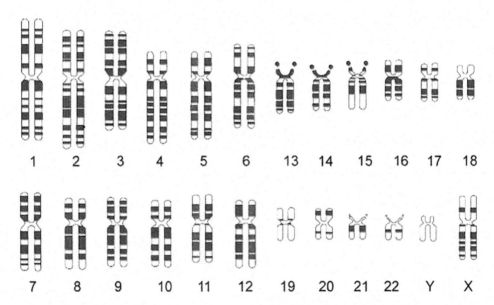

Figure 3.5 The 46 human chromosomes. There are 44 (22 pairs) of non-sex chromosomes, termed autosomes, plus a 23rd pair of X and Y chromosomes. Males possess an X and a Y chromosome and females possess two copies of the X chromosome (of which one is permanently inactivated). Number 1 is given to the largest chromosome and number 22 to the smallest. Note that visible chromosomes only form during the metaphase period of a cell cycle. In other phases of the cell cycle, individual chromosomes cannot be distinguished.

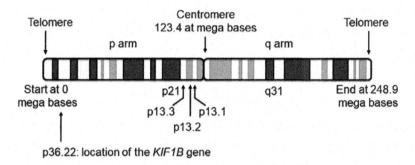

Figure 3.6 Schematic image (ideogram) of human chromosome 1 depicting the cytological bands and the physical length of the double helix in mega base pairs. Note that not all cytological bands are shown and labelled. Redrawn and adapted from Genetics Home Reference, National Library of Medicine, National Institutes of Health of the USA.

How are chromosomes structured?

Chromosomes have two arms and a central constriction which is termed the **centromere**. The short arm of a chromosome is denoted as **p** and the long arm as **q**. Each arm of the chromosome is subdivided into regions numbered consecutively from the centromere to the **telomere** which is the tip of each chromosome arm. Each **band** (i.e. the dark and light stripes of a chromosome seen in **Figure 3.5**) within a given region is identified by a number. With this nomenclature, it is possible to specify any chromosomal region by its "cytological address". For example, chromosome 1 is composed of about 249 million (mega, M) DNA base pairs (**Figure 3.6**). 1p22 refers to chromosome 1, p arm, region 2, band 2. Since the sequence of the DNA bases of the entire human genome is now available, it is possible to specify a physical position on a given human chromosome in terms of the exact base number in a sequence ranging from one to millions. For instance, there are 4,300 genes encoded on chromosome 1. The gene KIF1B which encodes kinesin family member 1B codes for a motor protein that transports vesicles within cells. It is located on 1p36.22 and extends from 10.21 to 10.38 M bases of DNA.

GENES, TRANSCRIPTION AND TRANSLATION

The name "gene" was introduced by the Danish botanist Wilhelm Johannsen and is derived from the term "genetics" which William Bateson had introduced in 1905. A simplified definition of a **gene** is **a DNA sequence that encodes one or more proteins**. The genome encodes the information needed to specify the sequence of amino acids of all proteins synthesized by the cellular machinery. This is important because organisms are either built from proteins (e.g. ≈70% of the dry mass of a muscle is protein) or from molecules that are made by a special class of proteins, termed enzymes.

How is a gene "read" to produce a protein? Francis Crick wrote in 1956 something in his notebook which he called the "central dogma". Today, this unpublished notion is widely known as the **central dogma of molecular biology**. It describes how the biological information flows in the "DNA → RNA → protein" direction. According to the dogma, DNA is equivalent to the instructions for the book of life. RNA is very similar to DNA, but it is single-stranded, whereas DNA is double-stranded (i.e. the double helix), and the sugar in RNA is a ribose, whereas the sugar in DNA is a deoxyribose. Also known as "messenger" RNA (mRNA), the RNA copies and delivers the DNA "message" to the protein-making machinery of the cell (in the ribosome) to make the protein. The making or synthesis of RNA from DNA

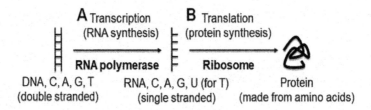

Figure 3.7 **(A)** Double-stranded DNA composed of C, A, G and T bases. During transcription, which is catalysed by RNA polymerase, one strand of DNA is transcribed into single-stranded RNA. RNA is similar to DNA but has ribose as a sugar and has uracil (U) instead of thymine (T). It is a template of the DNA which contains the instructions for a given protein. **(B)** Proteins are long, folded chains of amino acids. In DNA and RNA, 3 nucleotides encode 1 amino acid. The ribosome is an organelle made mainly from RNA which has the molecular machinery to read the RNA and synthesize a protein according to the nucleotide sequence of the RNA strand.

is termed **transcription** (RNA synthesis also described as gene expression) and the process of protein synthesis from RNA is termed **translation**. **Figure 3.7** illustrates Crick's central dogma. It is important considering recent advances in molecular biology to recognise that the central dogma is incomplete and that there are exceptions to the dogma.

According to the Genome Reference Consortium, we have 20,465 genes in our human genome. Most genes start with an ATG start codon (which is transcribed into AUG in RNA) that encodes methionine as the first amino acid of a protein. Genes also end with a stop codon which can be TAG, TGA or TAA in DNA or UAG, UGA and UAA when transcribed into RNA. Three DNA bases, termed a triplet, encode one amino acid within a protein. Since there are 64 possible combinations of 4 DNA bases in a 3-base code and only 20 common amino acids to specify, the genetic code is redundant, with most amino acids being encoded by more than one triplet. **Table 3.2** lists the codons encoding each amino acid.

A typical gene (**Figure 3.8**) consists of coding sequences termed **exons** (i.e. exons encode the amino acid sequence of a protein), interrupted by noncoding regions termed **introns**. Additionally, there are regulatory DNA sequences located upwards and downwards and at times far from the gene whose transcription depends on the regulatory DNA. The number of exons is highly variable, with a range from one (e.g. G-protein-coupled receptor genes, GPCRs, with no introns) to a few hundred exons, such as titin (gene symbol TTN), a gene with 363 exons.

Only one DNA strand of the double helix is transcribed into an RNA. Noncoding intron regions copied from the DNA are cut (spliced) out of the RNA and the exons fused together during its processing from "pre-RNA" to mature mRNA. Transcription begins at an initiation site along the DNA strand termed a promoter. Other regulatory DNA sequences are enhancer and silencer sequences. Enhancers are short DNA sequences that are activated when bound by proteins resulting in an increase in transcription (gene turned on). A gene sequence includes variable numbers of enhancers, and there is a relation between the number of activated enhancers and the rate of transcription and resultant amount of mRNA for that gene. In contrast, silencer sequences can bind repressor proteins that attenuate or suppress (silence) the transcription of a gene. Silencers just like enhancers are often located upstream of the gene-coding sequence, but they are also found downstream or far away from the gene. The latter can still impact the rate of transcription because the looping of the DNA within the nucleus brings the enhancer or silencer sites in close proximity to promoter(s).

Table 3.2 DNA codons and the amino acids that they encode

DNA codon(s)	Amino acid	Abbreviation*
ATT, ATC, ATA	Isoleucine	I, Ile
CTT, CTC, CTA, CTG, TTA, TTG	Leucine	L, Leu
GTT, GTC, GTA, GTG	Valine	V, Val
TTT, TTC	Phenylalanine	F, Phe
ATG	Methionine	M, Met
TGT, TGC	Cysteine	C, Cys
GCT, GCC, GCA, GCG	Alanine	A, Ala
GGT, GGC, GGA, GGG	Glycine	G, Gly
CCT, CCC, CCA, CCG	Proline	P, Pro
ACT, ACC, ACA, ACG	Threonine	T, Thr
TCT, TCC, TCA, TCG, AGT, AGC	Serine	S, Ser
TAT, TAC	Tyrosine	Y, Tyr
TGG	Tryptophan	W, Trp
CAA, CAG	Glutamine	Q, Gln
AAT, AAC	Asparagine	N, Asp
CAT, CAC	Histidine	H, His
GAA, GAG	Glutamic acid	E, Glu
GAT, GAC	Aspartic acid	D, Asp
AAA, AAG	Lysine	K, Lys
CGT, CGC, CGA, CGG, AGA, AGG	Arginine	R, Arg

*The one-letter and three-letter amino acid abbreviations are shown.

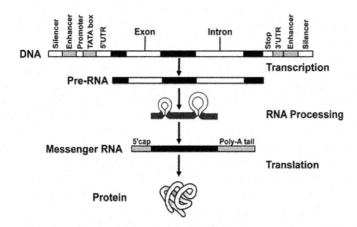

Figure 3.8 A simplified schematic structure of a protein-coding gene. A three-exon gene is illustrated. A partial regulatory region is depicted with elements on both sides of the coding sequence. The basic theorem of biology in which information flows from DNA to RNA to protein is shown with key RNA-processing steps including splicing intronic segments being illustrated.

CLAUDE BOUCHARD AND HENNING WACKERHAGE

Below is a specific example showing how five codons (i.e. 5 × 3 bases) are first transcribed by the enzyme RNA polymerase to RNA and then translated by ribosomes into five amino acids within a protein.

Protein-coding DNA strand:	**5'-ATG TTC ACT GGT GTG-3'**
	‖‖ ‖‖ ‖‖ ‖‖ ‖‖
Antisense DNA strand:	3'-TAC AAG TGA CCA CAC-5'
	↓ ↓ ↓ ↓ ↓
mRNA	5'-AUG UUC ACU GGU GUG … AAA-3'
	↓ ↓ ↓ ↓ ↓
Protein	Met Phe Thr Gly Val

You probably wonder about the 5' and 3' ends, and the "AAA" annotation at the end of the mRNA sequence. The 5' and 3' refer to the position of specific carbon atoms in the deoxyribose sugar of a DNA strand and provide information about the direction of the DNA strand. The "AAA" symbolizes the so-called poly-A tail added to each mRNA after transcription, which makes RNA more stable. Transcription of RNA takes place in the nucleus of the cell, whilst translation of proteins in the ribosomes occur in the cytoplasm. Once translated into a polypeptide, posttranslational modifications such as phosphorylation, methylation or acetylation may happen to generate a mature, functional protein or change how active the protein is.

With 3.2 billion pairs of nucleotides in the haploid human genome, about 20 million genes could be encoded. However, there are about 1,000 times fewer protein-coding genes than this estimate. But there are many more proteins than the 20,465 protein-coding sequences currently recognized in the human genome. The higher number of encoded proteins, for which the absolute number is still a matter of debate, is explained mainly by DNA-coding sequences producing more than one mRNA transcript, called a transcript variant. The disparity between the number of genes and gene transcripts results most frequently from alternative promoters and alternative splicing. As described above, splicing is the process by which introns are removed and exons sequences are fused together into an mRNA. Alternative splicing refers to a situation in which a single gene produces multiple messenger RNAs through different combinations of exons (**Figure 3.9**). Approximately 75% of the human genes with multiple exons have alternative splice sites. Alternative splicing may cause either inclusion or exclusion of one or several exons.

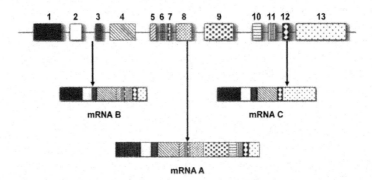

Figure 3.9 An example of a DNA sequence comprising 13 exons but giving rise to 3 mRNAs and eventually 3 proteins via a combination of alternative promoters and alternative splicing (not shown in detail here). mRNA (**A**) contains all 13 exons, mRNA (**B**) includes 9 exons and mRNA (**C**) only has 6 exons.

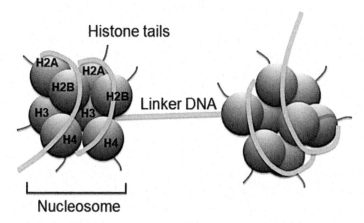

Figure 3.10 Schematic figure showing how DNA is wrapped around two nucleosomes, each of which is an octamer composed of 2 copies of each of the 4 histone proteins (H2A, H2B, H3 and H4). The linker DNA segment is composed of about 20 DNA base pairs.

How are 2 metres of human DNA packaged to fit within a tiny nucleus?

Linearly, the DNA of a diploid cell is about 2 m long and is packaged to fit into the tiny nuclei of our cells. How is this done? First, the linear DNA (primary structure) is twisted to form the double helix where one turn accommodates 10 nucleotides (**Figure 3.4**, secondary structure). The DNA double helix is then wrapped around protein complexes termed **nucleosomes**, which makes it appear like beads on a string (tertiary structure). A nucleosome contains 4 pairs of so-called histone (H) proteins (2 × H2A, 2 × H2B, 2 × H3 and 2 × H4) around which are wrapped 146 DNA base pairs of DNA (**Figure 3.10**). The DNA surrounding one nucleosome is joined to the DNA of the next nucleosome by about 20 linear DNA base pairs, known as linker DNA. Histones are not only the "curlers" of DNA but also regulate whether a stretch of DNA is open for biological interactions or tightly closed (more below on this topic).

Finally, the chromatin is further compacted to fit the DNA of all chromosomes of the human genome into a tiny cell nucleus (quaternary structure). Researchers using advanced microscopy to study how chromatin was actually packaged within the nucleus have shown that it is flexible and disordered and not neatly packaged (15).

How is the transcription of a gene regulated?

The regulation of gene expression is one of the most challenging issues in human biology. It is important because many adaptations to exercise occur when the transcription of a gene is increased or decreased overtime, in response to exercise stimuli (covered in detail in Chapter 7). Every cell with a nucleus contains a complete copy of the human genome. Some genes are expressed in most or all tissues because their RNA and protein are essential for normal cell functions ('housekeeping' genes). However, most genes are expressed only in specific tissues, and some are expressed when cells are stimulated, e.g. by calcium (Ca^{2+}), low glycogen or mechanical load that occur with exercise. To meet so many specific demands, gene expression must be closely controlled and coordinated. This is achieved mainly via networks of transcription factors (described below) with the contributions of alternative splicing, and multiple promoter, enhancer and silencer sequences.

The initiation of transcription requires the presence of a transcription factor that binds to specific DNA sequences (motifs) located both close and far away from the regulated gene (far-away DNA motifs can come close to the gene by DNA looping). The regulation of gene transcription is a coordinated mechanism involving 1,600 transcription factors that bind DNA and that allow RNA polymerase to transcribe the gene (16). Given there are over 20,465 transcribed genes and only 1,600 transcription factors implies that most transcription factors bind to the sequence motifs of multiple genes when activated, i.e. an average of about 13 genes per transcription factor. The number of transcription factors is an important feature of the biological complexity of an organism. Once a transcription factor, plus proteins known as co-activators are bound to the promoters or enhancers, an RNA polymerase binds to the transcription factor complex to begin transcribing the corresponding RNA encoded by the target DNA sequence. Whether enhancers, silencers and other response elements that are part of the regulatory sequences of a gene are activated or not will determine if transcriptional activity is increased, reduced, or inhibited. Once the DNA sequence of a gene has been transcribed into a messenger RNA, the primary RNA transcript (i.e. the full copy of the original template DNA) undergoes various posttranscriptional modifications. These include splicing and removal of the unwanted internal sequences (i.e. intronic sequences), fusion of the remaining exons, and capping at the initiation site (5 prime end) and polyadenylation at the terminal end (3 prime end) of the transcript. The splicing process is directed by specific nucleotide sequences at the exon–intron junctions.

EPIGENETICS: WHAT ARE DNA METHYLATION AND HISTONE MODIFICATIONS AND WHAT ARE THEIR IMPLICATIONS?

You will have probably heard the term epigenetics. What is it? Epigenetics means changes in gene expression that are due to chemical modifications of the DNA or of histones that are not due to variations in the DNA sequence (17). There are two main categories of epigenetic chemical reactions: the methylation of DNA and the addition of small chemical groups such as acetyl group to histones (**Figure 3.11**). DNA methylation refers to the addition of a CH_3 (i.e. methyl) group to a cytosine (C) that is followed by a guanine (G) in the same strand. This is referred to as a CpG where the "p" stands for the phosphate that connects two bases in the same strand. In contrast, CG refers to a C in one strand pairing with a G in the other strand. Regions with many methylated cytosines are also known as CpG islands. CpG islands are present across the whole genome but are often concentrated in promoter regions of genes.

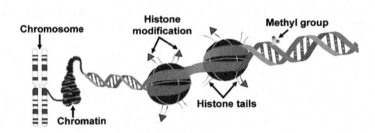

Methyl group added to cytosine in a CpG island

Acetylation, methylation or phosphorylation of a histone tail

Figure 3.11 Schematic representation of epigenetic modifications. Chemical acetylation, methylation or phosphorylation of histone tails of nucleosomes as well as methylation of a cytosine base on the double helix. The epigenome participates in the regulation of gene transcription/expression.

How does the CH_3 methyl group become added to some cytosines and what is the consequence? Cytosines are methylated by enzymes, called DNA methyltransferases. The resultant cytosine methylation then typically reduces gene transcription. This is because methylation reduces transcription factor binding to that region of DNA. Because CpG sites and CpG islands are located more frequently around promoter regions, reduced transcription factor binding after increased methylation suppresses gene transcription. In contrast, reduced methylation (demethylation or hypomethylation) of CpG sites in a promoter region can increase gene transcription, by allowing transcription factor binding. Exceptions are if increased methylation occurs in a silencer or repressor element where it can therefore increase transcription. CpG islands can be methylated (or demethylated) in the basal, natural state, but cytosine residues also become more or less methylated in response to nutrients, environmental stressors, exercise and other stimuli.

The second class of epigenetic events involves histones, the proteins that act as DNA curls. The double-stranded DNA of a chromosome is packed tightly in nucleosomes (tertiary structure), which consist of DNA wrapped around histone proteins. Chemical modifications, such as acetylation, methylation and phosphorylation, occur preferentially on the N terminus of histones, known as histone tails, with the net effect of altering the charge of the side chain and making the DNA available for binding of transcription factors. For example, in most cases, acetylation (addition of an acetyl group) to histones relaxes the chromatin structure and enables gene transcription, whereas reduced acetylation (deacetylation or removal of an acetyl group) results in more compact chromatin and is associated with reduced transcription. Overall, chemical histone modifications can change the conformation of chromatin and thereby change transcription activity.

In summary, the profile of DNA methylation and chemical modification of histone proteins constitutes the epigenome. Epigenetic modifications can occur at all ages as a result of exposure to environmental factors, such as nutrients, cellular insults and stress. The research on whether acute exercise, exercise training or inactivity are stimuli that can trigger epigenetic events is discussed later in this book (**Chapter 6**). Retained epigenetic modifications to DNA in skeletal muscle tissue have been identified following exercise training, even after a period of detraining, and therefore a so-called epigenetic memory or "epi-memory" of prior exercise has been proposed. Epigenetics of exercise and muscle "memory" are also discussed in detail later in this book (**Chapter 6**).

HOW WAS THE HUMAN GENOME SEQUENCED AND HOW MANY BASE PAIRS ARE IN A HUMAN GENOME?

One of the greatest scientific triumphs of the late 20th and early 21st centuries was the sequencing of the human genome and of the genomes of many other species. The starting point of all of these DNA sequencing projects was the development of DNA sequencing using the chain-termination method by the twice Nobel Laureate, Frederick Sanger (18). This method was first used to sequence all the 16,569 base pairs of the human mitochondrial genome (mitochondria have their own DNA termed mtDNA) (19). Following this, the state-funded multi-national Human Genome Project consortium ended up competing with a privately funded effort (Celera Corporation), the latter using so-called shotgun DNA sequencing. Both teams published their draft genome sequences in 2001, first the Human Genome Project consortium in *Nature* on the 15th of February (20) followed by Craig Venter and Celera Corporation one day later in *Science* (21).

Sequencing the human genome revealed its true size. The size depends on whether we focus on haploid gametes (i.e. sperm or oocyte) with 23 chromosomes or on diploid body cells with 46 chromosomes. The genome of human haploid cells has 3,609,003,417 (3.6 giga or billion) base pairs (Genome Reference Consortium Human Build 38), whereas diploid body cells have twice that number (7.2 giga bases).

WHAT ABOUT MITOCHONDRIAL DNA?

Each mitochondrion of a body (somatic) cell contains several copies of circular, double-stranded DNA molecules composed of 16,569 base pairs. The mitochondrial genome is very small compared to nuclear DNA. Mitochondrial DNA (mtDNA) was successfully sequenced in 1981 (19). mtDNA is able to replicate itself independently of nuclear DNA and has its own system of transcription and translation. The bulk of mtDNA is inherited from the mother through the egg cytoplasm at fertilization, but there is also evidence for paternal inheritance of a few copies of mtDNA. mtDNA codes for 37 RNA transcripts, 28 on the cytosine-rich light strand and 9 on the guanine-rich heavy strand. These 37 RNAs are processed into 13 polypeptides associated with the regeneration of ATP in the mitochondrion, two ribosomal RNAs and 22 transfer RNAs.

Since there are more than 1,000 proteins playing an active role in the mitochondria, mtDNA makes a critical but small contribution to mitochondrial biology. All other proteins and small molecules required for healthy mitochondria are encoded in the nuclear genome and are exported to mitochondria. The integrity of mtDNA is a great importance, as inherited or acquired anomalies in the mtDNA sequence can lead to mitochondrial dysfunction. Such dysfunctions are observed in some cases of exercise intolerance and a number of pathologies.

WHAT CAUSES VARIATION OF THE DNA SEQUENCE AND WHAT TYPE OF DNA VARIATIONS EXIST?

We have already mentioned that DNA sequence variants influence body height, strength, VO_{2max} trainability or disease risk. We will now discuss how variations in the DNA sequence occur, the different types of DNA variants and their frequency in human populations. But first we need to explain the vocabulary used to define DNA variants. A **mutation** is an event that changes a DNA sequence. The consequence of a mutation is a **DNA variant**. For example, a mutation may change a "…CTGT…" to a "…CTAT…" sequence resulting in a G/A DNA variant. Alleles are DNA variants of a given sequence. For example, assume that 20% of a population have a "CTGT" and 80% a "CTAT" DNA sequence in the myostatin gene (important in muscle mass regulation, covered in Chapter **4** and **8**); the CTGT variant would be the **minor** (frequency) **allele**, whereas the CTAT variant would be the **major** (frequency) **allele**. Alleles with a frequency of less than 1% are referred to as **rare alleles**, whilst those in the range of 1–5% are known as **low frequency alleles**. DNA variants with a minor allele carried by 5% and more of the population are labelled as **common alleles**. If a one-base substitution occurs in at least 1% of a population, then it is termed a **single-nucleotide polymorphism** and is abbreviated as SNP. SNPs are the most studied DNA variants.

The various types of single-nucleotide base substitutions are depicted in **Figure 3.12.** Many mutations are synonymous or silent substitutions; they do not change an amino acid in the final gene product because of the redundancy of the genetic code (see **Table 3.2**). They are the most frequently observed variants in coding DNA. Nonsynonymous substitutions result in an altered codon that specifies either a different amino acid (a missense mutation) or a termination codon (a nonsense mutation). A missense mutation can induce either a conservative or a nonconservative amino acid substitution. A conservative substitution refers to a DNA variant where the new amino acid is chemically similar to the old amino acid, whereas the amino acid introduced by a nonconservative substitution has different chemical characteristics. Thus, nonconservative substitutions are more likely to change the properties of the protein encoded by the gene than conservative substitutions. Insertions and deletions (**INDELs**) refer to the addition or removal, respectively, of one or a few nucleotides from the DNA sequence. These variations are relatively common in noncoding DNA. They are less frequent in exons where they could introduce minor or major shifts in the triplet reading frames of the mRNA resulting from transcription, thereby changing the final gene product (frameshift mutation). When INDELs occur in sequences contributing to the regulation of gene

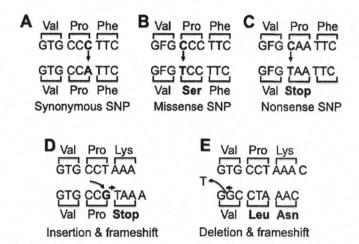

Figure 3.12 Single-base variants are illustrated. (**A**) The change of one base has no effect on the encoded amino acids (synonymous or silent), (**B**) it changes the amino acid, (**C**) it encodes a stop codon and (**D and E**) they illustrate the gain or loss of a single base, leading to a so-called frameshift mutation.

expression, they can have a significant effect on a phenotype. An example is the presence (insertion) or absence (deletion) of 287 base pairs in the angiotensin I-converting enzyme gene, encoding gene *ACE* (22) which was one of the first polymorphisms linked to endurance exercise-related traits (discussed in Chapter 5).

Other types of small DNA variants are **variable number of repeat sequences.** These are DNA variants where 2, 3 or more nucleotides are repeated multiple times. For instance, there is an abundance of CA dinucleotide repeats in the human genome. Some people may carry as few as 100 copies of the CA repeat at a given chromosomal site, whilst others may carry 10,000 or more copies of the repeat. Such repeat sequences are like a DNA fingerprint of an individual and are commonly used in forensic science to match DNA samples to an individual. One copy number variation occurs in the human *AMY1* gene which encodes the starch-digesting enzyme, amylase. Humans have a large variation in the number of copies of the amylase gene, and people of agricultural societies that eat more starch have on average more copies than e.g. hunter-gatherers, who eat less starch (23).

Finally, there are **structural DNA variants.** Here, hundreds, thousands or millions of bases are inserted, deleted or translocated from one chromosome to another. In some cases, a whole chromosome is gained or lost, resulting in **aneuploidy** (24). Some structural variants are illustrated in **Figure 3.13**.

Mutations have different effects depending on whether they occur in somatic cells or in sperm or oocytes

Mutations can occur in germ cells (i.e. sperm or oocyte) as well as in somatic cells (i.e. the other cells of the body). DNA variants in germ cells have potentially greater effects than the same DNA variants arising in a somatic cell during the course of the lifespan. This is because the latter will not be passed on to the offspring as they are not present in the sperm or oocytes. The majority of **mutations** occur when cells divide, and when the parent cell duplicates its 7.2 billion base pairs prior to cell division so that there is one complete genome for each of the two daughter cells, it has been estimated that 1 mutation occurs per 100,000 nucleotides during the **replication** of the genome. However, 99% of these mutations are corrected by proofreading mechanisms coming into play as part of the replication process (25). Whilst the true mutation rate post-DNA "proofreading" is low, it is not zero. Consequently, changes in the DNA

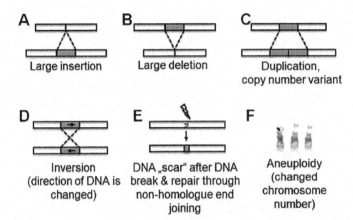

Figure 3.13 Schematic display of structural variations that involves large numbers of base pairs. This includes **(A)** large insertions, **(B)** large deletions of DNA and **(C)** a segment of DNA being duplicated. An example of the latter is the variable number of copies in the human salivary amylase (*AMY1*) gene which is associated with diet (23). **(D)** DNA fragments can be directionally inverted, or **(E)** due to imperfect DNA repair after a double-strand break. Finally, **(F)** during cell division, chromosomes can be transported in the wrong daughter cell, resulting in monosomy in one cell (only 1 chromosome) and trisomy (3 copies of a chromosome) in the daughter cell. The best-known example is trisomy 21 which causes Down syndrome.

sequence occur during each somatic cell division. Even more mutations can occur when cells become exposed to mutagens, such as ultraviolet radiation or tobacco smoke, which causes specific changes in the DNA sequence of the tissues exposed (26).

THE DEVELOPMENT OF GAMETES (OOCYTES AND SPERM) AND HUMAN GENETIC DIVERSITY

Mature sperms and oocytes are haploid cells, carrying only 23 instead of the normal 2 × 23 chromosomes in somatic cells. So how do we get from a normal cell with 46 chromosomes to a gamete? The answer is that this occurs during meiosis, a process by which female and male haploid gametes are generated. Spermatogenesis occurs in the testes and produces sperm in abundance. A subpopulation of these cells, spermatocytes, undergoes two meiotic divisions to form four haploid sperm cells. The process of female gamete production in the ovaries is called oogenesis. It begins during foetal development when thousands of primary oocytes are formed through mitosis. Primary oocytes enter meiosis but their meiotic progression is arrested until puberty with the onset of the menstrual cycle.

Two important events of meiosis contribute to human genetic diversity:

1. **Independent assortment of chromosomes**. Of each chromosome pair in somatic cells, gametes receive only one chromosome. Is it the mothers or fathers chromosome? This is a random process and so each spermatocyte or oocyte will have a random combination of chromosomes from maternal and paternal origins.
2. **Homologous recombination**. During meiosis, before chromosomes migrate to daughter cells, chromosomes cross over and exchange equivalent segments. For instance, a crossing over may occur between maternal and paternal chromosome 3, resulting in an exchange of DNA between the two

chromosomes. About 50–60 recombinations take place between all pairs of homologous chromosomes (i.e. pairs of chromosomes of maternal and paternal descent) during meiosis.

The independent assortment of chromosomes from either parent at meiosis, together with the homologous recombination occurring at several sites along each chromosome, augments genetic diversity and ensures that a new zygote and eventually offspring receives a random mix of chromosomes and DNA variants from each parent. In addition, new mutations occurring in sperm or oocytes have a probability to be present in a zygote and be passed on to the offspring. This is not a benign phenomenon as a new baby typically carries up to 100 single-nucleotide changes that were not present in the germline of his or her parents (27).

WHY DO SOME DNA VARIANTS BECOME MORE FREQUENT THAN OTHERS IN HUMAN POPULATIONS?

Within a population, DNA variants can be carried by "one in a million" or can be common. What factors determine whether a DNA variant will progressively become more frequent in a population over many generations? The answer is that evolution and specifically natural selection alters the frequency of mutations, depending on whether they are beneficial, neutral or detrimental for the carrier. The fact that evolution is the driving force for changes in allele frequency has been experimentally demonstrated in the unique **long-term evolution experiment** of Richard Lenski. On the 24th of February 1988, Lenski started 12 colonies of Escherichia coli (E. coli) bacteria in a medium that was low in the normal nutrient glucose but contained citrate, a nutrient that these bacteria could at the beginning only utilize in the absence of oxygen. Over three decades, his team cultured these 12 E. coli populations resulting in over 60,000 generations of E. coli. Lenski's team also took regular DNA samples to see if the genome was shaped by natural selection. They found that Darwinian fitness (as defined by the number of offspring) of the E. coli strains increased over time in the low glucose, high citrate medium even though the evolving E. coli strains could not utilize citrate for many generations. Then, on the 25th of June 2008, specifically after 31,500 generations, the growth of one E. coli population increased dramatically. The reason for the rapid growth was a random mutation that was carried by a so-called Cit+ variant that could metabolize citrate in the medium in the presence of oxygen. This variant quickly spread in that population through natural selection, as Cit+ E. coli could now grow faster and generate more offspring that those E. coli that still could not utilize citrate in the presence of oxygen (28). The Lenski team then additionally sequenced the DNA of E. coli after 2,000, 5,000, 10,000, 15,000, 20,000 and 40,000 generations. The 20,000th generation had 45 selected mutations when compared to the original E. coli colony, including 29 SNPs as well as 16 deletions, insertions or other polymorphisms (29). So why is this experiment so important? Lenski's long-term evolution experiment provides direct experimental evidence for evolution as it shows how specific, advantageous DNA variants are naturally selected, increase in frequency and eventually become fixed (i.e. all members carry the DNA variant) in a population. In humans, similar processes occur, but as the human generation time is about 20–30 years, it would take 200,000–300,000 years for just 10,000 generations as opposed to about 4 years in E. coli. The fact that Lenski et al. discovered 45 new DNA variants after 20,000 generations in E. coli (genome size of 4.6 million base pairs) in a challenging low-nutrient environment demonstrates that the selection of new, advantageous DNA variants can take a long time.

Back to our original question which was "why does the frequency of DNA variants change over time within a population?" The answer is that new or existing mutations are constantly subjected to natural selection. Whether the frequency of a new DNA variant increases or not over generations depends largely on the biological effects of the DNA variant as well as on random events:

1. **Deleterious DNA** variants cause a loss-of-function. An example would be loss-of-function mutations of the human DMD gene resulting in Duchenne muscular dystrophy. As the carriers are

not able to have children, the disease-causing DNA variants in the DMD gene are not transmitted to the next generation and do not spread in the human gene pool. This is known as **negative selection**.

2. **Advantageous DNA variants** increase the Darwinian fitness of individuals which means that their frequency is likely to increase over time in the population. Over many generations, advantageous DNA variants tend to become more common variants and some of them over time may become **fixed**, meaning that everyone carries the advantageous DNA variant. This phenomenon is known as **positive selection**.

3. **Neutral DNA variants** have no or only very minimal effects on Darwinian fitness. They may still spread or disappear from the genetic pool due to random genetic drift.

Whilst this seems straightforward, in practice the changes in the frequency of DNA variants are often difficult to understand. For example, genome-wide association studies (GWASs) have revealed thousands of common DNA variants that are positively or negatively associated with chronic diseases (30). How could potentially deleterious DNA variants become common variants and why were they not removed from the human gene pool by negative selection? In this case, the explanation probably has to do with the very small effect size of these GWAS common variants. A common variant associated with a chronic disease may also have become prevalent because the variant is simultaneously associated with another trait favouring reproductive fitness. Another example: high cardiorespiratory fitness (31) and high grip strength (32) are both associated with lower mortality. Why then don't we see strong evidence of positive selection of DNA variants that would result in humans having a VO_{2max} of about 80 ml/min/kg and Herculean strength? Perhaps we need to wait for a few thousand more generations to establish whether there is an ongoing selection trend that currently escapes us.

HOW MUCH DOES THE GENOME OF AN INDIVIDUAL DIFFER FROM THE HUMAN REFERENCE GENOME?

Improved DNA sequencing methods have allowed researchers to sequence thousands of whole human genomes, and this has informed us about how much genetic variability there is compared to the standard human reference genome. Whilst the sequencing of the first human genome took about 20 years, at a cost of $3 billion, today a human genome can be sequenced in a day for less than $1,000, thanks to the development of next generation DNA sequencing methods (33). These next generation sequencing methods were used to sequence the genome of James Watson (34) followed by the genomes of many others. Collectively, these sequencing experiments have shown that the typical human genome carries about 4–5 million DNA variants not found in the human reference sequence. About 20 million of the 3.6 billion base pairs, or 0.6% of the DNA sequence of an individual genome, differ when compared to the human reference genome (24). More than 99% of these DNA variants are SNPs and short INDELs. Additionally, a typical human genome has up to 2,500 structural or chromosomal variants which include about 1,000 large deletions and 160 copy number variants. Finally, an individual selected at random also carries from 200,000 up to 500,000 rare variants that have arisen in recent times.

DNA variants have different effects on the organism. Some variants alter regulatory DNA and through this the amount of proteins. Others modify biological function such as the activity of an enzyme, but others do not. So how many of the DNA variants within an individual genome influence physiological function? Only a partial answer to this broad question can be obtained at this time. For instance, humans typically carry at least 150 DNA variants that truncate mRNAs, which often leads to the partial loss or total knockout of protein function. Moreover, more than 10,000 DNA variants change the amino acid sequence of a protein, and around half a million DNA variants are outside the gene-coding sequence but reside in genomic regions that regulate the level of transcription of the target gene and eventually determine how much of the protein is synthesized (24).

HOW VARIABLE ARE THE DNA SEQUENCES OF GENOMES OF DIFFERENT HUMAN POPULATIONS?

Our ancestors in Africa, the region where our hominin species initially evolved, began to appear more than 200,000 years ago and then started to migrate to other parts of the world about 50,000 years ago (35). Long before our main forefathers and mothers, archaic humans predating Homo sapiens, such as the Neanderthalers and Denisovans, had left Africa about 400,000 years ago and lived in Europe and Asia before being extinguished after the arrival of modern humans. Interestingly, archaic humans and modern humans interacted and interbred. As a consequence, a small percentage of your genome will be Neanderthaler or Denisovan DNA (35).

Since the out-of-Africa migration of our main ancestors, there have been more than 2,000 human generations which is far less than the >60,000 E. coli generations of Lenski's long-term evolution experiment (36). This is a short time span for the selection process of DNA alleles to run its course. However, since humans have settled almost everywhere on Earth, from arctic to equator, strong selective pressures have led to the relatively quick selection of some DNA variants since the out-of-Africa migration. For example, the migration from Africa to colder climates with less sunlight has resulted in the selection of alleles related to diet, immunity, body height, and skin and eye colour (37). One example relates to alleles linked to the meat-rich diet of the Inuit in the Arctic, where very few edible plants such as berries only grow during the short summer period (38). Importantly, most common DNA variants have similar frequencies in different human populations of the world, suggesting that they have emerged and became frequent a long time ago (24). Less than 10% of DNA variants are found in specific racial or ethnic groups, whereas more than 90% of variants are found in all populations reflecting their common ancestry.

HOW CAN WE DISCOVER DNA VARIANTS THAT ARE ASSOCIATED WITH SPORT- AND EXERCISE-RELATED TRAITS?

The main goal of this chapter is to explain how to identify DNA variants associated with sport and exercise traits. Are these variants common or rare, are they SNPs, INDELs or structural variants? Do they affect the level of gene transcription/expression or the amino acid sequence of the protein or both? How important are epigenetic mechanisms? There are two major types of studies that are commonly reported on these topics. First, there are **genetic association studies** where researchers seek to identify all DNA variants that are associated with a trait. Second, researchers perform **mechanistic studies** to determine whether a DNA variant is causally related to a trait and define its underlying biology. To do so, the gene associated with the DNA variant is experimentally manipulated to investigate its effects on the trait. We will now explain these two types of studies.

How genotyping assays work and what are genetic association studies?

Identifying all DNA variants associated with sport- and exercise-related traits is a daunting task. Why? There are more than 80 million SNPs in the human genome plus large numbers of other DNA variants. Therefore, to try to identify DNA variants that influence phenotypes, one needs high-throughput genotyping assays. In early studies, sport and exercise geneticists typically used the polymerase chain reaction (PCR) to amplify a stretch of DNA followed by a genotyping reaction using, for example, a restriction enzyme to generate DNA fragments of variable length. The DNA bands visualized after electrophoresis on an agarose gel then identified the allele(s) carried by an individual. These assays only measured 1 or a few variants and were used in small cohorts to test whether specific candidate DNA variants were associated with sport- and exercise-related traits.

Examples of pioneering studies include associations between the ACE I/D variant with high-altitude performance and weight-lifting trainability (39) as well as between the ACTN3 R577X variant and sprinting performance (40), discussed in Chapters **five** and **four** respectively. However, most of the early studies had comparatively few participants and were statistically underpowered resulting in poor reproducibility. Moreover, even when an association between a common DNA variant and a sport- and exercise-related trait could be confirmed, it became clear that the DNA variant explained only a tiny fraction ($\approx$1%) of the trait variability. It took some time before researchers realized that there was not a single VO_{2max} or muscle contraction speed gene similar to a single gene trait such as the DMD gene in muscular dystrophy.

In 2004, the first papers reporting the development of SNP chips allowing the genotyping of thousands of DNA variants in a single experiment were published. In 2007, a research consortium used SNP microarrays to genotype half a million SNPs in 14,000 patients covering seven common diseases and compared the allele frequencies with those of 3,000 controls. This led to the identification of 24 SNP loci that were significantly associated with these diseases (41). Since then, hundreds of GWASs have been reported, and their findings can be accessed (https://www.ebi.ac.uk/gwas/). Typically, the global results of a GWAS are presented as a Manhattan plot, where the x-axis shows the genomic location (i.e. lists all the chromosomes) and the y-axis the association p-value of each SNP with the trait investigated. Importantly, because of the large number of statistical tests performed, the association between a SNP and a trait needs to reach a p-value of **5 x 10^{-8}** or less to be considered significant. **Figure 3.14** shows an example of a Manhattan plot from a study where the authors searched for SNPs associated with sleep duration and found one locus near the PAX8 gene on chromosome 2 that reached the significance threshold.

An important requirement of a GWAS is that the findings must be replicated before an association with a given SNP is confirmed and the finding published. For the study shown in **Figure 3.14**, the associations with DNA markers on chromosome 2 initially discovered in Caucasians were replicated in African Americans (42). Today, SNP microarrays genotype up to 2,000,000 SNPs (30) and associations between SNPs and a trait of interest are investigated in cohorts reaching, at times, one million individuals and more (43).

In sport and exercise genetics, the first GWAS report was published in 2009 with the aim to identify SNPs associated with adult leisure time exercise behaviour. The study identified a few potential SNPs but none

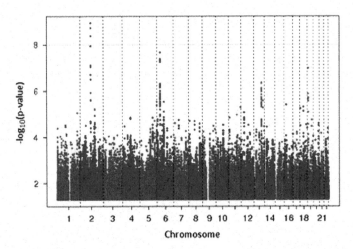

Figure 3.14 Manhattan plot for genome-wide association study results with usual sleep duration as the trait of interest. The significant SNPs are above the dotted line representing p-values of 5 × 10^{-8} and lower. Figure used with permissions from the author (42) Gottlieb DJ, et al. *Molecular Psychiatry* 20: 1232–9, 2015, Springer Nature. Where the author of this chapter (C Bouchard) was an author on the original manuscript.

reached the genome-wide p-value of 5×10^{-8} (44). Generally, GWAS focused on sport- and exercise-related traits have been unsuccessful mainly because sport- and exercise-related traits are influenced by hundreds or more DNA variants, which means that it is almost impossible to uncover associations with small effect sizes with sample sizes commonly used in such studies.

We will now briefly review two studies, one aimed at identifying DNA variants associated with altitude adaptation in Tibetans (45) and the second focused on SNPs associated with an exceptionally high VO_{2max} (46). In the altitude study, researchers used a case control design and a DNA variant chip to look for differences in 502,722 SNPs in only 35 Tibetans that lived at 3,200–3,500 metres (m) altitude with that of 84 Han Chinese that lived close to sea level. The researchers discovered SNPs that were significant at 5×10^{-8} near the EPAS1 gene which encodes the hypoxia regulator, hypoxia-induced factor 2α (45).

In contrast, in the VO_{2max} study, 1,520 endurance athletes with an average VO_{2max} of 79 ± 3 ml/kg/min were compared to 2,760 controls (mean of 40 ± 7 ml/kg/min). Using 43 times more elite athletes (n = 1,520) than the Tibetan (n = 35) study, and an exceptionally large difference in VO_{2max} between athletes and controls, should have led to the discovery of DNA variants that contribute to the large difference between endurance athletes and untrained controls. Surprisingly, the authors concluded that there is "*no evidence of a common DNA variant profile specific to world class endurance athletes*" (46). A subsequent meta-analysis of available studies revealed only one statistically significant marker near the GALNTL6 gene (46). Moreover, the gene GALNTL6 encodes the enzyme N-acetylgalactosaminyltransferase-like 6 for which there is currently no plausible mechanisms to VO_{2max}. It seems almost unfair that the comparison of less than 100 Tibetans and Han Chinese delivers a plausible explanation of the altitude tolerance trait in Tibetans, whereas comparing nearly 1,500 exceptional endurance athletes to nearly 3,000 controls with similar methodology does not reveal the genetic secret of an exceptionally high VO_{2max}.

Adaptation to the high altitude of the Tibetan plateau has been improved by strong positive selection of favourable DNA variants with apparently large effect sizes over a few generations. As a consequence, most Tibetans carry favourable altitude-related alleles of the EPAS1 gene, whereas few Han Chinese do because the same alleles do not confer advantage at or near sea level. In contrast, whilst a high VO_{2max} is associated with low mortality (31), a selection pressure for VO_{2max}-increasing DNA alleles has not, yet, been demonstrated. There are many potential reasons for this, including the following: such alleles do not impact reproductive fitness as they exert their influence on longevity well beyond the reproductive period; there are hundreds if not more of these alleles; such alleles are characterized by small effect sizes; and VO_{2max} as typically used in epidemiological studies is a composite of two uncorrelated traits, namely sedentary plus trained VO_{2max} levels. Another factor is that DNA variants may have variable effects in different populations or can be dependent on the global genetic background. This was nicely demonstrated in a paper published in *Science* in 2012. The researchers investigated the effect of 5,100 gene deletions in two strains of *saccharomyces cerevisiae* yeast (47). They found that in one strain, 44 genes were absolutely essential for viability, whilst in the other strain only 13 genes were essential. In humans too, the biological consequences of DNA variants can vary among people (48).

In summary, it has proved remarkably difficult to discover DNA variants associated with sport- and exercise-related traits, despite the substantial level of heritability for prominent phenotypes such as VO_{2max}, which is $\approx 50\%$ inherited. However, there are exceptions, including DNA variants in EPAS1 that became strongly enriched in Tibetans due to strong positive selection for survival at high altitudes. Other examples are rare and include a large deletion of DNA bases and ensuing truncation of several amino acids in the C-terminal region of the EPOR gene resulting in a much higher level of blood haemoglobin (49) or a mutation in the MSTN (myostatin) gene causing muscle hypertrophy (50). Based on the realization that studies performed over the last decade or so have not succeeded in defining the genotypes underlying sport- and exercise-related traits, a shift in research paradigm is underway with a focus on multi-centre studies, larger cohorts, genome-wide analyses of DNA variants, including epigenetic variants with additional support from gene expression and metabolomics technologies (4). We will discuss this further in the chapters on the genetics of endurance (**Chapter 5**) and strength/-muscle mass genetics (**Chapter 4**).

Mechanistic studies can explain the biology underlying a DNA variant of interest

The association of a DNA variant with a sport- and exercise-related trait does not automatically mean causation. To study causation in sport- and exercise-related genetics, researchers typically manipulate the relevant gene to test whether such manipulations affect the trait of interest. There are methods to reduce the expression of a gene, to make a gene dysfunctional, to overexpress it or to make the encoded protein more active. Generally, the aim is to either achieve a **loss-of-function** (also known as knockdown or knockout) or a **gain-of-function** (also known as knock-in) of the gene of interest either in cultured cells or in a living organism.

Usually, genes can be manipulated most easily in cell culture. Such experiments are termed *in vitro* (Latin *in glass*, as in a test tube). *In vitro* gene manipulation experiments can yield useful information about traits of interest (51) but are generally less productive for the study of more complex traits such as VO$_{2max}$. In these experiments, researchers use methods such as electroporation, transfection agents or viruses to get gene manipulation agents, such as small interfering RNA (siRNA), short-hairpin RNA (shRNA) or a gene construct into cells and their nuclei, where the gene manipulation agents can then block, reduce or increase the expression of a gene or manipulate the gene itself.

Alternately, experiments are often performed in animals *in vivo* where genes are either knocked in or knocked out to test whether the genetic manipulation causes a disease or alters a sport- and exercise-related trait. The most important animal model is the mouse. In 2007, Mario R. Capecchi, Martin J. Evans and Oliver Smithies received the Nobel Prize in physiology or medicine *"for their discoveries of principles for introducing specific gene modifications in mice by the use of embryonic stem cells"*. The procedure leads to transgenic mice and such genetically engineered animals have greatly helped our understanding of the connections between genes and phenotypes. Of relevance are the genes whose gain- or loss-of-function changes sport- and exercise-related traits such as muscle hypertrophy (52) or endurance performance (53).

Manipulations of genes used to be expensive, time-consuming and complicated but the **CRISPR-Cas** genome editing technology has made the manipulation of DNA much easier. How does CRISPR-Cas function? The method was inspired by an adaptive bacterial immune system that allows bacteria to detect and destroy the DNA of invading viruses. Researchers realized that this bacterial immune system could be re-purposed to target a single locus (fixed position on a chromosome), often a gene, within any DNA sequence of the genome. To achieve this, a so-called single-guide RNA (sgRNA) that specifically binds to the DNA locus of interest is generated. In the second step, Cas9 or another DNA-modifying enzyme finds the sgRNA and either knocks out the gene by introducing a double-stranded break and random mutations or modifies the sequence (54). CRISPR-Cas is schematically illustrated in **Figure 3.15**.

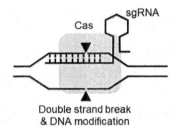

Figure 3.15 Schematic illustration of CRISPR-Cas genome editing. To selectively edit a gene or any given DNA sequence, a single-guide RNA (sgRNA) is designed and introduced into targeted cells or organisms. The sgRNA recruits Cas genes, illustrated as a grey box, to the sgRNA-bound site. The Cas enzyme cuts both DNA strands resulting in uncontrolled or controlled DNA modifications when compared to the original DNA sequence. CRISPR-Cas technology is used to knock out, edit or knock in genes.

CRISPR-Cas is commonly used to knockout or knock-in genes to test whether it affects a trait of interest in cells or organisms. CRISPR-Cas holds great promise for the treatment of Mendelian diseases such as Duchenne muscular dystrophy (55). On the downside, the comparatively easy usage of CRISPR-Cas can lead to misused and unethical applications. For instance, it could be used to create biological weapons, for personal biological enhancement or to modify genes for performance gain (i.e. **gene doping**, see below).

CAN ADVANCES IN SPORT AND EXERCISE GENETICS BE TRANSLATED INTO USEFUL APPLICATIONS?

Genetic tests aimed at consumers

Would it not be nice if we just had to take a mouth swab, send it away and later received an e-mail that informed us with high confidence how to train to maximize our individual genetic profiles. Where this test would inform us not only about our VO_{2max} trainability, potential for muscle hypertrophy and favourable changes in fibre-type distribution but also about having a talent for music, mathematics, our disease risk as well as our ancestry? Two decades after the first direct-to-consumer genetic test, a growing number of genetic tests are offered to consumers as summarized online in June 2018 by Scott Bowen and Muin J. Khoury at the US Centers for Disease Control and Prevention. In 2017 alone, more than 12 million consumers used these tests, mainly to learn about their own genome, health issues or genealogy.

What about direct-to-consumer genetic tests for sport- and exercise-related traits? Many sport- and exercise-related traits have a strong genetic component. It is therefore reasonable to expect that there will be measurable DNA variants allowing us to predict such traits. The topic was examined a decade ago by a group of experts from the British Association for Sport and Exercise Sciences (56) and more recently by a collective of international experts (57). The latest consensus is that the direct-to-consumer genetic testing aimed at talent identification in children or at personalized exercise prescription has no scientific merit in their currently marketed form. Whilst there are issues such as poor-quality genotyping and the lack of genetic counselling, the most important limitation of current commercial genetic tests is that they, at best, only explain a small fraction of traits in which the consumer is interested. There are currently almost 50 companies offering direct-to-consumer genetic testing services with a complete or partial focus on sport- and exercise-related phenotypes. None of these products have been properly validated in populations independent of the one from which they were originally derived. Discovery and then replication in an independent sample are essential for the proper derivation of sensitivity, specificity and other properties of a valid diagnostic test. However, considering that the variance explained by these commercially based genetic tests would be in the order of 5% of the trait variance or less, it is obvious that they have low sensitivity and poor specificity and, hence, no scientific merit as a classifier or diagnostic tool.

Gene therapy and gene doping

There are two types of gene therapies. The first covers methods where a defective gene is edited to make the DNA sequence normal in the genome of cells in a target tissue. The second type of therapy is one where a gene that is not functioning properly is silenced or knocked out. Whilst CRISPR-Cas has made such gene editing much easier (54), the challenge of delivering CRISPR-Cas or other DNA constructs in the correct target cells remains daunting but does not appear unsurmountable anymore. Typically, therapeutic DNA constructs are delivered with a so-called viral vector which has three limitations:

1. The viral vector may migrate to other types of cells in addition to the target cells, thus running the risk of harming healthy cells.
2. The viral vector may trigger immune reactions and inflammatory reactions, or the virus may itself cause a disease.
3. A low risk of off-target action resulting in the formation of tumours.

Gene replacement therapy is a highly regulated therapeutic field, and clinical trials based on the technology have been limited to diseases for which there is no known cure. However, gene therapy strategies may potentially be misused for **gene doping.** The World Anti-Doping Agency defines gene doping as the non-therapeutic use of genes or genetic elements with the intent of enhancing athletic performance. If the purpose of gene doping is to increase athletic performance, is this a realistic goal? How likely is it that some athletes engage in gene doping today?

It is clear from transgenic mouse studies that DNA can be manipulated in living organisms, for example, to increase muscle mass (52) or endurance performance (53). CRISPR-Cas has also been used in a very controversial way to correct a genetic defect in human embryos (58). Thus, it is technically possible to introduce performance-influencing alleles in human embryos. Thus, rogue athletes may attempt to use CRISPR-Cas to introduce performance-influencing mutations in the cells of a critical tissue or organ. However, whilst CRISPR-Cas provides the technology to edit the genome, say, to increase haematocrit or change mitochondrial content of muscles, the delivery of the CRISPR-Cas or other constructs to the right tissue remains a major problem as it requires viral vectors to deliver the constructs. Using such viruses means the same risks as discussed for gene therapy. Of course, we cannot exclude that some rogue athletes or coaches are already trying to manipulate DNA for performance enhancement purposes. However, given that the best molecular biologists still struggle to correct Mendelian diseases, such as Duchenne muscular dystrophy, using these techniques, it is highly unlikely in the near future that rogue biologists in an underground laboratory will be able to safely manipulate the DNA of an athlete to increase performance whilst ensuring that the gene editing constructs, or the viral vectors cannot be detected. Quite frankly, athletes or coaches who would use gene editing to improve sports performance would be undertaking highly dangerous and unethical behaviour.

SUMMARY

We hope that you agree at this point that there is strong evidence that the differences in our DNA sequence affect health and many sport- and exercise-related traits. This chapter aimed at providing a foundation for the understanding of the role of variation in DNA sequence and its implications for molecular exercise physiology. We briefly reviewed the main approaches of genetic epidemiology with an emphasis on methods based on twins, nuclear families and other relatives by descent or by adoption to evaluate whether sport- and exercise-related traits are influenced by a genetic component. Since the human genome is the causal pathway to biological inheritance, we described the DNA molecule, its organization and how the information flows from DNA to RNA and proteins. Subsequently, we explained the various types of DNA variants, their prevalence in people and populations with an emphasis on how these variants can lead to biological loss or gain of function potentially impacting phenotypes of interest for sports and exercise. Finally, we discussed some of the implications of advances in genomic science for practical issues such as genetic testing, gene therapy and gene doping.

Identifying the large numbers of DNA variants contributing to variability in sport- and exercise-related traits has proved to be an extremely difficult task (59). This challenge is not unique to molecular exercise physiology. Indeed, attempts to relate genotype characteristics to phenotypes have had limited success for complex human traits investigated to-date. Why is that so? In brief, the following three observations are at the origin of these difficulties. First, complex traits are influenced by genes and alleles with small effect sizes; second, the regulation of transcription, translation and other cellular processes is widely distributed and highly complex; and third, pervasive redundancy at gene, protein and pathway levels makes it more challenging to discover the link between genotype to phenotype. Of course, the task would in general be simpler if we were dealing with alleles with large effect sizes. Such alleles do exist (e.g. truncating mutation in the EPOR gene), but they are the exception rather than the rule. Globally, one should not think in terms of determinism when assessing the role of genetic variability in sport- and exercise-related traits but rather in terms of probability of small increments or decrements.

However, there are reasons to be optimistic. Advances in the human genome sequence (www.ncbi.nlm.nih.gov/Genbank or www.ebi.ac.uk.embl), DNA sequence variability (www.1000.genomes.org), the structure and functions of noncoding DNA (ENCODE) (www.encodeproject.org), the role of genotype on tissue specific gene expression (GTEx) (https://commonfund.nih.gov/gtex), high-throughput technologies, GWASs with large panels of SNPs, gene expression profiling, DNA methylation and histone profiling, and screening of the proteome and metabolome are giving more fire power to the efforts aimed at understanding the connection between genotype and phenotype. Moreover, computational biology and bioinformatics combined with the availability of online genomic resources provided by non-profit scientifically driven organizations are improving the odds of success to a considerable extent. The task is gigantic, which we did not realize in the beginning, but there are reasons to believe that progress is possible.

REVIEW QUESTIONS

* Explain why is it useful to study the heritability of exercise-related traits?

* Describe the central dogma of molecular biology and its relevance in the response to acute exercise and when adapting to the demands of exercise training.

* Describe what types of DNA sequence variants can potentially influence exercise capacity?

* Discuss the limitations of using genetic testing for talent identification and sports performance.

FURTHER READING

Bouchard C & Hoffman EO, Editors (2011). *Genetic and Molecular Aspects of Sports Performance. Encyclopaedia of Sports Medicine*, John Wiley & Sons.
Lightfoot JT, Hubal M, & Roth SM, Editors, (2020) *Routledge Handbook of Sport and Exercise Systems Genetics*, Routledge.
Pescatello LS & Roth SM (2011). *Exercise Genomics*, Springer.
Strachan T & Read A, (2020). *Human Molecular Genetics*, 4th Edition, Taylor & Francis.

REFERENCES

1. Silventoinen K, et al. *Twin Res: Off J Int Soc Twin Studies*. 2003. 6(5):399–408.
2. Gladwell M. *Outliers: The Story of Success*. 1st ed. New York: Little, Brown and Co.; 2008. p. 309.
3. Wood AR, et al. *Nat Genet*. 2014. 46(11):1173–86.
4. Bouchard C. *Br J Sports Med*. 2015. 49(23):1492–6.
5. Bouchard C, et al. *Med Sci Sports Exerc*. 1986. 18(6):639–46.
6. Simoneau JA, et al. *FASEB J*. 1995. 9(11):1091–5.
7. Bouchard C, et al. *Med Sci Sports Exerc*. 1998. 30(2):252–8.
8. Bouchard C, et al. *Compr Physiol*. 2011. 1(3):1603–48.
9. Hannon E, et al. *PLoS Genet*. 2018. 14(8):e1007544.
10. Cardon LR, et al. *Behav Genet*. 1991. 21(4):327–50.
11. Visscher PM, et al. *Nat Rev Genet*. 2008. 9(4):255–66.
12. Barbato JC, et al. *J Appl Physiol*. (1985). 1998. 85(2):530–6.
13. Bouchard C, et al. *J Appl Physiol (Bethesda, MD: 1985)*. 1999. 87(3):1003–8.
14. Watson JD, et al. *Nature*. 1953. 171(4356):737–8.
15. Ou HD, et al. *Science*. 2017. 357(6349):eaag0025.
16. Lambert SA, et al. *Cell*. 2018. 172(4):650–65.

17. Siggens L, et al. *J Intern Med.* 2014. 276(3):201–14.
18. Sanger F, et al. *Proc Natl Acad Sci.* 1977. 74(12):5463.
19. Anderson S, et al. *Nature.* 1981. 290(5806):457–65.
20. Lander ES, et al. *Nature.* 2001. 409(6822):860–921.
21. Venter JC, et al. *Science.* 2001. 291(5507):1304.
22. Rigat B, et al. *J Clin Invest.* 1990. 86(4):1343–6.
23. Perry GH, et al. *Nat Genet.* 2007. 39(10):1256–60.
24. Thousand-Genomes-Consortium. *Nature.* 2015. 526:68.
25. Pray LA. *Nat Educ.* 2008. 1(1):100.
26. Alexandrov LB, et al. *Science (New York, NY).* 2016. 354(6312):618–22.
27. Albers PK, et al. *bioRxiv.* 2018.416610.
28. Blount ZD, et al. *Proc Natl Acad Sci.* 2008. 105(23):7899–906.
29. Barrick JE, et al. *Nature.* 2009. 461(7268):1243–7.
30. Visscher PM, et al. *Am J Hum Genet.* 2017. 101(1):5–22.
31. Kodama S, et al. *JAMA.* 2009. 301(19):2024–35.
32. Celis-Morales CA, et al. *BMJ.* 2018. 361:k1651.
33. Margulies M, et al. *Nature.* 2005. 437(7057):376–80.
34. Wheeler DA, et al. *Nature.* 2008. 452(7189):872–6.
35. Nielsen R, et al. *Nature.* 2017. 541(7637):302–10.
36. Good BH, et al. *Nature.* 2017. 551:45.
37. Mathieson I, et al. *Nature.* 2015. 528(7583):499–503.
38. Fumagalli M, et al. *Science.* 2015. 349(6254):1343–7.
39. Montgomery HE, et al. *Nature.* 1998. 393(6682):221–2.
40. Yang N, et al. *Am J Hum Genet.* 2003. 73(3):627–31.
41. Wellcome-Trust-Case-Control-Consortium. *Nature.* 2007. 447(7145):661–78.
42. Gottlieb DJ, et al. *Mol Psychiatry.* 2015. 20(10):1232–9.
43. Evangelou E, et al. *Nat Genet.* 2018. 50(10):1412–25.
44. De Moor MH, et al. *Med Sci Sports Exerc.* 2009. 41(10):1887–95.
45. Beall CM, et al. *Proc Natl Acad Sci U S A.* 2010. 107(25):11459–64.
46. Rankinen T, et al. *PLoS One.* 2016. 11(1):e0147330.
47. Dowell RD, et al. *Science.* 2010. 328(5977):469.
48. Deplancke B, et al. *Science.* 2012. 335(6064):44–5.
49. de la Chapelle A, et al. *Proc Natl Acad Sci U S A.* 1993. 90(10):4495–9.
50. Schuelke M, et al. *N Engl J Med.* 2004. 350(26):2682–8.
51. Rommel C, et al. *Nat Cell Biol.* 2001. 3(11):1009–13.
52. Verbrugge SAJ, et al. *Front Physiol.* 2018. 9: 553.
53. Nezhad F, et al. *Front Physiology.* 2019. 10:262.
54. Knott GJ, et al. *Science.* 2018. 361(6405):866–9.
55. Nelson CE, et al. *Nat Med.* 2019. 25(3):427–32.
56. Wackerhage H, et al. *J Sports Sci.* 2009. 27(11):1109–16.
57. Webborn N, et al. *Br J Sports Med.* 2015. 49(23):1486–91.
58. Ma H, et al. *Nature.* 2017. 548:413.
59. Bouchard C. Exercise Genomics, Epigenomics, and Transcriptomics: A Reality Check! In: Lightfoot JT, Hubal MJ, Roth SM, editors. *Routledge Handbook of Sport and Exercise Systems Genetics.* New York: Routledge; 2019.

CHAPTER FOUR

Genetics of muscle mass and strength

Stephen M. Roth and Henning Wackerhage

LEARNING OBJECTIVES

At the end of the chapter, you should be able to:

1. Explain why muscle mass and strength are considered to be polygenic traits.
2. Discuss the ACTN3 R577X genotype and other polymorphisms that have been linked to muscle mass, strength and related traits.
3. Describe how a rare mutation in the myostatin gene is linked to increased muscle mass.
4. List and explain mutations in transgenic mice that affect muscle mass.
5. Explain how selective breeding and the analysis of inbred mouse strains have informed us about the genetics of muscle mass and strength.

INTRODUCTION

In weightlifting, Tatiana Kashirina holds the female world record for the combined snatch plus clean and jerk of 348 kg. For men, Lasha Talakhadze has lifted 477 kg for the combined exercises. What allows these athletes to lift such incredible amounts of weight? Certainly, years of training were needed to develop the muscle size and strength required to generate the force, as well as perfect the movement and lifting technique. Mental readiness for competition, nutrition and a passion for the sport no doubt contributed as well. But not everyone has the capacity to reach these records, even with training. Such an athlete must have the capacity to achieve these levels of muscle mass and strength, in addition to the years of training needed to actually achieve these records. As you will learn in the present chapter, muscle mass and strength are highly heritable traits, and it is very likely that both Tatiana and Lasha have a genetic advantage over others that allows them to excel in weightlifting competition.

Before reading this chapter, you should have read **Chapter 3** which introduces sport and exercise genetics. Also, **Chapter 8** covers the adaptation to resistance training which is closely related to the genetics of muscle mass and strength and therefore can be read in conjunction with the present chapter.

Muscle mass and strength are influenced by many limiting factors which depend on DNA sequence variants and on environmental factors such as resistance training. We will start this chapter by exploring the heritability of muscle mass and strength and discuss the findings of twin and family studies. After that, we will discuss DNA sequence variants that are common in the population, termed polymorphisms,

DOI: 10.4324/9781315110752-4

that have been linked to muscle mass and strength. As a specific example, we will discuss the *ACTN3* R577X polymorphism which is associated with speed and power performance and then summarize the results of other studies where muscle mass and strength-related polymorphisms have been identified. These studies do confirm the existence of polymorphisms which affect muscle mass and strength, but these polymorphisms typically only explain a small proportion of the variation of muscle mass and strength. It is currently unclear whether these polymorphisms are only the tip of the iceberg and whether there are many other polymorphisms that affect muscle mass and strength or whether there is a significant contribution of rare DNA sequence variants to these phenotypes.

Transgenic mouse models are discussed which show many 'candidate' genes where a 'loss or gain of gene function' affects muscle mass and/or strength. Genes where this has been demonstrated are termed **candidate genes**, reflecting the idea that these are 'candidates' for genes containing DNA sequence variation that may impact gene or protein function and thus impact the phenotype of interest, thus explaining some aspect of the heritability of that phenotype. Candidate genes are often identified based on known anatomy and physiology of the tissues underlying the trait of interest, or may be identified by genetic studies intended to find areas of the genome that appear to relate to a trait of interest. When examined in model organisms directly (e.g. in cultured muscle cells or 'in vitro', or in transgenic mice or other species in vivo), such candidate genes are commonly identified via **loss or gain-of-function** experiments where the gene is specifically manipulated to decrease or increase expression or function of that gene resulting in changes to the phenotype (explained in detail in **Chapter 3**). Many such candidate genes have been linked to mTOR and myostatin-Smad signalling that each has major effects on muscle mass and strength. One rare human DNA sequence variation with a large effect size on muscle mass has been reported so far. It is a DNA sequence variation in the first intron of the myostatin gene and a homozygous carrier had approximately a doubled muscle mass, which is in line with similar muscular phenotypes seen in animal species with myostatin gene variations. Finally, we will discuss new areas of genetic analysis, namely genome-wide association studies (GWASs) and next-generation sequencing (NGS), that will help highlight the contribution of new DNA variants in muscle mass and strength phenotypes.

MUSCLE MASS AND STRENGTH VARY GREATLY
AND ARE A POLYGENIC TRAIT

Humans not only come in all shapes and sizes, but they also differ greatly in their muscle mass, strength and power. For every Valkyrie and Hercules, there is a woman or man with low muscle mass and strength. It is striking by how much muscle mass and strength vary in the general human population. This is illustrated in **Table 4.1** where we show the means and ranges for the variation in muscle fibre numbers, muscle fibre size as well as elbow, hand grip and knee extension strength in young men.

The variation of muscle size and strength is especially evident from the estimated 95% confidence interval (i.e. 95% of individuals have values in between the lower and upper values shown). Here, for most variables, the upper value is twice as high as the lower value which means that the most muscular and strongest men have roughly two times the number of muscle fibres and two times as large muscle fibres. They are also typically twice as strong as the weakest men within the 95% interval. Therefore, the obvious question is: Where is this large variation in muscle size and function coming from?

Muscle mass and strength depend, like most other sport- and exercise-related traits, on many limiting factors which are summarized in **Figure 4.1**. The limiting factors that result in varying muscle mass and strength depend, in turn, both on environmental factors such as resistance training (**Chapter 8**) and nutrition (**Chapter 10**), epigenetics (**Chapter 6**) and on DNA sequence variants or genetics. The aim of this chapter is to focus on the latter, and review what we know about the genetics of muscle mass and strength and more specifically about the DNA sequence variants that contribute to such variation.

Table 4.1 Mean and variation values (standard deviation/SD and 95% confidence intervals) for muscle size and strength/function parameters in young men

Variable	Mean±SD	95% are in-between[3]	Reference
Fibre numbers[1]	648,000 ± 148,000	≈352,000 and 804,000 fibres	(1)
Type I CSA*	3554 ± 1214 μm²	≈1070–6038 μm²	(1)
Type II CSA*	3589 ± 1528 μm²	≈533–6645 μm²	(1)
Elbow flexion[2]	387 ± 84 N	≈218–556 N	(2)
Hand grip	616 ± 98 N	≈420–812 N	(2)
Knee extension	569 ± 118N	≈334–804 N	(2)

CSA – cross-sectional area, i.e. the size of muscle fibres: 1) Lexell et al. (1) analysed vastus lateralis muscles from men aged 18–22 years. 2) Silventoinen et al. (2) used data from Swedish army recruits aged 16–25 years. All calculated from the mean and SD values assuming a Gaussian distribution. 3) 95% are 'in-between', which means that you can be confident 95% of young men have a parameter (e.g. number of muscle fibres, fibre CSA and strength) that is within those ranges.

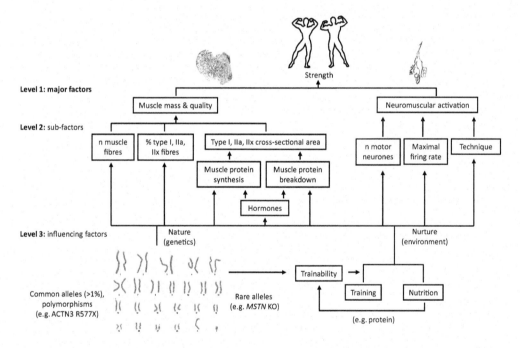

Figure 4.1 Muscle strength and mass depend on many limiting factors. Each of these limiting factors are influenced by DNA sequence variants, environmental factors of which, training and nutrition are the most important ones. Abbreviations: n – number, CSA – cross-sectional area (a measure of muscle fibre size), QTL – quantitative trait locus.

DNA sequence variants that effect muscle mass and strength can be common or rare and can have a large or small effect size. For example, the *ACTN3* R577X DNA variants (where one DNA variant encodes an arginine, abbreviated as R in position 577 and the other a stop codon, abbreviated as X – discussed later in this chapter in depth) are common but the differences between people that carry RR, RX or XX are small (3). In contrast, a mutation in the human myostatin (*MSTN*) gene is rare, but the aforementioned

Berlin boy carried such a mutation on each of the two copies of chromosome 2 that encode MSTN. However, the effect of being homozygous for the MSTN mutant was large as the boy seemed to have twice the muscle mass of children that were otherwise matched by age (4). Because muscle mass and strength depend, as we will show in this chapter, on DNA sequence variants in many genes, they are considered to be **polygenic traits**. Furthermore, the increase of muscle mass and strength in response to a resistance training programme (trainability to exercise) also varies greatly (5, 6). However, in contrast to VO_2max trainability, the heritability of mass and strength trainability has not yet been quantified but it seems likely that muscle mass and strength trainability are significantly inherited as is the case for endurance trainability (see **Chapter 5**). The factors that limit strength and muscle mass and their dependency on DNA sequence variation, training and nutrition are illustrated in **Figure 4.1**.

HERITABILITY OF MUSCLE STRENGTH AND MASS

How do we know that muscle mass and strength are partially inherited? The evidence of the role of genetic factors comes from twin and family studies which were introduced in **Chapter 3**. A popular way of estimating an individual's overall strength is the hand grip test, which is easy to perform, and correlates well with arm flexor and leg or knee extensor strength (2). In an extensive study, Silventoinen et al. (2) measured hand grip strength in over one million 16–25-year-old male subjects and found that it varied between 50 and 999 N. The hand grip strength of 100 of those subjects even exceeded the limit of the dynamometer used. Thus, the strongest men were at least 20 times stronger than the weakest. Because the researchers included a large number of monozygous ($_{MZ}$/identical) and dizygous ($_{DZ}$/non-identical) twins they were able to estimate the heritability of grip strength. The correlations coefficient r for hand grip strength for the monozygous twins was larger at 0.66 compared to the dizygous twins at 0.35. If the Falconer's formula, which we had discussed in **Chapter 3**, is used, then the heritability h^2 for grip strength can be calculated as 2 $(r_{MZ} - r_{DZ})$ or 2 $(0.66 - 0.35) = 0.62$ or 62%. This demonstrates that grip strength is significantly inherited.

Generally, the estimated heritability in strength varies depending on the type of strength measured and the volunteers investigated. Twin studies generally suffer from low subject numbers and assumptions about the environment, and thus the values should only be interpreted as a rough indication. An overview over the heritability of static, dynamic and explosive strength is given in **Table 4.2** (7).

Table 4.2 shows large variations in the heritability estimates for muscle strength which is probably due to the limitations of twin and family studies. Nonetheless, the studies demonstrate that strength is, like most other sport- and exercise-related traits, significantly inherited.

Muscle strength depends greatly on muscle mass, which cannot be measured directly in humans. However, it is possible to estimate muscle mass by measuring lean body mass which is body weight minus fat and bone mass which can be indirectly measured using various scanning techniques, including DXA, MRI or ultrasound scans. In men, $\approx 38\%$ and in women $\approx 31\%$ of the body mass is skeletal muscle

Table 4.2 Heritability estimates for static, dynamic and explosive strength (7)

	Twin studies	Family/sib-pair studies*
Static strength	14–83% (20 studies)	27–58% (5 studies)
Dynamic strength (isokinetic, concentric and eccentric)	29–90% (3 studies)	42–87% (concentric only; 2 studies)
Explosive strength or power (jump tests Wingate test	34–97% (7 studies)	22–68% (1 study)

For references, also see Peeters et al. (7). In this table, the heritability is given as a percentage rather than as a fraction of '1' as in the original paper.

(8), and thus muscle contributes roughly 50% to lean body mass in males and 40% in females but this varies a lot depending on the body composition. The heritability estimates of lean mass are high, in the order of 60–80% (9–11). Muscle cross-sectional area estimated from the circumference of the thigh is also highly inherited with an estimate of 91% (12). All these findings suggest that muscle mass, which is a major limiting factor for muscle strength, is highly inherited.

The mass of a given muscle depends on the number of muscle fibres and on the average size of the fibres within a muscle. Animal studies show that the number of fibres in the muscle is set during embryogenesis and generally changes little in the adult (13); however, that number can differ greatly between different individuals. For example, Lexell et al. investigated human cadavers and found that the muscle fibre number of the vastus lateralis varied in men aged 18–22 years between 393,000 and 903,000 fibres (1). Thus, there can be over a 2-fold difference in the number of fibres between individuals in a given muscle which explains why some untrained individuals have larger muscles than others. The heritability of fibre numbers per muscle is unknown but the large differences in untrained individuals and the small effect of training and nutrition on fibre numbers in adults strongly suggests that DNA sequence variation or heritability, probably together with the nutrition in the uterus and during early development, are major causes.

Muscle strength depends on muscle mass, which is significantly inherited as we have just shown and on neuromuscular activation or the ability of the nervous system to maximally innervate existing muscles. Unfortunately, little is known about the heritability of neuromuscular activation as there are no twin or family studies that seek to address this question. Thus, taken together, muscle mass and strength both depend significantly on DNA sequence variants or, in other words, are heritable. Strength is probably around 50% inherited although the heritability estimates vary greatly. Lean body mass and muscle mass are ≈60–90% inherited and the heritability of neuromuscular activation is unknown. Given that muscle mass and strength are significantly inherited, we will now review common and rare alleles or DNA sequence variants that are responsible for the heritability of muscle mass and strength.

TRANSGENIC MICE WITH A MUSCLE STRENGTH AND/OR PHENOTYPE

The identification of common polymorphisms or rare DNA sequence variants often depends on previous knowledge of genes that affect muscle mass and strength. Genes where this is demonstrated are termed **candidate genes**. Such candidate genes are commonly identified via gain or loss-of-function experiments in cultured muscle cells in vitro or in transgenic mice or other species in vivo. To identify genes whose transgenesis results in muscle hypertrophy, we systematically searched for publications that report that a gain or loss-of-function of a gene resulted in muscle hypertrophy in mice. This analysis revealed 47 genes that increased muscle size from 5 to 345% (**Figure 4.2**). Many of the mass mass-influencing genes belong to the myostatin-Smad, Igf-1-Akt-mTOR (dicussed in Chapter 8) and angiotensin-bradykinin pathways (14).

This analysis reveals that skeletal muscle is influenced by many genes of which some, like myostatin, reduce muscle mass, whereas others, like Akt1, stimulate muscle growth. To illustrate the effect, we have redrawn photos of transgenic mice with muscle hypertrophy in **Figure 4.3** in order to give a better idea about the phenotype.

So what have we learned? First, both the PKB/Akt-mTOR (15) and myostatin-Smad pathways (17, 18) can induce skeletal muscle hypertrophy when genes are knocked in or out both at birth or in the adult. Both pathways also affect fibre numbers when the genetic modification is present from birth and fibre size at all stages. Additional signalling pathways can also impact muscle size and strength, and a recent review reported that 47 different genes have been found to affect muscle mass in transgenic animal models (14). Only a small fraction of these have been studied in humans so far but as with the endurance

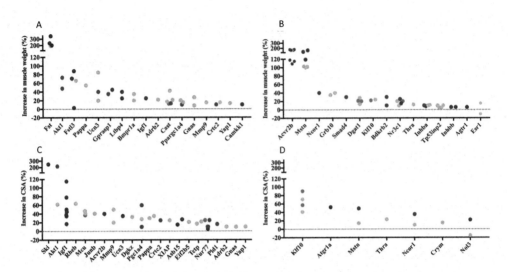

Figure 4.2 Genes whose gain (**A,C**) or loss-off-function (**B,D**) increases muscle weight (**A,B**) or cross-sectional area (CSA, **C,D**). Figure is taken from Verbrugge SAJ, et al. (14) as an open access (Attribution 4.0 International -CC BY 4.0, https://creativecommons.org/licenses/by/4.0/) article in *Frontiers in Physiology* where permissions are not required provided the work is properly cited.

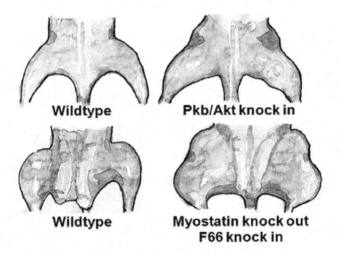

Figure 4.3 Top row - overexpression or knock-in of constitutively active PKB/Akt results in muscle fibre hypertrophy. We have redrawn wildtype and PKB/Akt knock in mice photos from Lai KM, et al. (15). Bottom row - a combined myostatin knockout and FLRG (F66) knock-in increases muscle mass by ≈4-fold, the most extreme example of a transgenic muscle hypertrophy mouse so far. We have redrawn wildtype and myostatin knock out/F66 knock in mice photos from Lee SJ (16).

genes (Chapter **5**), there are many variants of these genes in humans. Large-scale DNA-sequencing projects show that there are DNA sequence variations in exons that change the amino acid sequence or even cause a loss-of-function in almost all human genes. These databases are a 'goldmine' for sport and exercise geneticists (19, 20).

A second common observation is that muscle hypertrophy induced by targeting genes belonging to the PKB/Akt-mTOR (15) and myostatin-Smad (21) pathways can both reduce fat resulting in mice that are muscular and lean. Myostatin knockout mice have an increased myofibrillar protein synthesis rate per muscle (22) and activating mutations in the PKB/Akt-mTOR pathway increase protein synthesis as this is the major function of this pathway (23). Thus, in these mice, protein synthesis will rise above protein breakdown until a steady state with a presumably high protein turnover is reached. The energetic cost of protein synthesis or translation is the equivalent of 4 ATP for one peptide bond between amino acids. Using such information, the energy cost of protein synthesis has been estimated to be 3.6 kJ per gram of newly synthesized protein, respectively (24). Thus, to put it simply, an activation of the PKB/Akt-mTOR pathway or inhibition of the myostatin-Smad pathway will increase whole body protein synthesis and breakdown which, in turn, increases basal energy turnover. In line with this, the metabolic rate of myostatin knockout mice is significantly higher than that of wildtype mice (21). The high energetic costs of maintaining a high muscle mass also seems to be one plausible reason why DNA sequence variants that increase muscle mass do not enrich in the population as they will increase metabolic rate and thus limit survival time during periods where nutrients are limited.

SELECTIVE BREEDING AND INBRED MOUSE STRAINS

At the end of this chapter, we will discuss the results of selective breeding and inbred mouse strain studies as a non-biased strategy to identify genes that affect muscle mass and/or strength. Apart from farmers and horse breeders, geneticists have performed selective breeding experiments to (a) identify the cumulative effect of selecting, in the ideal case, all DNA sequence variants within a population that affect muscle size and (b) identify these DNA sequence variants. Selection studies for body weight have also led to an accumulation of genetic variants or alleles that increase or decrease muscle mass, as muscle mass is related to body mass. For example, the gastrocnemius weight in males of the so-called DUH mouse strain that have high body mass is ≈247 mg, whilst the gastrocnemius weight of mice selected for small body weight reaches only ≈66 mg (25). Among the selected alleles, there might be some that affect the growth of all cells such as genetic variations in the growth hormone system and alleles that affect muscle mass specifically such as those in the myostatin-Smad pathway.

Mice have also been helpful for understanding whether genetic factors affect number and size of muscle fibres. The soleus muscle of mice is a particularly useful muscle for such studies for several reasons. First, it is a small muscle where the number of fibres within a cross section can be counted reliably in reasonable time. Second, mouse solei consist almost only of slow-twitch type I and fast-twitch type IIa fibres which can be easily distinguished using ATPase-based fibre typing or immunohistochemistry using readily available antibodies. Finally, it does not, unlike other appendicular muscles, express type IIb fibres which are present in many other rodent but not human muscles (26). Such analyses have revealed that different inbred strains have ≈250 fibres in the soleus muscles of the Algerian mouse (m. spretus) (27) and between ≈500 and ≈800 fibres in the soleus of commonly used laboratory strains such as C57BL/6, C3H and DBA/2 (28). Finally, there are up to ≈1,250 fibres in the DUH strain selected for high body mass (25). This implies that DNA sequence variants determine much of the variation in the number of fibres in between these mouse strains. Importantly, however, larger muscle does not always mean more fibres (**Figure 4.4**). The LG/J and SM/J strains whose names originated from the fact that they were selected for large and small body weights, respectively, differ ≈2-fold in the size of the soleus and other muscles (25).

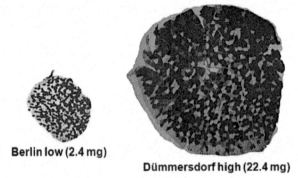

Berlin low (2.4 mg)

Dümmersdorf high (22.4 mg)

Figure 4.4 Differences in soleus muscle weight, fibre numbers and fibre size in two mice belonging to mouse strains selected for low and high body weights. In this example, the Berlin low mouse strain has 630 fibres in the soleus and its type I and IIa fibres are on average 700 and 750 µm², respectively. In contrast, the Dümmersdorf (DUH) high mouse strain has 1,223 fibres in the soleus and its type I and IIa fibres are 2,000 and 2,600 µm², respectively. This extreme example suggests that there are genetic variations that affect fibre numbers, fibre size and fibre proportions in a given muscle.

ASSOCIATION STUDIES AND THE ACTN3 R577X POLYMORPHISM

Many association studies have been carried out to try to identify common DNA sequence variants or alleles that affect muscle mass, strength and related variables in humans. Arguably, the most discussed polymorphism is the *ACTN3* R577X genotype and for this reason we will now review the research on this polymorphism in depth before summarizing the results of other association studies.

The actinins, abbreviated ACTNs, are actin-binding proteins located in the Z discs of sarcomeres within skeletal muscle. There are two ACTN isoforms of which ACTN2 is expressed in all muscle fibres, whereas ACTN3 is only expressed in fast type II muscle fibres. Initially, an ACTN3 deficiency was found in patients with muscular dystrophies but then a team led by Kathryn North demonstrated that the absence of ACTN3 is common in individuals with no muscle disease and caused by a so-called *ACTN3* R577X polymorphism (29). Using PCR and DNA sequencing, they identified a common C→T single nucleotide polymorphism (SNP) in exon 16 of the *ACTN3* gene. This one nucleotide difference in the DNA sequence greatly changes the ACTN3 protein because as a consequence, amino acid 577, which is normally an arginine (abbreviated as R), is changed to a stop codon, abbreviated as X. Hence, the polymorphism is abbreviated as *ACTN3* R577X.

The consequence of the premature stop codon in position 577 is that a shortened, non-functional version of ACTN3 protein is produced which is degraded. Thus, the 577X allele is comparable to a gene knockout for ACTN3.

Because DNA sequencing is expensive and time-consuming, the researchers developed a simplified PCR assay, which involves amplifying exon 16 of the ACTN3 gene using PCR followed by digestion of the PCR product with the restriction enzyme DdeI. This enzyme cuts the 577R allele into 205 and 85 base pair long fragments, whereas the 577X allele is cut into 108, 97 and 85 base pair long fragments (see **Figure 4.5A**). When the digested PCR product is then electrophoresed on an agarose gel, two bands indicate a homozygous 577R carrier, three bands a homozygous 577X carrier (who has no functional ACTN3) and a combination of the 205, 97 and 85 bands a heterozygous ACTN3 R577X carrier (30). Using this PCR assay, the researchers tested the DNA of individuals from many continents. They found that the frequency of those who were homozygous for the 577X knockout allele was <1% in an African Bantu population, whilst it was ≈18% in Europeans (3).

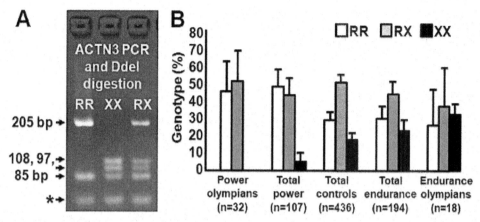

Figure 4.5 **(A)** Example result of an *ACTN3* R577X PCR genotyping test (3). The 290 bp product, which has been amplified by PCR, has been cut with DdeI either once into 205 bp and 85 bp fragments (ACTN3 577 RR) or twice into 108 bp, 97 bp and 85 bp fragments (ACTN3 577 XX). '*' refers to primer-dimers which are an artefact. **(B)** Association of *ACTN3* R577X genotype and athletic status. RR represents individuals that have the normal arginine in position 577 of the ACTN3 gene, XX individuals that have a premature stop codon in this position and RX represents heterozygous individuals. Redrawn from Yang N et al. (3). The key finding is that very few power athletes have an ACTN3 577 XX genotype, suggesting that such a genotype is detrimental for muscle power.

These observations raise some puzzling questions. Firstly, why did the 577X allele accumulate in some populations and not others? Secondly, why is there no obvious phenotype in individuals that are ACTN3 knockouts due to being XX carriers? One explanation for the lack of dramatic phenotype is that ACTN2, which is expressed in all muscle fibres, might compensate for the loss of ACTN3 in fast type II fibres. For this reason, the 577X allele might be neutral or even slightly beneficial allowing it to become enriched during human evolution at least in some populations. The research team also reasoned that as ACTN3 is expressed in fast fibres, the *ACTN3* R577X polymorphism might be associated with athletic performance. To investigate this, they obtained the genomic DNA from 107 power athletes from sports such as track, swimming, track cycling, judo, speed skating versus 436 controls and compared this with 194 endurance athletes that included long-distance swimmers, cyclists, runners, rowers and cross-country skiers. They even included fifty athletes who had competed in the Olympics. The researchers then performed the *ACTN3* R577X genotyping assay and found that the distribution of the R577X genotypes of the athletes differed between the different groups as is shown in **Figure 4.5B.**

Among 32 power Olympians, there was no subject with a XX 'knockout' genotype, although if the distribution was the same as in the controls, one would expect ≈6 individuals with a XX genotype in this cohort. In contrast, the XX genotype was higher in endurance athletes than in the controls suggesting that it might be beneficial for endurance performance.

So, is the *ACTN3* R577X genetic test useful to identify future elite power athletes? The test has been marketed as such a talent identification test and is commercially available. The answer, however, is no. The reasons are as follows: First, ≈80% of people will have an RR or RX genotype which indicates that they carry a genotype that is typical for both power and endurance Olympians albeit at slightly different frequencies. Only ≈18% in the population will be XX ACTN3 knockout carriers. These XX carriers might have a tiny advantage in endurance sports, and it is likely that they may miss one of the many factors that are required to be a speed or power Olympian. Thus, the test is at best an exclusion test for ≈18% of the population. We see no major ethical issues with performing the test unless it is done using the DNA of minors or embryos.

Follow-up studies have generally confirmed the high frequency of RR carriers in speed and power athletes especially in Europeans (31). More mechanistic studies showed that ACTN3 knockout mice displayed a fast-to-slow fibre-type shift consistent with decreased power performance (32). In association studies in humans, it was shown that XX carriers had 5% less type IIx fibres than RR carriers, again consistent with the hypothesis that the XX genotype reduces power performance (33). To conclude, the ACTN3 R577X genotype is associated with sport performance, though the effect is small and may only be seen at elite levels in individuals with the XX genotype.

Many other association studies have been performed in order to test whether specific candidate genetic variants are associated with muscle mass or strength. The results of several of these studies are shown in **Table 4.3**.

Table 4.3 demonstrates that there are common DNA sequence variants or polymorphisms in the human population that explain a fraction of the variability of muscle strength, mass and related traits seen in humans. It is important to point out, however, that the sample sizes of these studies are very small, and the results should be viewed with caution. A recent review by leaders in the field of exercise and sport genetics (47) offers a caveat that association studies such as these are fraught with problems, including small sample sizes, complex phenotypes and variation in phenotype measurement techniques that make their results challenging to replicate or compare across studies. Better approaches include both GWAS, outlined below, or meta-analysis of multiple studies.

Given that relatively few DNA sequence variants have been identified that clearly contribute to muscle mass or strength, a key question is whether the so far unexplained heritability of muscle strength and mass is due to undiscovered polymorphisms or whether rare DNA sequence variants play an important role. The answer to this question awaits the large-scale application of whole genome sequencing to sport- and exercise-related questions (48). Also, the fact that each of the polymorphisms listed in **Table 4.3** only explains a small fraction of the heritability of strength or mass implies that, for example, a single

Table 4.3 Polymorphisms or common DNA sequence variants that have been associated with muscle mass and strength-related traits in humans

Gene	Trait affected and study design	Number of subjects	Reference
Activin receptor 1B (ACVR1B)	Strength (QTL mapping study)	500, 266	(34)
ACTN3 R577X	Muscle power trainability, association study	157	(35)
CNTF SNPs	Muscle strength	494	(36)
IGF-1 repeat promoter polymorphism	Strength trainability, association study	67	(37)
IGF2 'ApaI' SNP	Hand grip strength, association study	693	(38)
IGF2 'ApaI' SNP	Fat free mass, strength and sustained power; association study	579	(39)
IL15RA SNPs	Muscle mass trainability, association study	153	(40)
Myostatin K153R	Muscle power, association study	214	(41)
Myostatin pathway genes	Strength (QTL mapping)	329	(42)
TNFα promoter SNPs	Muscle mass, association study	1050	(43)
UBR5 SNP	Muscle fibre cross-sectional area, fibre type	357	(44)
Vitamin D receptor poly A repeat	Muscle strength, association study	175	(45)
Vitamin D receptor SNPs	Muscle strength, association study	109	(46)

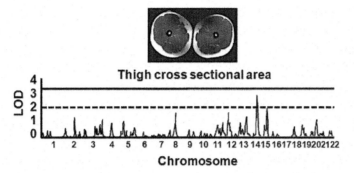

Figure 4.6 The top image shows an MRI scan of a human leg at mid-thigh level. The thigh muscles are seen in grey. Below is the result of a genome-wide linkage scan for genome regions (quantitative trait loci, or QTLs) that are associated with cross-sectional area of the thigh. An LOD score of 2.2 (dotted line) suggests linkage, whilst an LOD score of 3.3 is considered to be significant linkage. Redrawn from De Mars G, et al. (50).

polymorphism test will yield little useful information about the muscle strength of the individual. Given that polymorphisms have an additive effect, one possible strategy might be to measure a panel of muscle strength and mass-influencing polymorphisms and to calculate a genetic score from that. This approach may be a better indicator of an individual's genetic muscle strength and mass potential, but this will only be effective once these DNA sequence variants are conclusively identified.

Some genome-wide SNP analyses have been conducted for muscle strength. An experimental multi-gene approach was used in several studies by a Belgian team. The team first performed a 'linkage analysis' in 367 male siblings (49) and after that in 283 male siblings a genome-wide SNP-based multipoint linkage analysis (50). A linkage analysis aims to identify DNA variants that are close together on chromosomes that are inherited together and associated with a specific trait, such as muscle strength or CSA. The technique relies on the fact that DNA is inherited in chromosome segments and many of those segments are shared within families. So, correlations can be studied between chromosome regions and a trait of interest in multiple families, with those chromosome regions containing candidate genes that can be studied individually. Linkage analysis has recently given way to GWAS, which allows a more refined map of genome regions associated with a trait of interest. The linkage analysis in the first Belgian study showed linkage for some genes in the myostatin pathway, albeit not myostatin itself, with muscle strength (49). In the second study, they found several peaks of suggestive linkage and significant linkage on chromosome 14q24.3 (**Figure 4.6**).

The researchers used a linkage analysis to further refine the regions identified in their previous studies and so far, have identified the activin receptor 1B (ACVR1B) related to muscle strength (34). Examination of this gene in a transgenic mouse model would be a possible next step to better understand the implications of variation in this gene for muscle traits, but only limited work in this area has been performed to date. This gene is related to the myostatin pathway which we discuss in **Chapter 8** as a potential regulator of the adaptation to resistance exercise.

A MYOSTATIN KNOCKOUT MUTATION AS AN EXAMPLE OF A RARE DNA SEQUENCE VARIATION WITH A LARGE EFFECT SIZE

We have just reviewed common DNA sequence variants or polymorphisms that are associated with muscle mass and strength. However, such polymorphisms only explain a fraction of the heritability, for example, the ≈20-times difference in hand grip strength that is seen in the human population (2). As discussed below, performing GWAS experiments in large groups that are well phenotyped for muscle mass and strength will help to uncover more polymorphisms that explain the inherited variability of muscle mass and strength.

Additionally, it is likely that there are rare DNA sequence variants which are only present in individuals or families which affect muscle mass and strength. As we have stated earlier, the whole genome sequencing of large cohorts will eventually allow us to determine how important common and rare DNA sequence variants are for muscle mass and strength (48). In this section, we will discuss in detail one rare DNA sequence variation that has a large effect size on muscle mass. This proof-of-principle example is a knockout mutation in the myostatin gene which has been associated with a double muscling phenotype in humans. This is similar in principle to the transgenic mouse models discussed previously where the knock-in or knockout of a gene has an effect on muscle mass and strength.

We discuss myostatin as a major regulator of muscle mass in **Chapter 8**. Here, we discuss the case of a boy who was born to a former athlete where stimulus-induced involuntary twitching was observed after birth triggering further investigations. During these investigations, it was noted that the boy 'appeared extraordinarily muscular, with protruding muscles in his thighs and upper arms' (4). Ultrasonography showed that the quadriceps muscle mass was 7.2 standard deviations above the mean for age- and sex-matched controls. Anecdotal evidence about unusually strong family members emerged, and the research team decided to test whether the mutation was due to a knockout mutation in the myostatin gene, as such mutations could lead to double muscling in mice and other species (17, 51). Thus, a mutation of the myostatin gene was a candidate cause for the phenotype and the researchers started to test this hypothesis.

The myostatin gene comprises three exons and the researchers designed primers to amplify all exons and the introns in between followed by DNA sequencing of the PCR products. The researchers found no unusual DNA sequence in the exons but noted a DNA sequence variation in the first intron where a G was mutated to an A. The authors abbreviated this DNA sequence variation as **IVS1+5 G→A**. This indicates a DNA sequence variation in intron 1 (IVS stands for intervening sequence which means the same as 'intron') where after five base pairs, a guanine (G) was changed to an alanine (A).

The researchers now had to demonstrate that the IVS1+5 G→A was a rare DNA sequence variant, as otherwise other carriers should also have the unusual double muscling phenotype. First, they developed a simplified genotyping assay in order to test for the IVS1+5 G→A variation in a larger cohort. They developed a restriction fragment length polymorphism (RFLP) PCR assay which involved, in a similar way to the ACTN3 R577X assay, a PCR reaction to amplify a 166 bp PCR product followed by digestion of the PCR product with a restriction enzyme AccI into 135 base pair and 31 base pair fragments, whilst the mutated IVS1+5 G→A fragment of the boy was left uncut. After running the digested PCR product on an agarose gel, the rare, uncut IVS1+5 A allele appeared as a 166 bp band, whereas the common IVS1+5 G allele appeared as a 135 bp band (**Figure 4.7**; note that the common 31 bp band is not shown). Using this assay, the authors demonstrated that the boy was homozygous for this mutation, whilst the mother was heterozygous and that all controls were homozygous for the common IVS1 + 5 G allele which is in line with the hypothesis that the IVS1+5 A allele was a rare DNA sequence variation.

But at this stage it still was unknown whether the IVS1+5 A allele had an effect on the serum (blood) concentration of myostatin. To study this, the authors obtained serum from animals and patients who are known to produce normal myostatin, performed an immunoprecipitation to concentrate myostatin then performed a Western blot to detect the different forms of myostatin to which the antibody binds. The Western blot demonstrated that the myostatin propeptide was absent in the boy, whilst it was present in other individuals. This was strong albeit not perfect evidence that the IVS1+5 G→A mutation resulted in no myostatin propeptide as a likely cause for the double muscling phenotype.

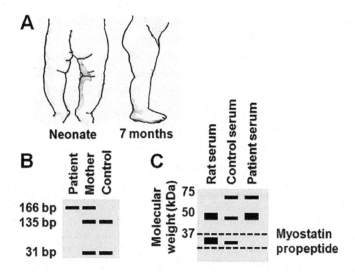

Figure 4.7 Top left. (A) Gross leg phenotype of a toddler showing pronounced skeletal muscle hypertrophy. The researchers hypothesized DNA variations in the myostatin gene as a cause and identified an **IVS1+5 G→A** mutation in the myostatin gene in the patient using PCR and Sanger sequencing. **(B)** Schematic result of a **IVS1+5 G→A** PCR assay. After PCR amplification of the part of the myostatin gene with the DNA sequence variation, the restriction enzyme AccI is used to cut the wildtype PCR product into a 135 bp and 31 bp piece, whilst the mutated DNA remains uncut and is 166 bp. This schematic of the gel shows that the patient is homozygous for the mutation, the mother is heterozygous and the control is homozygous for the normal DNA sequence. **(C)** Schematic demonstrating the analysis of serum from a patient, rat and human control with a JA16 anti-myostatin antibody which detects several forms of myostatin. The key band is the myostatin propeptide, which is absent in the toddler. Photos and gels redrawn from Schuelke M, et al. (4).

This research identified the first rare DNA sequence variation with a large effect on muscle mass and strength and reemphasizes the important role of myostatin. This fits in well with the literature showing that a knockout of the myostatin gene results in skeletal muscle hypertrophy in mice (17). Similarly, a naturally occurring knockout mutation of the myostatin gene has been associated with a large muscle mass in Piedmontese and Belgian Blue cattle breeds (51). In two species, the myostatin gene has been linked to athletic performance suggesting that not only muscle mass depends on myostatin genotypes. First, it has been shown that 'bully' whippet race dogs are heterozygous for a naturally occurring myostatin knockout mutation (52). Second, researchers performed a GWAS in thoroughbred race horses and found that myostatin genotype was a good predictor of race distance (53). All the studies in the myostatin field demonstrate that myostatin knockout mutations can occur naturally and that they can lead to a large increase in muscle mass. However, in some cases, a myostatin knockout mutation only increases muscle mass, whereas strength does not rise proportionally (54), possibly a consequence of myostatin's role in tendon structure (discussed in Chapter 8).

Taken together, this research suggests that the existence of rare DNA sequence variants in genes such as myostatin may sometimes have a large effect size on muscle mass and strength. Whole genome sequencing will accelerate the discovery of such rare DNA sequence variants to inform researchers, whereas rare DNA sequence variants contribute significantly to the variation of muscle mass and strength seen in humans and other species (48).

GWAS AND NGS STUDIES IN RELATION TO LEAN BODY MASS, MUSCLE MASS AND STRENGTH

As outlined so far, genetic studies have yet to identify the major contributors to the heritability of muscle mass and strength. Only a small number of common polymorphisms and rare variants have demonstrated, with any confidence, to have an influence on muscle phenotypes. Researchers are now moving to newer genetic analysis techniques that hold the promise of identifying important DNA sequence variants, namely GWASs and NGS studies.

Candidate gene studies focus on genes that reside in signalling pathways known to be important for muscle, like we have seen in the Smad-myostatin pathway. Those studies are limited to looking for genes that fit within what we already know about physiology, though discovering genes or DNA regions outside of these known pathway genes requires a different approach, namely GWAS. Here, rather than simply looking within a single gene or small number of known genes, many thousands of polymorphisms are examined widely across the entire genome allowing identification of DNA sequence variants in genes that we may not realize are important based on our knowledge of signalling pathways alone. GWASs for muscle mass and strength are beginning to be completed. As with candidate gene studies, sample size is an important consideration and GWASs must have many subjects in order to have adequate statistical power. As such, GWASs generally focus on phenotypes that are measured using standard approaches, allowing thousands of subjects to be measured. For example, a GWAS examining appendicular lean mass was performed in over 85,000 middle-aged men and women, which identified over 130 genome regions of interest, with 38 candidate genes identified (55). A similar GWAS for hand grip strength was performed in over 195,000 individuals and identified 16 genome regions associated with strength (56). In addition, because GWASs often use standard approaches, multiple GWASs can be combined and compared using the 'meta-analysis' technique, adding additional power to the method. One such recent meta-analysis combined the results of 53 different GWASs examining lean body mass (as an indirect measure of muscle mass) (57). Each of those GWASs had their own conclusions, but the meta-analysis approach allowed researchers to examine all of the findings and look for similarities. Across all of those studies, the researchers identified five genes that were associated with lean body mass (*HSD17B11*, *VCAN*, *ADAMTSL3*, *IRS1* and *FTO* genes) and only one of those (*FTO*) had previously been identified using candidate gene approaches. Little is known about how most of these genes may influence muscle mass, but these GWAS findings provide strong evidence for an in-depth analysis for each of these genes.

In addition to GWAS, researchers are using NGS studies to identify rare DNA sequence variants that may impact muscle mass and strength. Whilst GWAS remains focused on common DNA polymorphisms, NGS allows for analysis of rare variants, carried by only one or few individuals. Effectively, NGS is the rapid sequencing of the entire genome, a process that was once prohibitively expensive but now is generally <$1,000 USD per sample, when many samples are analysed. Whilst such studies for lean mass are just now beginning to be performed, genetics researchers have developed NGS analysis techniques needed to identify not only the specific DNA sequence variants that contribute to a trait but also learn more about the proportion of rare versus common DNA sequence variants underlying the heritability of a phenotype.

SUMMARY

Muscle mass and strength are limited by the number of muscle fibres, the size of muscle fibres and neuromuscular activation. These limiting factors depend on DNA sequence variants which can be common or rare, and also depend on environmental factors such as resistance training and nutrition. Lean body mass, muscle mass and strength both vary greatly (hand grip strength varies more than 20-fold in humans) and are significantly inherited. The heritability estimates vary considerably, and it is impossible to give a reliable overall estimate. Muscle mass and/or strength depends on both common DNA sequence variants and polymorphisms such as the ACTN3 R577X genotype and on rare DNA sequence variants such as the myostatin IVS1+5 G→A allele. It is currently unclear how much of the muscle mass and strength variation in human populations is due to common polymorphisms and rare DNA sequence variants. Transgenic mouse models and inbred mouse strains show that the variation of muscle mass and strength depends on DNA sequence variants that affect the number of muscle fibres within a given muscle and on the size of muscle fibres. Also, mouse studies demonstrate that a combination of gain- or loss-of-function mutations can increase muscle mass at least 4-fold compared to wildtype mice. Key candidate genes are found within the PKB/Akt-mTOR and myostatin-Smad signal transduction pathways which have also been implicated in the adaptation to resistance exercise, covered in Chapter **8**. GWAS and NGS studies are now beginning to be performed for muscle phenotypes that will further advance our understanding in the coming years.

REVIEW QUESTIONS

• Draw a diagram to illustrate what factors limit muscle mass and strength. Ensure to include common and rare DNA sequence variations as one of the causative factors.

• Discuss what is known about the heritability of human muscle size and strength?

• Describe the discovery of the *ACTN3* R577X polymorphism. Can an *ACTN3* R577X genetic test alone be used to identify, with good likelihood, someone who has the potential to become a world class sprinter?

• Describe the experimental strategy that researchers have used to identify a mutation in the myostatin gene as a rare DNA sequence variation responsible for doubling muscle mass in a toddler.

• Explain and compare two transgenic mouse models where a transgene in either the PKB/Akt-mTOR or myostatin-Smad pathway has increased muscle size. What is the maximal muscle size increase that has been achieved in a transgenic mouse model when compared to the wildtype (controls)?

• How do GWASs and NGS studies improve our ability to identify DNA sequence variants related to muscle mass and strength?

FURTHER READING

Bouchard C & Hoffman EP (2011). *Genetic and Molecular Aspects of Sports Performance: 18 (Encyclopedia of Sports Medicine)*, John Wiley & Sons.

Peeters MW, Thomis MA, Beunen GP, & Malina RM (2009). Genetics and sports: an overview of the pre-molecular biology era. *Med Sport Sci* 54, 28–42.

Pescatello LS & Roth SM (2011). *Exercise Genomics*, Springer.

Schuelke M, Wagner KR, Stolz LE, Hubner C, Riebel T, Komen W, Braun T, Tobin JF, & Lee SJ (2004). Myostatin mutation associated with gross muscle hypertrophy in a child. *N Engl J Med* 350, 2682–8.

Yang N, MacArthur DG, Gulbin JP, Hahn AG, Beggs AH, Easteal S, & North K (2003). ACTN3 genotype is associated with human elite athletic performance. *Am J Hum Genet* 73, 627–31.

REFERENCES

1. Lexell J, et al. *J Neurol Sci.* 1988. 84(2–3):275–94.
2. Silventoinen K, et al. *Genet Epidemiol.* 2008. 32(4):341–9.
3. Yang N, et al. *Am J Hum Genet.* 2003. 73:627–31.
4. Schuelke M, et al. *N Engl J Med.* 2004. 350(26):2682–8.
5. Ahtiainen JP, et al. *Age.* 2016. 38(1):10.

6. Hubal MJ, et al. *Med Sci Sports Exerc*. 2005. 37(6):964–72.
7. Peeters MW, et al. *Med Sport Sci*. 2009. 54:28–42.
8. Janssen I, et al. *J Appl Physiol*. 2000. 89:81–8.
9. Hsu FC, et al. *Obes Res*. 2005. 13(2):312–9.
10. Souren NY, et al. *Diabetologia*. 2007. 50(10):2107–16.
11. Bogl LH, et al. *J Bone Miner Res*. 2011. 26(1):79–87.
12. Huygens W, et al. *Can J Appl Physiol*. 2004. 29:186–200.
13. Ontell MP, et al. *Dev Dyn*. 1993. 198(3):203–13.
14. Verbrugge SAJ, et al. *Front Physiol*. 2018. 9:553.
15. Lai KM, et al. *Mol Cell Biol*. 2004. 24(21):9295–304.
16. Lee SJ. *PLoS One*. 2007. 2(8):e789.
17. McPherron AC, et al. *Nature*. 1997. 387(6628):83–90.
18. Whittemore LA, et al. *Biochem Biophys Res Commun*. 2003. 300:965–71.
19. Lek M, et al. *Nature*. 2016. 536(7616):285–91.
20. Karczewski KJ, et al. *Nature*. 2020. 581(7809):434–43.
21. McPherron AC, et al. *J Clin Invest*. 2002. 109:595–601.
22. Welle S, et al. *Am J Physiol Endocrinol Metab*. 2011. 300(6):E993–1001.
23. Proud CG. *Biochem Biophys Res Commun*. 2004. 313(2):429–36.
24. Hall KD. *Br J Nutr*. 2010. 104(1):4–7.
25. Lionikas A, et al. *J Anat*. 2013. 223(3):289–96.
26. Smerdu V, et al. *Am J Physiol*. 1994. 267(6 Pt 1):C1723–8.
27. Totsuka Y, et al. *J Appl Physiol (1985)*. 2003. 95(2):720–7.
28. Nimmo MA, et al. *Comp Biochem Physiol A Comp Physiol*. 1985. 81(1):109–15.
29. North KN, et al. *Nat Genet*. 1999. 21(4):353–4.
30. Mills MA, et al. *Hum Mol Genet*. 2001. 10:1335–46.
31. Alfred T, et al. *Hum Mutat*. 2011. 32(9):1008–18.
32. MacArthur DG, et al. *Hum Mol Genet*. 2008. 17(8):1076–86.
33. Vincent B, et al. *Physiol Genomics*. 2007. 32(1):58–63.
34. Windelinckx A, et al. *Eur J Hum Genet*. 2011. 19(2):208–15.
35. Delmonico MJ, et al. *J Gerontol A Biol Sci Med Sci*. 2007. 62(2):206–12.
36. Roth SM, et al. *J Appl Physiol*. 2001. 90(4):1205–10.
37. Kostek MC, et al. *J Appl Physiol*. 2005. 98:2147–54.
38. Sayer AA, et al. *Age Ageing*. 2002. 31:468–70.
39. Schrager MA, et al. *J Appl Physiol*. 2004. 97:2176–83.
40. Riechman SE, et al. *J Appl Physiol*. 2004. 97:2214–9.
41. Santiago C, et al. *PLoS One*. 2011. 6(1):e16323.
42. Huygens W, et al. *Physiol Genomics*. 2004. 17(3):264–70.
43. Liu D, et al. *J Appl Physiol*. 2008. 105(3):859–67.
44. Seaborne RA, et al. *J Physiol*. 2019. 597(14):3727–49.
45. Grundberg E, et al. *Eur J Endocrinol*. 2004. 150:323–8.
46. Wang P, et al. *Int J Sports Med*. 2006. 27(3):182–6.
47. Lightfoot JT, et al. *Med Sci Sports Exerc*. 2021. 53(5):883–7.
48. Wheeler DA, et al. *Nature*. 2008. 452(7189):872–6.
49. Huygens W, et al. *Physiol Genomics*. 2005. 22(3):390–7.
50. De Mars G, et al. *J Med Genet*. 2008. 45(5):275–83.
51. McPherron AC, et al. *Proc Natl Acad Sci U S A*. 1997. 94(23):12457–61.
52. Mosher DS, et al. *PLoS Genet*. 2007. 3(5):e79.
53. Binns MM, et al. *Anim Genet*. 2010. 41(Suppl 2):154–8.
54. Amthor H, et al. *Proc Natl Acad Sci U S A*. 2007. 104(6):1835–40.
55. Hernandez Cordero AI, et al. *Am J Hum Genet*. 2019. 105(6):1222–36.
56. Willems SM, et al. *Nat Commun*. 2017. 8:16015.
57. Zillikens MC, et al. *Nat Commun*. 2017. 8(1):80.

CHAPTER FIVE

Genetics of endurance

Stephen M. Roth and Henning Wackerhage

LEARNING OBJECTIVES

At the end of the chapter, you should be able to:

1. Explain why endurance performance is considered to be a polygenic trait and describe the heritability of VO_2max, VO_2max trainability and fibre-type percentages.
2. Describe transgenic mouse models where endurance capacity and related traits are affected.
3. Discuss the effect of selective breeding and inbreeding on genetic variations that affect endurance capacity.
4. Discuss the *ACE* I/D polymorphism in relation to endurance and the limitations of endurance exercise-focused association studies.
5. Discuss the results of gene score and genome wide association studies (GWASs) in relation to endurance performance.
6. Discuss how a rare EPO receptor mutation increases endurance capacity.
7. Explain the relevance of genetic testing for sudden cardiac death in young athletes.

INTRODUCTION

When it comes to endurance performance, the racehorse Frankel is one of the all-time greats. Frankel has won each of its 14 races and was awarded 140 points in the World's Best Racehorse Ranking as the highest ranked racehorse. After his racing career, Frankel mated 133 mares in his first season as a stallion to transfer many of his endurance DNA variants into a new generation of racehorses. This has produced some winning horses such as Cracksman who won Ascot's Champion Stakes in 2017. Frankel's 'stud fee' was reported to be £125,000 and has increased to £175,000 in 2018 (http://www.bbc.com/sport/-horse-racing/41823871). Frankel is a result of generations of selective breeding for endurance running ability, being the offspring of both a championship-winning sire and championship-winning dam, both with their own champion pedigrees.

What about humans? Untrained humans not only differ greatly in their VO_2max (1), muscle fibre distribution (2), haemoglobin concentration and body weight (all factors that limit endurance capacity) but also adapt differentially to the same endurance training programme (3). Twin and family studies imply that much of this variation is explained by variants in the DNA sequence that differ between individuals or families. Unfortunately, despite much research, we still know little about the DNA sequence variants that result in elite endurance performance.

DOI: 10.4324/9781315110752-5

LEARNING OBJECTIVES

At the end of the chapter, you should be able to:

1. Describe the heritability of endurance and the factors that limit endurance performance.
2. Discuss the DNA sequence variants that affect human endurance performance.

Before reading this chapter, you should have read **Chapter 3** which introduces sport and exercise genetics. Also, reading **Chapter 9** in conjunction with the present chapter is recommended, as it covers the molecular regulators of adaptation to endurance training.

Research questions related to endurance exercise were a major research focus of many exercise physiologists from the time when exercise physiology emerged as a sub-discipline of physiology. Researchers have been keen to identify the limits of human endurance performance, as well as what distinguishes elite athletes from average performers. Indeed, the VO_2max was already being measured in the 1920s by Archibald Vivian (A.V.) Hill (4) and others. Classical endurance research is focused on the function of the cardiovascular and muscular organ systems during exercise and on the effect of environmental factors such as training and nutrition. However, whilst VO_2max and VO_2max trainability are both ≈50% inherited (3), only a small proportion of VO_2max and endurance training research has been directed at identifying the genetic variations that are responsible for the large variation in endurance capacity and trainability in the human population. Nonetheless, as the ability to study genetic influences on endurance traits improves, the number of exercise physiologists engaged in genetic research on endurance is increasing.

We start the present chapter by reviewing factors that may limit performance in endurance events, such as maximal cardiac output, blood oxygen transport capacity and skeletal muscle fibre-type variability. Endurance capacity, and probably most of its limiting factors are significantly inherited and polygenic. We know that genetic factors are important from the study of transgenic mouse models that have an increased endurance capacity, perhaps most strikingly in the case of the PEPCK mouse (5). Such transgenic mice highlight 'candidate genes' where common or rare DNA sequence mutations may have an effect on endurance-related traits. We also discuss selective breeding for endurance performance in rodents and racehorses and discuss the variation of endurance-related traits in inbred mouse strains. Studying these genetic factors in humans is more difficult, with both common and rare genetic variants contributing to endurance traits, and both of those types of variants being hard to conclusively identify. In this context, we discuss the *ACE* I/D genotype which was the first common polymorphism linked to endurance-related traits. After that, we will discuss more sophisticated genome-wide association studies (GWASs) and outline how the field is moving forward to more conclusively identify the genetic underpinnings of endurance performance. We will then discuss a rare DNA sequence variant in the erythropoietin (EPO) receptor found in the family of the cross-country skier and treble Olympic gold medal winner Eero Antero Mäntyranta (6). This suggests that some rare DNA sequence variants may have a large effect size on endurance-related traits. Finally, we discuss sudden death in athletes and outline the basis for which genetic testing may someday be commonplace as a predictive mechanism to help prevent these tragedies.

ENDURANCE: A POLYGENIC TRAIT

Endurance performances such as a Marathon race depend on many limiting factors or **quantitative traits** (QTs) such as the VO_2max, the % of VO_2max at lactate threshold (%VO_2max at LT) and movement efficiency which are summarized in **Figure 5.1** (7). The variations of these factors are explained by both environmental factors such as endurance training (Chapter 9) and nutrition (Chapter 10), and genetic factors, including DNA sequence variants. Because endurance is dependent on many limiting factors

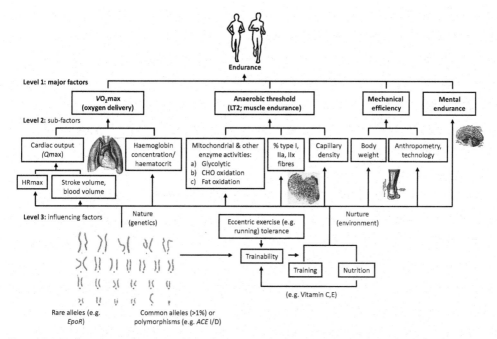

Figure 5.1 Endurance performance depends on many factors which, in turn, are influenced by common and rare DNA sequence variations or quantitative trait loci and environmental factors such as training and nutrition. Abbreviations: Qmax – maximal cardiac output. HRmax – maximal heart rate. CHO – Carbohydrates. ACE I/D –angiotensin converting enzyme insertion/deletion variant. EpoR – erythropoietin receptor. Brain, lung and foot images are taken from Grey's Anatomy and do not require permissions under the Creative Commons CC0 License, https://creativecommons.org/publicdomain/zero/1.0/; or Creative Commons Attribution-ShareAlike License, https://creativecommons.org/licenses/by-sa/4.0/.

which are affected by genetic variation, it is clear that endurance itself is, as we will show throughout this chapter, a polygenic trait.

HERITABILITY OF VO_2MAX, VO_2MAX TRAINABILITY

AND FIBRE-TYPE PERCENTAGES

The earliest studies on the genetics of human endurance performance were twin and family studies. These studies estimated heritability of VO_2max and fibre-type proportions. The heritability of VO_2max was estimated to be between 40% (8) and 93% (9). More recently, the heritability of VO_2max has been estimated to be 50% (1), whilst the heritability of VO_2max trainability has been estimated to be 47% (3). Thus, keeping the limitations of the methods for measuring heritability in mind (see **Chapter 3**), it is close enough to assume that VO_2max and VO_2max trainability are both ≈50% inherited. Similarly, many of the traits underlying VO_2max are similarly heritable, including maximal heart rate, stroke volume and cardiac output.

After the VO_2max, the fibre-type proportion is the second major factor which affects endurance performance. In humans, adult skeletal muscle comprises slow type I, intermediate type IIa and fast type IIx fibres, as discussed in **Chapter 9**. Humans carry the gene for myosin heavy chain IIb but it is not normally expressed in human skeletal muscle (10). Within individuals, the proportions of fibre types differ from muscle to muscle. They range from 15% of slow type I fibres in eye muscles to 89% in the soleus (calf) muscle (11). From a genetics perspective, the most important finding is that the fibre proportions in a given muscle differ considerably in the population. For example, a quarter of North American Caucasians have either less than 35% or more than 65% type I fibres (2). Extremes of fibre-type proportions for locomotor muscles are found in speed/power versus endurance athletes, with speed/power athletes having a high percentage of fast type II and endurance athletes having a high percentage of slow type I fibres (12, 13). In this research, Costill and Saltin already concluded that 'these measurements confirm earlier reports which suggest that the athlete's preference for strength, speed, and/or endurance events is in part a matter of genetic endowment' (12). The heritability estimates for fibre types range from no significant genetic effect (14) to 92.8–99.5% for females and males, respectively (15), which highlights the limitations of such studies. Today, the consensus estimate for fibre-type proportions is ≈45% heritability (2).

Taken together, the key endurance-limiting factors of VO_2max, VO_2max trainability and fibre-type proportions are ≈50% inherited. In the following text, we will now discuss common and rare DNA sequence variations that contribute to the variation of endurance traits seen in the human population.

WHAT DO TRANSGENIC ANIMAL AND ANIMAL BREEDING STUDIES TELL US ABOUT GENES WHOSE DNA VARIANTS AFFECT ENDURANCE?

The identification of polymorphisms or rare mutations depends on previous knowledge of genes that affect endurance-related traits. Genes where this has been demonstrated are termed **candidate genes**, reflecting the idea that these are 'candidates' for genes containing DNA sequence variation that may impact gene or protein function and thus impact the phenotype of interest, thereby explaining some aspect of the heritability of that phenotype. Candidate genes are often identified based on known anatomy and physiology of the tissues underlying the trait of interest or may be identified by genetic studies intended to find areas of the genome that appear to relate to a trait of interest. When examined in model organisms directly (e.g. in cultured muscle cells in vitro, or in transgenic mice or other species in vivo), such candidate genes are commonly identified via gain or loss-of-function experiments where the gene is specifically manipulated to increase or decrease expression or function of that gene to study the resulting changes to the phenotype (explained in general in **Chapter 3** and in the context of muscle size and strength in Chapter **4**).

Much information on endurance-related candidate genes has been obtained by creating transgenic mice and by testing their endurance capacity. In these studies, the investigators either knockout or overactivated genes. To gain a more complete overview over genes whose manipulation alters endurance in mice, we have systematically searched for publications that report this. We found 31 genes whose transgenesis increased endurance performance by up to 1800% illustrated in **Figure 5.2**.

The most striking transgenic mouse model had a mutation of the Pck1 gene which encodes the gluconeogenic enzyme PEPCK-C. This enzyme is normally expressed especially in the liver but its overexpression in skeletal muscle changes energy metabolism. The authors found that the PEPCK-C-overexpressing mice ran up to 6 km at 20 m.min^{-1}, whilst the wildtype controls ran only ≈200 m at that speed to fatigue (**Figure 5.3**).

The PEPCK mouse demonstrates the large effect that some individual genes can have on endurance-related traits.

Does the DNA sequence of these genes vary in humans and does this explain the variation in human exercise capacity? First, the genome-wide sequencing of exons, termed exomes, in 60,706 individuals revealed that all genes are affected by DNA variants that change the amino acid sequence of the encoded proteins and often result in a knockout of the human gene. Some of these knockouts are even homozygous so that there are knockout humans for some genes (17). For example, when searching for

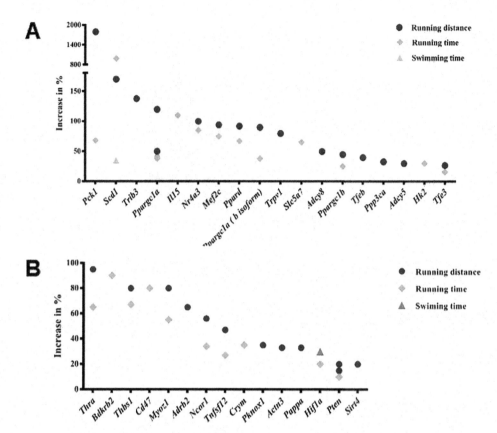

Figure 5.2 Genes whose gain or loss-of-function increases endurance performance in mice (16). Figure is taken from Yaghoob Nezhad F, et al. 2019 (16) as an open access (Attribution 4.0 International -CC BY 4.0, https://creativecommons.org/licenses/by/4.0/) article in *Frontiers in Physiology* where permissions are not required provided the work is properly cited.

DNA variants for *PPARGC1A*, the gene that encodes the mitochondrial biogenesis regulator protein PGC-1α, the genome aggregation database browser reveals 385 missense SNPs and 5 loss-of-function SNPs, albeit none of these are homozygous (17, 18). This suggests function-modulating DNA sequence variation in genes whose increased or decreased function increases endurance performance, at least in mice. It is therefore a mystery as to why these genes are not highlighted in human GWAS studies where the investigators searched for SNPs that were associated with endurance markers such as VO₂max (19).

SELECTIVE BREEDING FOR ENDURANCE AND ENDURANCE-RELATED TRAITS OF INBRED MOUSE STRAINS

Towards the end of this chapter, we will discuss the results of selective breeding and inbred mouse strain studies as a non-biased strategy to identify genes that affect endurance-related traits. Selective breeding for endurance-related traits has been used for a long time to breed race dogs and horses. The aim of selective breeding for endurance is to accumulate DNA sequence variations found in the founder

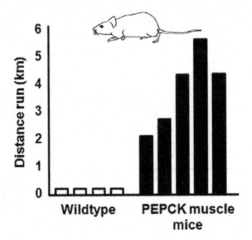

Figure 5.3 Effect of skeletal muscle-specific expression of the PEPCK gene in skeletal muscle on running distance at 20 m.min⁻¹. PEPCK muscle mice can run more than 10 times the distance at 20 m.min⁻¹ when compared to wildtype controls. Redrawn from Hakimi P, et al. 2007 (5) as an open access (Attribution 4.0 International -CC BY 4.0, https:// creativecommons.org/licenses/by/4.0/) article in *Journal of Biological Chemistry* where permissions are not required provided the work is properly cited.

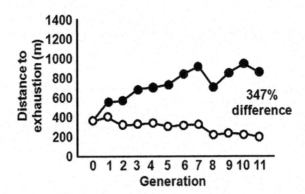

Figure 5.4 Selective breeding for high (high-capacity runners, HCR) and low (low-capacity runners, LCR) exercise capacities in rats over 11 generations. The study shows that such breeding selects, amongst other factors, for high expression levels of known regulators of mitochondrial biogenesis and fibre type, e.g. PGC-1α and PPAR-γ. Redrawn from Wisloff U, et al. (20).

population that increase endurance capacity. In a scientific experiment, researchers have selectively bred rats for low and high running ability (20). After 11 generations, the running distance to exhaustion differed by 347% between the high- and low-capacity runners (**Figure 5.4**).

Genetically, this has led to the enrichment of DNA sequence variants that are favourable for running capacity. The researchers additionally measured the expression of known regulatory factors of endurance and found that, for example, PGC-1α and PPAR-γ are expressed at significantly higher levels in the high-capacity compared with the low-capacity runners. This could be due to DNA sequence variants in the promoter or enhancer regions of these genes or due to variants that affect the expression or activity of factors which increase the expression of PGC-1α and PPAR-γ.

A different model is inbred mice. In this case, the genetic variations are fixed in a species by interbreeding family members for at least 20 consecutive generations. Via this procedure, the animals become genetically more or less identical and have similar phenotypes and thus retain exercise-related traits such as muscularity or fibre-type percentages. Such inbred mice strains are a powerful tool for molecular sport and exercise geneticists because the DNA of many of these inbred mouse strains have now been sequenced across the genome. Researchers can thus measure the phenotype or QTs such as voluntary running, muscle size, number of muscle fibres for a given muscle or fibre-type proportions. Then, by using linkage analysis, they can identify the regions that harbour variations in the DNA sequence or **quantitative trait loci** (QTLs) that are responsible for the variation of QTs in between inbred strains. New DNA-sequencing technologies can then be used to directly sequence those regions and identify the specific DNA sequence variants underlying the traits.

Inbred mouse strains have been used to search for QTLs that affect voluntary wheel running that also depends on psychological factors. **Figure 5.5A** shows the large difference in average wheel running between the 41 inbred mouse strains studied and the small variation within each strain. This suggests two things: first, the large variation between strains suggests that there are DNA sequence variants in the

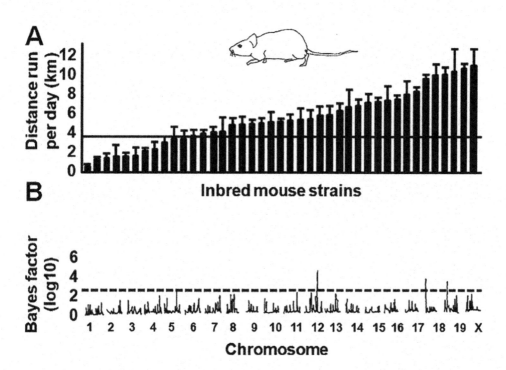

Figure 5.5 **(A)** Average running wheel distance (mean ± SD) per day for different inbred mouse strains. **(B)** The x-axis shows the location on the chromosomes of mice and the y-axis shows a measure of the likelihood that variation in that genomic location has an influence on average running distance. Redrawn from Lightfoot JT, et al. (21).

heterogenous founder population that affect voluntary wheel running. Second, the small variation within a group suggests that the effect of these DNA sequence variants is large. **Figure 5.5B** then shows QTLs in the mouse genome that are associated with the differences in voluntary wheel running distance. The highest QTL is located on chromosome 12.

Researchers can now search more selectively for the exact DNA sequence variants which cause the differences in voluntary wheel running that are seen in inbred mouse strains.

ASSOCIATION STUDIES AND THE ACE I/D POLYMORPHISM

Until recently, updates on genes that were associated with sport- and exercise-related traits were published in a journal article in *Medicine in Science and Sports and Exercise* (22). These articles give a good overview of the results from association and other genetic studies in relation to endurance performance and other sport- and exercise-related traits. In this section, we will first focus in detail on the angiotensin converting enzyme (ACE) insertion/deletion (I/D) polymorphism as a putative endurance-related polymorphism, and then list other polymorphisms that have been linked to endurance-related traits. Importantly, sport and exercise genetics are changing quickly, and single-gene polymorphism studies are becoming increasingly obsolete, instead being replaced by GWAS and more complex molecular genetic studies.

The *ACE* gene arose as a possible candidate for endurance performance, given its central role in the renin-angiotensin system and the regulation of blood pressure. The *ACE* gene contains a 287-base pair insertion/deletion (I/D) in intron 16 where both alleles are common in many populations, which is why this DNA sequence variant is considered to be a polymorphism (23). A report in 1997 showed that the *ACE* I/D genotype was associated with growth of the left ventricle in the heart in response to physical training in military recruits, with D/D genotype carriers having greater cardiac growth than I/I carriers (24). This led to a flood of studies aimed at determining the importance of this DNA sequence variant on endurance-related traits. Cross-sectional studies of athletes followed, several of which showed a greater proportion of high-level endurance athletes carrying the I/I genotype than would be expected in the general population. However, the results were not all consistent, as several studies found no or only little association with endurance-related traits and especially VO_2max (25) (**Figure 5.6**).

There are two main explanations for the inconsistent results. First, the effect size of the *ACE* I/D polymorphism is relatively small, and for this reason hundreds or even thousands of subjects are required for sufficient statistical power to detect true associations (27, 28). As is typical in most exercise genomics studies, these large sample sizes have not been achieved in most studies on the *ACE* I/D polymorphism; thus, the studies are likely underpowered to confidently show small effects. Second, the *ACE* I/D polymorphism may not be associated with VO_2max but instead with skeletal muscle endurance. Skeletal muscle has a local, tissue-specific renin-angiotensin system whose function might be altered by the *ACE* I/D polymorphism. Therefore, studies emerged examining metabolic efficiency and skeletal muscle fibre type and here researchers were able to find more meaningful associations. *ACE* I/I genotype has been associated with both enhanced metabolic efficiency and a higher proportion of type I muscle fibres in comparison to the D/D genotype, both of which could contribute to improved endurance performance. Interestingly, the D/D genotype has been associated with strength and power-related traits, but we will not discuss this further in this chapter.

Overall, the *ACE* I/D polymorphism appears to have a small influence on endurance performance such that I/I individuals have a slight advantage. But this advantage must be put in context: there are elite endurance athletes without the I/I genotype and many endurance athletes with the I/I genotype never reach elite status as the effect size of the *ACE* I/D polymorphism is small. In other words, the *ACE* I/D polymorphism is just one of many contributors to endurance performance and should be considered within this context.

Over the years, many other polymorphisms have been linked to endurance-related traits. We give a brief overview in **Table 5.1** of some key genes studied in relation to endurance performance traits. Note, however, that none have a sample size over >1,000 subjects, which many would argue is a minimum for a quality cross-sectional genetic analysis.

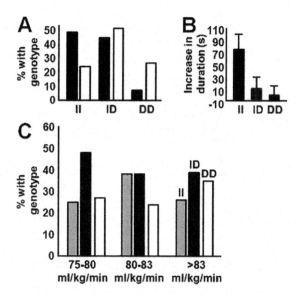

Figure 5.6 **(A)** Frequency of the *ACE* I/D polymorphism in 25 British mountaineers/climbers (black bars) and 1,906 healthy British men (white bars) which served as controls. **(B)** Improvement of duration of repetitive elbow flexion after 10 weeks of physical training amongst British army recruits in relation to the *ACE* I/D polymorphism. **A** and **B** were redrawn from Montgomery HE, et al. (26). **(C)** No association between the *ACE* I/D genotype and VO_2max in a study of elite endurance athletes. Redrawn from Rankinen T, et al. (25).

Overall, these studies demonstrate the presence of common DNA sequence variants or polymorphisms that explain a small proportion of the variability of endurance-related traits such as VO_2max or endurance trainability. They do, however, not explain all the heritability of such traits and it is currently unclear whether the remaining heritability is related to currently unknown polymorphisms or whether rare DNA sequence variants have a major effect. The whole genome sequencing of large cohorts that are reliably phenotyped for endurance-related traits will eventually yield an answer to this question.

MULTIPLE POLYMORPHISMS: GENE SCORE AND GENOME-WIDE ASSOCIATION STUDIES

Association studies for single-gene polymorphisms with small effect size are not very informative and testing athletes for one polymorphism generally yields little information about an individual's genetic potential for endurance sport. For these reasons, researchers have started to analyse several polymorphisms together to calculate 'genetic scores' in order to better predict the genetic endurance potential of an individual (45, 46). GWASs have also been performed as a non-biased way to search for multiple polymorphisms that are associated with sport- and exercise-related traits (47).

- In one 'genetic score' study, 1,423 Russian athletes and 1,132 controls were genotyped for multiple endurance-related polymorphisms (46). The authors analysed their data for multigene effects and found that individuals that carry ≥ 9 endurance alleles are significantly more likely to be an endurance athlete, suggesting that endurance-associated polymorphisms contribute to talent for endurance sports. However, the practical usefulness of this genetic score test is still limited because 38% of the controls also had ≥ 9 endurance alleles, suggesting that the test does not allow researchers to reliably identify endurance athletes based on the presence of several alleles associated with endurance

Table 5.1 DNA sequence variations affecting endurance-related traits in humans (key studies only)

Gene	Trait affected and study design	Number of subjects	Reference
PPARGC1A	Predicted VO$_2$max	599	(29)
PPARGC1A	Case control of endurance athletes versus controls	204	(30)
PPARGC1A	Case control of endurance athletes versus controls	395	(31)
ADRB2	VO$_2$max	63	(32)
ADRB2	Case control of endurance athletes versus controls	600	(33)
ADRB2	VO$_2$max	62	(34)
VEGFA	VO$_2$max	146	(35)
NOS3	Triathlon competition performance	443	(36)
ADRB1	VO$_2$peak, exercise time	263	(37)
ADRB1	VO$_2$peak	892	(38)
HIF1A	VO$_2$max and training response of VO$_2$max	125	(39)
BDKRB2	Muscle efficiency	115	(40)
BDKRB2	Case control of endurance athletes versus controls	346	(36)
CKM	VO$_2$max and training response of VO$_2$max	240	(41)
AMPD1	Training response of VO$_2$max	400	(42)
ATP1A2	Training response of VO$_2$max	472	(43)
PPARD	Training response of VO$_2$max	264	(44)

(**Figure 5.7**). These early genetic score analyses are reliant on candidate genes studies that are often underpowered, thus somewhat also limited in their applicability.

The next step-up from multigene or 'genetic score' studies is to search the whole genome for common DNA sequence variants or polymorphisms that are associated with endurance-related traits. Here, we discuss two such studies. In the first study, Timmons et al. (48) obtained muscle biopsies from individuals where the response of the VO$_2$max to a standard endurance training programme was measured which is known as VO$_2$max trainability. The researchers then used a sophisticated strategy in order to find DNA sequence variants that predicted VO$_2$max trainability and a so-called expression QTL strategy. Their strategy was as follows (48):

- Use microarrays to measure the expression of mRNAs in resting muscle.

- Use statistics to determine those mRNAs whose expression correlates best with VO$_2$max trainability. These mRNAs were termed **quantitative molecular classifiers**.

- In another experiment measure SNPs within the gene sequence of the **quantitative molecular classifiers** and analyse the SNP data for association with VO$_2$max trainability.

The important finding of this experimental strategy was that the researchers were able to identify 11 SNPs that predicted 23% of the total variance in VO$_2$max trainability (**Figure 5.8**).

Given that VO$_2$max trainability is ≈50% inherited, (3), the described method predicts roughly half of the inherited variation of VO$_2$max trainability based on common DNA sequence variations (48). The results do require validation in other cohorts of subjects to ensure the results can be generalized beyond the original study. If the results are validated, explaining such a large proportion of VO$_2$max trainability would be useful for identifying potential low and high responders prior to an endurance training programme, which may then help inform the prescription of the training regime.

In another VO$_2$max trainability study, the researchers performed a GWAS (47). In this study, gene chips were used to measure 324,611 SNPs spread over all chromosomes in 473 individuals from 99 families. The association between all SNPs and VO$_2$max trainability was then calculated using a

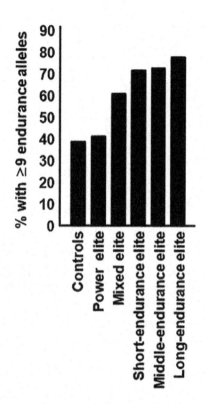

Figure 5.7 Percentage of subjects with a high number of 'endurance alleles'. The study suggests that more endurance athletes carry a high number of endurance alleles (i.e. ≥ 9 endurance alleles). Redrawn from Ahmetov, II, et al. (46).

sophisticated statistical analysis. At the end of the procedure, the research team identified 21 SNPs that predicted 49% of the variation of VO_2max trainability (47) which arguably explains approximately all of the genetic variation, given the $\approx 50\%$ heritability of VO_2max trainability. Given that these gene chips measure common DNA sequence variants, it also implies that nearly all of the genetic variation of VO_2max trainability is due to common as opposed to rare alleles. Clearly, these results will need to be replicated, but both studies show how newer approaches can improve our ability to find DNA sequence variations that explain much of the variability of endurance-related traits.

Because GWAS focuses on common DNA sequence variants, researchers must also account for rare variants that may be significantly influencing endurance traits in individuals and families but are too rare to be predictive for the population (an example is described in the next section). As such, researchers are moving towards using whole genome sequencing, often called next-generation sequencing (NGS), to identify rare DNA sequence variants that may impact endurance-related traits. NGS allows a relatively inexpensive method (compared to older methods) and rapid sequencing of the entire genome, a process that was once prohibitively expensive. Such studies will allow researchers to better understand the contributions of rare versus common DNA sequence variants to endurance performance phenotypes.

In recognizing that sample size is a primary limitation in the analysis of the genetic impact on sport performance. Research groups around the world are coming together to combine cohorts and data to achieve sample sizes that are significantly more statistically powerful than those that could be achieved by any research group alone. The highest profile of these groups is known as the Athlome Consortium (49)

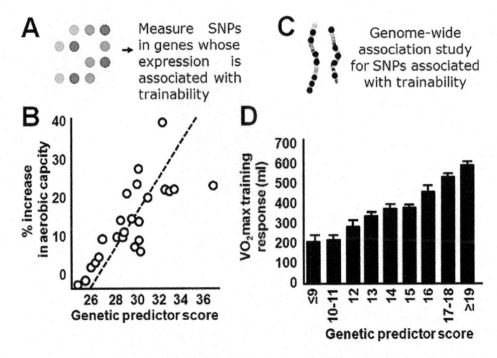

Figure 5.8 (A) The authors used mRNA expression analysis to identify mRNAs of the so-called classifier genes, whose expression predicts VO_2max trainability. The DNA of the classifier genes was then checked for SNPs that also predict VO_2max trainability. Eleven of these SNPs were used to calculate a gene predictor score. **(B)** Relationship between the gene predictor score and VO_2max trainability. According to this study, the gene predictor score explains 23% of VO_2max trainability. Redrawn from Timmons JA, et al. (48). **(C)** GWAS study of SNPs (each dot in this schematic drawing represents one SNP; in reality, many more SNPs were measured) that are associated with VO_2max trainability. The authors found that variation in 21 SNPs accounted for 49% of the variance of VO_2max trainability. **(D)** A predictor SNP score calculated from 21 SNPs predicts VO_2max trainability well. Redrawn from Bouchard C, et al. (47).

which has combined data from 15 different research groups from around the world. The large and high-quality sample sizes combined with GWAS and NGS technologies will finally allow researchers to more confidently identify the genetic factors underlying endurance performance traits.

EERO ANTERO MÄNTYRANTA AND THE EPO RECEPTOR MUTATION AS AN EXAMPLE OF A RARE DNA SEQUENCE VARIATION WITH A LARGE EFFECT SIZE

In the previous sections, we mostly discussed common endurance alleles or polymorphisms. However, exceptional endurance performance may also partially depend on the presence of one or more rare DNA sequence variants with a large effect on endurance performance. The best example of such a rare DNA sequence variant is one contained in the erythropoietin receptor (EPOR) gene which was found in a Finnish family. In this section, we will review this study in some detail.

Many studies have demonstrated that removal or reinfusion of red blood cells or erythrocytes immediately reduces or increases both haematocrit and VO_2max, respectively (50). Thus, blood doping or the administration of EPO, which stimulates erythrocyte production by the bone marrow and thus increases haematocrit, can have a large effect on VO_2max and endurance performance (50). To control blood doping and EPO, the cycling world governing body (Union Cycliste Internationale/UCI) has stated that a haematocrit >50% and haemoglobin levels >170 g.l^{-1} are defined as abnormal and athletes that exceed these thresholds are subject to disciplinary action. This is a good point to start the story of the Finnish cross-country skier and treble Olympic gold medal winner, Eero Antero Mäntyranta, and his family. Eero was a member of a family where a rare erythrocyte-related phenotype was segregating. In this family, the affected males had haemoglobin concentrations of between 183 and 231 g.l^{-1} (6) which are all abnormal levels according to the UCI and are considered detrimental to human health, with possible adverse cardiovascular complications. The fact that this trait was segregating in a family suggested that it was inherited. Furthermore, the researchers found normal bone marrow, low or normal EPO values and the plasma of the individuals did not stimulate erythroid progenitor cells. Thus, the problem was unlikely to be a problem of the bone marrow or of EPO or another serum hormone that had an effect on the production of red blood cells.

Therefore, what could be the cause of the high haematocrit? The researchers found that bone marrow and blood cells from affected family members formed more erythrocyte colonies at low EPO concentrations, even in the absence of EPO, where controls do not form colonies. This suggested that something within the cells was responsible for the increased erythrocyte production. The researchers reasoned that the EPO receptor (EPOR) might be a candidate gene. They designed primers to cover the EPOR gene and used the genomic DNA of affected family members and controls to amplify the EPOR gene. The amplified EPOR gene was then sequenced, and a G-to-A mutation was found in position 6002 of the gene which was abbreviated as EPOR 6002 G→A. The mutation results in a premature stop codon in the DNA which gives rise to a shortened EPOR mRNA and protein. Normally, incompletely transcribed proteins are degraded or function less, but in this case, it seemed as if the truncated version of the EPOR protein was more active. The most likely explanation is that the last bit of the EPOR protein must somehow inhibit EPOR activity.

Because DNA sequencing was expensive at the time of this study, the authors used the restriction enzymes NcoI and StyI to digest the PCR product of the EPOR gene. In this assay, the PCR product of control subjects is cut by NcoI and StyI into two bands (see **Figure 5.9**) but not in the affected family members (one band; heterozygous individuals display three bands) (6). The figure shows that there are no individuals with only one band suggesting that there are no homozygous carriers. As there was no individual that was homozygous for the EPOR mutation in the whole study, it may be that the mutation is lethal if homozygous.

The researchers then also checked DNA from 50 random Finnish blood donors to ensure that this mutation was not present in the normal population, which was confirmed by this experiment.

Now back to Eero Antero Mäntyranta. As a heterozygous carrier of the mutation, his haematocrit and haemoglobin were so high that he would probably be banned from competing in cycling today. Essentially, he had the advantage of what is equivalent to EPO or blood doping which increased his VO_2max and endurance performance. Given that he was competing at a time when EPO doping was unknown and blood doping was unlikely, he had a genetic advantage over his competitors which partially explains why he was able to win one gold medal in 1960 in Squaw Valley and two gold medals in 1964 in Innsbruck.

However, this poses an important question: If he was a cyclist today, should he be banned from competing as his haematocrit is too high? Of course not, because talent for sport is all about being a carrier of advantageous common and rare DNA sequence variants (45). Thus, individuals who carry a DNA sequence variation that results in a conflict with anti-doping rules should be allowed to use genetic testing in order to prove that such a mutation is natural and be allowed to compete if it can be shown that an abnormal result is due to the mutation. To summarise, the EPOR 6002 G→A DNA sequence variant

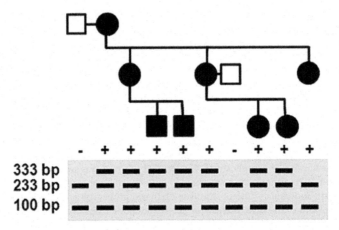

Figure 5.9 Schematical drawing showing a pedigree for the Finnish family affected by the rare EPOR DNA sequence variation and the result of a genetic test on an agarose gel. Black symbols and '+' represent affected family members, white symbols and '–' represent unaffected members. Redrawn from de la Chapelle A, et al. (6). Unaffected family members, who do not carry the mutation, have a 233 bp and a 100 bp fragment as their EPOR PCR product is cut into two pieces by NcoI or StyI restriction enzymes. Heterozygous, affected family members have three bands: 233 bp and 100 bp for the normal EPOR allele and 333 bp for the mutated allele. Family members that are homozygous for the mutated form would have just one 333 bp band but none were present suggesting that being homozygous for this genotype may be lethal.

demonstrates as a proof-of-principle for the presence of rare DNA sequence variants with a large effect on endurance performance.

GENETICS OF SUDDEN DEATH IN SPORT

Another aspect of understanding the genetic factors underlying endurance performance is identifying the causes of tragic sudden death in sport, particularly that seen in otherwise healthy young athletes participating in strenuous activity. The vast majority of these deaths are related to cardiac arrhythmias, coronary artery malformations and myocardial hypertrophy (51). However, they are idiopathic which means that the athlete has no prior symptoms or knowledge of a problem or increased risk. The majority of deaths in young athletes are attributed to Sudden Arrhythmic Death Syndrome (SADS) in which the heart and coronary artery structure are normal, but the electrical rhythms of the heart are not. These abnormal heart rhythms are exacerbated by physical exertion.

Whilst the underlying causes of SADS and related sudden death are often identifiable by autopsy, being able to effectively predict risk for such an event would be useful to help combat SADs. However, sophisticated cardiac screening of all athletes is cost-prohibitive at this time, which means that every year athletes with underlying cardiac abnormalities participate in sport and experience adverse events. Genetic factors, likely rare DNA sequence variants, are contributing to the majority of these abnormalities and thus genetic testing can be envisioned as a future screening tool to predict risk for sudden cardiac death due to exertion. Whilst specific DNA sequence variants have been identified to underlie specific cardiac disorders, genetic testing for risk prediction is not yet effective for broad athlete screening as the full extent of such variants have not yet been discovered. Nevertheless, researchers continue to add to our knowledge base for the genetic factors that contribute to an increased risk of sudden death in young athletes, and we can envision a day when such predictive genetic tests become more mainstream.

SUMMARY

Endurance performance as a complex physiological endpoint is a polygenic trait. VO_2max is a key variable and both VO_2max and VO_2max trainability are ≈50% inherited (3), whilst fibre-type proportions are ≈45% inherited (2). Transgenic mice display changes in cardiac performance, muscle fibre-type percentages, mitochondrial content and actual running performance. The most striking example is a mouse where the PEPCK enzyme has been overexpressed in skeletal muscles. Such transgenic mouse studies highlight candidate genes where DNA sequence variants may affect endurance-related traits. Selective breeding has been used to breed race dogs and horses but also in scientific studies for high and low running capacity (20). Inbred mouse strains differ amongst each other in endurance capacity but within an inbred mouse strain the difference in capacity is small, suggesting that underlying genetic variants explain these differences (21). As the genomes of many inbred mouse strains have been sequenced, genetic variants that are responsible for high or low endurance capacity can be identified (21). In humans, early association studies have suggested that common DNA sequence variants or polymorphisms such as the ACE I/D polymorphism are associated with endurance-related traits (26). However, these common polymorphisms have a small effect size on endurance and thus large sample numbers are needed to conclusively demonstrate the effect of such polymorphisms (27, 28). In many of the published studies, the sample sizes are too small which explains the divergent, inconsistent results. The trend in newer studies is therefore towards GWAS, whole genome sequencing, large consortia and other sophisticated study designs that can better predict genetic contributions to endurance-related traits (47, 48). However, rare DNA sequence variants also contribute to the variation of endurance-related traits, with a proof-of-principle being the EPOR 6002 G→A allele which results in a premature stop codon and hyperactive EPOR and increased haematocrit. The importance of rare DNA sequence variants for endurance-related traits is currently unknown and only whole genome sequencing (52) of individuals phenotyped for endurance traits will eventually yield an answer. Once common and rare DNA sequence variants are conclusively identified that contribute meaningfully to endurance performance, it seems likely that genetic tests will be developed on the basis of such studies to predict performance of young athletes and potentially personalize exercise training programmes (53). Similarly, this type of genetic testing will likely become more prevalent for the prediction of risk variants for sudden cardiac death in young athletes.

REVIEW QUESTIONS

- Drawing a diagram to illustrate what factors limit endurance performance. Ensure to include common and rare DNA sequence variations as some of the causative factors.

- Discuss the heritability of key traits underlying endurance performance.

- Describe how transgenic mice and inbred mouse strains can be used to improve our understanding of the genetic underpinnings of endurance performance and endurance-related traits?

- Explain one polymorphism that has been linked to endurance performance. What are the limitations of polymorphism studies?

- Describe the current approaches to improve candidate gene studies for identifying the genetic underpinnings of endurance performance traits in humans.

- What genetic advantage contributed to the success of Eero Mäntyranta and what specific trait was impacted and how?

FURTHER READING

Bouchard C & Hoffman EP (2011). *Genetic and Molecular Aspects of Sports Performance: 18 (Encyclopedia of Sports Medicine)*, John Wiley & Sons.

Bray MS, Hagberg JM, Perusse L, Rankinen T, Roth SM, Wolfarth B, & Bouchard C (2009). The human gene map for performance and health-related fitness phenotypes: the 2006–2007 update. *Med Sci Sports Exerc* 41, 35–73.

de la Chapelle A, Traskelin AL, & Juvonen E (1993). Truncated erythropoietin receptor causes dominantly inherited benign human erythrocytosis. *Proc Natl Acad Sci U S A* 90, 4495–9.

Hagberg JM, Rankinen T, Loos RJ, Perusse L, Roth SM, Wolfarth B, & Bouchard C (2011). Advances in exercise, fitness, and performance genomics in 2010. *Med Sci Sports Exerc* 43, 743–52.

Peeters MW, Thomis MA, Beunen GP, & Malina RM (2009). Genetics and sports: an overview of the pre-molecular biology era. *Med Sport Sci* 54, 28–42.

Pescatello LS & Roth SM (2011). *Exercise Genomics*, Springer.

REFERENCES

1. Bouchard C, et al. *Med Sci Sports Exerc*. 1998. 30(2):252–8.
2. Simoneau JA, et al. *FASEB J*. 1995. 9:1091–5.
3. Bouchard C, et al. *J Appl Physiol*. 1999. 87(3):1003–8.
4. Hill AV, et al. *Q J Med*. 1923. 16:135–71.
5. Hakimi P, et al. *J Biol Chem*. 2007. 282(45):32844–55.
6. de la Chapelle A, et al. *Proc Natl Acad Sci U S A*. 1993. 90(10):4495–9.
7. Bassett DR, Jr., et al. *Med Sci Sports Exerc*. 2000. 32(1):70–84.
8. Bouchard C, et al. *Med Sci Sports Exerc*. 1986. 18(6):639–46.
9. Klissouras V. *J Appl Physiol*. 1971. 31(3):338–44.
10. Smerdu V, et al. *Am J Physiol*. 1994. 267(6 Pt 1):C1723–8.
11. Johnson MA, et al. *J Neurol Sci*. 1973. 18(1):111–29.
12. Costill DL, et al. *J Appl Physiol*. 1976. 40(2):149–54.
13. Costill DL, et al. *Med Sci Sports*. 1976. 8(2):96–100.
14. Bouchard C, et al. *Can J Physiol Pharmacol*. 1986. 64:1245–51.
15. Komi PV, et al. *Acta Physiol Scand*. 1977. 100:385–92.
16. Yaghoob Nezhad F, et al. *Front Physiol*. 2019. 10:262.
17. Lek M, et al. *Nature*. 2016. 536(7616):285–91.
18. Karczewski KJ, et al. *Nature*. 2020. 581(7809):434–43.
19. Rankinen T, et al. *PLoS One*. 2016. 11(1):e0147330.
20. Wisloff U, et al. *Science*. 2005. 307(5708):418–20.
21. Lightfoot JT, et al. *J Appl Physiol*. (1985). 2010. 109(3):623–34.
22. Sarzynski MA, et al. *Med Sci Sports Exerc*. 2016. 48(10):1906–16.
23. Rigat B, et al. *J Clin Invest*. 1990. 86(4):1343–6.
24. Montgomery HE, et al. *Circulation*. 1997. 96:741–7.
25. Rankinen T, et al. *J Appl Physiol*. 2000. 88:1029–35.
26. Montgomery HE, et al. *Nature*. 1998. 393:221–2.
27. Altmuller J, et al. *Am J Hum Genet*. 2001. 69(5):936–50.
28. Studies N-NWGoRiA, et al. *Nature*. 2007. 447(7145):655–60.
29. Franks PW, et al. *Med Sci Sports Exerc*. 2003. 35(12):1998–2004.
30. Lucia A, et al. *J Appl Physiol*. (1985). 2005. 99(1):344–8.
31. Eynon N, et al. *Scand J Med Sci Sports*. 2010. 20(1):e145–50.
32. Moore GE, et al. *Metabolism*. 2001. 50:1391–2.
33. Wolfarth B, et al. *Metabolism*. 2007. 56(12):1649–51.
34. McCole SD, et al. *J Appl Physiol*. (1985). 2004. 96(2):526–30.

35. Prior SJ, et al. *Am J Physiol Heart Circ Physiol.* 2006. 290(5):H1848–55.
36. Saunders CJ, et al. *Hum Mol Genet.* 2006. 15(6):979–87.
37. Wagoner LE, et al. *Am Heart J.* 2002. 144(5):840–6.
38. Defoor J, et al. *Eur Heart J.* 2006. 27(7):808–16.
39. Prior SJ, et al. *Physiol Genomics.* 2003. 15(1):20–6.
40. Williams AG, et al. *J Appl Physiol (1985).* 2004. 96(3):938–42.
41. Rivera MA, et al. *Med Sci Sports Exerc.* 1997. 29(11):1444–7.
42. Rico-Sanz J, et al. *Physiol Genomics.* 2003. 14(2):161–6.
43. Rankinen T, et al. *J Appl Physiol.* 2000. 88:1571–5.
44. Hautala AJ, et al. *Am J Physiol Heart Circ Physiol.* 2007. 292(5):H2498–505.
45. Williams AG, et al. *J Physiol.* 2008. 586(1):113–21.
46. Ahmetov, II, et al. *Hum Genet.* 2009. 126(6):751–61.
47. Bouchard C, et al. *J Appl Physiol.* 2011. 110(5):1160–70.
48. Timmons JA, et al. *J Appl Physiol.* 2010. 108(6):1487–96.
49. Pitsiladis YP, et al. *Physiol Genomics.* 2016. 48(3):183–90.
50. Cooper CE. *Essays Biochem.* 2008. 44:63–83.
51. Finocchiaro G, et al. *J Am Coll Cardiol.* 2016. 67(18):2108–15.
52. Wheeler DA, et al. *Nature.* 2008. 452(7189):872–6.
53. Roth SM. *J Appl Physiol.* 2008. 104(4):1243–5.

CHAPTER SIX

Epigenetics of exercise

Daniel C. Turner*, Robert A. Seaborne* and Adam P. Sharples

LEARNING OBJECTIVES

At the end of this chapter, you should be able to:

1. Describe the fundamental principles of epigenetics.
2. Explain the various types of common epigenetic modifications.
3. Discuss the role of DNA methylation, histones modifications and miRNAs during endurance exercise.
4. Discuss the role of DNA methylation, histone modifications and miRNAs during resistance exercise.
5. Discuss how epigenetics contributes to muscle memory.

INTRODUCTION

In **Chapters 3 to 6**, the theory of genetics in sports and exercise, as well as the role of inherited genes in muscle mass and endurance were discussed. In this chapter, the role of **epigenetics** in exercise will be discussed. Epigenetics (meaning 'above' genetics) is the topic surrounding the interaction between environmental 'stressors' (e.g. physical activity and exercise) and our inherited DNA that ultimately influences the acute molecular response and overall adaptation to exercise. For example, exercise can create molecular modifications to our inherited DNA code that subsequently affects how genes are turned on and off following exercise. While epigenetics is not necessarily a new field, researchers have only really begun investigating epigenetics within the context of exercise in the past 10 years. In **Chapter 1**, we provided a brief history of the emergence of epigenetics within molecular exercise physiology and in **Chapter 3**, we introduced the important epigenetic modifications that occur in the DNA and histone proteins. In **Chapter 7**, we include epigenetics within the 'signal transduction theory' that covers the fundamental steps involved in the molecular response to exercise. The focus of the present chapter, however, is to provide an insight into the emerging field of molecular exercise **epigenetics** and to gain a more in-depth understanding of the main epigenetic modifications that occur in response to exercise. The chapter will also cover the effects of performing divergent types of exercise (i.e. endurance and resistance exercise) that cause these epigenetic modifications. Finally, given that epigenetic modifications can also be retained following exercise, epigenetics has been linked with playing a role in muscle 'memory' or a so-called 'epi-memory' (1). Therefore, this chapter will end by discussing the role of epigenetics in muscle memory and why this phenomenon is of interest to molecular exercise physiologists.

---

*These authors contributed equally.

DOI: 10.4324/9781315110752-6

WHAT IS EPIGENETICS?

The formal study of epigenetics (*epi* in Greek means 'on top') can trace its origin back as far as the 1940s, when Conrad Waddington described this topic as a means by which genetic components (genotype) interact with their surrounding environment to create a 'phenotype', thereby bridging the gap between 'genotype and phenotype'. Since then, and with the development of pioneering technologies, our understanding of epigenetics and its definition has advanced significantly. Indeed, epigenetics is now more commonly known to be the field of study interested in changes in gene expression that are not attributable to underlying variations within inherited DNA (2). In essence, epigenetic research is interested in the biological/biochemical modifications that regulate chromatin, histones and DNA and therefore gene expression rather than changes to the underlying genetic sequence due to genetic variations or mutations (discussed in **Chapters 4 and 5**). More plainly, we now know that the turning on or off of genes (gene expression) can be controlled 'epi'-genetically via modifications to the DNA itself or by modifications to proteins such as histones that encase genomic DNA (gDNA). These modifications involve adding or removing small chemical groups such as methyl or acetyl groups to specific sites on the DNA or histones. Within molecular exercise physiology, we would therefore be interested in what modifications to DNA are caused by exercise, which genes are involved (and turned on or off) and if these play a role in a molecular response that leads to exercise adaptation. Epigenetic modifications can be dynamic in that they can change, influence gene expression and then return to normal. However, some can also be retained for much longer, and some modifications can even be 'enhanced' if they have been encountered before. For example, epigenetic modifications have been shown to be remembered during later detraining and retraining following an earlier training period (3, 4).

What causes epigenetic modifications? The full repertoire of epigenetic modifications and the enzymes that create or erase these modifications are not completely known (reviewed in (5)). Currently, more than 200 modifications and accompanying enzymes have been identified (6, 7). However, the most common epigenetic modifications include SUMOylation, phosphorylation, ubiquitination, acetylation and methylation, with the latter two being the most studied to date.

As you may now begin to appreciate, the field of epigenetics encompasses an enormous amount of highly complex biological processes that stretch far beyond the scope of this chapter. We will therefore focus primarily on a small subset of modifications, including **DNA methylation, histone methylation** and **acetylation,** and non-coding micro RNAs **(miRNAs),** all of which are considered the most comprehensively studied in the context of molecular exercise physiology. But what are these modifications, where do they take place and how do they occur?

EPIGENETIC MODIFICATIONS: WHAT, WHERE AND HOW?

DNA methylation

There are two primary sites where biochemical epigenetic modifications occur. The first of these sites is located on the 'linker' gDNA that joins nucleosome complexes.

Within the cell nucleus, approximately 150 bp of double helical DNA is tightly wrapped around four pairs of histone proteins (H2A, H2B, H3 and H4) which acts as a scaffold to enable the condensed packaging of DNA. These protein complexes are termed the **nucleosomes** and are often referred to as 'beads on a string' owing to their appearance (see **Chapter 3**, **Figure 3.11**). Each nucleosome complex is separated by short strands (~20–90 bp) of 'linker' DNA that is often subjected to epigenetic modifications, specifically **DNA methylation**. As described in **Chapter 3**, DNA is composed of four different nucleotides; the purine base nucleotides, adenine (A) and guanine (G), and the pyrimidine base nucleotides, thymine (T) and cytosine (C). Traditionally, it is the 'C' nucleotides within the DNA that are susceptible to methylation modification, which involves the biochemical attachment of a covalent methyl (CH_3) chemical group to the 5th position of the pyrimidine ring of the 'C' nucleotide, resulting in

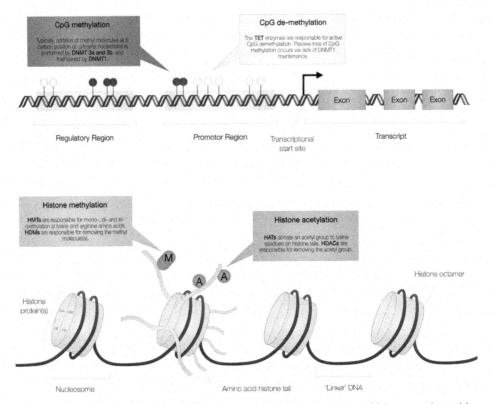

Figure 6.1 Overview of epigenetic modifications located across the DNA and histone amino acid tail. Red circle = methylated CpG sites, White circle = demethylated (hypomethylated) CpG sites, A = acetylation, M = methylation, TET = ten-eleven translocation enzymes, DNMT = DNA methyltransferase, HMTs = histone methyltransferases, HATs = histone methyltransferases, HDACs = histone deacetylases.

5-methylcytosine (5mC) (see **Figure 6.1**). Methylation of these cytosine nucleotides occurs in the context of CpG dinucleotide positions, also known as CpG sites. CpG sites are characterised by the presence of a cytosine nucleotide followed by a guanine nucleotide that are linked by a phosphate ('p') group within the same strand of DNA. This process therefore yields increased methylation, which is more commonly referred to as **'hyper'-methylation** when looking at whether exercise changes (or differentially methylates) the methylation status of a CpG site.

In contrast, the removal of a CH_3 methyl group from cytosine nucleotides results in 5-hydromethylcytosine (5-hmC) and therefore results in reduced methylation, also known as **'hypo'-methylation** or demethylation. Such biochemical processes are catalysed by several key enzymes that determine whether CpG sites are either methylated or demethylated. For example, specific enzymes are required to either promote, maintain or remove methyl groups and are sometimes referred to as 'writers', 'readers' and 'erasers' respectively (see (5) for a detailed review). The DNA methyltransferases (DNMTs), DNMT3a, DNMT3b and DNMT1, are the main family of enzymes that promote increased DNA methylation (**Figure 6.1**) (8). The former two enzymes (DNMT3a and DNMT3b) are essential for establishing *de novo* or 'new' methylation, whereas the latter DNMT1 enzyme is most responsible for maintaining methylation during cell division, enabling daughter cells to preserve methylation profiles that would otherwise be lost without DNMT1 enzyme activity (9, 10). Indeed, a lack of DNMT1 enzyme activity results in the passive loss of DNA methylation during DNA replication. However, another set of important enzymes, known as the ten-eleven translocation (TET) enzymes (TET1, 2 and 3), directly catalyse the removal of methyl groups and this is therefore considered an 'active' mechanism responsible for reduced methylation/hypomethylation (11, 12) (**Figure 6.1**). The extent to which DNMT and TET enzymes are differentially activated to modulate the DNA methylation response after endurance and resistance exercises is discussed below in this chapter.

It is estimated that the human genome contains ~28 million CpG sites (~ less than 1% of the genome), with a large proportion (~70–80%) of these sites being methylated (13, 14). Clusters of CpG sites are termed CpG islands and are often located within gene promoter, enhancer and silencer-specific regions of DNA that are defined and illustrated in **Chapter 3** (15). As these specific DNA sequences (gene promoter, enhancer and silencer-specific regions) are responsible for the initiation, elevation and suppression of messenger RNA (mRNA) expression, respectively, it is possible that alterations in the methylation status of these gene regions can determine the level of gene expression. One mechanism responsible for this phenomenon in the methylated state (or after an increase in methylation) is via the recruitment of CpG methyl-binding proteins that inhibit binding of the transcriptional 'machinery' (i.e. RNA polymerase and transcription factors) and therefore block the initiation of gene transcription (16), resulting in reduced gene expression. Moreover, DNA methylation leads to the recruitment of proteins that remodel or tighten the adjacent chromatin (termed **heterochromatin**), also inhibiting the gene transcriptional process (17). Conversely, where there is reduced/hypomethylated DNA, binding of the transcriptional apparatus and/or loosening of the chromatin (referred to as **euchromatin**) can occur, enabling and increasing gene transcription. This is most likely to be the case if changes in methylation occur in gene promoter or enhancer regions. Therefore, a general rule is that increased/hypermethylated DNA typically (but not always) results in reduced gene expression, whereas decreased/hypomethylated DNA typically (and also not always) results in increased gene expression. However, the opposite is more likely to be true if this occurs in gene-silencing regions, where a positive correlation is more likely. For example, within gene-silencing regions, increased methylation may lead to an increase in gene expression and reduced methylation may lead to reduced gene expression. It is also important to note that methylation within promoter, enhancer or silencer regions is not solely associated with the regulation of gene expression, as there is also evidence of associated gene body methylation and reduced transcription of the corresponding genes (18–20). As we focus on DNA methylation in the context of exercise in the following sections, it is worth noting that when assessing DNA methylation across the genome (or the 'epigenome'), it is more appropriate to refer to this as the **methylome**. In this chapter, we will therefore use this description when referring to studies that have used microarray or bisulphite sequencing (described in **Chapter 2**) to assess genome-wide DNA methylation after exercise.

Histone modifications

Histones are octamer structures made up of two copies of four individual histone proteins (H2A, H2B, H3 and H4). These histones tightly bind gDNA into chromatin structures which helps to condense our genetic material down and package it into our nucleus. Protruding from these individual histone proteins are amino acid tails (**Figure 6.1**). These amino acid tails are of great interest from an epigenetic perspective, because depending on the exact amino acid in question and the histone for which the amino acid tail belongs, it becomes a target for several different epigenetic modifications. For example, the 4th amino acid that sits along the histone tail protruding from histone 3 (e.g. H3K4) is a site that is susceptible to a methylation modification. On the same histone tail, the lysine residue at the 27th amino acid position (H3K27) is liable to both methylation and acetylation modifications, where only one modification can occur at any given time on these amino acids (e.g. they cannot be co-epigenetically modified). The four main modifications that occur at histone amino acid tails include methylation, acetylation, ubiquitination and phosphorylation, with all of these modifications requiring enzymes that are able to create, recognise and remove these modifications – commonly referred to as writers, readers and erasers, respectively.

At the histone level, the enzymes responsible for inducing and removing epigenetic modifications are complex, and for a full exploration of the enzymes, we suggest reading previous review papers (21, 22).

Briefly, several families of proteins (SET domain and the Mixed Lineage Leukaemia family [MLL]) are largely responsible for creating methylation on histone lysine residues, known as histone methyltransferases (HMTs). These enzymes are able to replace each hydrogen of the NH3+ group of these lysine amino acids with a methyl molecule, with each lysine residue able to accept up to three methyl molecules leading to three 'levels' of methylation (e.g. mono, di- or tri-methylation). The removal of these methyl molecules (or hypomethylation) on lysine residues is orchestrated by several enzymes, referred to as histone demethylases (HDMs). For example, the Jumonji C (JmjC) domain-containing proteins are a large family of protein enzymes that target and demethylate sites such as H3K9 and H3K4 (23). Typically, histone methylation is usually associated with a repressive/supressed transcriptional state resulting in reduced gene expression. However, this is not always the case, and is very much dependent on the specific amino acid that is modified and the 'level' of this modification (e.g. mono-, di- or tri-methylation). For example, if we take the amino acid residues highlighted above, tri-methylation of H3K27 (H3K27me3) is strongly associated with inactivated gene promoters and attenuation of gene transcription. However, mono methylation of this same histone and amino acid position (H3K27me1) has been suggested to localise to active gene promoters; thus, it is associated with increased gene expression (24). The **histone mark** ('mark' refers to the modification of a specific residue extending from a particular histone protein) H3K4, is also a target for methylation, but irrespective of whether its mono-, di- or tri-methylated, all marks are strongly associated with activating gene expression (24). The complexity of histone methylation is therefore clear, and great care must be taken when interpreting data and research papers in this area.

By contrast, acetylation modifications that occur along histone tails are relatively more straightforward to understand. Indeed, the enzymes responsible for histone acetylation are commonly referred to as histone acetyltransferases (HATs; three main protein families, GNAT, MYST and P300/CREB families) and histone deacetylases (HDACs; for which there are sub-categories of class I, class II, class III and class IV HDACs). More comprehendible is that histone acetylation only exists on one level, unlike histone methylation that has three levels as described above. Of particular interest for molecular exercise physiologists is the class III HDACs, as this class of enzymes contains the sirtuins which are commonly implicated in processes such as ageing, stress resistance and low-calorie insults (25). The sirtuins are $NAD^+/NADH$ sensors that detect the signal from the alterations in mitochondrial reduction/oxidation (redox state) following the breakdown of nutrients required to fuel exercise, as discussed in general in **Chapter 7** and in response to endurance exercise in **Chapter 9**. As a general rule, when a histone is acetylated, it creates a chromatin state that helps to activate and increase transcription of genes near those loci (site or location of a specific gene). This is due to the new acetyl group changing the electrical charge between histones and DNA, repelling their tightly bound association and creating a more accessible chromatin structure which allows the relevant machinery to perform the processes required for gene transcription (26).

Non-coding RNAs as epigenetic modifications

Small non-coding RNA species, particularly miRNAs, are the third main epigenetic regulator. The single-stranded RNA, consisting of only ~18–25 nucleotides, bind directly to the 3' UTR region of mRNA molecules, thereby disrupting the mRNA strand just before it's required to be translated into protein. Hence, this is known as a post-transcriptional epigenetic modification as it occurs after the mRNA has been transcribed (27). Despite being so small in nature, these RNA species have tremendous capacity to modify the post-transcriptional processes of mRNA, with individual miRNAs reported to target hundreds of mature mRNA molecules (28), and with estimates suggesting that more than 30% of human genes can be regulated by miRNAs (29, 30). RNA polymerase II transcribes miRNAs in the cell nucleus to produce primary miRNA (pri-miRNA) (**Figure 6.2**). The pri-miRNA molecule is then processed by a ribonuclease (Drosha) and protein (DGCR8/Pasha) complex to form a small hairpin RNA structure of about 60–100 nucleotides (nt) in length, known as the precursor miRNA (pre-miRNA) (**Figure 6.2**). The pre-miRNA is subsequently exported out of the nucleus and is further processed by a different endonuclease known as Dicer, forming the 18–25 nt long miRNAs (31). These small miRNAs can then join with the protein Argonaute-2, to create an RNA-inducing silencing complex (RISC) which guides this newly formed

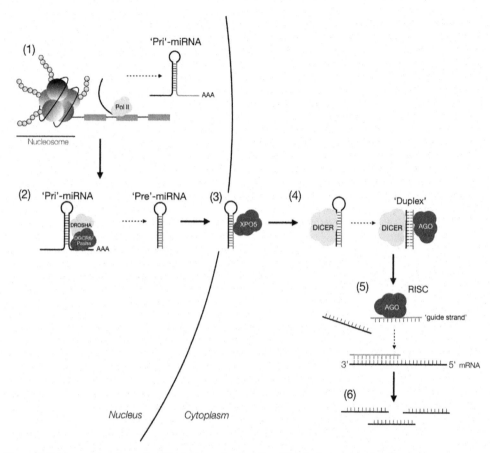

Figure 6.2 The mechanisms of miRNA biosynthesis. **(1)** miRNA is first transcribed by RNA polymerase II (POL II), resulting in primary miRNA ('pri'-miRNA) of ~60–100 nucleotides (nt) in length. **(2)** Pri-miRNA is then cleaved by the DGCR8/Pasha and DROSHA protein complex to become precursor miRNA ('pre'-miRNA). **(3)** The resultant pre-miRNA molecule is then transported out of the nucleus by Exportin-5 (XPO5). **(4)** Within the cytoplasm, pre-miRNA is processed by DICER to produce a 'duplex' of both 'passenger' (black) and 'guide' (grey) miRNA strands (~18–25 nt). The Argonaute-2 (AGO) protein binds to the duplex to create the RNA-induced silencing complex (RISC). **(5)** After processing, only the guide strand is retained in order to direct the RISC-miRNA complex to the target mRNA transcript. **(6)** This typically results in mRNA degradation and the prevention of protein translation that would have occurred in the ribosome.

complex to specific mRNAs (**Figure 6.2**). Commonly, binding of this miRNA complex to the mRNA of specific genes results in reduced gene expression, either by direct degradation of the targeted mRNA molecule or by suppressing its translation into protein (27). While this is the case when the RISC-miRNA complex binds to the 3' UTR region of the specific mRNA, some research has also shown that when bound to the coding region (or the 5' UTR locus of the mRNA), this can result in activation and an upregulation in protein translation (32).

Intriguingly, miRNAs have been shown to interact with both histone and DNA methylation to create an miRNA-epigenetic feedback loop. The key histone and DNA epigenetic enzymes (e.g. DNMTs, HDACs and

HMTs) have been shown to be targeted by miRNAs that then regulate the expression of these key enzymes (33). Conversely, the expression of miRNAs themselves has been shown to be under the regulation of the local epigenetic landscape for which they are transcribed, where, for example, it has been shown that hypermethylation of the promoter region of specific miRNAs is sufficient to reduce their expression (34, 35).

THE EPIGENETIC REGULATION OF ENDURANCE EXERCISE

The specific cellular and molecular responses to exercise are considerably determined by the type of exercise performed, where high-forceful resistance exercise (RE) elicits increased size and strength of myofibres, whereas increased mitochondrial density, oxidative capacity and improved fatigue resistance are largely mediated by the less-forceful, high-frequency continuous contractions experienced during aerobic/endurance exercise (extensively covered in **seven to nine**). Interestingly, the specific genes that are differentially expressed after exercise that eventually lead to such contrasting phenotypes are also regulated by various epigenetic modifications. Among the two divergent types of exercise, the epigenetic regulation of endurance exercise has been more extensively studied compared to RE. The next section of this chapter will therefore focus on some of the key epigenetic modifications (i.e. DNA methylation, histone methylation/acetylation and miRNA changes) that are associated with the acute and chronic responses to endurance exercise. After which we will go on to focus on the epigenetic modifications identified following acute and chronic resistance exercise.

Acute endurance exercise and DNA methylation

As introduced in **Chapter 1**, peroxisome proliferator-activated receptor γ co-activator 1α (PGC-1α) is a gene that is significantly increased after endurance exercise (36, 37), and its increase is considered as a 'hallmark' in the molecular response to endurance exercise and in the adaptation of mitochondria and improvements in oxidative capacity (38, 39). Its full role in the molecular response and adaptation to exercise are discussed at length in **Chapters 7 and 9**. Briefly, PGC-1α acts as a transcriptional co-activator through the recruitment, binding and co-regulation of multiple transcription factors, including nuclear respiratory factor 1 (NRF-1), NRF-2, myocyte enhancer factor 2 (MEF2) estrogen-related receptor α (ERRα) and mitochondrial transcription factor A (TFAM). This process ultimately regulates the mRNA expression of metabolic and mitochondrial genes associated with endurance-type exercise (40). Given its well-established role in exercise, it is no surprise that some of the first studies investigating promoter DNA methylation, and changes in gene expression after exercise in skeletal muscle, also focussed on PGC-1α. In a non-exercise context, studies first identified that there was an inverse relationship between the degree of PGC-1α promoter methylation and how much the gene was expressed in skeletal muscle tissue (41). To show this, the authors profiled this transcriptional co-activator in the muscle of those who were healthy (glucose-tolerant) and those with type-II diabetes (glucose-intolerant). The healthy individuals demonstrated the lowest methylation levels and largest gene expression compared with type-II diabetics who had higher promoter methylation and reduced gene expression. In a separate study, a similar pattern was also demonstrated after a period of nine days bed rest that increased PGC-1α DNA methylation levels while reducing corresponding gene expression in the muscle (42). Given endurance exercise is generally known to increase gene expression of PGC-1α, it could therefore be hypothesised that exercise may be able to prevent the increased DNA methylation and reduced gene expression in bedridden individuals to levels seen in healthy muscle. However, despite a trend towards reduced DNA methylation and increased gene expression of PGC-1α after 4 weeks of exercise (30 minutes/day cycling at 70% VO_{2max}, 6 days/week), methylation and gene expression were not completely 'rescued' to pre-bed rest levels (42). The first study to demonstrate that exercise itself could alter PGC-1α DNA methylation was performed in healthy individuals that undertook an acute bout of aerobic exercise (cycling at 80% VO_{2peak} until 1674 kJ/400 kcal was expended). This study demonstrated that PGC-1α, alongside other mitochondrial-related genes (TFAM, PDK4 and PPAR-δ), were significantly hypomethylated and displayed significant increases in gene expression (43). The exercise regime was also generally a hypomethylating stimulus to the entire DNA, where they performed an assay that showed exercise reduced global DNA methylation. Another fascinating finding from this study, was that these changes only occurred when performing higher-intensity (80% VO_{2peak}) compared with lower-intensity exercise (40% VO_{2peak}) that was matched for the amount of energy expended (1,674 kJ/400 kcal). This suggested that exercise intensity was perhaps an important regulator of altered DNA methylation in skeletal muscle. However, other studies have also shown steady state cycling (120 minutes cycling at ~60% VO_{2peak}) in competitive endurance-trained individuals results in changes of DNA methylation in key metabolic genes (FABP3, COX4I1) with associated changes in gene expression (44). Although PGC-1α gene expression increased after exercise in this study, DNA methylation was not assessed, which would have perhaps complimented the current literature examining both PGC-1α gene expression and DNA methylation. A study in mice has however shed more light onto DNA methylation of PGC-1α after acute exercise. The researchers profiled methylation of PGC-1α's canonical (Exon 1a) and alternative (Exon 1b) promoters after progressive rotarod exercise (35 rpm for 20 min, 40 rpm for 30 min and 45 rpm for 10 min) (45). Of relevance, it has been previously shown that the canonical promoter (A) is expressed more highly at rest than the alternative promoter (B), but that the alternative promoter B is more responsive to increases in gene expression following exercise, cold-water immersion, and exercise under low glycogen conditions (46, 47). Interestingly, the mouse study demonstrated that progressive exercise caused DNA hypomethylation in the A promoter yet with no increase in gene expression (perhaps due to higher resting levels as observed before). Further, while authors did not see any change in methylation of the alternate B promoter, they reported increased tri-methylation of lysine 4 on histone 3 (H3K4me3) in the B promoter together with a large increase in promoter B gene expression (45). As referred to in the above section, this epigenetic modification is associated with increasing the accessibly of DNA and therefore a more permissive (i.e. euchromatin) DNA state to allow gene expression to occur more readily (48). Therefore, these studies showing that the B promoter of PGC-1α is more prone to larger increases in histone methylation and increased gene expression in response to exercise, indicate that histone modifications may be more important than DNA methylation for increasing PGC-1α gene expression that is observed after endurance exercise. However, more studies would be required to confirm this in humans, and if this still occurs at different exercise intensities.

Despite some epigenetic alterations in key metabolic genes after acute exercise, very little is known about the vast majority of genes in pathways that are associated with the endurance exercise response, such as those in the AMPK or p38MAPK pathway introduced in **Chapter 1** and discussed in detail in **Chapters 7 and 9**. Briefly, AMPK is an important sensor of ATP turnover [ATP]/[ADP][P$_i$] that is required to fuel exercise, and p38 MAPK an important sensor of oxidative stress, reactive oxygen species (ROS) and glycogen breakdown following exercise. The reason for this lack of knowledge is because there are little to no studies using array or sequencing technology to profile DNA methylation across the 'methylome' after acute endurance exercise in healthy human muscle. An exception is a recent study that used DNA microarray technology to profile around 850,000 CpG sites (methylation array principle and methodology are described in **Chapter 2**) after straight line running compared with change of direction running exercise in human skeletal muscle. While the distance, speed of running and the number of accelerations/decelerations were the same in both trials, simply by introducing changes of direction changing direction in the protocol (compared with running in a straight line) evoked greater physiological (heart rate), metabolic (lactate), exertion and movement (measured using GPS) load compared with straight line running. At the methylation level, change of direction exercise evoked hypomethylation of the DNA and this hypomethylation was 'enriched' (or more prominent) in genes within the exercise responsive pathways of AMPK, MAPK and insulin compared to the straight line running exercise (49). Moreover, change of direction exercise resulted in greater promoter region-specific hypomethylation of important angiogenic genes such as Vascular Endothelial Growth Factor (VEGFA) and metabolic transcription factors genes such as Nuclear Receptor Subfamily 4A1 (NR4A1). Interestingly, recent sprint running transcriptome analysis and exercise transcriptome meta-analyses, VEGFA and NR4A1 were genes that are ranked most highly as being associated with corresponding changes in PGC-1α gene expression in human skeletal muscle (50, 51). While there was no significant change in PGC-1α methylation detected in this study, there was an increase in PGC-1α, VEGFA and

NR4A1 genes after change of direction exercise compared to straight light running (49), suggestive of an important connection between some of these genes in terms of changes in DNA methylation and gene expression after exercise. For example, it is known that PGC-1α co-activates estrogen-related receptor α (ERRα), a transcription factor known to target VEGFA and therefore stimulate angiogenesis/capillary formation (discussed in detail in relation to endurance exercise in **Chapter 9** and in different environmental conditions in **Chapter 11**). Overall, this study extends earlier work showing that acute exercise induces a global reduction in DNA methylation, and that this was the case using a genome-wide analysis, that also identified genes within exercise-responsive pathways such as AMPK, MAPK and insulin signalling to demonstrate these hypomethylated signatures. Another important finding was that the hypomethylation signatures identified were more prominent in these pathways after 30 minutes of exercise in skeletal muscle compared with after 24 hours, suggesting that methylation changes are extremely rapid and dynamic following exercise and that they likely precede changes in gene expression that may typically peak between 3 and 6 hours after exercise (discussed at length in **Chapter 7**). However, these studies require confirmation following exercise that is more representative of endurance exercise given that this study was investigating sprint interval exercise that likely incorporated a considerable anaerobic stimulus in already well-trained individuals. Overall, more studies will be required in the future to investigate methylome changes after acute endurance exercise and whether these lead to similar alterations at the gene expression and importantly at the protein level, as well as the temporal synchronisation of these molecular responses.

Chronic endurance training and DNA methylation

Compared with acute exercise, there have been more studies investigating genome-wide DNA methylation after chronic endurance training, with the caveat that some of these have not been in healthy populations. The first exercise training study that undertook DNA methylome analysis in skeletal muscle used early DNA methylation array technology that profiled around 29,000 CpG sites. The study included 6 months of supervised aerobic training (3 days/week with no indication of exercise intensity, although VO_{2max} increased post-training suggesting an improvement in aerobic fitness) in 28 males with an average age of 37.5 years (52). Fifteen of those men were categorised with, and 13 categorised without, a first-degree relative who had type-II diabetes (52). As with acute exercise described above, chronic exercise training across both groups (i.e. regardless of family history of type-II diabetes) was able to evoke a predominant hypomethylated signature, where the number of significantly differentially methylated genes identified after the training period compared with pre-training was 2,051 hypomethylated compared to 766 hypermethylated. The hypomethylation post-training was more frequently observed in gene pathways, including insulin and calcium signalling pathways, as well as in carbohydrate and retinol metabolism. Using larger coverage microarray technology (that profiled around 450,000 CpG sites) after progressive endurance training in obese type-II diabetics (3 days/week of 40–60 minutes cycling at 65–85% of heart rate reserve for a total 16 weeks), Rowlands et al. also reported a greater reduction in DNA methylation after training, where 386 CpG sites were hypomethylated compared to 169 CpG sites that were hypermethylated (53). This differential methylation was enriched in gene pathways involved in carbohydrate and lipid metabolism, metabolic disease, cell death and survival, and cardiovascular system development/function (53). Therefore, both of the aforementioned studies identified a predominance of DNA hypomethylation after chronic endurance training. Studies published the same year further confirmed that altered methylation profiles occurred after endurance training in gene pathways associated with metabolism and oxidative phosphorylation (54). In this particular study, healthy untrained males and females completed 3 months of progressive unilateral endurance training (45 minutes of unilateral knee-extension exercise, 4 days/week for 3 months), where in slight contrast to those studies above, Lindholm et al. reported a similar number of hypomethylated and hypermethylated CpGs after training (54). However, when performing a complimentary transcriptome-wide gene expression analysis (which the other training studies had not performed) in tandem with a genome-wide methylation analysis, the authors were able to demonstrate that, out of the 4,919 differentially methylated CpG sites, there were slightly more genes that were upregulated (i.e. increased gene expression) and hypomethylated (273)

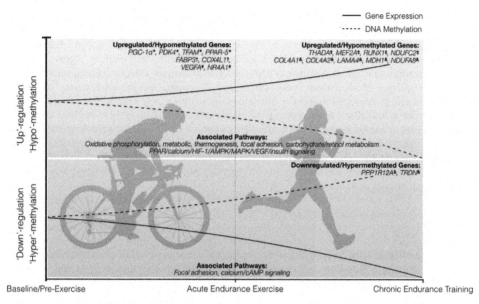

Figure 6.3 Schematic summarising the regulation of DNA methylation and gene expression after acute and chronic endurance exercise. The schematic demonstrates the more prominent inverse association of DNA methylation and gene expression, either upregulated/hypomethylated (green upper panel) or downregulated/hypermethylated (red lower panel) after exercise. The relevant exercise-related pathways are also displayed. * *genes identified in* (43); # *genes identified in* (49); † *genes identified in* (44); ‡ *genes identified in* (52); & *genes identified in* (54).

compared with the number of genes that were hypermethylated and downregulated (255) (i.e. decreased gene expression). Furthermore, the majority of differentially methylated genes demonstrated an inverse compared to a positive relationship with gene expression, where 203 genes were upregulated and hypermethylated (compared with 273 upregulated/hypomethylated) and 70 genes were downregulated and hypomethylated (compared with 255 upregulated/hypomethylated). An additional interesting finding from Lindholm et al.'s study was that most of the training-induced alterations in methylation occurred in the so-called enhancer regions and less frequently in promoter regions (54). However, the authors noted that when looking at both the methylation and associated gene expression levels, there was a significant inverse correlation between gene expression and DNA methylation of several interesting genes associated with muscle structure (COL4A1, COL4A2, LAMA4), actin-myosin interaction (PPP1R12A), oxidative metabolism (MDH1, NDUFA8) and calcium release (TRDN) (outlined in **Figure 6.3**), where the methylation changes also occurred in gene promoters, transcription start sites or within the 1st exon (54). Therefore, even if alterations in promoter methylation were less frequent in response to endurance training, these findings suggest that those CpG sites that did change in promoter regions were perhaps more likely to affect corresponding gene expression levels. Despite these studies broadly presenting agreeable results, it is worth noting that another study that was conducted a few years later using the same coverage DNA methylation arrays found little change in methylation after endurance training. Indeed, Robinson et al. undertook 12 weeks of alternating endurance exercise sessions (3 days/week of 4 × 4 minute intervals high-intensity exercise at >90% VO_{2peak} each separated by 3 minutes rest; 2 days/week 45-minute continuous lower-intensity incline walking exercise at lower 70% VO_{2peak}) and reported that training resulted in less than a 10% change in CpG methylation compared with pre-training (55), which is in contrast to the studies described above demonstrating differential methylation after both acute endurance exercise and training. Perhaps one explanation for not identifying a significant change in methylation in this study was that the authors used stringent statistical 'cut offs' for identifying changes in methylation with relatively small numbers of participants. This was not necessarily incorrect as the statistical cut-off is often important to reduce the number of false positive results, and others have also used similar statistical cut-offs. However, due to demanding training interventions, together with the expense of performing such high-throughput DNA methylation analysis, exercise training studies are often undertaken in low-moderate numbers of participants. Therefore, DNA methylation array or sequencing technology allows researchers to simultaneously analyse a large number of CpG sites across the entire genome in a smaller cohort, enabling the identification or 'discovery' of methylated sites that are more consistently altered across all participants. This approach can then inform the researcher of potential interesting differential methylation in genes that can be further validated in future experiments. For example, targeted sequencing of gene regions can be performed to quantitatively confirm the changes in methylation seen in the high-throughput experiments (if arrays have been performed), as well as looking whether such alterations in DNA methylation are also associated with changes in gene expression or even alterations in protein levels. Therefore, helping to validate the 'discovery' data with more targeted analysis of individual genes e.g. in larger sample sizes. Additionally, future studies can also confirm or refute those analyses by examining the genes of interest across different experimental models. One other discrepancy you may have noticed is that quite a few studies vary in terms of the number of differentially methylated sites that are discovered. For example, studies report changes in methylation in the range of hundreds to thousands of CpG sites. This large variation is also, in-part, due to the varying number of subjects recruited and the different statistical cut-offs used.

One final important study worth mentioning in regard to DNA methylation and endurance training used the most recent and comprehensive microarray technology to examine genome-wide DNA methylation in a much larger number of CpG sites (over 850,000 CpG sites – see **Chapter 2** for methods) after endurance training (4 days/week, 10 weeks of progressive intensity) (56). Within this particular study, Stephens et al. analysed skeletal muscle from type II diabetics, and categorised these participants as either responders or non-responders to exercise according to their phosphocreatine (PCr) recovery rates (56). Five hundred and thirty-three CpG sites were differentially methylated between the two groups following the training period, with non-responders displaying reduced promoter methylation for genes associated with glutathione metabolism, insulin signalling and mitochondrial metabolism pathways (56). Collectively, this studies suggested that exercise training predominantly leads to hypomethylation, rather than hypermethylation, of DNA. Yet, it's perhaps important to note that it may have been more plausible to hypothesise that hypomethylation would have been more prominent in exercise responders and not those classed as non-responders to exercise. One explanation might be that exercise responders could have potentially had reduced methylation of their DNA prior to training compared to non-responders; however, this presumption would require further investigation to make this conclusion. It has been demonstrated that increased physical activity levels are associated with a more hypomethylated DNA profile in skeletal muscle (57, 58), which would provide indirect evidence to support this hypothesis under the general assumption that responders might be individuals who may have undertaken more exercise in the past. However, this perhaps would not be the case for those who are genetically predisposed to be exercise responders (see **Chapters 4 and 5**). Furthermore, whether PCr recovery rates determine if the exercisers then go onto become responders to the training across other molecular and physiological systems may also influence the interpretation of these results. In summary, endurance training seems to evoke differential methylation in genes within metabolic pathways. Furthermore, hypomethylation and increased gene expression of corresponding genes seem to occur more frequently in response to endurance exercise training.

Finally, it is worth also mentioning that most studies to date in this area have been associative in nature. For example, the 'association' between DNA methylation and gene expression. Other studies have attempted to address the question as to whether DNA methylation changes are causative of altered gene expression after exercise by 'knocking out' (or experimentally preventing gene expression) in the DNA methyltransferase 3a (DNMT3a) enzyme in mice and then looking at the effects on energy metabolism and exercise capacity (59, 60). Given the findings from the previous studies outlined above, and the fact that DNMT3a is responsible for *de novo* or 'new' methylation, it may have been hypothesised that removing DNMT3a would cause an even greater hypomethylated DNA profile than that typically observed after training alone. However, two separate studies reported contrasting results, with one study showing no effect of muscle-specific DNMT3a knockout on energy metabolism and exercise capacity in mice (59), whereas the other study demonstrating impaired aerobic capacity, oxidative capacity, mitochondrial respiration coupled with a larger increase in ROS production (60) (**Figure 6.2**). Therefore, these contrasting results require further investigation. To further extend this work, it may be pertinent for future studies to overexpress the TET enzymes which are directly associated with reducing methylation to bolster the exercise effect, rather than taking away the ability for de novo or 'new' methylation to occur. Therefore, these are important studies that require future attention in molecular exercise physiology.

Endurance exercise, histone modifications and chromatin dynamics

The previous section discussed how performing endurance exercise effects the methylation signatures of DNA and how this is associated with changes in gene expression. The present section of this chapter will focus on some of the key epigenetic modifications that occur at the histone (rather than DNA) level after endurance exercise and training in skeletal muscle, and how these can also remodel the chromatin, and therefore influence the transcriptional state of our genes. Histone methylation and acetylation have gained significant interest in the molecular exercise physiology field, given these particular modifications are affected by exercise and are also associated with altering the behaviour of exercise-responsive genes. While this section will focus on the work conducted in skeletal muscle, there is also a fascinating body of research focussed on histone modifications that occur in other tissues after exercise. Therefore, interested readers are also advised to read an excellent review by Mcgee SL and Walder KR (61).

Some of the first studies to investigate histone acetylation in response to endurance-type exercise were performed in rodent skeletal muscle. In one early study, rats underwent acute endurance exercise (5 × 17 minutes of interval swimming with an additional 3% body weight load) and the acetylation of histone 3 (H3) in the promoter region of the glucose transporter, GLUT4, was assessed via chromatin immunoprecipitation (or 'ChIP' for short, as described in **Chapter 2**) (62). This is an important region in the GLUT4 gene, where the muscle-specific transcription factor, MEF2, binds to its promoter and turns the gene on. GLUT4 is critical for skeletal muscle glucose uptake in response to muscle contraction with exercise. MEF2 is also a transcription factor that is regulated by PGC-1α in order to promote an increase in its gene expression. Moreover, MEF activity is activated by either continuous (through CaMKII) or high-intensity (through AMPK) endurance exercise (described in detail in **Chapter 9)**. In this study, the authors reported an increased binding of MEF2A to the GLUT4 promoter and increased GLUT4 gene expression after exercise in rats (62). The authors also reported increased H3 acetylation levels of the MEF2 binding site on the GLUT4 gene, which suggested that the exercise-induced increase in acetylation would enable a permissive (or 'relaxed') chromatin state to help MEF2A to bind to the GLUT4 gene promoter. Interestingly, experimentally inhibiting CaMKII activity, that is normally elevated by increased calcium for muscle contraction, reduced GLUT4 expression and MEF2 binding, but also prevented the increase in acetylation (62). Overall, this study helped confirm that histone acetylation is important to enable the binding of MEF2 to the GLUT4 gene promoter and increase GLUT4 expression which is important for the uptake of glucose into the working muscle to support exercise-induced muscle contraction. In corroboration with these findings, others have also demonstrated increased H3 acetylation near different genes that also result in altered gene expression after endurance exercise. In this study, Joseph et al. (63) employed a similar exercise model in rats (5 × 17 minutes of interval swimming over a longer period of five consecutive days) to that described in the study above. They demonstrated that exercise led to increased acetylation of H3 in the MEF2A promoter region where the transcription factor NRF-1 binds, and also increased NRF-1 and MEF2A gene expression (63). The authors also demonstrated that inhibiting CaMKII, using the same method as the previous study, prevented the exercise-induced increase in H3 acetylation and associated increases in MEF2A and NRF-1 gene expression (63). It is also worth noting that Smith et al. also demonstrated increased CaMKII phosphorylation after an acute bout of swimming described above (62). Collectively, these studies indicate that increased histone acetylation following exercise can modulate the transcriptional status of genes involved in glucose transport via altering calcium/CaMKII signalling. Calcium signalling in exercise and muscle fibre phenotype after endurance exercise is covered in more detail in **Chapter 9**. As with the rodent studies described above, studies have also investigated histone acetylation after endurance exercise in human skeletal muscle. McGee et al. were the first to analyse histone acetylation levels in human muscle and found that a single bout of endurance exercise (60 minutes cycling at 70–80% VO_{2max}) significantly increased global H3 acetylation levels, particularly on lysine residue 36 (H3K36), which is associated with the elongation step of gene transcription (64). As histone acetylation signatures are mediated by the activity of acetylating promoting (i.e. HATs) and suppressing (i.e. HDACs) enzymes, the authors investigated the activity of some of the key deacetylating HDAC class IIa enzymes such as HDACs, 4 and 5. Interestingly, the abundance of HDAC4 and HDAC5 within the nucleus significantly reduced after exercise which suggested that the exercise stimulus caused these enzymes to be transported (or 'exported') out of the nucleus so that they could no longer induce their repressive effects on gene expression. Despite the authors not measuring gene expression in this study, other studies have demonstrated that HDAC4 and HDAC5 are able to interact with, and therefore suppress the activity of muscle-specific genes (including MEF2 described above), impair myogenesis (i.e. the formation of skeletal muscle fibres/myotubes discussed in more detail in **Chapters 2 and 13**) (65) and even suppress the skeletal muscle adaptive response typically observed after endurance exercise (i.e. slow-fibre formation and enhanced running exercise capacity – see **Chapter 9**) (66). Another interesting collective finding is that the nuclear export of HDAC is dependent on well-characterised kinases that are associated with endurance exercise adaptation, including those observed in the previous rodent studies. Indeed, McGee et al. also showed increased phosphorylation levels of CaMKII (~2-fold) after exercise, where increased activity of this kinase has also been associated with nuclear export of HDACs in muscle cells (67), and reduced HDAC5 mRNA and protein expression after endurance exercise in the rat hippocampus (68). One explanation for this is that HDAC4 protein contains a domain that is recognised by CaMK, causing an interaction between the two proteins that can also affect HDAC5 (69). McGee et al. also reported increased AMPK (~4-fold) phosphorylation levels after exercise (64). Although not necessarily a surprising result, given the well-characterised role of this kinase during endurance exercise adaptation (see **Chapter 9**), in a non-exercise context, the same group showed that experimentally increasing AMPK activity (using the AMPK activator, AICAR) in cultured human skeletal muscle cells, resulted in increased HDAC phosphorylation, reduced HDAC5 nuclear abundance, as well as increased H3 acetylation (on lysine 9 and 14) and GLUT4 gene expression (70). Taken together, these studies have shed light on the possible key mechanisms responsible for exercise-induced alterations in histone acetylation that occur because of phosphorylation and subsequent nuclear export of HDAC4/5 deacetylating enzymes, resulting in increased H3 acetylation.

More recent studies have investigated the H3 acetylation levels near regions of other exercise-responsive genes and across different muscle groups. In one study, Masuzawa et al. examined total H3 and H3K27 acetylation levels within different regions of the PGC-1α gene in both slow (soleus) and fast (plantaris) muscles after acute aerobic exercise (20 minutes treadmill running at 24 metres/min) in rats (71). As mentioned above, PGC-1α is of particular interest, given this gene is extremely responsive to endurance exercise and regulates the increase in the size and number of mitochondria (termed 'mitochondrial biogenesis') within skeletal muscle after endurance training (see **Chapter 9**). In this study, exercise led to an increase in total H3 and H3K27 acetylation levels that also corresponded to an increase in total and isoform-specific PGC-1α gene expression, which suggested there was a critical role for histone acetylation in regulating gene signatures favourable for metabolic adaptation (71). An interesting finding from this study was that these changes in histone acetylation predominantly occurred in the slow oxidative soleus muscle that is more responsive to endurance exercise compared to the fast glycolytic plantaris muscle. In relation to longer-term endurance training, the same lab also investigated the effects of both the total duration and amount of daily exercise performed on histone acetylation. In this particular study, Ohsawa et al. subjected rodents to three different endurance training regimes, all of which consisted of treadmill running at a speed of 24 metres/minute, however, the different regimes were altered for duration of exercise per day and performed over different time periods (4–8 weeks)

(71). The authors reported increased total H3 acetylation after training in two out of the three regimes and in one of the training protocols identified assembly of the H3.3 histone variant into the nucleosome (71). The role of H3.3 here is intriguing, given its nucleosomal incorporation is associated with histone marks such as H3K4me3 that positively affect gene expression, and if you recall from earlier in this chapter, H3K4me3 also increases after endurance exercise in the PGC-1α alternative B promoter, and corresponds to increased gene expression of PGC-1α (45, 72, 73). Indeed, in the present study rodents demonstrated the largest increases in gene expression in the endurance regime that identified assembly of the H3.3 histone variant into the nucleosome (71).

Diverting our attention towards histone methylation, Ohsawa et al. also looked at the methylation levels of different histone proteins across the different training regimes. The authors specifically focussed on the tri-methylation of histone 4 on lysine 20 (H4K20me3) as this particular modification is associated with 'condensed' chromatin (i.e. heterochromatin) and therefore reduced gene expression (74). Interestingly, in the regime that demonstrated increased total levels of H3.3 described above, the levels of H4K20me3 significantly decreased (71). These findings may therefore suggest that the endurance exercise-induced increase in the total level of H3.3 variants was able to mitigate the condensed chromatin state typically induced by the tri-methylation of H4K20 that would otherwise result in reduced gene expression (71). In contrast to the inhibitory effects of methylated H4K20 on gene transcription, methylation of other histone proteins (particularly histone 3) is oppositely associated with increased gene transcription, such as tri-methylation of lysine 4 on histone 3 (H3K4). Therefore, this modification has attracted the attention of researchers to study this mark in the context of exercise. In one study, also described earlier in this chapter, mice that performed progressive endurance exercise did not demonstrate reduced DNA methylation levels of the PGC-1α alternative B promoter, rather, the authors reported an increase in H3K4me3 methylation that corresponded to a large increase in PGC-1α alternative B promoter gene expression (45). Given that the authors also observed hypomethylation of DNA in PGC-1α's canonical promoter A after exercise (that did not affect gene expression) whereas increased H3K4me3 of PGC-1α's alternative promoter B corresponded with increased gene expression, collectively indicates that histone methylation (together with H3K27 acetylation (46)) maybe key epigenetic modifications underpinning increased PGC-1α gene expression following endurance exercise. However, contrasting results that identified associated promoter DNA hypomethylation and increased gene expression of PGC-1α after endurance exercise in humans (43) make it difficult to generalise these finding across species.

Overall, in the field of molecular exercise physiology and epigenetics, histone acetylation and methylation modifications and their effects on transcriptional regulation after exercise have shed light onto some of the key epigenetic mechanisms that underpin the molecular response to endurance exercise.

Endurance exercise and non-coding miRNA

miRNAs are capable of post-transcriptionally regulating the expression of target mRNA transcripts and are thus considered an important post-transcriptional epigenetic modification involved in the molecular response to exercise. Moreover, miRNAs are tissue-specific, where those exclusively expressed in muscle are typically referred to as myomiRs. Analysis of myomiRs has identified a family of miRNAs that seem to be crucial for muscle physiology. This family, known as the muscle-specific miR-1 family, consists of miR-1, miR-206, miR-133a and miR-133b, all of which are abundantly expressed in both cardiac and skeletal muscles (75), with the exception of miR-206 which is believed to be skeletal muscle specific (76). The importance of this family of miRNAs for muscle tissue physiology has previously been highlighted, with several of them being implicated in processes of cardiac muscle development, myogenesis, muscle regeneration and cell fate determination (reviewed in (77)). As suggested earlier in this chapter, in general, miRNAs are typically associated with repressing transcriptional activity of their target genes, although there are some exceptions (30). In the context of molecular exercise physiology, several studies have reported differentially altered miRNA expression in response to divergent types of exercise, revealing some fascinating insights into the mechanisms that control post-transcriptional modifications to mRNA after exercise. There have been studies identifying alterations in circulating (in the blood) miRNAs following exercise (78–80), however in this chapter we will focus on the miRNA response to endurance exercise in skeletal muscle tissue.

One of the first studies to look at the miRNA response after endurance exercise was conducted in Professors Mark Tarnopolsky's lab. In this study, the authors showed that acute endurance exercise (treadmill running at 15 metres/minute for 90 minutes) in adult mouse muscle culminated in differential expression of several miRNAs that are associated with altered myogenesis and substrate metabolism (81). Specifically, the expression of miR-107 and miR-181, together with muscle-specific miR-1 (that is associated with regulating myogenesis), all increased 3 hours post-exercise. Interestingly, the largest alteration in miRNA expression was seen in miR-23 that significantly reduced by more than 80% (81). Moreover, in line with the reduction in miR-23, PGC-1α and its downstream targets, ALAS, cytochrome c and citrate synthase all increased at the mRNA expression level after exercise. Given that miR-23 is considered a putative negative regulator of PGC-1α, and that increased PGC-1α protein expression after exercise was negatively correlated with reduced miR-23, these findings suggest an associated role for this miRNA and the metabolic response after endurance exercise (81). In corroboration with these findings, other studies have also reported associated changes in miRNAs and PGC-1α expression. In one particular study, Aoi et al. used microarray technology to simultaneously reveal the expression levels of several miRNAs after endurance training (starting from 18 metres/minute for 20 minutes which increased to 32 metres/minute for 60 minutes, 5 days/week) in mice (82). Following the 4 weeks of training, expression of one miRNA (miR-21) increased, whereas three miRNAs (miR-696, miR-709 and miR-720) decreased (82). Interestingly, reduced levels of miR-696 after training were oppositely affected by physical inactivity or disuse of the muscle, where expression levels increased after five days of hindlimb immobilisation. The authors therefore went on to investigate the predicted mRNA targets of miR-696 and found that PGC-1α was amongst the 827 potential gene targets (82). To confirm these predictions, overexpression of miR-969 in cultured muscle cells in vitro was able to reduce expression of PGC-1α and its downstream target genes, PDK4 and COXII (82). Together, both the in vivo and in vitro data derived from these studies suggest that miR-696 post-transcriptionally regulates PGC-1α and therefore the molecular response to exercise training.

Other miRNAs have been shown to target different genes that promote mitochondrial biogenesis. Indeed, Yamamoto et al. performed experiments in both skeletal muscle cells and tissue to identify and characterise the miR-494 species (83). Microarray analysis identified miR-494 as the only miRNA that significantly reduced during myoblast differentiation, that also coincided with increased mitochondrial DNA (mtDNA) content, suggestive of an increased number of mitochondria. Follow-up loss and gain of function experiments in cultured muscle cells/myoblasts showed that knocking down miR-494 induced an increase in mitochondrial content together with the protein expression of transcription factor A (mtTFA) and forkhead box j3 (Foxj3), whereas overexpressing miR-494 oppositely reduced all these variables (83). To determine the role of miR-494 during mitochondrial adaptation to exercise, miR-494 was also assessed after seven days of swimming (7 × 15 minutes intervals, each separated by a 5-minute rest period) in mice. As expected, given the results from muscle cell culture experiments, miR-494 led to a significant reduction in mtDNA content, mtTFA and Foxj3 protein expression, alongside an exponential increase in PGC-1α gene expression (83). Taken together, these results clearly highlight another important miR, miR-494, that is capable of regulating mitochondrial content in skeletal muscle after exercise.

While the aforementioned studies have investigated the miRNA response to endurance exercise in mouse skeletal muscle, studies have also investigated the miRNA response to chronic endurance training in human skeletal muscle. The first of these by Nielsen et al. analysed the expression of muscle-specific myomiRs, miR-1, miR-133a, miR-133b and miR-206, in the human quadriceps muscle after a bout of endurance cycling exercise (60 minutes at 65% of max power output), performed both before and after 12 weeks of endurance training (3 days/week of continuous cycling interspersed with high-intensity interval sessions on the remaining days/week) (84). Like that seen in mouse skeletal muscle (81), expression of miR-1, alongside miR-133a expression, increased after acute exercise, which was also supported by a later study that demonstrated miR-133a decreased after seven days of physical inactivity due to bed rest in human muscle (85). Interestingly however, the abundance of these miRNAs remained unchanged after the 12 weeks of training (84). Furthermore, the reduced resting miRNA expression levels, induced by the exercise training, were attenuated following 2 weeks of detraining (i.e. cessation of exercise) where basal miRNA expression returned to pre-training levels (84). These results were quite surprising, as one may have hypothesised that the enhanced miR-1 and miR-133a expression, reported after acute exercise, may also result in increased expression after more chronic endurance training. However, the authors speculated that the unaltered miRNA expression observed after endurance training could have been due to the possible transient behaviour of these miRNAs in response to physiological stress, as expression levels only increased immediately after acute exercise and returned to pre-exercise levels after 3 hours. In support of these observations after acute endurance exercise, Russell et al. also reported increased expression of miR-1, miR-133a, as well as miR-133b and miR-181a after 60 minutes of cycling exercise of moderate-intensity ($\sim$70% VO_{2peak}) in human muscle (86). Moreover, mRNA expression levels of several components that are critical for the production (or the biosynthesis) of miRNA molecules, including the endoribonuclease 'cleaving' enzymes, Drosher, Dicer and the nuclear transport protein, Exportin-5 (**Figure 6.2**), all increased after acute exercise (86). Also, after ten days of repeated exercise of alternating intensity (75%-100% VO_{2peak}) and duration (30–90 minutes), the authors reported increased and reduced expression of miR-1 and miR-133a, respectively, which partially supports but also differs from the findings observed by Nielsen et al. (84), where expression of both miRNAs was unaltered after 12 weeks of endurance training. This discrepancy in miR-1 expression could perhaps be due to the difference in training status of the recruited participants between the two studies. Where it is plausible to suggest that the less trained participants in the study by Russell et al. (determined by a lower VO_{2peak} and peak power output) may have experienced a greater physiological stress response, despite exercising at similar intensity and duration to those participants in the Nielsen et al. (84) study, perhaps resulting in a more sustained elevation of miR-1 levels after training (86). In support of this hypothesis, the endurance training period significantly differed between the two studies (12 weeks versus 10 days), where the more trained participants in the Nielsen et al. study exercised for much longer and may have therefore elicited a different adaptive response to the endurance training programme. Interestingly, Keller et al. performed a microarray analysis to identify exercise-responsive miRNAs after 6 weeks of endurance exercise (4 × sessions/week of 45 minutes cycling at 70% of pre-training VO_{2max}) in human skeletal muscle and found that miR-1 was specifically among the 14 miRNAs that decreased, rather than increased (total of 7 miRNAs) after training (87). In the same study, the authors identified several miRNA target-binding sites (areas on genes where miRNA can bind and influence gene expression) on a number of transcription factor genes which all also increased at the gene expression level after endurance exercise. Indeed, the transcription factors, RUNX1, PAX3 and SOX9, possessed miRNA-binding sites for 4 (miR-101, miR-144, miR-92, miR-1) out of the 14 miRNAs that decreased after exercise, whereas these genes did not have target sites for any of the 7 miRNAs that increased after exercise. Overall, suggesting that the reduction in these specific miRs (miR-101, miR-144, miR-92, miR-1) were important for the increase in gene expression observed following endurance training. Of particular interest was perhaps miR-1 that reduced after exercise and has been shown previously to target the paired box 3 (PAX3) transcription factor that is crucial for myogenesis (see **Chapters 2 and 13**) (87). Despite the disparities in miR-1 expression profiles reported in the literature, it is evident that miR-1 is consistently altered across the different studies in response to physical activity and inactivity in skeletal muscle. However, the contrasting miR-1 expression profiles observed across the different studies make it difficult to conclusively define the precise role for this particular miRNA in the response to exercise.

Overall, research to date indicates that small non-coding miRNAs play an important role in the molecular responses to endurance exercise. Given a number of these miRNA are oppositely affected by physical inactivity, such as immobilisation and bed rest, further supports the idea that they are important epigenetic regulators in the molecular responses to endurance exercise in skeletal muscle.

THE EPIGENETIC REGULATION OF RESISTANCE EXERCISE

By comparison to work in aerobic/endurance exercise, research examining the interplay between Resistance Exercise (RE) training, modifications to the epigenome and what consequential effect this has on physiological adaptations is slightly more under-developed. Indeed, only within the last few years have we begun to see concerted efforts to unravel the epigenomic landscape following both acute and chronic RE, and the role these DNA modifications play in regulating various physiological adaptations. Nonetheless, the research that has been produced has revealed some remarkable findings, making this area of molecular exercise physiology incredibly exciting.

Resistance exercise and DNA methylation

To date, most of the epigenetic analysis following RE has looked at DNA methylation, with the first of these studies only published in 2014. In this work, Rowlands et al. trained a group of obese, type-II diabetic human participants for 16 weeks of RE (6–8 reps until failure, including 8 exercises targeting all major muscle groups 3 times per week) before using a genome-wide methylation array (Infinium HumanMethylation 450 BeadChip) to examine what affect this would have on the human muscle methylome. The researchers reported more than 500 CpG sites that showed statistically significant changes in their DNA methylation patterns, with most of these sites (> 400 CpG sites) demonstrating a reduction in their methylation profile (i.e. hypomethylation) (53). Of these analysed CpG sites, the authors identified the genes for which they were localised to, and the role these genes play in biological processes, where they identified tissue morphology, cellular development and cellular assembly/ organisation as key enriched processes. This work was therefore the first to suggest that following chronic RE training, the muscle methylome in humans is preferentially hypomethylated, and that these changes may be associated with key processes of tissue adaptation. What is more, the same study compared 16 weeks of RE to 16 weeks of endurance exercise, identifying a similarly hypomethylated profile following both exercise modalities, but crucially, that the genes differentially methylated were located in distinct pathways between the divergent exercise types. Suggesting that DNA methylation in human muscle is differentially responsive to the specific exercise that it is exposed to (53). The finding that RE creates a hypomethylated change in the human methylome has since been supported by several prevailing studies. First, in white blood cells (leukocytes) found in the circulation, 12 weeks of chronic RE (first 2 weeks of 40–50% 1RM [4 sets of 15 reps], and final 10 weeks of 70% 1RM [3–4 sets of 10– 12 reps]) globally reduced DNA methylation levels (88). In a separate study also in leukocytes, looking at a targeted group of genes, Denham et al. (2016) identified a significant reduction in DNA methylation levels, with a corresponding increase in gene expression of both growth hormone-releasing hormone (GHRH) and fibroblast growth hormone (FHG1) following RE (3 sets of 8–12 reps of 80% 1 RM, 3 times a week for 8 weeks; (89)).

With the development of more advanced array technology (see methodologies, described in **Chapter 2**), researchers were able to examine the human muscle methylome following RE more extensively. In 2018, Seaborne et al. provided the most comprehensive examination of the human muscle methylome following both acute and chronic RE (3, 4). At the acute level (+30 minutes post-RE cessation), these authors reported 10,284 CpG sites to display a significant reduction in their methylation profile, compared to pre-exercise levels. This number of hypomethylated CpG sites exceeded that of CpG sites that increased in methylation following an acute RE stimulus (7600 CpG sites), again suggestive of a preferentially hypomethylated human methylome following acute RE (4). What was more interesting in this research, however, was the findings these authors report in their chronic RE training programme. Using a 7-week training, detraining, retraining model they were able to examine what happened to the muscle methylome during an earlier encounter with chronic RE (training), a period of exercise cessation (detraining) and a secondary encounter with chronic RE training (retraining). Muscle phenotype data collected from this work reported significant increases in lower limb lean mass following the initial chronic RE training period, which returned to pre-exercise/basal levels following the ensuing 7-week period of detraining. Authors subsequently reported the largest increase in lower limb lean mass following the secondary period of 7 weeks of retraining, which, even when normalised to pre-7-week training levels, was superior to the increase observed following the first original RE training block. In summary, authors report an enhanced ability for lower limb lean mass acquisition during a later, secondary period of RE retraining compared to an initial period, a suggestion further supported by similar trends observed in muscle strength measured by isometric knee extensor maximal voluntary contractions. When examining muscle tissue following their RE programme (using Illumina's MethylationEPIC '850K' BeadChip array), the authors identified a large 'remodelling' of the muscle methylome following both the initial 7-week RE training period and the subsequent RE cessation detraining period, where more than 17,000 CpG sites were significantly differentially regulated at both time points, with a tendency for a greater hypomethylated profile following RE training, than following detraining. The most interesting observation, however, was made following the secondary period of RE retraining. Indeed, Seaborne et al. identified a similar number of significantly hypermethylated CpG sites as were seen following the initial training and detraining periods, but they reported a doubling in the number of CpG sites that were hypomethylated during retraining compared to the previous initial training period. In essence, this study reported that after a secondary bout of chronic RE retraining, where the largest increase in lower limb lean mass and strength was observed, they also found a significant increase in the number of CpG sites that were hypomethylated compared with the earlier training period (3, 4). This data initially indicated to the authors that if there were more DNA sites being hypomethylated during retraining, then this was perhaps as a consequence of encountering RE training earlier, and the authors went on to ask the question of whether some of the DNA sites may have been retaining this hypomethylated signature after the training period (even during detraining when muscle mass was lost again), and also if the DNA sites simply experienced an increased frequency of hypomethylation and/or enhanced (greater) hypomethylation on the same sites after later retraining. Ultimately, this is what the authors found, fascinatingly, some genes even retained their hypomethylated and gene turned on signature during detraining (even when muscle was completely lost back to pre-exercise levels) following the first training period. This suggested that the DNA retained this methylated signature over a significant amount of time, suggestive of a molecular 'memory' at the DNA methylation level. Some other genes during retraining also demonstrated enhanced hypomethylation and increased gene expression, also demonstrating an epigenetic retention or memory of the earlier training period. Epigenetic muscle memory and its implications for molecular exercise physiology are discussed in a dedicated section, later in this chapter.

Understanding the consequence of changes in epigenetics is important, as without it, the true meaning of these changes is hard to interpret. To partially address this, Turner et al. (58) performed large-scale bioinformatic analyses of publicly available transcriptome data sets following both acute and chronic RE in healthy human subjects and overlapped this with previously published methylome data sets performed at similar post-RE time frames (3, 4). This explorative work identified the expression levels of 866 genes to be upregulated following RE at the acute level, with 270 of these genes being hypomethylated (reduced methylation). Conversely, they also showed that 936 genes were down regulated immediately post-RE, with 216 of these displaying hypomethylation (58). However, in those that were upregulated a larger proportion of hypomethylated sites occurred in promoter regions, which would therefore be more likely to influence gene expression. From these analyses, it suggests an association between DNA methylation and gene expression levels immediately following acute RE, with a preferential hypomethylated profile, as previously suggested. This trend was further corroborated when the same analysis was performed following chronic RE training. Where authors reported that of the 2,018 genes transcriptionally upregulated following chronic RE, 592 were hypomethylated, and of the 430 transcriptionally downregulated genes, 98 displayed a hypermethylated profile. Importantly, this study also identified the wider pathways that demonstrated enriched hypomethylated and gene turned on signatures after acute and chronic RE. These epigenetic signatures tended to be enriched in growth, extracellular matrix, actin structure and remodelling as well as mechano-transduction gene pathways. While these pathways are already known to alter at the gene expression level after RE due to their role in muscle mass remodelling and regulation, this study was the first to suggest that there was a strong association with the genes in these pathways' methylation levels and their corresponding gene expression changes in response to RE and training.

While this study is the first to investigate and identify a tandem relationship between fluctuations in DNA methylation state and gene expression across the genome following both acute RE and RE training, it is clear that DNA methylation modifications may not account for all observed changes in transcriptomic behaviour. This maybe in part due to this data set being performed on a number of publicly available data sets, for which the RE protocols, time and anatomical location of biopsy as well as handling of material differed. For example, the transcriptome data examined for the acute post-RE time point included all publicly available data sets where biopsies were taken at any point up until 24 hours post-RE, whereas the methylome data set was performed on muscle biopsies obtained as early as 30 minutes post-RE. While this type of study is perhaps highly powered due to a large number of participants' data, this is a common problem when performing cross-comparative analyses on publicly available data sets. To address this, albeit expensive to perform, future research should seek to perform more well-controlled multi-omic analyses of the post-acute and chronic RE periods within the same participants.

While the previously highlighted studies made great strides in expanding our understanding for the effects that RE has on the human methylome over both acute and chronic time frames, they are not without limitation. Most notably, the BeadChip array technologies employed in these studies, while powerful, are limited in their ability to examine certain areas of our genome such as regions associated with ribosomal DNA regions or mtDNA (as discussed in **Chapter 2**). Two recent studies have however, employed sequencing-based chemistry to examine the epigenetic state of skeletal muscle following RE, which combats some of the shortfalls in array methods.

Using reduced representative bisulphite sequencing (RRBS; see methods, described in **Chapter 2**), recent work explored the differential methylome of myonuclei following overload treatment in mice, a protocol that is commonly used to mimic human skeletal muscle hypertrophy in vivo. This work is interesting, as most previous research up to this point has examined skeletal muscle as a whole tissue homogenate, with this homogenate likely to contain different proportions of other cell types (such as fibroblasts). However, using a genetically engineered mouse model, labelling of myonuclei (nuclei contained in muscle fibres only) and fluorescence-activated cell sorting, works from Dr. Murach's lab were able to purify myonuclei from muscle fibres and perform reduced representation bisulphite sequencing to investigate the DNA methylome between sham control and overload treated mice (90). In keeping with previous work, these authors identified a hypomethylated remodelling of the myonuclei methylome following overload, where more than 11,000 CpG sites were hypomethylated compared with less than 3,500 hypermethylated (90). Pathway analysis of their RRBS data sets implicated key regulatory factors involved in skeletal muscle growth, development and quality control. Indeed, the authors identified the pathways of PTEN, PIP3 and TP53 as being hypomethylated following overload stimulus (90), with constituent parts of these pathways influencing the mechanistic target of rapamycin (mTOR), regarded as crucial regulator of skeletal muscle growth (91) (discussed in detail in Chapter **8**). Elaborating on these findings further, using RRBS, the same authors examined how CpG methylation modifications are affected after 8 weeks of training using progressive weighted wheel running in mice (92). Authors also undertook elaborate 'nuclei' labelling experimentation to provide the methylation status of both myonuclei and interstitial nuclei. This is because a muscle tissue biopsy, as explained above, includes both myonuclei (nuclei within the muscle fibres) and nuclei from other cells such as satellite cells (**Chapter 13**) and other non-muscle cell types (e.g. fibroblasts). Therefore, albeit a simplified comparison, the aim was to identify if methylation changes after resistance-type training occur in the muscle fibres themselves or other muscle-derived cells within the muscle tissue niche (with one caveat that is explained below). Indeed, Wen et al. identified that promoter hypomethylation slightly outweighed hypermethylation signatures in myonuclei after training, and promoter hypomethylation was also predominant in interstitial nuclei (92), with these data like those signatures observed in whole human muscle tissue after the first resistance training period, described in detail above (4). At the pathway level in myonuclei, there was enriched hypomethylation of promoter CpGs particularly in Wnt signalling and growth-related pathways, with hypomethylation and upregulation of gene expression in growth-related pathways also observed after RE in humans (58). Interestingly, Wen et al. were also able to demonstrate a differential regulation of the Wnt pathway in interstitial nuclei, where they revealed opposite profiles to myonuclei, i.e. hypermethylation of Wnt pathway genes in interstitial nuclei compared with hypomethylation in the myonuclei (92). This demonstrated the importance of such experiments differentiating between nuclei of different muscle-derived cell types to determine which pathways are epigenetically regulated after exercise training (in mice) in the muscle fibre itself. The only caveat is that the interstitial cells included satellite cells, and given that satellite cells ultimately fuse into the muscle fibres and become myonuclei during training (see Chapter **13**), it meant that the contribution of these important muscle stem cells to methylation profiles after training could not be distinguished within these studies. We will discuss this paradigm again in the epigenetic muscle memory section below, as the authors also undertook repeated training (training, detraining, retraining) in mice to investigate epigenetic muscle memory and we will compare results with repeated training, detraining and retraining methylation studies in human skeletal muscle (4).

More recent work has also taken better advantage of the benefits of using sequencing-based technology to examine the methylome following RE. The mtDNA is one of the most intriguing organelles of the human genome. As introduced in Chapter 3, it is not part of the traditional chromosomal genome, instead existing as a separate ~16kb circular genome entity with a separate and distinct evolutionary origin. Despite being the first significant part of the human genome to be fully sequenced (93), incorporation of its sequence into array-based technologies has proven difficult and thus, to date, all BeadChip array platforms do not have probes that recognise the mtDNA genome. However, using the benefits of the restriction enzyme digests of the RRBS workflow (see **Chapter 2**), recent work has examined how RE in elderly untrained human subjects remodels the mtDNA methylome (94). Following six weeks of whole body RE training, these subjects report a significant reduction in DNA methylation of the 16kb mtDNA region, with 159 of 254 interrogated CpG sites showing a significant reduction in methylation from pre- to post-six weeks training. Interestingly, they identified the ~1.2kb D-loop/control region of the mtDNA genome as displaying an enriched hypomethylated profile following RE in these subjects. This region is crucial for regulating replication and transcription of the whole ~16kb unit, with analysis in expression and protein levels of several genes on both the H- and L-strands of the mtDNA confirming this. This study helps corroborate previous findings of a preferentially hypomethylated remodelling of DNA following RE in human subjects, but, importantly, is the first to examine these effects in areas outside of the traditional nuclear genome (94). Given these changes occur with RE, while you might expect greater changes to mitochondria with aerobic exercise, future studies may therefore want to examine the changes in mtDNA methylation after endurance exercise.

The modifications of DNA methylation following periods of both acute and chronic RE have seen a surge in interest in recent years, with this work showing a clear and potentially important regulation in the molecular responses to exercise. Crucially, more work needs to be done to further our understanding of how integral modifications at the CpG methylation level are, in orchestrating the adaptive response of skeletal muscle following chronic RE training.

Resistance exercise, histone modifications and chromatin dynamics

As outlined above, histone modifications are regarded as one of the hallmark epigenetic regulators of the mammalian organism and a crucial dictator of cellular transcriptomic behaviour. Despite this, there is little work examining how RE training reprogrammes histone marks and what consequential affect this has on gene expression.

Recently, Lim et al. (2020) produced the most comprehensive study to date, examining various histone modifications following RE in human skeletal muscle (95). Authors first conducted a transcriptome-wide RNA sequencing analysis, in which they found more than 150 genes to be upregulated in expression following acute RE (6 reps at 60% of 1 RM for back squat, single leg lunge and deadlift, repeated twice). Following chronic RE training (3 sets of leg press, leg extension and leg curl three times per week, for 10 weeks), the authors identified far fewer genes to be regulated at the expression level (4 genes up-, 5 genes downregulated) (95). Focussing on the post-acute RE period time point and in transcript locations that were shown to be significantly upregulated following acute RE, Lim et al. reported intriguing findings relating to the distribution of histone 3, and its variants. First, authors reported a drastic reduction in total H3 levels in genes transcriptionally activated following acute RE. These findings were further confirmed by a reduction in the histone 3 variant, H3.3, in absolute terms, which, when relative to the levels of H3 (e.g. to relativise for the observed reduction following acute RE), maintained significance. Authors speculated that these data were suggestive of nucleosome dissembling following acute RE. This is intriguing, given that nucleosome incorporation of H3.3 is widely thought to be associated with accumulation of gene-activating chromatin modifications (96), where, for example, the same authors found that endurance training increased the incorporation of H3.3 into the nucleosome (71).

When examining further, authors also reported a significant increase in tri-methylation of H3K27 in absolute terms, and relative to levels of H3. They also reported an increase in acetylation of H3 and mono methylation of H3K4, when relativised to levels of total H3 (95), where previous in vitro work reported H3K27me3 and H3K4me1 to be negatively and positively correlated with gene transcription, respectively, in a wider range of cell types (24). In a muscle-specific context, both these marks seem to be highly conserved to the myogenic basic helix loop helix (bHLH) transcription factor, MyoD, in both myoblasts and myotubes, leading to muscle-specific gene transcription (97) (MyoD and its role in myogenesis/ satellite cells with exercise are covered in **Chapter 13**). Despite these interesting results, very few other studies have characterised the regulation of histone modifications following RE training; thus, it remains an under-developed area of molecular exercise physiology research. It is, however, a tantalising avenue of research and is surely poised to see a flurry of exciting, explorative, and illuminating work in the near future.

Resistance exercise and non-coding RNA

Work elucidating the behaviour of myomiRs after RE, and specifically the miR-1 family, has identified some interesting, albeit ambiguous findings. Early work from Karyn Essers' lab first highlighted the role of miR-1 and miR-133a in the post-hypertrophic stimulus phase of skeletal muscle remodelling (98). Using a synergistic ablation model (see **Chapter 1**) in mice to provide a seven-day chronic overload/ hypertrophic stimulus to the plantaris muscle, they showed a significant increase in the expression of pri-miRNA levels of miR-1 and 133a and a huge increase (~18.5-fold increase) in miR-206 (95). In contrast however, the expression levels of the mature version of these miRNAs significantly reduced by ~50% (95). They analysed the expression of the key components of the miRNA processing pathway (**Figure 6.2**), Drosha and Dicer, but also found both to have an increased expression of ~50%, suggesting that the discrepancy between primary and mature levels of myomiRs in skeletal muscle following hypertrophy could not be explained by a dysregulation in the processing of the miRNAs (95). In a separate study analysing the miR-1 family of myomiRs following an acute anabolic stimulus (resistance exercise plus essential amino acid ingestion), Drummond et al. also demonstrated a reduction in miR-1 expression (99). In humans (young versus old participants), the authors sampled skeletal muscle at baseline, 3 and 6 hours post-acute RE and identified a reduction in pri-miR-1, pri-miR-133a at 6 hours post-exercise in young participants. Conversely, but in keeping with the earlier findings from Karyn Essers group (98), they also found an increase in pri-miR-206 across all post-exercise time points in both the young and aged participants (99). It is important to note that these authors also found pri-miR levels of miR-1 and mi-133a to be elevated in older compared to young individuals at baseline. At the mature miRNA level, Drummond et al. found miR-1 to be significantly reduced (in young participants only) at both the 3- and 6-hour post-RE time points, compared to baseline controls, in keeping with results found in mice (99). MiR-1, specifically, has been shown to target and directly bind to various factors within the IGF-I/Akt signalling pathway (100), a crucial pathway implicated in the post-hypertrophic stress response/adaptation period (91, 101, 102), discussed in **Chapter 8**. While further work is required to consolidate and expand these findings, it provides proof-of-principal that miRNA expression may play a key role in the molecular response to anabolic and hypertrophic stimuli, and subsequently, skeletal muscle adaptation.

Finally, it would be important to note that interesting work in the vastus lateralis muscle of high or low responders to 12 weeks of RE training in humans suggested a differential miRNA response pattern between these groups. Sampling the top (responders) and bottom (low responders) ~20% of this RE cohort, Davidsen et al. profiled the expression of 21 different miRNAs in the vastus lateralis muscle. Of these array analyses, they reported miR-26a, 29a and 378 to be significantly reduced compared to levels seen before training in the low responder group only, with no change observed in the high responder group (103). The authors further correlated the levels of miR-378 expression change (relative to pre-training control levels) with the changes in lean body mass, identifying a positive correlation between the two, suggesting that maintenance of miR-378 expression maybe crucial for lean body mass adaptations to RE (103). The role of miR-378 in skeletal muscle adaptation is supported by in vitro data reporting this miRNA as a key target of the myogenic repression protein, MyoR. In this context, miR-378 acts to attenuate the repressive characteristics of MyoR, thereby promoting myoblast differentiation (104, 105).

There is evidence for a role of myomiRs in the molecular paradigm that orchestrates molecular responses to the post-hypertrophic and RE periods. Nonetheless, the field requires more explorative research to fully understand and interpret the role these small non-coding RNAs play in these responses.

FUTURE PERSPECTIVES IN THE FIELD OF EXERCISE AND EPIGENETICS

Muscle memory

Thus far, the present chapter has focussed on key epigenetic modifications that are associated with changes in gene expression that occur after both acute and chronic endurance or resistance exercise. We will conclude this chapter by introducing the emerging concept of 'muscle memory' and discuss the key mechanisms underlying this phenomenon. Recent developments in the field have implicated that epigenetics, specifically the preservation of epigenetic marks or 'imprints', underpins the molecular phenomenon by which 'muscle memory' is orchestrated. This is because epigenetic modifications can be retained following exercise and have therefore been linked with playing an important role in muscle memory or a so-called 'epi-memory' (1). However, in order to understand the role of epigenetics in muscle memory, we first need to define what muscle memory is, and highlight other important cellular mechanisms contributing to this phenomenon.

So what is muscle memory? Whenever we come across the term 'memory', the recollection of a specific moment or event that we have encountered in our lives that is stored in the brain (or central nervous system) often springs to mind. However, there has been a notion that certain genetic material may allow individual cells, from many bodily organs, to possess an internal memory capacity, where, for example, it has recently been described that there is a cellular and epigenetic memory residing in muscle cells. Skeletal muscle memory has therefore been defined as 'the capacity of skeletal muscle to respond differently to environmental stimuli in an adaptive (positive) or maladaptive (negative) manner if the stimuli have been encountered previously' (1). In the context of exercise, muscle typically responds in an adaptively advantageous manner. More so, it is now becoming evident that these advantageous molecular and phenotypic responses to exercise training are accentuated when a period of similar type exercise has previously been performed. For example, when muscle hypertrophies as a result of performing resistance training, the muscle can grow faster and larger in response to a second period of resistance training, even after a prolonged period of reduced activity or 'detraining' where resistance training is ceased and muscle returns to its pre-exercise state. Following studies in the 1990s demonstrating that resistance training is capable of 'priming' muscle for a greater adaptive response to later 'retraining' following several months of reduced activity (106), advanced mechanistic studies that have been conducted over the past decade have begun to unravel some of the key underlying mechanisms responsible for muscle possessing a memory of exercise at both the cellular (i.e. myonuclei) and epigenetic (i.e. DNA methylation) levels.

Cellular muscle memory

Work by Professor Kristian Gundersons' lab first provided evidence that skeletal muscle is capable of 'remembering' anabolic stimuli at the cellular level. In one study, Egner et al. treated mice with testosterone for 14 days which led to an increase in muscle size and myonuclei number acquired from the fusion of satellite cells into the existing muscle fibres (see **Chapter 13**) (107). After a 3-week testosterone 'washout' period, muscle size reduced to baseline/pre-testosterone levels; however, the additional myonuclei gained from the testosterone treatment were retained. Most importantly, when mechanical overload (a stimulus that evokes hypertrophy introduced in Chapter 1) was undertaken after a longer 3-month testosterone washout period, the animals that had received an earlier encounter with testosterone exhibited a 31% increase in muscle cross-sectional area versus control untreated mice that showed only a non-significant 6% increase in growth in the same period of time (107). This therefore suggested that the animals who had previously received testosterone in the past were able to grow their muscle more quickly when encountering exercise later (**Figure 6.4**). The same group also demonstrated that mechanical loading alone was sufficient for the accretion of myonuclei in growing EDL mouse muscle that was also not lost during denervation-induced muscle atrophy (108). However, in these studies, it was not possible to determine whether the newly acquired myonuclei from loading could accentuate the hypertrophic response following repeated loading due to the methodological limitations of the synergistic ablation and denervation models used in this study. To overcome this, a recent study utilised a more physiologically relevant RE protocol to assess whether muscle grew larger after future retraining when previously exposed to earlier exercise training, and whether this was facilitated by an increase and retention of myonuclei after the initial training and detraining, respectively. In this particular study, mice underwent 8 weeks of progressive loaded ladder climbing (3 sets × 5 reps starting from 50% body weight and gradually increased to 300% body weight, twice per day, every third day for 8 weeks), followed by 20 weeks of detraining and another 8 weeks of retraining (109). As with the previous studies, myonuclei were accrued with training and persisted throughout detraining. Interestingly, despite no larger increases in all measures of muscle size after retraining (i.e. CSA and absolute mass), relative muscle mass was enhanced further after repeated exercise (109). Collectively, these studies demonstrate that anabolic stimuli such as testosterone and resistance exercise/mechanical loading are able to promote the acquisition of new myonuclei that can be retained and allow muscle to 'remember' its previous encounters with growth.

It is important to consider that while the general consensus of these studies is that myonuclei number is increased after testosterone, RE/mechanical overload, John McCarthy and Charlotte Peterson's groups reported a normal hypertrophic response when genetically removing satellite cells, suggesting that myonuclei addition acquired from satellite cells may not necessarily be essential for load-induced hypertrophy (110) (discussed in more detail in **Chapters 9** and **13**). Furthermore, other more physiologically relevant models of exercise-induced hypertrophy such as progressive weighted wheel running (PoWeR) training in rodents have also reported no significant retention of myonuclei into the detraining period (111). However, this model of weighted wheel running evoked mainly hypertrophy of oxidative fibre types and therefore may not be fully representative of fibre type changes that occur to faster fibre types after higher load resistance exercise. It is also worth mentioning here that all the studies mentioned thus far have all been performed in rodent muscle.

An important question is therefore, does human skeletal muscle accumulate more myonuclei in response to RE, and are these newly acquired nuclei retained throughout detraining as seen in some models of rodent muscle hypertrophy? Another study by Niklas Psilander and Kristian Gundersen's labs attempted to address this question by assessing muscle size and function, as well as myonuclei number, in human quadriceps muscle before and after 10 weeks of resistance training (3 times per week of unilateral leg extension and leg press exercise at 70–85% 1RM with the non-exercise leg serving as a contralateral control limb), post-20 weeks of detraining and following another 5 weeks of bilateral (rather than unilateral in the first 10 weeks) retraining (112). In contrast to the rodent data, myonuclei number and fibre volume were not altered after the first training period, despite an increase in muscle size and strength. Moreover, muscle size and strength did not significantly differ between the training only versus repeated training (retrained) leg, suggesting that the initial training stimulus did not 'prime' the muscle for an enhanced response after retraining, perhaps owing to the lack of altered myonuclei number during the initial training period (112). These intriguing findings together with the accessibility of the raw data meant that others were able to reanalyse the data set and publish their own interpretations of this study as a viewpoint in the *Journal of Applied Physiology* (113). The analysis of individual participants (rather than the mean of all participants for each condition at each time point) and genders demonstrated the somewhat heterogenous response of human muscle to RE, where most participants had gained myonuclei after training which were not retained after detraining. However, a small subset of participants did retain their myonuclei into detraining. The disparities in the results gained further traction across the scientific community where several expert muscle scientists published their comments in the same issue (114). While we will not necessarily highlight all of these within this chapter, it seems that there is still some controversy as to whether humans retain myonuclei after resistance training. However, given the variable response between individuals, additional studies across a larger population of human participants in both genders, and following different training parameters are required to investigate this further.

Epigenetic muscle memory

As we have now seen, the retention of newly accreted myonuclei after resistance training seems to be important for the accentuated muscle response to future exercise in rodent muscle, but this is still to be fully elucidated in humans. Interestingly, the role of epigenetics has emerged as another important mechanism for enabling muscle to remember previous exercise in humans. A recent landmark study in this area demonstrated that DNA methylation signatures that are modulated in response to RE may be essential for the muscle memory effect of exercise. In this particular study, Professor Adam Sharples' group performed genome-wide methylome (850K CpG sites) and targeted gene expression analyses on the DNA and RNA, respectively, of human muscle tissue from the quadriceps that had undergone 7 weeks of resistance training (3 days/week), 7 weeks of detraining (where participants returned to normal habitual activity) and another 7 weeks of training (or retraining) (4). The authors demonstrated that the number of hypomethylated (reduced) CpG sites was slightly larger than the number of hypermethylated (increased) sites after the initial 7 weeks of resistance training-induced increases in lean leg mass. Interestingly however, the number of both hypomethylated and hypermethylated CpG sites remained stable during detraining when lean muscle mass returned to baseline (pre-exercise levels). After retraining however, where the participants had gained the largest increase in lean leg mass, the number of hypomethylated CpG sites substantially increased, doubling in number compared to training and detraining, whereas the total number of hypermethylated sites remained constant throughout the 21-week intervention (4). When interrogating the specific genes that were epigenetically modified and showed corresponding changes in gene expression (i.e. reduced methylation and increased gene expression), the authors identified 2 distinct gene signatures. The first signature included genes that were hypomethylated with an increased gene expression after the initial training period. Where interestingly hypomethylation of these genes was maintained throughout detraining, and fascinatingly, gene expression also remained elevated, despite the participants undertaking no exercise and lean leg mass returning to pre-exercise levels. Moreover, hypomethylation of these genes continued throughout retraining, where some genes had an even larger hypomethylation and enhanced gene expression. Therefore, this gene profile indicated that muscle possessed an epigenetic memory at the DNA methylation level of previous training-induced hypertrophy due to the retention of DNA methylation imprints throughout detraining that led to larger hypomethylation and enhanced gene expression after later retraining. The second gene signature in another subset of genes also demonstrated hypomethylation and enhanced gene expression following training. However, both DNA methylation and gene expression returned to baseline/pre-exercise levels during detraining. Interestingly however, these genes displayed an even greater hypomethylation and further enhanced gene expression after retraining compared to the initial 7 weeks of training (see **Figure 6.4**) (4). Consistent with these findings, a recent paper looked at the methylation signatures using a similar model of training, detraining and retraining in rodents and also reported a greater

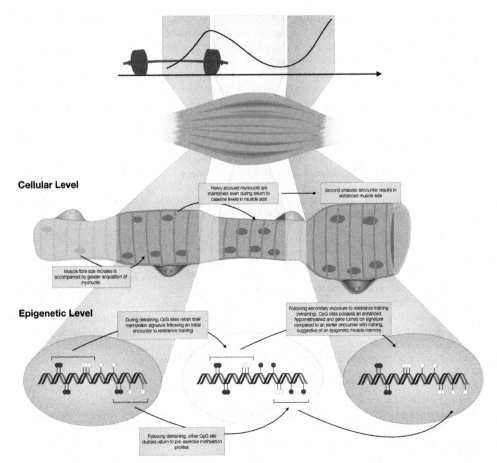

Figure 6.4 Schematic representation of the two main theories of skeletal muscle memory. A cellular memory and an epigenetic memory.

hypomethylation and retention of DNA methylation signatures into detraining following an earlier training period. Importantly, the authors extended this research by investigating the methylation signatures specifically in the myonuclei compared with other interstitial nuclei residing in the mouse plantaris muscle (92). In this study, Wen et al. used a progressive weighted wheel running (PoWeR) protocol to exercise mice for 8 weeks, followed by 12 weeks of detraining and 4 weeks of retraining. The group then measured DNA methylation and gene expression in both myonuclei and interstitial nuclei using RRBS and RNA-sequencing, respectively (92). After the initial training period, the authors reported slightly greater promoter hypomethylation versus hypermethylation within the myonuclei, whereas the interstitial nuclei predominantly demonstrated hypomethylation, corroborating the methylation signatures reported in human muscle after an initial period of resistance training (4). The methylation profiles of genes also differed across the divergent nucleus types where genes associated with Wnt signalling and growth-related pathways demonstrated gene promoter hypomethylation within the myonuclei, whereas these genes oppositely showed hypermethylation in promoter regions within the interstitial nuclei (92). These differences clearly highlighted the importance of conducting nuclei-specific experiments for differentiating the epigenetic response between different muscle-derived cell types, and inform us of the muscle memory response that occurs within the nuclei of the muscle fibres themselves, at least in rodents.

The methylation profiles reported during the detraining period also confirmed the overarching key findings from the previous human studies described above (4). Indeed, the methylation signatures observed after initial training persisted throughout the detraining period, further indicating an epigenetic memory of earlier muscle growth. Specifically, hypomethylation was predominantly maintained throughout detraining in the interstitial nuclei, whereas the hypermethylation was mainly retained in the myonuclei (92). Therefore, the retention of hypomethylation seen in whole human muscle homogenates during detraining (4) was more representative of the hypomethylated signatures observed in the rodent interstitial cells rather than the hypermethylation of myonuclei. However, it's important to note that in the rodent analyses, there was still some retention of hypomethylation observed for some genes within the myonuclei after detraining, just less than the number of genes hypermethylated. Finally, despite reporting greater hypertrophy after 5 weeks of retraining, it was not possible to analyse the epigenetic response after retraining in this particular rodent study due to various practical reasons (92). Therefore, the authors were unable to identify the second epigenetic memory signature described in the human studies above, where methylation was decreased after training and then further enhanced after later retraining. Therefore, this epigenetic profile remained untested in the rodent study and would be important to follow up in later studies using this experimental protocol, especially to identify if this memory profile occurs in the myonuclei and/or interstitial nuclei. Nonetheless, these differences between myonuclei and interstitial nuclei provide clear evidence of a cell-specific epigenetic response to training and detraining and significantly contribute new knowledge within the epigenetic muscle memory field. One important consideration in this field for future research is that the interstitial nuclei in this study would have included satellite cell nuclei, and therefore given that satellite cells would also fuse into the muscle fibres and become myonuclei during training, it means that the contribution of satellite cells to the methylation signatures after training could not be distinguished within these studies. Therefore, future studies should perhaps be designed to separate out alterations of the methylome in satellite cells, then compare this with data derived from the myonuclei in fibres before and after the fusion of satellite cells occurs due to training, detraining and retraining. This is also important, as satellite cells (unlike muscle fibres that cannot divide due to their multinucleated make-up) are single cells that do have the capacity to undergo cellular division (proliferation) and fuse with the muscle fibres to contribute to increases in the number of myonuclei (discussed in detail in **Chapters 8 and 14**). Therefore, it is perhaps possible that satellite cells can pass on the retained epigenetic information to their daughter cells that are ultimately incorporated into muscle fibres as myonuclei to enable a longer-term muscle memory.

Overall, future cell nuclei-specific studies are required in humans and complementary rodent models to broaden our understanding of the epigenetic response to resistance training. Despite there being more research that has investigated the epigenetic response to acute and/or endurance exercise compared with RE, there is a considerable lack of research looking at DNA methylation in response to endurance training, detraining and retraining. Unpublished data from Simone Porcelli and Adam Sharples laboratories suggest that there are over 1,000 CpG sites on the human genome that demonstrate a retained hypomethylated profile into detraining after an earlier higher intensity endurance training period. However, further analysis is required to identify the specific gene pathways (and if these are different to the memory genes identified with RE) and if changes in methylation result in altered gene expression.

The present chapter has specifically focussed on a 'positive' memory (hypertrophy) that is apparent following anabolic steroid use or exercise training. However, it is also possible that muscle may also possess a 'negative' memory after muscle loss associated with injury/immobilisation, metabolic disease, cancer cachexia and ageing (i.e. sarcopenia). Indeed, studies show that muscle cells derived in culture remember these negative encounters, such as impaired lipid metabolism and insulin sensitivity, when obtained from T2D and obese individuals (115, 116). Additionally, high-dose exposure of the inflammatory cytokine, TNF-α, in muscle cells impairs myotube size and alters DNA methylation that is retained long after TNF-α is removed (117). Ageing also results in greater hypermethylation versus hypomethylation (the opposite to that seen after exercise as described above) in human skeletal muscle tissue (118). Interestingly, muscle taken from elderly individuals who have undertaken regular exercise throughout their life displays a more hypomethylated profile compared to elderly individuals who have been less active throughout their life (57), suggesting that exercising in earlier life may protect the muscle from a negative memory. Moreover, it has recently been shown that ageing also hypermethylates the mitochondrial (mtDNA) genome in skeletal muscle, where resistance training was able to rejuvenate the hypermethylated mtDNA signature towards a more hypomethylated state that was observed in trained young adults (94). This is suggestive that the negative epigenetic signatures associated with ageing skeletal muscle might be counteracted by undertaking exercise training.

In summary, skeletal muscle has a memory of previous encounters from testosterone and exercise, resulting in an enhanced exercise response when similar exercise is performed again in the future. So far, there have been two mechanisms proposed for this muscle memory phenomenon: (1) a cellular memory (i.e. increased and retained myonuclei number) and (2) an epigenetic memory (retained DNA methylation). It is not currently known how long cellular or epigenetic memory lasts and this question is likely to be the focus of future research in this area. It is also likely that other epigenetic modifications such as those that occur on histone proteins and other factors such as the three-dimensional chromatin configuration could also be retained after exercise and lead to an enhanced genomic response to future exercise. Therefore, these other epigenetic modifications are also likely to be important areas of muscle memory research in the future. Finally, could altering the exercise frequency, intensity or timing of training lead to a memory of exercise that lasts for longer? Hypothetically, addressing these important questions would allow the incorporation of this type of 'optimised muscle memory' training into athletes or recreationally active persons periodised training programmes. This may help reduce the total volume or intensity of training required, meaning that more time could be spent recovering (therefore reducing injury) or on skill-specific tasks for a given event/sport. For example, if athletes could train at higher volume/intensity less frequently and still have the same adaptation, this would be beneficial in freeing up more time for recovery. However, more studies that are designed to test these specific exercise variables throughout training, detraining and retraining are required to answer these questions.

REVIEW QUESTIONS

- Explain and describe what the different types of common epigenetic modifications are and how they influence gene expression. Discuss why these modifications might be important in the molecular responses to exercise.

- Describe the most important changes in DNA methylation, histones modifications and miRNAs following endurance exercise, and discuss how these might be involved in the molecular response to acute exercise and exercise adaptation to training.

- Describe the most important changes in DNA methylation, histone modifications and miRNAs following resistance exercise, and discuss how these might be involved in the molecular response to acute exercise and exercise adaptation to training.

- Explain the two main theories (cellular and epigenetic) contributing to muscle memory and describe the overarching epigenetic modifications that occur in skeletal muscle in response to training, detraining and retraining. In your answer, discuss what future studies investigating muscle memory may wish to focus on in order to further explain how epigenetics is associated with muscle memory.

FURTHER READING

Seaborne RA & Sharples AP (2020). The interplay between exercise metabolism, epigenetics, and skeletal muscle remodeling. *Exerc Sport Sci Rev* 48(4), 188–200.

Sharples AP, Stewart CE, & Seaborne RA (2016). Does skeletal muscle have an 'epi'-memory? The role of epigenetics in nutritional programming, metabolic disease, aging and exercise. *Aging Cell* 15(4), 603–16.

McGee SL & Walder KR (2017). Exercise and the skeletal muscle epigenome. *Cold Spring Harb Perspect Med* 7(9), a029876.

Sharples AP & Seaborne RA (2019). Exercise and DNA methylation in skeletal muscle. In: Debmalya Barh and Ildus I. Ahmetov, editor. *Sports, Exercise, and Nutritional Genomics*. Academic Press, Cambridge, MA. pp. 211–29.

REFERENCES

1. Sharples AP, et al. *Aging Cell*. 2016. 15(4):603–16.
2. Siggens L, et al. *J Intern Med*. 2014. 276(3):201–14.
3. Seaborne R., et al. *Sci Data*. 2018. 5:1–9.
4. Seaborne R., et al. *Sci Rep*. 2018. 8(1):1–17.
5. Seaborne RA, et al. *Exerc Sport Sci Rev*. 2020. 48(4):188–200.
6. Duan G, et al. *PLoS Comput Biol*. 2015. 11(2):1–23.
7. Minguez P, et al. *Mol Syst Biol*. 2012. 8(599):1–14.
8. Suzuki MM, et al. *Nat Rev Genet*. 2008. 9(6):465–76.
9. Trasler J, et al. *Carcinogenesis*. 2003. 24(1):39–45.
10. Denis H, et al. EMBO Rep. 2011. 12(7):647–56.
11. Tahiliani M, et al. *Science (80-)*. 2009. 324(5929):930–5.
12. Ito S, et al. *Nature*. 2010. 466(7310):1129–33.
13. Lister R, et al. *Nature*. 2009. 462(7271):315–22.
14. Ziller MJ, et al. *Nature*. 2013. 500(7463):477–81.
15. Bird AP. *Nature*. 1986. 321(6067):209–13.
16. Bogdanović O, et al. *Chromosoma*. 2009. 118(5):549–65.
17. Jones P., et al. *Nat Genet*. 1998. 19:187–91.
18. Anastasiadi D, et al. *Epigenetics Chromatin*. 2018. 11(1):1–17.
19. Ball MP, et al. *Nat Biotechnol*. 2009. 27(4):361–8.
20. Brenet F, et al. *PLoS One*. 2011. 6(1):e14524.
21. Greer EL, et al. *Nat Rev Genet*. 2012. 13(5):343–57.
22. Torres IO, et al. *Curr Opin Struct Biol*. 2015. 35:68–75.
23. Marmorstein R, et al. *Biochim Biophys Acta – Gene Regul Mech*. 2009. 1789(1):58–68.
24. Barski A, et al. *Cell*. 2007. 129(4):823–37.
25. Lee IH. *Exp Mol Med*. 2019. 51(9): 1–11.
26. Eberharter A, et al. *EMBO Rep*. 2002. 3(3):224–9.
27. Lujambio A, et al. *Nature*. 2012. 482(7385):347–55.
28. Selbach M, et al. *Nature*. 2008. 455(7209):58–63.
29. Valinezhad Orang A, et al. *Int J Genomics*. 2014. 2014(June 2013): Article ID 970607.
30. Vasudevan S, et al. *Science (80-)*. 2007. 318(5858):1931–4.
31. Yi R, et al. *Genes Dev*. 2003. 17(24):3011–6.
32. Ørom UA, et al. *Mol Cell*. 2008. 30(4):460–71.
33. Kwa FAA, et al. *Drug Discov Today*. 2018. 23(3):719–26.
34. Jin C, et al. *J Neurooncol*. 2017. 133(2):247–55.
35. Shao L, et al. *Mol Cancer Res*. 2018. 16(4):696–706.
36. Baar K, et al. *FASEB J*. 2002. 16(14):1879–86.
37. Pilegaard H, et al. *J Physiol*. 2003. 546(3):851–8.
38. Puigserver P, et al. *Cell*. 1998. 92(6):829–39.
39. Wu Z, et al. *Cell*. 1999. 98:115–24.
40. Martínez-Redondo V, et al. *Diabetologia*. 2015. 58(9):1969–77.
41. Barrès R, et al. *Cell Metab*. 2009. 10(3):189–98.
42. Alibegovic AC, et al. *Am J Physiol – Endocrinol Metab*. 2010. 299(5):752–63.
43. Barrès R, et al. *Cell Metab*. 2012. 15(3):405–11.
44. Lane SC, et al. *J Appl Physiol*. 2015. 119(6):643–55.

45. Lochmann TL, et al. *PLoS One*. 2015. 10(6):1–16.

46. Masuzawa R, et al. *J Appl Physiol*. 2018. 125(4):1238–45.

47. Allan R, et al. *Eur J Appl Physiol*. 2020. 120(11):2487–93.

48. Schuettengruber B, et al. *Nat Rev Mol Cell Biol*. 2011. 12(12):799–814.

49. Maasar MF, et al. *Front Physiol*. 2021. 12(February):1–17.

50. Rundqvist HC, et al. *PLoS One*. 2019. 14(10):1–24.

51. Pillon NJ, et al. *Nat Commun*. 2020. 11(1).

52. Nitert MD, et al. *Diabetes*. 2012. 61(12):3322–32.

53. Rowlands DS, et al. *Physiol Genomics Exerc Heal Dis Multi-omic*. 2014. (46):747–65.

54. Lindholm ME, et al. *Epigenetics*. 2014. 9(12):1557–69.

55. Robinson MM, et al. *Cell Metab*. 2017. 25(3):581–92.

56. Stephens NA, et al. *Diabetes Care*. 2018. 41(10):2245–54.

57. Sailani MR, et al. *Sci Rep*. 2019. 9(1):1–11.

58. Turner DC, et al. *Sci Rep*. 2019. 9(February):1–12.

59. Small L, et al. *PLoS Genet*. 2021. 17(1):1–24.

60. Damal Villivalam S, et al. *EMBO J*. 2021. 40(9):1–15.

61. Mcgee SL, et al. *Cold Spring Harb Perspect Med*. 2017. 7(9):1.

62. Smith JAH, et al. *Am J Physiol – Endocrinol Metab*. 2008. 295(3):698–704.

63. Joseph JS, et al. *Biochem Biophys Res Commun*. 2017. 486(1):83–7.

64. McGee SL, et al. *J Physiol*. 2009. 587(24):5951–8.

65. Lu J, et al. *Proc Natl Acad Sci U S A*. 2000. 97(8):4070–5.

66. Potthoff MJ, et al. *J Clin Invest*. 2007. 117(9):2459–67.

67. McKinsey TA, et al. *Nature*. 2000. 408(6808):106–11.

68. Gomez-Pinilla F, et al. *Eur J Neurosci*. 2011. 33(3):383–90.

69. Backs J, et al. *Mol Cell Biol*. 2008. 28(10):3437–45.

70. McGee SL, et al. *Diabetes*. 2008. 57(4):860–7.

71. Ohsawa I, et al. *J Appl Physiol*. 2018. 125(4):1097–104.

72. Wirbelauer C, et al. *Genes Dev*. 2005. 19(15):1761–6.

73. Hake SB, et al. *J Biol Chem*. 2006. 281(1):559–68.

74. Lu X, et al. *Nat Struct Mol Biol*. 2008. 15(10):1122–4.

75. Chen JF, et al. *Nat Genet*. 2006. 38(2):228–33.

76. Hak KK, et al. *J Cell Biol*. 2006. 174(5):677–87.

77. Horak M, et al. *Dev Biol*. 2016. 410(1):1–13.

78. Polakovičová M, et al. *Int J Mol Sci*. 2016. 17(10).

79. Silva GJJ, et al. *Prog Cardiovasc Dis*. 2017. 60(1):130–51.

80. Van Guilder GP, et al. *Am J Physiol Heart Circ Physiol*. 2021. 320(6):H2401–15.

81. Safdar A, et al. *PLoS One*. 2009. 4(5):19–22.

82. Aoi W, et al. *Am J Physiol – Endocrinol Metab*. 2010. 298(4):799–806.

83. Yamamoto H, et al. *Am J Physiol – Endocrinol Metab*. 2012. 303(12):1419–27.

84. Nielsen S, et al. *J Physiol*. 2010. 588(20):4029–37.

85. Ringholm S, et al. *Am J Physiol – Endocrinol Metab*. 2011. 301(4).

86. Russell AP, et al. *J Physiol*. 2013. 591(18):4637–53.

87. Keller P, et al. *J Appl Physiol*. 2011. 110(1):46–59.

88. Dimauro I, et al. *Redox Biol*. 2016. 10:34–44.

89. Denham J, et al. *Eur J Appl Physiol*. 2016. 116(6):1245–53.

90. Von Walden F, et al. *Epigenetics*. 2020. 15(11):1151–62.

91. Bodine SC, et al. *Nat Cell Biol*. 2001. 3(11):1014–9.

92. Wen Y, et al. *Function*. 2021. 2(5):1–11.

93. Andersen S, et al. *Nature*. 1981. 290:457–65.

94. Ruple BA, et al. *FASEB J*. 2021. 35(9): e21864.

95. Lim C, et al. *PLoS One*. 2020. 15(4):1–18.

96. Loyola A, et al. *Trends Biochem Sci*. 2007. 32(9):425–33.

97. Blum R, et al. *Genes Dev*. 2012. 26(24):2763–79.
98. McCarthy JJ, et al. *J Appl Physiol*. 2007. 102(1):306–13.
99. Drummond MJ, et al. *Am J Physiol – Endocrinol Metab*. 2008. 295(6):1333–40.
100. Elia L, et al. *Circulation*. 2009. 120(23):2377–85.
101. Glass DJ. *Curr Opin Clin Nutr Metab Care*. 2010. 13(3):225–9.
102. Glass DJ. *Nat Cell Biol*. 2003. 5(2):87–90.
103. Davidsen PK, et al. *J Appl Physiol*. 2011. 110(2):309–17.
104. Gagan J, et al. *J Biol Chem*. 2011. 286(22):19431–8.
105. Lu J, et al. *Proc Natl Acad Sci U S A*. 1999. 96(2):552–7.
106. Staron RS, et al. *J Appl Physiol*. 1991. 70(2):631–40.
107. Egner IM, et al. *J Physiol*. 2013. 591(24):6221–30.
108. Bruusgaard JC, et al. *Proc Natl Acad Sci U S A*. 2010. 107(34):15111–6.
109. Lee H, et al. *J Physiol*. 2018. 596(18):4413–26.
110. Mccarthy JJ, et al. *Development*. 2011. 138(17):3657–66.
111. Murach KA, et al. *J Cachexia Sarcopenia Muscle*. 2020. 11(6):1705–1722.
112. Psilander N, et al. *J Appl Physiol*. 2019. 126(6):1636–45.
113. Murach KA, et al. *J Appl Physiol*. 2019. 127(6):1814–6.
114. Miller Pereira Guimarães, et al. *J Appl Physiol*. 2019. 127:1817–20.
115. Kase ET, et al. *Biochim Biophys Acta – Mol Cell Biol Lipids*. 2015. 1851(9):1194–201.
116. Bakke SS, et al. *PLoS One*. 2015. 10(3):1–17.
117. Sharples AP, et al. *Biogerontology*. 2016. 17(3):603–17.
118. Turner DC, et al. *Sci Rep*. 2020. 10(1):1–19.

CHAPTER SEVEN

Signal transduction and exercise

Brendan Egan and Adam P. Sharples

LEARNING OUTCOMES

At the end of this chapter, you should be able to:

1. Outline the molecular responses to acute exercise in the context of the regulation of adaptation to exercise training.
2. Describe the molecular response to exercise, explain the 'signal transduction hypothesis' and illustrate how these processes are central to adaptation to exercise in skeletal muscle.
3. Distinguish between aerobic, resistance and high-intensity intermittent types of exercise and their associated molecular responses to acute exercise.
4. Discuss the specificity of signal transduction after divergent exercise modalities and the interference effect after concurrent exercise.
5. Discuss the practical relevance and translational potential of the study of signal transduction in skeletal muscle, including the exercise mimetic concept.

INTRODUCTION

How does skeletal muscle adapt to exercise training? This question has puzzled and enthused exercise physiologists for over half a century since the seminal work by John Holloszy demonstrated remarkable adaptive changes in mitochondrial content in rodent skeletal muscle in response to intense exercise training; see **Chapter 1** (1). This phenomenon is further illustrated by the remodelling of muscle structure and function, with respect to muscular size, force, endurance and contractile velocity as a result of changes in functional demand induced by exercise training (2). In other words, exercise training makes muscles bigger, stronger, fitter and faster depending on the type of training performed. The changes are also of benefit to non-athletes as demonstrated by the ability of regular exercise to prevent or ameliorate pathophysiological disease states to which physical inactivity and skeletal muscle insulin resistance contribute (3).

One might also ask the question: how does skeletal muscle adapt in a divergent manner to specific *types or modes* of exercise training? To the sport and exercise scientist, the breadth of this divergence in adaptations to exercise training is most obviously illustrated by the vastly different morphological appearance, energy utilisation and performance parameters of well-trained marathon runners compared to elite short-distance sprint athletes. However, adaptations are not always so skewed in one direction for endurance versus strength and power. For example, elite decathletes excel in events requiring wide-ranging contributions from both endurance and strength domains, whereas professional rugby players seek to

DOI: 10.4324/9781315110752-7

combine a high degree of muscularity with an ability for repeated sprint performance in a sport with relatively high demands for aerobic fitness.

In simple terms, exercise training-induced adaptations are a consequence of the training stimulus, which is largely determined by the volume of training (as a function of frequency, intensity and duration of sessions) and the type(s) (e.g. aerobic, resistance, interval and sprint) and mode(s) (e.g. running, cycling and rowing) of training performed. However, the question remains as to how variations in these parameters of exercise prescription have the potential to produce somewhat divergent adaptive responses. The contemporary view is that a network of molecular responses activated by exercise occurs in a manner that, broadly speaking, induces adaptive responses following repeated exercise sessions, and that these networks are sensitive enough to the specifics of the training stimulus to ultimately explain divergence in training adaptations.

In this chapter, we aim to answer questions such as 'how and why do we adapt to exercise?' and 'how do similar and divergent adaptations to different types of exercise occur?'. We begin by describing the nature of the exercise stimulus and its implications for molecular responses to exercise in skeletal muscle. The consequent molecular adaptations to exercise training are described in more detail in **Chapters 8** (resistance exercise) and **9** (endurance exercise). Next, we describe how molecular responses to exercise and associated exercise-induced signal transduction are encompassed in the **'signal transduction hypothesis'**, and how these responses are proposed to lead to adaptation in skeletal muscle. This model is currently the most widely accepted explanation for how molecular responses to acute exercise determine how skeletal muscle adapts to exercise training. Delving more deeply into this model, we discuss how the exercise **'stimulus'** generates molecular **'signals'** that are **sensed** by **'sensor' proteins**, computed and conveyed by **'signal transduction' proteins**, and how **'effector' proteins**, in turn, **regulate transcription, translation or protein synthesis, protein degradation and other cellular processes**. All these processes provide the molecular basis for how muscle (and perhaps other organs) adapt to exercise over time. Examples of some of the key players in skeletal muscle that are currently thought to be regulators of adaptation to exercise are provided, as well as consideration for the emerging techniques to identify novel molecular players. Importantly, while molecular responses do exhibit some specificity in response to divergent exercise types, at present, these responses do not fully explain the divergent adaptive responses to exercise training, and this will be discussed in this chapter. Lastly, we consider the practical significance and translational potential for the investigation into this level of molecular regulation, and specifically delve into the concept of exercise mimetics as an illustrative example.

To prepare for this chapter, you should read **Chapter 1**, which provides a historical overview of molecular exercise physiology and introduces the signal transduction theory as well as the emergence of some of the best-studied regulators such as the AMPK/PGC-1α pathways for aerobic exercise, and mTOR pathways for resistance exercise; **Chapter 2**, which introduces molecular exercise physiology and basic wet laboratory research methods through which many of the findings in this chapter have been revealed; and **Chapter 3**, which introduces sport and exercise genetics and explains concepts such as gene expression, transcription and translation. Each chapter covers core concepts that are important for the understanding of pathways and processes described in this chapter.

WHY DO WE ADAPT TO EXERCISE? THE INTERPLAY BETWEEN THE CONCEPTS OF HOMEOSTASIS AND OVERLOAD

Physiological, or phenotypical, adaptations are changes that occur within individuals in response to external factors such as exercise training, and other environmental factors, such as altitude. For example, a larger muscle mass can be a phenotype caused by performing progressive resistance exercise training, or improved oxygen-carrying capacity through increased haemoglobin mass can be a phenotype after performing aerobic exercise training at altitude. In the history of adaptation research, one early idea is the overload concept proposed by Julius Wolff, who linked the mechanical loading of bones to their adaptation in an 1892 book entitled 'Das Gesetz der Transformation der Knochen' [The Law of Bone Remodelling; translated by Maquet and Furlong (1986) (4)]. The original hypothesis is now known as Wolff's law. This principle can be applied to other organs such as skeletal muscle, if the meaning of the term overload is extended beyond mechanical overload to include the many perturbations to homeostasis that occur with the onset of, and recovery from, exercise. The principle of overload in the context of sport and exercise is indeed one of the core principles of exercise training. Overload in this case refers to an exercise stimulus that perturbs the stability of the internal environment (i.e. homeostasis), and with repeated exercise stimuli produces physiological adaptations, which we call 'training effects'.

Unsurprisingly therefore, this concept of homeostasis is central to the why and how we adapt to exercise. As classically defined, homeostasis is the ability to maintain relative constancy or uniformity in the internal environment in the face of significant changes in the external environment. To the exercise physiologist, this can be reframed as the ability of the whole body (broadly) and/or the skeletal muscle (locally) to maintain constancy under the challenge of exercise. Of notable relevance here is that the very definition of exercise acknowledges this fact, i.e. exercise is 'any and all activity involving generation of force by the activated muscle(s) that results in disruption of a homeostatic state' (5). Exercise induces perturbations in homeostasis in virtually every organ system but particularly so in skeletal muscle (6, 7), and activates acute responses and long-term adaptations that act as compensatory mechanisms to re-establish and/or preserve homeostasis as required. This chapter therefore takes a skeletal muscle-centric view of signal transduction and its role in mediating the adaptive response to exercise.

As such, at a cellular and regulatory level in skeletal muscle, adaptive responses to exercise training take many forms, including changes in abundance, regulation and/or maximal activity of key proteins involved in energy provision, the remodelling of cellular components such as contractile proteins and the extracellular matrix, and the biogenesis of organelles such as ribosomes and mitochondria (8–12). Consequent to these cellular changes are alterations at the level of tissues and systems such as angiogenesis (the development of new blood vessels), muscle hypertrophy and altered substrate utilisation. The teleological understanding of these coordinated changes in skeletal muscle form and function is that they occur to minimise perturbations to cellular homeostasis, with this better maintenance of cellular homeostasis likely contributing to improved fatigue resistance during future sessions of exercise and thereby improving performance by maximising substrate delivery, mitochondrial respiratory capacity and contractile function during exercise (2, 13).

How do we adapt to exercise? The signal transduction hypothesis

While the principle of overload states that regular and progressive exercise training induces repeated perturbations to homeostasis, which are necessary stimuli for adaptations that defend perturbations to homeostasis in future sessions of exercise (i.e. **why** we adapt to exercise), this principle does not explain the mechanisms by which skeletal muscle or other organs respond and adapt (i.e. **how** we adapt to exercise). Therefore, while there is little doubt about the effects of exercise training to produce wide-ranging adaptations within skeletal muscle (for more details, see **Chapter 8** for size/strength adaptations, and **Chapter 9** for endurance adaptations), the mechanistic bases for how these changes occur remain a topic of much investigation. In the corresponding chapter in the First Edition of *Molecular Exercise Physiology*, Burniston et al. (2014) (14) summarised these mechanistic bases under a model that they termed 'the signal transduction hypothesis of adaptation', which we have revised and updated for this chapter in the Second Edition (**Figure 7.1**). We present this model as six sequential steps from the onset of the exercise stimulus to a change in protein abundance and/or activity (which parallels the flow of genetic information in a cell), but as described later, there are temporal and molecular overlaps between the various steps (**Figure 7.1**).

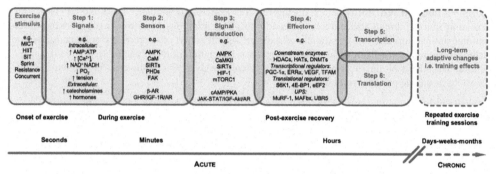

Figure 7.1 Schematic representation, as sequential steps, of the molecular response to exercise and associated exercise-induced signal transduction that are encompassed in the 'signal transduction hypothesis' that underpins adaptation to exercise in skeletal muscle. Examples, but a far from exclusive list, of the respective signals, sensors, signal transduction proteins and effectors are provided. Further details for each step and key molecular players are provided in the main text. Abbreviations: 4E-BP1, eukaryotic translation initiation factor 4E-binding protein 1; Akt, protein kinase B; AMP, adenosine monophosphate; AMPK, AMP-activated protein kinase; AR, androgen receptor; ATP, adenosine triphosphate; β-AR, beta-adrenergic receptor; $[Ca^{2+}]_i$, intracellular calcium concentration; CaM, calmodulin; CaMK, Ca^{2+}/calmodulin-dependent protein kinase; cAMP, cyclic AMP; DNMT, DNA methyltransferase; eEF, eukaryotic translation elongation factor; ERR, estrogen-related receptor; FAK, focal adhesion kinase; FOXO, forkhead transcription factor O box subfamily; HAT, histone acetyltransferase; HDAC, histone deacetylase; HIF, hypoxia-inducible factor; HIIT, high-intensity interval training; IGF, insulin-like growth factor; JAK-STAT, Janus kinase-signal transducer and activator of transcription; MAFbx, muscle atrophy F-box; MICT, medium intensity continuous exercise; mTORC1, mechanistic target of rapamycin complex 1; MuRF, muscle RING finger 1; NAD+, oxidised form of nicotinamide adenine dinucleotide; NADH, reduced form of NAD; PGC-1, peroxisome proliferator-activated receptor γ coactivator 1α; PHD, prolyl hydroxylase domain; PKA, protein kinase A; PO_2, partial pressure of oxygen; S6K1, ribosomal protein/p70 S6 kinase; SIRT, sirtuin; SIT, sprint interval training; TFAM, mitochondrial transcription factor A; UBR5, E3 ubiquitin-protein ligase UBR5; UPS, ubiquitin-proteasome system; VEGF, vascular endothelial growth factor.

Signal transduction generally refers to stimuli, stressors and signals originating from the either outside or inside of a cell resulting in the activation of the intracellular transfer of signals, usually by kinase or phosphatase cascades, or other posttranslational mechanisms and signalling processes, to protein targets in various cellular locations (e.g. cytosolic and nuclear) that regulate downstream processes. In the present context, the onset of exercise is the **stimulus** for a myriad of acute responses occurring at multiple systemic and cellular levels, many of which are related to the supply of blood, oxygen and nutrients, the rate of ATP turnover and substrate flux, and the mechanics of muscle contraction and loading themselves. These responses produce **molecular signals (Step 1 – 'signals')** such as the reduction in blood oxygen and reduced intracellular **partial pressure of oxygen (PO_2)** producing the signal of local tissue hypoxia; changes in ATP turnover and mitochondrial reduction/oxidation (redox state) producing the signals of altered **$[ATP]/[ADP][P_i]$** and **$NAD^+/NADH$**, respectively; muscle contraction requiring an increase in **calcium ions (Ca^{2+})** in the sarcoplasmic reticulum (SR) as well as increased and intermittent **tension** through the muscle. Signals originating outside of the contracting muscle include the increasing concentration of **circulating hormones**, often in a manner related to the type and intensity of exercise. For example, increases in testosterone, growth hormone and insulin-like-growth factor 1 (IGF-I) occur most prominently with resistance exercise, and increases in catecholamines such as adrenaline and noradrenaline occur most prominently with aerobic/endurance exercise. Similarly, endogenous and exogenous **nutrient availability** before, during and after exercise can influence signalling via changes in circulating glucose, amino acid and free fatty acid (FFA) concentrations. Moreover, many acute responses continue in the hours after the cessation of an exercise session as part of the recovery of homeostasis, and include inflammatory and anti-inflammatory responses, restoration of fluid balance, lactate oxidation and removal, resynthesis of muscle glycogen and intramuscular triglycerides oxidised during exercise, and elevations in muscle protein synthesis (MPS).

Collectively, these responses during and after an exercise session are sensed by a variety of **sensor proteins (Step 2 – 'sensors')** that detect alterations in the molecular signals described above. For example, AMP-activated protein kinase **(AMPK)** and sirtuins (e.g. **SIRT1**) are intracellular sensors of the aforementioned changes in **[ATP]/[ADP][P$_i$]** and **NAD$^+$/NADH**, respectively. **Calmodulin** (CaM) is a sensor of the changes in Ca^{2+} flux in the SR that are required for muscle contraction. **'Mechanosensors'** are a diverse set of proteins that can sense changes in mechanical load (tension) through muscle, and downstream pathways often converge on **mTORC activation.** Sensor proteins also include **receptors present on the cell membrane** that can sense specific peptide hormonal signals arising from changes in circulating concentrations. Examples include growth hormone, IGF-I and adrenaline and their activation of the growth hormone receptor (GHR), IGF-I receptor (IGF-IR) and β-adrenergic receptor (β-AR), respectively. However, receptors as sensor proteins can also include intracellular receptors for steroid-based hormone signals such as the androgen receptor (AR), which is generally located in the cytoplasm but can translocate to the nucleus upon activation and thereby regulate gene expression. AR is a specific receptor for testosterone, which is notable for the fact that these types of lipid-based hormones can pass directly through the cell membrane, rather than relying on the activation of a receptor on the cell membrane.

This array of sensor proteins acting as, and in concert with, primary and secondary messengers then transduce these signals arising from metabolic, mechanical, hormonal and neuronal stimuli through complex **signal transduction networks (Step 3 – 'signal transduction')**. This signal transduction of molecular information mainly includes protein-protein interactions and posttranslational modifications (e.g. phosphorylation and acetylation) that lead to the activation and/or repression of an array of proteins that relay the sensed molecular signals inside the cell and are ultimately coupled to a myriad of **effector proteins (Step 4 – 'effectors')**. These effector proteins, such as transcription factors and coregulators, and translation initiation and elongation factors, regulate **transcriptional processes (Step 5 – 'transcription')**, and the regulation of **protein translation (Step 6 – 'translation')**, respectively.

The six steps described occur in a temporal manner, such that homeostatic perturbations due to the exercise stimulus produce molecular signals (Step 1) that are sensed by sensor proteins (Step 2), and which, in turn, induce signal transduction (Step 3) and pre-transcriptional/translational regulation via effector proteins (Step 4) occurs during exercise and in the early phase of recovery (minutes to hours). Changes in messenger RNA (mRNA) and protein abundance (Steps 5 and 6) then occur in the hours and day(s) that follow (**Figure 7.1**). Changes that ultimately result in functional improvements in exercise capacity and performance occur in the following days, weeks and months consequent to cumulative effect of frequent, repeated sessions of exercise (15, 16). Hence, while each individual session of exercise is necessary as a stimulus for adaptation, the long-term adaptation to exercise (i.e. exercise training) is the result of the progressive and cumulative effects of each acute exercise session, thereby leading to a new functional threshold.

This working model for molecular regulation of skeletal muscle adaptation to exercise has been in existence since the turn of the century (17, 18). With the greater adoption of cellular and molecular analytical techniques by sport and exercise scientists (see **Chapters 1 and 2**), much new information has been discovered in the interim, and as a result there have been several extensive, more recent reviews (8, 9, 12, 19, 20). This model remains the most widely accepted explanation of the molecular events that link the acute responses to exercise with long-term adaptation. The fundamental molecular physiology concepts, key players and processes will now be discussed in the remainder of this chapter. Indeed, Steps 1-6 of the signal tranduction hypothesis are also described in more detail below, but before this, it is important to understand the nature of the exercise stimulus and how this influences the molecular responses to exercise.

THE NATURE OF THE EXERCISE STIMULUS AND ITS CONSEQUENCES FOR MOLECULAR RESPONSES TO ACUTE EXERCISE

Because the molecular responses to exercise and associated exercise-induced signal transduction are encompassed in the signal transduction hypothesis (that is often proposed to explain the specificity of adaptation to exercise training), it is important to first consider the nature of the exercise stimulus and its consequences for molecular responses to acute exercise. At present, exercise type and intensity have been the independent variables most often explored in this context, and the effect of duration is less well-defined. However, broadly speaking, divergent types (e.g. aerobic versus resistance) and divergent intensities (e.g. low versus high) induce both overlapping and distinct physiological demands on skeletal muscle, the extent of which is largely dependent on factors such as force (load), velocity, duration, frequency and number of contractions. Aside from determining the physiological demands of an exercise session, these factors also influence the recruitment profile of different fibre types in skeletal muscle. Fibre type is likely to have an important role in the molecular response to exercise, but for the purposes of this chapter, we will focus almost exclusively on the responses in mixed muscle.

For didactic purposes, it can be useful to categorise exercise under three broad types: (1) aerobic (or endurance) training; (2) resistance (or strength) training and (3) various intermediate combinations, including, but not limited to, concurrent aerobic and resistance training, circuit training, high-intensity interval training (HIIT), sprint interval training (SIT) and short-sprint training (**Figure 7.2**). In fact, we have previously placed aerobic and resistance exercise on opposite ends of a spectrum of exercise types to conceptualise the breadth of differences in the acute metabolic and molecular responses, and chronic adaptive changes (8, 21). However, as discussed in a later section of this chapter, while this conceptualisation is useful in theory, it must also be said that for several reasons, it is too simplistic of a framework to fully explain the specificity and continuity between acute molecular responses and physiological and functional adaptations indicated by endurance and hypertrophy phenotypes.

The consideration of intermediate types of exercise is exemplified by the intensified interest in high-intensity intermittent exercise (HIIE) models of exercise in the past two decades, and the ability of this style of training to produce training effects that are characteristic of both aerobic and resistance exercise training. The various forms of HIIE can produce improvements that resemble the traditional endurance or strength/power phenotypes, which should not be surprising because it is well-established that even short-sprint training [generally recognised as brief (<10 second efforts) but intense exercise with the goal to develop speed and power] can produce a broad range of adaptive responses (22). Again, the specifics of the adaptive responses in skeletal muscle are highly dependent on the duration of sprint efforts, recovery between sprints and total volume within sessions (22).

The traditional view of aerobic exercise, in contexts so far described in this chapter, is of moderate-intensity continuous exercise (MICE) or training (MICT). However, intermittent or 'interval training' has also been well-established for many decades as a type of 'aerobic' exercise training. The more recent emergence of, and interest in, HIIE models of exercise training has centred largely on the so-called HIIT and SIT (**Figure 7.2**). In their comprehensive review on the topic, MacInnis and Gibala (2017) (23) define HIIT as 'near maximal' efforts generally performed at an intensity that elicits $\geq 80\%$ of maximal heart rate, whereas SIT is characterised by efforts performed at intensities equal to or greater than maximal oxygen uptake (VO_{2max}), and are therefore often characterised by 'all-out' or 'supramaximal' efforts. Again, this general classification scheme is imperfect, given that HIIE could involve a range of exercise modes, including those used for aerobic (e.g. cycling and running) and resistance exercise (e.g. weightlifting and plyometrics) as well as novel forms such as bodyweight exercise or resisted sled sprinting, or may target sport-specific repeated sprint ability characterised by short sprints (≤ 10 seconds), often including a change of direction, and interspersed with brief recovery periods (usually ≤ 60 seconds).

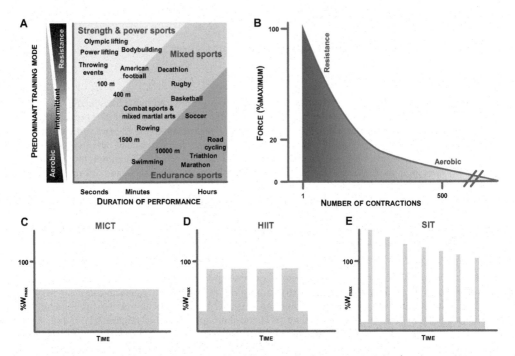

Figure 7.2 The nature of the exercise stimulus. **(A)** Conceptual representation of various sports along continuums of the predominant training undertaken and the duration of performance in competition. **(B)** The main distinction between aerobic and resistance exercise is described largely by the force of each contraction and the number of contractions that are performed during typical training sessions, or that can be performed at a given intensity before fatigue. **(C–E)** Representations of moderate-intensity continuous training (MICT), high-intensity interval training (HIIT) and low-volume sprint interval training (SIT). The intensity is depicted as a percentage of the maximal power output (%W_{max}) achieved during a typical incremental exercise test to assess VO_{2max} on a cycle ergometer. A similar representation could be made using the percentage of velocity at VO_{2max} (vVO_{2max}) instead of %W_{max} for testing and training using running.

Despite consisting of a small number of repeated, intermittent sets of high (force) power output, all-out sprint activity (e.g. repeated all-out 30-second Wingate test efforts interspersed with a few minutes of recovery), SIT induces molecular responses similar to those associated with aerobic exercise (24). Moreover, it is long known that high-volume SIT demonstrates a remarkable capacity to produce an endurance phenotype in skeletal muscle (25), which can occur relatively quickly (e.g. within two weeks) even with low-volume SIT (23).

The training status of the individual is also an important consideration for the nature of the exercise stimulus. The effect of exercise training on metabolic, mechanical, hormonal and neuronal demands and molecular responses can be inferred by comparing responses between groups of trained and untrained individuals (cross-sectional designs), or by studying the same group of individuals before and after a defined exercise training intervention (pre-post designs). For cross-sectional designs, it is most common to assess responses based on comparisons made from exercise performed at the same *relative* intensity [e.g. groups matched on %VO_{2max} or percentage of one repetition maximum (%1RM)], but due to the training effect, this often means that the absolute intensity (e.g. power output or load lifted) is much greater in the trained group. For pre-post designs, it is most common to assess responses based on comparisons made from exercise performed at the same absolute intensity before and after the period of exercise training, but again due to the training effect, this often means that acute response at post-training is being assessed at a lower *relative* intensity compared to pre-training.

The effect of training status or a period of exercise training on the molecular responses to acute exercise has been the subject of a handful of studies of various designs, including comparisons at the same relative intensity, same absolute intensity, responses to maximal effort and comparison of diverse training backgrounds, i.e. aerobic versus resistance exercise-trained athletes. The available evidence suggests that the magnitude of response in terms of outcomes such as activation of signal transduction pathways or changes in mRNA abundance are attenuated in moderate- to well-trained individuals (26–28), or after exercise training interventions (29–31). This observation succinctly illustrates that why and how we adapt to exercise are inextricably linked; perturbations to homeostasis are essential stimuli and signals to activate and/or repress the molecular pathways that induce adaptive responses to exercise, but as 'fitness' improves, by definition, the body is better able to maintain homeostasis in the face of the challenge provided by exercise. Ultimately, over time with repeated exercise sessions, perturbations to homeostasis are less pronounced (e.g. change in AMP and ADP, depletion of muscle glycogen and degree of muscle damage), and therefore, the molecular response to acute exercise is diminished. Another viewpoint that has recently emerged is that the diminished molecular response after training is also reflective of better efficiency of signal transduction and gene expression processes. One such example is that a trained muscle can retain what can be thought of as 'muscle memory' through epigenetic changes to genes known as hypomethylation (covered in detail in **Chapter 6**). These epigenetic changes mean that a change in gene expression may occur more efficiently, and sometimes to an even greater magnitude, when returning to training after a period of detraining (32, 33).

In summary of this section, the characteristics of divergent types of exercise differ in terms of the force (load), velocity, duration, frequency and number of contractions. These parameters are important determinants of the metabolic, mechanical, hormonal and neuronal stimuli acting on skeletal muscle, and therefore, in turn, the perturbations to homeostasis induced by any single session of exercise. As will be seen throughout the remainder of this chapter, these fundamental characteristics of an exercise session therefore have implications for the molecular responses to the various types of exercise in each step of the signal transduction model.

STEP 1 – SIGNALS: MOLECULAR SIGNALS IN RESPONSE TO PERTURBATIONS TO HOMEOSTASIS INDUCED BY THE STIMULUS OF ACUTE EXERCISE

In Step 1, the onset and continuation of exercise, and ensuing post-exercise recovery period, result in both intrinsic and extrinsic responses to the exercising skeletal muscles that act as important signals initiating the molecular response to exercise. Returning to the concept of homeostasis, in a resting myofibre, homeostasis is a function of competing processes, including, but not limited to, substrate flux though the various metabolic pathways, ion distribution across the plasma membrane, Ca^{2+} sequestration in the SR and cycling of contractile proteins. However, the stimulus of acute exercise induces a myriad of challenges to the maintenance of homeostasis, including:

- electrolyte imbalances across cell membranes,

- changes in muscle cell and tissue volume,

- changes in the regulators of the various energy-producing pathways and ATP turnover, such as Ca^{2+} release, metabolites related to the cytosolic phosphorylation potential ($[ATP]/[ADP][P_i]$) and the mitochondrial reduction/oxidation (redox) state ($NAD^+/NADH$),

- declining concentrations of muscle glycogen,

- declining pH,

- reduced partial pressure of oxygen,

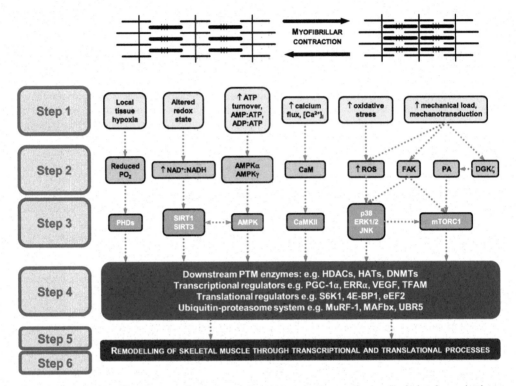

Figure 7.3 Overview of major exercise-induced signal transduction pathways in skeletal muscle that arise from signals intrinsic to the contracting muscle. Abbreviations as listed in Figure 7.1 in addition to: ADP, adenosine diphosphate; DGK, diacylglycerol kinase; ERK, extracellular signal-regulated kinase; JNK, c-Jun N-terminal kinase; PA, phosphatidic acid; ROS, reactive oxygen species.

- increased oxygen free radical production,

- increased muscle temperature and

- increased mechanical load/tension and associated sarcolemmal disruption.

Those listed above are responses intrinsic (intracellular) to the contracting muscle, but extrinsic (circulating) factors that influence the molecular responses to exercise, including the prevailing hormonal and substrate milieu such as concentrations of catecholamines (e.g. adrenaline and noradrenaline) and hormones (GH, IGF-I, testosterone), and/or concentrations of glucose, amino acids and FFAs. Together, these intrinsic and extrinsic factors represent many, but unlikely all, of the responses that constitute perturbations to homeostasis induced by the stimulus of acute exercise, which, in turn, lead to the initiation of signal transduction pathways (**Figure 7.3**).

STEP 2 – SENSORS: SENSING OF EXERCISE-INDUCED SIGNALS

In Step 2, perturbations to homeostasis are linked to subsequent signal transduction pathways by a panoply of **sensor** proteins found within cells and on cell membranes. For the purposes of this model, these sensor proteins can be further divided into **small molecule sensors, transmembrane receptors**

and other cellular sensors/receptors. Small molecule sensors are proteins that contain domains capable of binding molecules such as Ca^{2+}, or AMP and ADP, and this enables the cell to sense changes in the concentration of these molecules through sensor proteins such as CaM and AMPK, respectively (**Table 7.1**). **Transmembrane (or cell surface) receptors** are typically single proteins (e.g. G protein-coupled receptors), enzyme-linked receptors (e.g. receptor tyrosine kinases) or protein complexes (e.g. ligand-gated ion channels) that are embedded in the cell membrane in a transmembrane orientation that allows for the transduction of signals arising outside the cell into processes inside the cell, i.e. coupling external signals to internal regulation of metabolism. Activation of a receptor occurs when a ligand such as a hormone, e.g. insulin, or a neurotransmitter, e.g. acetylcholine, binds to the relevant receptor on the cell surface, which, in turn, leads to a cascade of downstream events. A subclass of hormone receptors are receptors for steroid hormones such as testosterone, which typically reside in the intracellular (cytoplasmic) space and are referred to as nuclear receptors. These receptors, such as the AR, typically have a strong influence on gene expression due to their function as transcription factors with an ability to bind directly to DNA and thereby influence transcription of target genes. **Other cellular sensors/ receptors** typically include sensors of physical properties rather than the presence of metabolites or ligands, with examples being sensors of stretch or load through a cell, or of the physical interactions between cells.

We will use AMPK as an example to illustrate the function and role of **small molecule sensors**. AMPK is a type of protein known as a serine/threonine kinase and exists as a protein complex that comprises three different proteins (termed a 'heterotrimer'). Through its action as a kinase, AMPK modulates cellular metabolism through phosphorylation of metabolic enzymes, and modulates transcription and translation through phosphorylation of transcription factors and other signalling proteins. The activity of AMPK as a kinase is influenced by binding of ADP and AMP (activation), and glycogen (inhibition). An increase in AMPK phosphorylation and enzymatic activity by exercise has been shown many times, and this is unsurprising, given the cellular energy deficit caused by exercise, and the increase in ADP and AMP as a consequence of ATP hydrolysis and resynthesis in the process of muscle contraction. Additionally, activation of AMPK often occurs in an intensity-dependent manner, which probably reflects intensity-dependent effects of exercise on ATP turnover and adenine nucleotide concentrations, as well as the greater reliance on, and therefore depletion of, muscle glycogen with increasing intensity of exercise. The role of muscle glycogen concentration in the activation of AMPK by exercise is notable because of the recent findings that training in a state of reduced carbohydrate availability or low muscle glycogen leads to greater activation of AMPK and may lead to greater adaptive changes in skeletal muscle with exercise training over time (34). Overall, AMPK activation acts to protect ATP concentrations by stimulating catabolic pathways to restore cellular energy stores while simultaneously inhibiting biosynthetic pathways and anabolic pathways; in other words, its activation by a perturbation to homeostasis ultimately acts to restore homeostasis – a recurring theme in this chapter. In fact, in experimental models that utilise the repeated and chronic activation of AMPK (acting as a surrogate for exercise training), observe adaptive changes in skeletal muscle that are similar to those seen with aerobic exercise training, e.g. an increased number and size of mitochondria. This again nicely illustrates the coupling of the response to acute exercise as a means to produce adaptive changes that lead to a more robust defence of homeostasis to later challenges of exercise.

Cells also sense signals from the superior endocrine and nervous systems via **transmembrane receptors, including hormone and neurotransmitter receptors**. An example of a hormone receptor is the β-adrenergic receptor (β-AR), which is a G protein-coupled receptor (GPCR) that is activated by the catecholamines adrenaline and noradrenaline. Binding of adrenaline to the extracellular domain of the receptor changes the intracellular region, which activates the so-called G-proteins. This instigates a cascade of events, including opening of membrane Ca^{2+} channels and stimulation of adenylate cyclase, which is the enzyme that produces adenosine 3′,5′-cyclic monophosphate (cAMP) from ATP. This has immediate effects on muscle metabolism such as the increased rate of glycogenolysis, i.e. the rate of breakdown on muscle glycogen. In this regulation, an increase in cAMP activates protein kinase A, which, in turn, leads to the activation of glycogen phosphorylase from its more inactive *b* form to its more active

Table 7.1 Selected sensor proteins and receptors, their function and role in skeletal muscle, and their associated downstream targets, pathways and processes relevant to exercise-induced signal transduction

Sensor protein or receptor	Regulation	Downstream targets, pathways & processes	Suggested further reading
Sensors of intrinsic/intracellular signals			
AMP-activated protein kinase (AMPK)	• A heterotrimeric (catalytic α, and regulatory β and γ subunits) Ser/Thr kinase • Senses energy status of the cell via AMP/ATP and Cr/PCr ratios, and depletion of muscle glycogen	• Modulates metabolism acutely through phosphorylation of metabolic enzymes • Modulates adaptive changes through regulation of the activity of transcriptional (e.g. HDACs, PGC-1a, CREB, and FOXO1) and translational (e.g TSC2) regulators	Kjøbsted, Hingst (36)
Ca2+/Calmodulin-dependent protein kinases (CaMKs)	• Ser/Thr kinases activated in a CaM-dependent • Elevations in $[Ca^{2+}]_i$ are decoded by CaM, a multifunctional signal transducer that activates downstream kinases and phosphatases	• Modulates adaptive changes through regulation of the activity of transcriptional regulators, including HDACs, MEF2, CREB, and SRF	Chin (37)
Sirtuins (silent mating type information regulation 2 homolog) (SIRTs)	• A family of NAD+-dependent protein deacetylases • Sense fluctuations in NAD+ as well as the ratio of NAD+/NADH • Increases in NAD+ increase the enzymatic activity of SIRTs	• Catalyse lysine deacetylation of both transcriptional regulators (SIRT1) and mitochondrial enzymes (SIRT3) • Couple redox state to gene expression and enzyme function	Philp and Schenk (38)
Prolyl hydroxylases (PHDs)	• Enzymes that sense intracellular PO_2 • Low intracellular PO_2 concentrations inhibit the hydroxylase activity of PHDs • Hydroxylation of HIF-1α marks HIF-1α for proteasomal degradation	• Declining intracellular PO_2 concentrations and consequent inhibition of PHD activity result in declining hydroxylation of HIF-1α and upregulation of HIF-1-dependent gene expression	Lindholm and Rundqvist (39)
Mitogen-activated protein kinases (MAPKs)	• Three main MAPK subfamilies in human skeletal muscle: (i) ERK1/2, (ii) JNK, and (iii) p38 MAPK • Sense changes in growth factors, cytokines, and cellular stress, including ROS	• Regulate transcriptional events by phosphorylation of diverse substrates localised in the cytoplasm or nucleus, including transcriptional regulators • Proposed to regulate protein translation through mTORC1-independent pathways	Kramer and Goodyear (40)

Sensor protein or receptor	Regulation	Downstream targets, pathways & processes	Suggested further reading
Focal adhesion kinase (FAK)	• A class of transmembrane receptors that act as protein tyrosine kinases in the integrin signalling pathway • Sense forces arising inside or outside of a cell, most obviously in the form of tension in skeletal muscle	• Facilitate the phosphorylation of downstream targets in mechanotransduction pathways, including mTORC1, leading to enhanced protein synthesis	Boppart and Mahmassani (41)
Mechanistic target of rapamycin complex (mTORC)	• Ser/Thr kinase that exists in two functionally and structurally distinct multiprotein complexes, mTORC1 and mTORC2 • Senses a variety of inputs, including growth factors, insulin, amino acids, and factors relating to mechanotransduction	• Most focus in skeletal muscle and exercise has been on mTORC1 activity • Acting through the canonical mTOR signalling pathway, especially 4E-BP1 and S6K1 in the control of protein translation and cellular growth	Goodman (42)
Sensors of extrinsic/extracellular signals			
β-adrenergic receptors (β-AR)	• Members of the large family of GPCRs, and have a seven-transmembrane helix topology • Sense and transduce signals encoded in catecholamine hormones and neurotransmitters to activate intracellular signalling • Signal via G protein-dependent pathways to control key physiological functions	• Acutely activates adenylate cyclase to convert ATP to cAMP, then activating several different cAMP-dependent protein kinases that phosphorylate target proteins to alter cell processes, including substrate utilisation and force potentiation • Chronic activation associated with β-agonists associated with skeletal muscle hypertrophy	Cairns (43)
Androgen receptor (AR)	• Member of the steroid receptor superfamily and a ligand-dependent nuclear transcription factor • Senses concentrations of androgens/steroid hormones, including testosterone	• Regulate transcription by testosterone binding to ARs and acting as a transcription factor by binding to specific androgen response elements • Androgen-mediated increase in protein synthesis proposed through mTORC1 via IGF-I/Akt and/or ERK1/2 • ARs are expressed in satellite cells and may also activate and increase the number of satellite cells for growth and regeneration of muscle	Rossetti (44)

(Continued)

Sensor protein or receptor	Regulation	Downstream targets, pathways & processes	Suggested further reading
Growth hormone receptor (GHR)	• Transmembrane receptor that belongs to the class I cytokine receptor family • Senses circulating concentrations of growth hormone	• Direct effects exerted by binding of growth hormone to skeletal muscle GHRs activating JAK-STAT signalling and increasing protein synthesis through mTORC1 • Indirect effects exerted through increased production of IGF-I from the liver and consequent IGF-IR signalling in skeletal muscle	Dehkhoda (45)
Insulin-like growth factor 1 receptor (IGF-IR)	• Transmembrane receptor that belongs to the large class of tyrosine kinase receptors • Activated by IGF-I in circulation and/or IGF-I and MGF acting in autocrine/paracrine manner	• Binding of IGF-1 to IGF-IR results in phosphorylation of the intracellular adaptor proteins Shc or insulin receptor substrate 1 (IRS-1), which results in the activation of MAPK and PI3K/Akt signalling, respectively • Downstream effects are to stimulate protein synthesis largely through mTORC1 signalling	Schiaffino (46)

a form. Glycogen phosphorylase is the key enzyme in the breakdown of muscle glycogen to glucose-1-phosphate prior to its entry into the glycolytic pathways as glucose-6-phosphate. Therefore, the regulation of glycogenolysis within contracting skeletal muscle is partly regulated by signals arising outside the muscle in the form of circulating adrenaline and noradrenaline acting through a receptor on the cell surface. Intracellular regulation during exercise is also likely to occur via changes in sarcoplasmic Ca^{2+} concentrations, and greater accumulation of ADP, AMP and inorganic phosphate (P_i) (34).

Finally, there are **other cellular sensors/receptors** that sense signals other than small molecules or the input from the nervous and endocrine systems. Such sensors measure stretch/length, tension/force, electrical potentials, and interactions with proteins and molecules on the surface of other cells. In the context of skeletal muscle, one example of such receptors is a class of proteins known as 'mechanoreceptors', with their sensory role currently being investigated as 'mechanosensory' regulation of signal transduction (also 'mechanotransduction'), especially in the context of resistance exercise and the regulation of muscle protein synthesis MPS (12). The process of MPS and the marked increase in the rate of MPS during recovery from exercise, and its augmentation by the presence of aminoacidemia through protein ingestion, is considered to be central to inducing muscle hypertrophy in response to resistance exercise training (35). In other words, repeated transient increases in MPS through exercise and appropriate nutrition lead to the accumulation of myofibrillar proteins, and thereby increased size in the trained muscle (see **Chapter 8**). In the acute regulation of MPS, one example of mechanotransduction involves focal adhesion kinase (FAK) proteins, a class of transmembrane receptors that act as protein tyrosine kinases. FAK proteins are key elements for the transmission of contractile force through the skeletal muscle architecture, and thereby can also act as sensors of contractile activity. High force contraction can result in conformational changes and activation of FAK phosphotransferase activity, which activates MPS through mechanisms related to the mechanistic target of rapamycin complex (mTORC), ribosomal protein/p70 S6 kinase (S6K1) and several downstream pathways that can be both mTORC-dependent and -independent (12). For more details on mTORC and other potential mechanosensors for resistance exercise, see **Chapter 8**.

These three examples illustrate some of the most important sensor proteins in skeletal muscle in an exercise context, but many others exist and continue to be discovered. An overview of selected sensor proteins and their relevance to exercise-induced signal transduction is provided in **Table 7.1**.

STEP 3 — 'SIGNAL TRANSDUCTION': THE TRANSDUCTION OF EXERCISE-INDUCED SIGNALS

In Step 3, the exercise stimulus, consequent homoeostatic perturbations and associated signals interact with cellular sensors resulting in the activation and/or repression of **signal transduction** pathways that amplify and/or dampen the initial signals. Much of the focus of research on this step has been on 'classical' signal transduction (i.e. protein-protein interactions, posttranslational modifications such as phosphorylation and acetylation, and protein translocation), particularly as it relates to modulating the activities of downstream transcriptional and translational regulators (**Figure 7.3**) (8, 9, 14). These regulators, or effector proteins, and their associated processes, constitute Step 4 of the model as described below. Prior to that discussion, given that signal transduction is the 'engine room' of the model, it is pertinent to describe to classical features of intracellular signal transduction in more detail.

Signal transduction pathways compute and convey the sensed information using several mechanisms, including:

- **protein-protein interactions:** binding between proteins and other molecules allows information to be conveyed from one protein to the next;

Table 7.2 Protein modifications, enzymes that catalyse such reactions and the amino acids (residues) within a protein that are modified

Modification (Chemical group that is added or removed)	Modifying enzyme(s)	Amino acids (Residues) modified
Phosphorylation, dephosphorylation (PO_4^{3-} or P_i)	Kinases, phosphatases	Ser, Thr, Tyr (all have an OH group)
Ubiquitination • **the addition of an 8.5 kDa peptide termed 'ubiquitin'**	Ubiquitin ligases and deubiquitinases	Lys
Sumoylation • **the addition of a 12 kDa peptide termed 'Sumo'**	Sumo ligases and Sumo-specific proteases	Lys
Acetylation and deacetylation (CH3CO)	Acetylases and deacetylases	Lys
Methylation and demethylation (CH_3)	Methyltransferases and demethylases	Lys, Arg
Glycosylation • **the addition of a glycan such as O-GlcNAc to a protein**	Glycosyltransferases	Ser, Thr, Asn
Fatty acylation and prenylation (myristate, palmitate, farnesyl and geranylgeranyl)	Fat-adding and -removing enzymes (e.g. prenylases)	Small amino acids such as Gly or Cys

- **protein modifications:** covalent modifications (including phosphorylation, acetylation, glycosylation and ubiquitination among others – see **Table 7.2**) change the activity of proteins by causing changes in the shape of the protein;

- **translocation:** movement of signal transduction proteins between the nucleus, cytoplasm, membrane, organelles and from inside to outside the cell and vice versa allows signals to be transported through intracellular spaces;

- **synthesis and degradation:** changes in the concentration of signalling proteins can amplify or terminate intracellular signals.

The above mechanisms are linked such that several of the individual mechanisms contribute to each step of signal transduction. For example, protein-protein interaction is necessary for one protein to bind and modify the next. The change in protein shape due to modification may then expose a localisation signal that causes the protein to translocate. In some cases, posttranslational modification causes translocation to the nucleus and affects gene transcription. In the case of ubiquitination, modification causes translocation to the proteasome where the protein is degraded, and this terminates the signalling event. For the sake of clarity, we will discuss the transduction processes individually in more detail below with specific examples of molecular regulation in exercising skeletal muscle.

Protein-protein and other molecule interactions

Proteins need to be in contact with each other so that they can modify one another and convey information. Furthermore, many cellular functions rely on multimeric protein complexes, such as AMPK being a heterotrimeric complex as mentioned above. Another example of a multimeric protein complex is mTORC, which consists of several proteins and is a key regulator of ribosomal translation implicated in nutrient-, growth factor- and exercise-induced MPS. In fact, mTORC can assimilate as either mTORC1 or mTORC2 depending on conformation and composition of the various proteins that make up each complex, with each complex then serving somewhat different roles in cellular metabolism.

Protein-protein interactions are often based on relatively weak chemical bonds and generally involve protein-binding domains, which function like the interfaces of 'Lego' bricks. One such example is WW domains ('W' is the single-letter code for the amino acid tryptophan) on some proteins interacting with proline-rich motifs (e.g. PPxY) on other proteins. Another example of protein-protein interactions is that in several cases, transcription factors, which are proteins that bind to DNA to enhance gene transcription, need to join together in order to recognise and bind DNA sequences. For example, the peroxisome proliferator-activated receptor γ coactivator 1α (PGC-1α) acts as a transcriptional coactivator through recruitment, binding and coregulation of multiple proteins known as transcription factors, including nuclear respiratory factor 1 (NRF-1), NRF-2, myocyte enhancer factor 2 (MEF2), estrogen-related receptor α (ERRα) and mitochondrial transcription factor A (TFAM), which regulate skeletal muscle gene expression, particularly the expression of metabolic and mitochondrial genes in response to exercise (47). PGC-1α has often been considered a 'master' or 'hallmark' regulator of mitochondrial biogenesis, and therefore central to exercise training-induced adaptations in skeletal muscle and is explored in more detail in **Chapter 9**. While we use PGC-1α and its associated transcriptional coregulators as an example of protein-protein interactions in signal transduction, these factors are also key examples of effector proteins in the regulation of transcriptional processes described in Steps 4 and 5 below. Their activity as effector proteins is regulated by upstream signal transduction pathways, and thus they serve dual roles as signalling molecules and effector proteins, which serves as a useful illustrative example of the overlapping nature of the steps, factors and processes in the molecular response to exercise.

Protein modifications

Signalling proteins are modified by the addition or removal of 'side groups' to specific amino acid residues. This usually occurs via strong covalent bonds and changes the shape (or conformation) of the protein. Several small chemical groups can be attached to different amino acids of a protein. Often, a particular modification may be prerequisite to the next, and several residues may need to be modified in a sequential manner to change the function of a given protein. There are probably more than 200 different types of protein modification, although not all of them have well-described roles in signal transduction. We have listed the most commonly studied protein modifications, the enzymes that catalyse these modifications and the amino acids that are modified in **Table 7.2**.

The most studied protein modification is phosphorylation. In the 1950s, Edmond Fischer and Edwin Krebs were the first to make the link between the phosphorylation of a protein and stimulation of its enzymatic activity. Their work demonstrated that reversible protein phosphorylation is a key regulatory mechanism for the enzymatic activity of the aforementioned glycogen phosphorylase, and was awarded the Nobel Prize in Physiology or Medicine in 1992. Proteins are phosphorylated on serine (Ser) or threonine (Thr) residues by Ser/Thr **kinases** (such as AMPK) and on Tyrosine (Tyr) residues by Tyr kinases (such as FAK) (see **Table 7.2**). Ser, Thr and Tyr contain a hydroxyl (OH) moiety to which an inorganic phosphate taken from ATP is bound: ATP + protein ↔ ADP + protein-P_i. Note that P_i stands for inorganic phosphate or PO_4^{3-}, which is not just a phosphorus atom. At physiological pH, P_i is a negatively charged group and this modification changes the conformation (shape) of the protein by pushing away other negative charges and attracting positive charges on amino acids. Such conformational changes explain why phosphorylation affects the function of a protein, which is often the activity of an enzyme.

At least one-third and possibly up to one-half of all proteins are estimated to be phosphorylated, which gives an indication of the importance of this modification as a regulatory mechanism. Indeed, over 500 human genes encode protein kinases, which represents about 2% of the entire human genome. Phosphorylation is a reversible modification and phosphorylated proteins can be dephosphorylated by serine/threonine or tyrosine **phosphatases**. The number of tyrosine kinases is roughly the same as the number of tyrosine phosphatases. However, the number of serine/threonine phosphatases (~40) is much less than the number of known serine/threonine kinases. This difference in the number of phosphatases to kinases is likely due to their different mechanisms of action.

Kinases typically are highly selective, which is based on their ability to recognise a particular amino acid motif on the phosphorylated protein. In contrast, phosphatases can usually dephosphorylate many proteins, but they target individual proteins for dephosphorylation by interacting with regulator subunits. For example, protein phosphatase-1 (PP1) interacts with the PP1 regulatory subunit 3A to specifically dephosphorylate muscle glycogen phosphorylase.

Translocation of signal transduction proteins

Signal transduction also depends on the controlled movement of signal transduction proteins within the cell. One of the most important transport events is the bidirectional shuttling (i.e. translocation) of signal transduction proteins between the cytosol and the nucleus. In some cases, such movement depends on the activation of a nuclear localisation signal or sequence (NLS) on a protein. NLSs are recognised by proteins that transport protein cargo through nuclear pores from the cytosol into the nucleus of a cell. Usually, the activation of NLS involves protein modification or a change in protein-protein interaction, which exposes the NLS. For example, NF-κB is bound to its inhibitor, IκB, when it is in the cytosol. This is because IκB masks the NLS of NF-κB, which prevents it from transiting into the nucleus. In a similar manner, binding of class IIa histone deacetylases (HDACs) with the chaperone protein 14-3-3 masks the NLS and exposes the nuclear export sequence (NES) resulting in nuclear export and cytosolic retention of HDAC4. Phosphorylation of three serine residues (Ser^{246}, Ser^{467} and Ser^{632}) plays a key role in modulating HDAC4 translocation by increasing 14-3-3 binding and leading to nuclear export and the de-repression of gene transcription.

Class IIa HDACs represent a particularly good example of the complexity of signal transduction in response to exercise. HDACs regulate gene transcription by contributing to the balance between acetylation and deacetylation status of histones and consequent effects on chromatic architecture as described later in this chapter, and in more detail in **Chapter 6**. As described in **Table 7.2**, deacetylation via the enzymatic activity of HDACs is a posttranslational modification of lysine residues on target histone proteins, and largely results in transcriptional repression. Dynamic nucleocytosolic shuttling as described above is a fundamental mechanism regulating the function of class IIa HDACs, and is itself regulated by posttranslational modifications such as serine phosphorylation by upstream kinases such as AMPK and CaMKII, which modulate HDAC binding with the 14-3-3 family of proteins. However, class IIa HDACs can also exert transcriptional repression by direct interaction and inactivation of specific target transcription factors, in addition to recruitment of several distinct corepressors and/or protein-modifying enzymes that switch target transcription factors to their inactive form. Therefore, several features of classical signal transduction (protein-protein interactions, protein modifications and translocation) are concurrent and interdependent in the regulation of HDAC function. Lastly, a single session of aerobic exercise has been demonstrated to increase class IIa HDAC phosphorylation, increase nuclear export, increase histone acetylation and reduce HDAC-mediated repression of exercise-responsive genes such as GLUT-4 and PGC-1α by reducing HDAC-MEF2 protein interactions, and therefore class IIa HDACs are seen as a central component of the molecular response to exercise (19).

Synthesis, degradation and stabilisation of signalling proteins

The last of the classical signal transduction mechanisms relevant to exercise is the modulation of transduction by changing the synthesis or degradation rates and/or stabilisation of signal transduction proteins. This concept can be illustrated by the example of the regulation of hypoxia-inducible factor (HIF) because of the reduction in the intracellular partial pressure of oxygen (PO_2) that occurs during exercise, covered in detail in **Chapter 11**. Briefly, HIF is a heterodimeric transcription factor composed of two subunits, HIF-1α and HIF-1β. Under normoxic conditions, hydroxylation of HIF-1α occurs by prolyl hydroxylase (PHD) enzymes, and triggers degradation of HIF-1α (39). PHDs act as sensors of cellular oxygen tension, such that during hypoxia or conditions of reduced PO_2, the hydroxylase activity of PHD enzymes is inhibited, allowing stabilisation of HIF-1α. Consequently, HIF-1α translocates to the nucleus to form an active complex with HIF-1β, resulting in the transcription of target genes involved in erythropoiesis, angiogenesis, glycolysis and energy metabolism (39). Unsurprisingly, HIF-1α protein abundance is increased during acute aerobic exercise, accumulates in the nucleus and shows enhanced DNA binding as part of its role in transcriptional regulation (48).

Another example specific to the role of protein degradation in the termination of signal transduction is with the myogenic regulatory factors MyoD, Myf5, myogenin and MRF4, which regulate muscle development or myogenesis (discussed in detail in Chapter 2 and Chapter 13). These signals are terminated by selective degradation of the transcription factors. For example, the MyoD signal is stopped when the ubiquitin ligase MAFbx ligates ubiquitin to MyoD, and marks MyoD for degradation by the 26S proteasome (49). Like the example of PGC-1α above, this example encompasses both signal transduction via posttranslational modification and the activity of transcriptional regulators acting as effector proteins described in the next steps below.

STEP 4 – 'EFFECTORS': EFFECTOR PROTEINS AND PROCESSES REGULATING EXERCISE-INDUCED ADAPTATION

The first three steps of the signal transduction model ultimately involve the activation and/or repression of an array of intracellular pathways. In turn, these pathways are coupled to a myriad of downstream effector proteins (**effectors**) involved in the regulation of the transcriptional and translational processes that constitute Steps 4–6 of the model. The consequences of these last three steps are changes in abundance, regulation and/or maximal activity of key proteins involved in energy provision, the remodelling of cellular components such as contractile proteins and the extracellular matrix, and the biogenesis of organelles such as ribosomes and mitochondria (8–11). In other words, those changes are the final processes by which exercise changes our skeletal muscles via adaptation. For the purposes of this chapter, we will consider the regulation of transcription and translation as *acute* responses to exercise downstream of the signal transduction pathways described so far. A table of selected regulators of transcription and translation that have been associated with exercise training-induced changes in skeletal muscle is provided in **Table 7.3**, many of which are described in more detail in **Chapters 8** and **9** for resistance and endurance exercise, respectively, and some of which have been used as illustrative examples throughout this chapter.

Historically, the activity of transcription factors, coactivators and repressors was the subject of much interest. This interest was driven by the observation that changes in tissue form and function as an adaptive process are driven by acute and chronic changes in mRNA abundance prior to changes in proteins that provoke gradual structural remodelling and long-term functional adjustments (9). In other words, the diffusible gene copies (mRNA) produced provide the message for the instruction of muscle tissue remodelling via translation and assembly of the encoded proteins. Based upon this relationship, it was hypothesised that the systematic exploration of differences in mRNA abundance relative to functional adjustments arising from acute exercise sessions and chronic exercise training would reveal the mechanisms underlying muscle adaptation. The transcription of exercise-responsive genes remains central to the model of exercise training-induced adaptations in skeletal muscle (50), and in the last decade the discovery of novel roles for pre- and post-transcriptional processes particularly as they relate to epigenetics, namely histone modifications (19), DNA methylation (20) and microRNA (miRNA) (51), has added further and intriguing complexity to the model (see **Chapter 6**).

However, the regulation of protein translation has been viewed as having increasing importance in this step, primarily given the repeated observation of acute activation by exercise of canonical pathways involved in protein synthesis, namely mTORC, S6K1 and several downstream targets involved in various steps in translation (12, 52). The maintenance of cellular homeostasis is a dynamic process, and therefore to optimise survival, a cell must be able to rapidly respond to intrinsic or extrinsic changes in its environment. Thus, proteins that control these responses must be able to increase or decrease in short time frames, by altering the rate of translation of existing mRNA rather than necessary relying on

Table 7.3 Selected effector proteins and processes proposed as regulators of exercise training-induced adaptations in skeletal muscle

Protein	Function & regulation
Transcriptional regulators	
Estrogen-related receptors (ERRs)	• ERRα and ERRγ are constitutively-active orphan nuclear receptors with high expression in oxidative tissues
	• regulate OXPHOS, fatty acid oxidation and mitochondrial DNA genes, and angiogenesis
	• ERRα is regulated by PGC-1α; necessary for PGC-1α-induced mitochondrial biogenesis and VEGF-induced angiogenesis
Forkhead transcription factors, O-box subfamily (FOXOs)	• regulated by PTMs, including acetylation and phosphorylation
	• FOXO1 regulates genes involved in energy metabolism and shifts in fuel selection, often in concert with ERRα
	• FOXO3 drives muscle atrophy through upregulation of the muscle-specific ubiquitin ligases MAFbx and MuRF1
Histone acetyltransferases (HATs) and deacetylases (HDACs)	• antagonistic enzymes regulating gene expression by altering histone acetylation status
	• transcriptional activators (HATs) and repressors (HDACs) by catalysing acetylation and de-acetylation, respectively
	• rate of transcription determined by balance between HATs and HDACs
Hypoxia-inducible factor (HIF)	• heterodimeric transcriptional factor composed of HIF-1α and HIF-1β
	• normoxia: HIF-1α is hydroxylated PHDs and degraded
	• hypoxia: PHD activity inhibited allowing stabilisation of HIF-1α, activation in concert with HIF-1β
	• activated HIF-1 induces transcription of target genes involved in erythropoiesis, energy metabolism and angiogenesis
Mitochondrial transcription factor A (TFAM)	• nuclear-encoded transcription factor essential for the replication, maintenance and transcription of mitochondrial DNA and normal mitochondrial function
	• expression regulated by PGC-1α coactivation of NRF-1 and -2 binding to its promoter
Myocyte enhancer factor (MEF) 2	• MADS-box transcription factor involved in muscle remodelling through binding sites present in promoters of a wide range of metabolic and myogenic genes
	• association with class II HDACs represses MEF2 activity
	• works in concert with PGC-1α to enhance transcription
Nuclear respiratory factor (NRF) 1 and 2	• nuclear-encoded transcription factors linked to regulation of many mitochondrial genes
	• NRF-1 regulates genes of all five electron transport chain complexes
	• regulates mitochondrial DNA transcription through TFAM activation
	• coactivation by PGC-1α with NRF-2 and ERRα regulates their own and NRF-1 expression
Peroxisome proliferator-activated receptors (PPARs)	• family (PPARα, β/δ and γ) of ligand-dependent nuclear hormone receptors
	• regulate transcription by dimerising with the retinoid X receptor and binding to PPAR response elements (PPREs)
	• regulatory roles in lipid metabolism and whole-body fuel turnover (Continued)

Protein	Function & regulation
PPARγ coactivator (PGC) family	• family of transcriptional coactivators: PGC-1α, PGC-1β and PGC-1α-related coactivator (PRC)
	• regulated by PTMs in response to environmental cues governing pathways of thermogenesis, gluconeogenesis, muscle differentiation and cell growth
	• orchestrators of cellular metabolism through network of transcriptional activators and repressors
	• dominant effects observed in nucleo-mitochondrial regulation of mitochondrial biogenesis
	Translational regulators
Ribosomal protein/p70 S6 kinase (S6K1)	• substrate of mTORC1 whose kinase activity is increased by mTORC1 activation leading to phosphorylation of downstream targets, including ribosomal protein S6 (rpS6)
	• activation of S6K1 is proposed to result in increased translation initiation through eIF4B, and translation elongation through eEF2K
	• phosphorylation of S6K1 and rpS6 are often used as indicators of mTORC1/anabolic signalling
Eukaryotic initiation factors (eIFs)	• eIFs are group of factors regulating translation initiation through the recognition of mRNA, priming of ribosomal subunits and the creation of pre-initiation and initiation complexes
	• eIF3 and eIF4 are bound to the signalling proteins S6K1 and 4E-BP1, respectively, which are phosphorylated after mTORC1 activation, resulting in the detachment of eIF3 and eIF4 from S6K1 and 4E-BP1 and steps towards the assembly of ribosomes and translation initiation
Eukaryotic translation initiation factor 4E-binding protein 1 (4E-BP1)	• translation initiation factor that is another major target of phosphorylation by mTORC1
	• removal of 4E-BP1 from eIF4 allows the pre-initiation complex to be recruited to the mRNA strand before translation initiation
	• phosphorylation of 4E-BP1 is often used as an indicator of mTORC1/anabolic signalling
Eukaryotic elongation factors (eEFs)	• eEFs aid the translation of mRNA to protein by enhancing the recruitment of loaded transfer RNAs to the ribosome and accelerating the shift of the ribosome to the next codon once a peptide bond has been formed
	• the phosphorylation status of eEF2 is primarily regulated by eEF2 kinase (eEF2K), which is phosphorylated and inhibited by S6K1 (downstream of mTORC1 activation), thereby activating eEF2
	• may also be regulated by mTORC1-independent mechanisms

exercise-induced changes in mRNA abundance. Once protein synthesis has been stimulated above basal rates by a session of exercise, translation of specific proteins is primarily dependent on translational efficiency and capacity. Both increased translational efficiency (protein synthesis per unit of RNA) and elevated translational capacity (total RNA content per unit of tissue) because of ribosome biogenesis have emerged as important regulators of the adaptive response to exercise (10). Ultimately, regardless of whether there is a change in mRNA abundance (increase or decrease) in response to exercise, phenotypic and functional consequences of exercise are entirely dependent on protein, be that in the form of altered abundance, maximal activity, sensitivity to the regulatory mechanisms determining activity or function and so on.

Step 5 – 'Transcription': Regulation of gene expression

The full details of the transcription of a gene and the regulation of this process are beyond the scope of this chapter and are covered in detail in **Chapter 3**, but briefly here, important aspects of the control of transcription include:

- the unwinding of the tightly packed DNA, which is termed chromatin remodelling, which makes genes accessible;

- the recruitment of RNA polymerase II to the start of a gene;

- the rate at which RNA polymerase II transcribes the DNA of the gene into RNA and

- the termination of transcription and recycling of RNA polymerase II.

Most, if not all, aspects of the control of transcription have been demonstrated to be altered in skeletal muscle by exercise, and we will now discuss some key events during transcriptional regulation, and the impact of exercise or other signals on those events.

After a stretch of DNA has opened up by chromatin remodelling, it can be accessed by transcription factors to affect the expression of genes. A stretch of DNA called the core promoter is found close to the start of genes and serves as a docking site for RNA polymerase II as well as proteins known as basic transcription factors that together form the transcription pre-initiation complex. The assembly of the transcription pre-initiation complex is sufficient for low rates of gene transcription. However, exercise and otherwise-activated signal transduction pathways modulate transcription via sequence-specific transcription factors that bind proximal promoter and enhancer regions and greatly enhance the rate of transcription.

The proximal promoter is a region of DNA up to a few hundred base pairs upstream from the core promoter, whereas enhancer regions can often be hundreds of thousands of base pairs away from the transcription start site. Loops in the DNA bring distal enhancer regions close to the proximal promoter. Indeed, it is very rare for an enhancer to affect genes that are adjacent to it in the DNA sequence. Current predictions suggest that there are ~70,000 regions with promoter-like features and ~400,000 regions with enhancer-like features in the human genome (53), which gives an indication of the complexity of this process. To give an example from skeletal muscle, the myogenic regulatory factor MyoD binds to >20,000 DNA sites in muscle cells (54).

Additionally, estimates suggest that there are ~1,800 DNA-binding transcription factors in the human genome (55). Of these, the regulatory transcription factors have a DNA-binding domain that recognises a specific DNA motif. Some common types of DNA-binding domain include the C2H2 zinc-finger, homeodomain and basic helix-loop-helix (55). Often, it is necessary for transcription factors to form homo- or heterodimers (protein-protein interactions) in order to create a correct DNA-binding motif. These transcription factors are further regulated by the binding of co-factors, such as the abovementioned PGC-1α, and its interaction with the transcription factors NRF-1, NRF-2, MEF2, ERRα and TFAM in the regulation of skeletal muscle gene expression. Moreover, a single session of aerobic exercise alters the DNA-binding activity of a variety of transcription factors, including MEF2 (56), NF-κB (57), and NRF-1 and NRF-2 (58).

Chromatin remodelling or epigenetic regulation: the opening up or packaging of DNA

Epigenetic regulation of gene expression is briefly introduced in **Chapter 3** and a full chapter in this textbook is dedicated to epigenetics of exercise (**Chapter 6**), but given the importance of chromatin remodelling and modulation of DNA methylation as part of the signal transduction model, we will briefly cover the topic in this chapter too. The DNA in the human genome has been estimated to be almost 2 m long. In order to fit into a tiny nucleus of only 10 micrometres, it must be very tightly packaged. This DNA packaging is achieved by wrapping DNA around complexes which are built from eight histone proteins. DNA together with histone complexes is termed chromatin and the DNA wrapped around one histone complex is called a nucleosome.

The tight packaging of DNA in chromatin must be unravelled before a gene can be transcribed. The packaging and unpackaging of DNA are known as chromatin remodelling and histone modifications are key mechanisms in the regulation of this process. Mapping of packaged and unpackaged DNA on a genome-wide scale has revealed that histone tail modifications are cell-specific and mark genes, transcription start sites and stretches of regulatory DNA, via which gene expression is regulated (53). Indeed, signal transduction pathways modulate chromatin remodelling by methylating (CH_3, methyl group), acetylating (CH_3CO, acetyl group) and phosphorylating histone proteins, especially in the tail regions of histones H3 and H4 (19, 20). The enzymes that catalyse these modifications include histone methyltransferases and histone demethylases as well as histone acetyltransferases and HDACs. The resultant modifications are abbreviated stating first the histone number, second the amino acid which is modified and finally the type of modification (ac stands for acetylation, me1, me2, m3 for methylation, dimethylation and trimethylation, respectively). For example, H3K27me3 refers to the trimethylation of lysine 27 (K is the one-letter abbreviation for lysine) of histone 3.

Epigenetic modifications are one of the many regulatory events that contribute to the molecular response and adaptation to exercise. Specifically in relation to acute aerobic exercise, transient changes in histone modifications (59), and gene-specific changes in DNA methylation status (60), are rapidly induced, and precede changes in gene expression in the post-exercise period (59, 60). For example, exercise can induce an increase in histone acetylation by inhibition of the activity of HDACs (59), which typically results in increased accessibility of chromatin and activation of transcription. Similarly, increased accessibility of chromatin and activation of transcription occurs with hypomethylation of GC-rich consensus-binding sequences of DNA in specific genes and has been observed for various exercise-responsive genes (60).

More recently, it has been observed that the same pattern of hypomethylation induced by acute exercise occurs in response to both resistance exercise (32, 33) and change-of-direction repeated sprint exercise (61). At the genome-wide level, there was a considerable overlap between genes that were transiently hypomethylated soon after exercise and increases in mRNA abundance in the several hours of recovery that follow (32, 33). Importantly, many of the genes identified as being hypomethylated are those that are established as important for the response and adaptation to different types of exercise, but there are also identified novel epigenetically regulated genes associated with the exercise response and the epigenetic memory of skeletal muscle in response to training, detraining and retraining (32, 33, 61).

The molecular networks that link exercise to epigenetic regulation are presently a matter of intense research interest, with a large number of enzymes regulating acetylation/deacetylation and methylation/demethylation. Many of these enzymes are linked to metabolic pathways and signal transduction pathways that are established as being exercise-responsive (e.g. AMPK and CaMKII) and again therefore link exercise-induced signal transduction in skeletal muscle to regulators of epigenetic modifications (19, 20).

Post-transcriptional regulation of mRNA

The primary transcript produced by RNA polymerase II undergoes post-transcriptional processing to become mRNA that can be translated into protein. Most human genes consist of multiple exons, which is the part of the gene that encodes the protein interspersed by introns or intervening sequences. Introns are removed by spliceosomes and, depending on which exons are retained, different splice variants of the gene can be created. A well-known example relevant to exercise physiology is the alternative splicing of the IGF-I gene to create mechano-growth factor (MGF, also referred to as IGF-IEc in humans or IGF-IEb in rodents), which was discovered by Geoffrey Goldspink's group in the late 1990s (62). The activity of spliceosomes is in part regulated by proteins that recognise and mark different splice sites, but as yet it is unclear as to how exercise regulates alternative splicing.

Post-transcriptional processing can also be modulated by small RNA species, typically ~21–26 nucleotides long, known as small interfering RNA (siRNA) and miRNA (63). miRNAs and siRNAs are produced in different ways but have similar functions on selective mRNA degradation. After their synthesis, miRNA and siRNA form part of the RNA-inducing silencer complex (RISC). The miRNA and siRNA confer specificity to the RISC complex by binding complementary sequences in their target mRNA, which results in either the degradation of the target mRNA or inhibition of mRNA translation, which ultimately results in changes in protein abundance. More than 2,000 miRNA transcripts have been so far identified; it is estimated that up to 30% of human gene transcripts may be affected by this type of regulation. miRNA-induced alterations in protein abundance are subtle (often less than 10% change in abundance), but a single miRNA can alter the abundance of tens to hundreds of diverse proteins, meaning that changes in miRNA in response to stimuli such as exercise can have important implications when it comes to adaptive changes. To add further complexity, while one miRNA can have many targets, simultaneously one transcript may be targeted by several miRNAs. Changes in miRNA abundance occur in response to both aerobic (64) and resistance (65) exercises, and so miRNA may be involved in modulating the adaptive response to exercise training (51) (alterations in miRNA with endurance and resistance exercise are discussed in detail in Chapter 6).

How should we interpret a measured change in mRNA abundance in response to acute exercise?

The example of post-transcriptional regulation of mRNA illustrates an important point for the interpretation of acute responses to exercise: gene expression can be controlled at various points beyond transcription, so the magnitude by which a protein will change in abundance or function in response to an acute stimulus cannot always be predicted from the magnitude of change in mRNA abundance (66). Moreover, relative mRNA abundance does not always accurately reflect the abundance of a target protein, and can give no information regarding features such as their posttranslational modifications or enzymatic function where relevant (67). Gene expression in its broadest sense therefore refers to a process encompassing the activation of transcriptional regulators to the synthesis of a functional protein (68). Increments in mRNA abundance are often referred to as increases in gene expression, but this is technically incorrect, because increased gene expression and its associated phenotypic/functional manifestations do not take place until there is an increase in the concentration of the protein encoded by the gene.

A well-established example is in yeast wherein the correlation between mRNA and protein expression patterns was insufficient to predict protein abundance from quantitative mRNA data (69). Indeed, for some genes while the relative expression of respective mRNAs was similar, the abundance of the respective protein targets varied by more than 20-fold. Conversely, steady-state abundances of certain proteins were observed; yet, the respective mRNA abundance varied by as much as 30-fold in the same time frames. Therefore, when interpreting alterations in transcriptional regulation, although mRNA abundance is both a fundamental precursor and a determinant of protein abundance, this relationship is neither simple nor linear. The study of a model organism such as yeast (*Saccharomyces cerevisiae*) allows for the creation of hundreds of data points with serial samples, provided a high degree of temporal resolution for exploring mRNA and protein kinetics. Studying similar concepts in human skeletal muscle in molecular exercise physiology is a much more challenging proposition and is ultimately often limited by what is feasible with participant numbers and sampling timepoints. The latter is a point worth emphasising; the temporal resolution provided by muscle biopsies is a limitation because muscle biopsies only provide a static picture of a dynamic process, in addition to the half-lives and kinetics of change in mRNA and protein being quite variable. By way of example, the expression of PGC-1α mRNA is robustly increased for ~1 to 8 hours after a session of aerobic exercise, with a peak usually between 2 and 4 hours post-exercise, whereas the half-life of PGC-1α protein is relatively short (~2.3 hour), but can be stabilised by phosphorylation by upstream kinases such as AMPK and MAPK, leading to a tripling of its half-life and an increase in observed PGC-1α protein abundance (70). Therefore, any consideration for the measurement of PGC-1α mRNA and protein in human skeletal muscle would need to be cognisant of these features, especially if the aim is measurement of both mRNA and protein abundance in response to acute exercise.

Aside from acute exercise, one might also consider the examples of changes in mRNA and protein abundance in response to altered loading states or exercise training. For example, a gene of particular interest in recent years has been the E3 ligase UBR5 (E3 ligases and their actions are discussed later in this chapter). When UBR5 expression is measured at the level of its DNA methylation, mRNA and protein levels, there can be strong agreement between mRNA and protein expression patterns across several species and in response to chronic hypertrophy and atrophy stimuli that produce differential expression patterns of UBR5 (32, 71, 72). However, care still needs to be taken in identifying the timepoints for detecting changes in mRNA and protein abundance when large proteins (like UBR5 which is over 300 kDa) perhaps require more time to be translated from mRNA to protein after an acute bout of exercise. Another example is a study comprising 14 consecutive days of aerobic exercise training for 60 minutes per day at ~80%VO$_{2peak}$, with skeletal muscle biopsies on the morning following (+16 hours after exercise) the first, third, seventh, tenth and fourteenth training sessions in order to explore mRNA and protein kinetics (16). For genes such as ERRα and cytochrome c, the expected pattern emerged whereby measurable increases in mRNA abundance were present after just one session of exercise and maintained throughout the training period, whereas increases in protein abundance were not observed until after the third and seventh training sessions, respectively. Conversely, an increase in PGC-1α protein abundance occurred before the training-induced increase in mRNA abundance, whereas GLUT4 protein abundance increased in the absence of any measurable increase in mRNA abundance during training. These results underscore the need to consider gene-specific mRNA and protein kinetics, and interested readers are referred to the work of Booth and Neufer (2012) (66) for a more detailed discussion of these aspects of gene expression.

Awareness of these aspects is also important when interpreting the outcomes of studies that measure gene expression. To give one specific example, an often-studied target in post-exercise muscle biopsy samples is pyruvate dehydrogenase kinase 4 (PDK4). PDK4 is of interest because of its role in the regulation of the pyruvate dehydrogenase (PDH) complex. The PDH complex is the key step regulating the complete oxidation of glucose proceeding from glycolysis to the tricarboxylic acid cycle. Inactivation of this complex via phosphorylation by PDK4 attenuates the conversion of pyruvate to acetyl CoA, resulting in allosteric inhibition of glycolysis and suppression of glucose oxidation leading to the diversion of imported glycogen precursors to storage (73). The mRNA abundance of PDK4 is rapidly and robustly increased in response to acute exercise and fasting, but it is incorrect to say that such changes in PDK4 mRNA abundance lead to 'sparing of glucose' or some similar metabolic consequence. Simply put, by measuring only mRNA, there is no evidence that this change results in altered PDK4 protein content nor is there any evidence that its enzymatic activity, i.e. leading to phosphorylation of the PDH complex, is increased. A more conservative statement is that the transcriptional regulation of PDK4 was altered (by exercise, for example) and this response may form part of a coordinated metabolic response that contributes to glucose sparing and increased fat oxidation.

More broadly speaking, the observation that a specific mRNA was significantly changed during and/or after acute exercise indicates a regulatory effect of muscle contractions upon the transcriptional regulation of that gene. However, the finding that the protein content does not subsequently change may demonstrate that acute exercise does not necessarily produce a *detectable* change at the level of protein translation; albeit the aforementioned issues with the temporal resolution provided by muscle biopsies, the half-lives and kinetics of change in mRNA and protein, as well as issues with the sensitivity of protein measurement (**Chapter 2**), must be factored into the final interpretation of the data. These results may simply reflect the transient nature of changes in mRNA and protein abundance, whereby the sampling points for muscle biopsies may conflict with measurable and functional changes in mRNA and protein due to variations in respective half-lives or stability.

Step 6 – 'translation': regulation of mRNA translation to synthesise proteins

Mature mRNA is translated into protein by ribosomes. Ribosomes consist of a small 40S and a large 60S subunit of rRNA (ribosomal RNA) that join together to create an 80S ribosome. Ribosomes are controlled by ribosomal proteins, which are regulated especially by the mTOR signalling pathways. In the late 1990s, it was noted that translational regulators were modulated in hypertrophying muscles (74), and subsequent research has shown that activation of translation via the mTOR pathway is a key mechanism by which exercise increases MPS (in a later section, we provide more discussion on the role of MPS in aerobic and resistance exercise adaptations). Protein translation occurs in three stages:

1. initiation, where the ribosome and mRNA are assembled;
2. elongation, where the mRNA is read by the ribosome and translated into an amino acid chain; and
3. termination.

While all three processes are tightly regulated, the rate-limiting step for translational control occurs at the initiation step. Translation initiation is a multi-step process regulated by eukaryotic initiation factors (eIFs) and culminates in formation of the 80S initiation complex. Prior to translation, the subunits of the ribosome are separate and the 40S subunit must be 'primed' by binding of a transfer RNA (tRNA). The tRNA contains the complementary sequence for the AUG start codon of mRNA and is also attached to the amino acid methionine. Proteins are synthesised beginning from the amine N-terminal and ending with a carboxyl C-terminal; therefore, all proteins begin with a methionine residue, although this is often removed after translation. The priming of the 40S ribosome is regulated in part by eIF3 and creates a 43S pre-initiation complex. Next eIF4 guides the 5'-end (named 5'-cap) of mRNA into the 43S complex and the ribosome proceeds along the mRNA until the start codon is found. Other eIFs then recruit the 60S subunit to create the 80S initiation complex, and synthesis of the polypeptide chain begins.

When inactive, the initiation factors eIF3 and eIF4 are bound to the signalling proteins S6K1 and 4E-BP1, respectively. Resistance exercise, in particular, nutrients such as essential amino acids, and hormones such as insulin, all activate mTORC1, which, in turn, phosphorylates S6K1 and 4E-BP1 (and various other proteins in the mTOR signalling pathway). As a consequence, eIF3 and eIF4 detach from S6K1 and 4E-BP1 and contribute to the assembly of ribosomes.

Elongation follows the assembly of the 80S initiation complex. The process of elongation is controlled by **eukaryotic elongation factors** (eEFs) and uses the codons of the mRNA as a template to recruit the correct sequence of tRNA. Elongation proceeds in a cycle involving (i) binding of activated tRNA (i.e. tRNA bound with its respective amino acid), (ii) peptide bond formation and (iii) release of the inactive tRNA. The ribosome moves along the mRNA in this manner until a stop codon is reached, at which point the process is terminated by **eukaryotic release factors** (eRFs). Beyond the acute regulation of protein translation, the biogenesis and activity of organelles, including the ribosomes and lysosomes (autophagy, relevant to protein degradation as described later), and the importance of skeletal muscle satellite cells for adaptation (Chapter **13**), are some of the emerging areas of interest in this field (10).

Regulation of protein degradation

Signal transduction pathways also regulate the lifespan of proteins by controlling their degradation rate. Protein degradation can occur via at least three processes:

- ubiquitin-proteasome system;
- autophagy-lysosomal pathways and/or
- cytosolic proteolytic systems.

The cytosolic proteolytic systems include the caspase and calpain proteases, which are associated with apoptotic cell death and Ca^{2+}-activated proteolysis, respectively. These systems are not strongly implicated in protein turnover, so, for the remainder of this section, we will focus on the roles of the ubiquitin-proteasome and the autophagy-lysosomal pathways relevant to exercise and skeletal muscle tissue.

The majority of intracellular proteins are degraded by the ubiquitin-proteasome system (UPS), which involves ubiquitination (a protein modification mentioned in **Table 7.1**) of target proteins followed by their degradation in the 26S proteasome. Proteins are selected for degradation by attaching ubiquitin to the ε-amino group of a lysine residue, or in some cases to the N-terminus of the protein. Ubiquitin itself is a small protein with 76 amino acid residues and it is so named because it is ubiquitous in all tissues. Ubiquitination is an energy-requiring process carried out in series by three types of enzymes, named E1, E2 and E3 enzymes. Ubiquitin is first of all 'primed' by an E1-activating enzyme and then transferred to one of several E2-conjugating enzymes, which, in turn, interact with numerous E3 ligases. The actual ubiquitination of target proteins is performed by E2-E3 pairs, but it is the E3 ligase that confers specificity to the process. Two muscle-specific E3 ligases are muscle RING finger 1 (MuRF1) and muscle atrophy F-box (atrogin-1/MAFbx), which were first discovered in models of muscle atrophy (75) and are key regulators of skeletal muscle proteolysis under catabolic conditions. However, many other E3 ligases also regulate muscle protein degradation, and there are more than 600 different E3 ligases in the human genome. Ubiquitinated proteins are then digested via the 26S proteasome, which is a barrel-shaped complex built from more than 30 proteins. The role of the UPS in skeletal muscle atrophy is well-established, but whether this pathway has an important role in the adaptive response to exercise is not as clear. For example, the expression of MuRF1 and MAFbx specifically have been observed to either increase, not change or decrease in response to various forms of acute exercise (76). However, the general trend that MuRF-1 expression is induced by both acute aerobic and acute resistance exercises, and the logical interpretation that proteolysis would be involved in skeletal muscle remodelling is broadly suggestive of a role for activation of the UPS by exercise-induced signal transduction pathways.

Interestingly, recent work has identified the E3 Ubiquitin ligase, UBR5, as a novel regulator of skeletal muscle mass (32, 71, 72). Of particular note is that UBR5 activity is in contrast to MuRF1 and MAFbx in that its expression is associated with anabolism and hypertrophy, is correlated with skeletal muscle mass and muscle-specific UBR5 knockout attenuates anabolic signalling resulting in muscle atrophy (32, 71, 72). These findings suggest that the role of the UPS in the regulation of skeletal muscle adaptation is more complex than simply being a protein 'breakdown'/atrophy pathway. Instead, dual roles are likely to exist for the E3 ligases via degradation of positive and negative regulators of muscle mass resulting in muscle atrophy and hypertrophy, respectively. This is discussed in more detail in Chapter 8 covering molecular adaptation to resistance exercise.

In a similar manner to proteolysis, autophagy is likely to be involved in skeletal muscle remodelling in response to exercise. Autophagy occurs through various mechanisms that differ in the way they capture proteins or organelles and deliver them to the lysosome for degradation. Macro-autophagy involves entire regions of the cytosol or specific organelles and protein complexes being engulfed by a vacuole known as an autophagosome, which then fuses with the lysosome. Micro-autophagy involves the direct uptake of cytosolic components into lysosomes. In addition, more selective types of autophagy, known as chaperone-mediated autophagy and chaperone-assisted selective autophagy, are able to degrade specific proteins. Like the concept of protein degradation by the UPS, on first impression, autophagy may seem highly destructive, but autophagic processes are initiated by a single session of exercise (77), and in fact may be required for adaptation to exercise training (78). Conceptually, this can be thought of acute exercise activating signal transduction pathways that increase the turnover of skeletal muscle proteins (i.e. activation of both degradation and synthesis), and this increase in turnover is essential for the remodelling of cellular components such as contractile proteins and the extracellular matrix that occurs with exercise training.

INTEGRATIVE ASPECTS AND CONTEMPORARY PERSPECTIVES ON EXERCISE-INDUCED SIGNAL TRANSDUCTION

In the text above, we have described in detail the molecular mechanisms that regulate changes in transcription and translation, and how these mechanisms can be activated and/or repressed by a single session of exercise, which over time with repeated exercise can result in adaptive changes known as training effects. We have also provided a smattering of examples of the currently known regulators of these processes in skeletal muscle. We acknowledge that it is easy to get lost because of all the detail stated, and indeed some of the detail that is not stated here but described in other chapters (see **Chapters 6, 8 and 9**). For these reasons, we will now summarise the most important points, as well as aim to integrate what is known and what remains to be discovered. In fact, while there are now hundreds of proteins implicated in the regulation of adaptation to exercise, it is probable that many more remain to be discovered, and this is the focus of several recent pieces of 'omics' research described below. More detailed discussion of the specific pathways and proteins involved in adaptation to exercise in skeletal muscle is available in several other reviews that interested readers may pursue (8, 9, 12, 19, 20).

In summary, the working model for exercise-induced signal transduction and its role in adaptation to exercise begins when a myriad of perturbations to homeostasis occurs, broadly within our bodies and specifically within our muscles, including important changes in energy turnover and muscle tension. All these perturbations act as **'signals'** (Step 1) to be sensed by molecular **'sensors'** (Step 2) or sensor proteins that are then computed and conveyed by **'signal transduction'** pathways and/or networks (Step 3). Such signalling works by protein-protein binding, phosphorylation and other modifications, and signalling proteins move, for example, in between the cytosol and nucleus. Finally, the exercise-induced downstream signalling proteins regulate **'effectors'** (Step 4), which are effector proteins and processes that include transcription factors, regulators of protein synthesis and protein breakdown and can be thought of as regulators of gene **'transcription'** (Step 5) and protein **'translation'** (Step 6) that over time determine adaptations to exercise in skeletal muscle, described in later chapters.

Specificity of signal transduction pathways

The pathways of signal transduction are clearly very complex, with some pathways complementing each other's activity, and others potentially interfering with, and blunting, each other's activity. Moreover, the relative activation/repression, contribution and magnitude of the described pathways and downstream targets are largely dependent on the intensity, duration and type of the exercise stimulus, that are also dependent on other imposed environmental variables such as nutrient and oxygen availability. An obvious question might be – why is there such complexity in this regulation? One answer is that a multiple signal transduction control system has inherent redundancy (i.e. not all pathways are active or necessary at all times) and therefore, the potential advantage of allowing for fine-tuning of adaptive responses to multiple metabolic and physiological stimuli is required in the context of exercise training. In fact, redundancy and compensatory regulation are key characteristics of biological systems that act to preserve physiological responses and adaptations to a variety of challenges to homeostasis. While both aerobic and resistance exercise share the same fundamental characteristics of exercise in muscular contraction, elevated energy expenditure and disruption to homeostasis; the metabolic, mechanical, hormonal and neuronal demands of different types of exercise can be divergent and are reflected to a certain extent in the nature of adaptations to prolonged aerobic exercise training compared to resistance exercise training when performed in isolation. Therefore, one postulate of the signal transduction hypothesis (as the process that underpins adaptation) has been that the type of exercise stimulus and the contemporaneous environmental conditions are reflected in the specificity and divergence of the molecular signatures that are induced.

Early work did indeed suggest that the divergence between the broad categorisations of exercise type at the level of frequency, force and duration of contraction resulted in divergent molecular signatures in response to single sessions of aerobic versus resistance exercise. For example, Atherton et al. (2005) (79) isolated skeletal muscles from Wistar rats and subjected them to electrical stimulation that was either continuous low-frequency (10 millisecond pulse duration at 10 Hz for 180 minutes) to mimic aerobic exercise, or intermittent high-frequency (6 sets of 10 × 3 second contractions at 100 Hz over ~20 minutes) to mimic resistance exercise. A point of note here is that 'frequency' in the context of electrical stimulation does not refer to the number of contractions, but instead refers to the stimulation pulses delivered per second, which, in turn, is proportional to the force of contraction, i.e. high-frequency muscular electrical stimulation is the equivalent of high force contractions. When the molecular signature consequent to each contraction profile was determined, the results demonstrated that there was selective activation of pathways associated with anabolic signalling and muscle growth (Akt/mTOR signalling) after the resistance exercise model, and selective activation of pathways associated with mitochondrial adaptation (AMPK/PGC-1α pathways) after the aerobic exercise model. Moreover, aerobic exercise-like stimulation inhibited Akt/mTOR and its downstream targets leading the authors to postulate that an 'AMPK/Akt master switch' and the selective activation of either AMPK/PGC-1α or Akt/mTOR could explain the divergent adaptations associated with aerobic and resistance exercise training (79).

This study illustrated that it is possible to induce quite divergent molecular signatures in response to different profiles of muscle contraction that characterise the divergent exercise types. However, since that study, there has been little data from human exercise studies to support the hypothesis of a simple switch to explain specificity of training adaptation. Therefore, the concept remains somewhat theoretical that these divergent signatures are coupled to a functional outcome, i.e. the specific molecular responses are contiguous with divergent physiological and functional adaptations to the respective exercise modalities as indicated by the endurance and hypertrophy phenotypes (2). However, this is not to dismiss the concept of specificity outright; in the section below describing the practical significance of these investigations, there are more convincing studies around the specificity of signal transduction as it relates to exercise performed in hypoxia or with reduced carbohydrate availability.

'Interference effect' between signal transduction pathways

A related concept to the discussion of specificity of signal transduction is that of the interference between signal transduction pathways. When first proposed, the hypothesis was that the activation of AMPK by aerobic exercise antagonises the activation of mTORC-dependent pathways, attenuating pathways that dominate the adaptation to resistance exercise, and thereby suggesting that the conflicting molecular reprogramming that underlies the respective adaptations could produce an **interference effect** (80, 81). This hypothesis was a molecular basis to explain a phenomenon known as the 'concurrent training' or 'interference' effect that had first been observed in the early 1980s (82). Simply stated, training for both endurance and strength outcomes through concurrent aerobic and resistance exercise training can result in a compromised adaptation to resistance exercise training in terms of strength, power and hypertrophy when compared with training using resistance exercise alone. Data from elite decathletes confirm as much, suggesting that it is possible to excel at disciplines that require predominantly endurance or predominantly strength/power, but not both (83). Thus, phenomenon had been proposed to be explained by the specificity in signal transduction pathways modulated in response to aerobic compared to resistance exercise. However, in practice, the concurrent training effect is not widely observed in exercise training studies (84), nor does the inhibition of mTOR signalling by activation of AMPK provide a complete explanation of the phenomenon when it is observed (85).

This discussion also serves to highlight that the conceptualisation of aerobic and resistance exercises at opposite extremes of the continuum is useful in theory, but as stated above, it must also be said that this is too simplistic of a framework for two obvious reasons. The first reason is that ultimately there are several obvious overlaps in the molecular networks that regulate skeletal muscle adaptation to exercise, such that it is often impossible to distinguish or isolate mechanisms for divergent acute responses or adaptive changes, at least with our currently available methods and tools for investigation. In fact, in the time since the work of Atherton et al. (2005) (79), many studies have failed to observe divergent molecular signatures in response to aerobic and resistance exercises. For interested readers, these concepts have been reviewed in detail elsewhere (81, 85). The second reason is practical; there are many 'intermediate' types of exercise that fall between the two extremes, and yet produce similar adaptive responses to aerobic and resistance exercise. Similarly, the agreement between the specific demands of an exercise session and the induced molecular responses is not absolute. For example, as described above, SIT consists of a small number of repeated, intermittent sets of high (force) power output, all-out sprint activity, yet demonstrates a remarkable capacity to produce an endurance phenotype in skeletal muscle and improve endurance performance (23). Such outcomes are consistent with SIT-inducing molecular responses largely similar to those associated with aerobic exercise (86–88). Another contradiction is that aerobic exercise training can produce modest hypertrophy of muscle fibres (89), whereas resistance exercise training can produce modest improvements in mitochondrial function (90) and whole-body aerobic fitness measured by VO_{2max} (91). Lastly, many studies in molecular exercise physiology using skeletal muscle biopsies from human participants have analysed mixed muscle homogenates, rather than isolating muscle fibre types to facilitate the analysis in a fibre type-specific manner. However, the analysis of mixed muscle homogenates may result in the 'masking' or 'dilution' of results that are seen specifically in type I compared to type II fibres as recently illustrated in terms of the utilisation of muscle glycogen during exercise (92), and the adaptive response of the skeletal muscle proteome to exercise (93).

These points highlight that while there are some signals, sensors, signal transduction proteins and pathways, and downstream targets and processes that exhibit somewhat differential responses to these divergent types of exercise stimuli, a simplified view of exercise training adaptations being drawn along extremes of the exercise continuum because of discrete and specific signal transduction pathways is inaccurate. In fact, while pathways in this chapter have largely been depicted as linear (**Figure 7.3**), these pathways ostensibly demonstrate some degree of dependence, crosstalk, interference and redundancy in their regulation, making the exact contribution of each pathway to measured changes in gene expression difficult to isolate. Antithetical to this complexity has been the discussion of 'master regulators' and 'master switches' as determinants of adaptive changes in skeletal muscle such as PGC-1α in the context of mitochondrial adaptation to aerobic exercise, or mTORC1 in the context of muscle hypertrophy in response to resistance exercise. While the concept of master regulation can be seductive, based on current knowledge and indeed much of what has been described in this chapter, it is highly improbable that any one protein or pathway is capable of such regulation. In this context, it is worth noting that many basic science discoveries about regulators of skeletal muscle phenotype are made utilising transgenic and knockout mouse models and/or cell culture experiments in vitro. While crucial in identifying and mechanistically exploring the importance of these genes, the change in abundance or activity of a protein or pathway using the models can sometimes range from complete ablation to several hundred-fold increase, whereas physiological changes in response to exercise are, most often, orders of magnitude lower than such changes.

For illustration, as stated above, PGC-1α has often been considered a master regulator of mitochondrial biogenesis, and therefore central to exercise training-induced adaptations in skeletal muscle (47). However, first, the observational nature of human exercise studies means that no cause-and-effect relationship can be established for changes in skeletal muscle phenotype. Second, the observations in mice that demonstrate either whole-body (94) or muscle-specific (95) PGC-1α knockout does not impair aerobic exercise training-induced changes in skeletal muscle indicate that networks that are independent of PGC-1α also contribute to adaptive responses in skeletal muscle. For interested readers, more detailed arguments against PGC-1α being a so called 'master' regulator of adaptation are outlined elsewhere (11, 96). Of course, this discussion is not to suggest that PGC-1α is not an important regulator of skeletal muscle phenotype – it is perhaps just not a 'master' regulator (if they exist). Similarly, the emergence of evidence of mTORC-independent pathways regulating muscle hypertrophy also speaks against mTORC as a 'master' regulator of this process (42).

The role of stimulation of muscle protein synthesis

Another misconception that in our opinion has become pervasive over the past couple of decades and is related to the concept of master regulators and/or switches is that aerobic exercise adaptation depends on the regulation of transcriptional processes, whereas resistance exercise adaptation depends on the regulation of translational processes. Therefore, we will briefly address what we believe to be the root of that misconception here.

The control of muscle mass is proposed to be determined by the balance between processes of muscle protein synthesis MPS and muscle protein breakdown (MPB), with MPB being largely under the control of the protein degradation pathways described above. Muscle hypertrophy occurs when accumulated rates of MPS exceed accumulated rates of MPB for an extended period of time (35). Because MPB is challenging to measure and likely to be less consequential than MPS in the regulation of resistance exercise training-induced muscle hypertrophy, there has been much focus on understanding the regulation of MPS in the context of resistance exercise. As a result, molecular investigations to date have predominantly centred on the activation of critical regulators of MPS, namely mTORC, S6K1, 4E-BP1 and related downstream targets. However, a myriad of other nodes exist in the regulation of muscle hypertrophy, including ribosome biogenesis, satellite cell function (activation, proliferation, differentiation, survival) and myonuclear domain size and myonuclei number (10), in addition to circulating factors such as growth factors and cytokines (97).

However, given the marked changes in mitochondrial size, number and function with aerobic exercise training, much research has focussed on the process of mitochondrial biogenesis and its regulation by transcription factors and transcriptional processes (11). Again, despite the dominant focus on MPS and muscle hypertrophy in response to resistance exercise training, it is worth noting that resistance exercise has also marked effects on the skeletal muscle methylome and transcriptome in both acute and chronic contexts (31–33, 50, 98). In recent years, in addition to mitochondrial biogenesis, several other molecular processes have emerged as regulators of mitochondrial adaptation in skeletal muscle, including mitochondrial fission-fusion dynamics, the mitochondrial unfolded protein response and mitochondrial quality control through mitophagy (11). Given that there are immediate and transient changes in these processes in the post-exercise period, and as described above in relation to the UPS and the processes of autophagy, the current model suggests that in addition to transcriptional regulation, exercise-induced signal transduction pathways initiate turnover of the mitochondrial pool within skeletal muscle in a coordinated process of removal of dysfunctional mitochondria, in collaboration with the activation of biogenesis.

An important point in this discussion is the delineation of the term 'MPS'. In the broadest sense, this term refers to mixed MPS. While it is true that muscle hypertrophy will require the accretion of a greater volume of muscle protein per se, the process of adaptation to any kind of exercise, be that aerobic, resistance or SIT or any variation thereof, will most likely require an increase in protein synthesis, i.e. the synthesis of new proteins from existing, or exercise-induced changes in, mRNA abundance. Thus, focus on MPS as a process as being only relevant to resistance exercise is erroneous; for example, SIT is a potent stimulus to promote MPS and skeletal muscle anabolism in a general sense (52). Another notable point is that resistance exercise-induced increases in MPS acutely over a few hours after a single exercise session do not always correlate with predicted changes in muscle size with prolonged exercise training (99, 100). Rather, the specificity of exercise adaptation to divergent types of exercise is most likely to reside at the level of differential responses of different protein fractions (myofibrillar, sarcoplasmic and mitochondrial) (101) as well as individual proteins (35), with 'mixed' MPS producing results that combine all these fractions. Of course, it is also worth noting that increases in strength can be dissociated from increases in muscle size or cross-sectional area, i.e. neural adaptations in the absence of hypertrophy (102), and increases in specific force or muscle quality (force per unit of cross-sectional area) but the molecular basis for how signal transduction pathways could influence neuromuscular changes in skeletal muscle is largely unexplored.

The emergence of 'omic' approaches

Many of the pathways and targets thus far rely heavily on the data from the interrogation of 'known' pathways, especially those with well-established canonical roles in cell physiology, in the context of the molecular response to acute exercise. The limitations of methods such as qPCR or Western blotting (**Chapter 2**) are that only one or a few mRNA or protein targets are measured at a time, and that the choice of these targets is somewhat subjective. An alternative approach is to measure many if not all genetic variations, metabolites, histone modifications, DNA methylations, mRNAs and proteins in an 'unbiased' approach. Methods such as metabolomics, epigenomics, gene expression profiling or transcriptomics, and proteomics are examples of unbiased approaches for assessing the changes in metabolites, epigenetic modificatons, mRNA and protein, respectively. Each of these methods generally results in long lists of data that are then analysed and presented using bioinformatics. The above methods can be abbreviated as 'omic' approaches. The historical emergence of omics approaches in molecular exercise physiology is described in **Chapter 1** and the methods themselves have been covered in more detail in **Chapter 2**, but here to provide an illustrative example for the advantages of omic approaches, we will briefly cover the relevance of transcriptomics and proteomics in the context of signal transduction pathways.

Investigating changes in mRNA abundance by transcriptomics (i.e. gene expression profiling using expression microarrays or RNA-seq) has provided enormous insight into the effect of acute exercise on the transcriptional profile of skeletal muscle, as well as effects on skeletal muscle induced by inactivity or more chronic exercise training. Recently, data collected from human studies of aerobic and resistance exercise, including acute exercise sessions and chronic exercise training, were integrated using meta-analysis methods to create an online open access database known as MetaMEx (www.metamex.eu), alongside an interface to readily interrogate the database (50). This database at the time of writing contained 66 transcriptomic datasets from human skeletal muscle, including 13 studies of acute aerobic exercise, and 8 studies of acute resistance exercise. In one interrogation of the database using gene ontology and pathway analyses on the expression profiles, a transcription factor NR4A3 (nuclear receptor 4A3; also known as NOR-1) was identified as one of the most exercise- and inactivity-responsive genes, with the expression of this transcription factor being induced by both aerobic and resistance exercises (50). Notably in the spirit of the discovery that is a goal of unbiased omics approaches, NR4A3 had not been previously identified by authors of the individual published datasets, despite this transcription factor having regulatory roles in skeletal muscle consistent with producing an endurance phenotype (103). With these findings in mind, in an independent study, it was subsequently demonstrated that marked changes in NR4A3 mRNA abundance 3 hours after change-of-direction repeated sprint running, with these changes coinciding with hypomethylation of the NR4A family of genes (61).

More specifically in relation to the focus on this chapter on signal transduction, proteomics is the key omics approach used by molecular exercise physiologists to study acute and chronic responses to exercise in unbiased approaches. The difficulty with proteins, in contrast to RNA extraction and gene expression profiling, is that it is practically impossible to isolate all proteins using a single extraction nor can proteins be amplified, and the proteome is much more diverse than the transcriptome. However, much progress has been made, especially with proteome mining techniques, to catalogue all proteins that can be detected by mass spectrometry. For example, already over a decade ago, Parker et al. (2009) (104) identified over 2,000 proteins from biopsy samples of human skeletal muscle. In addition to abundant myofibrillar proteins and metabolic enzymes, the proteins identified included signalling proteins that are mentioned elsewhere in this chapter, including sensing proteins such as AMPK and CaMKs as well as numerous eIFs, proteasome subunits and E3 ligases (104). More recently, Deshmukh et al. (2021) (93) described a new approach to the proteomic analysis of skeletal muscle in a fibre type-specific manner, and with deeper coverage of the skeletal muscle proteome than other studies to date, and with this method detected over 4,000 proteins. This approach allowed for comparison between fibre types at rest and resulted in identifying a remarkable 471 proteins being different between type I and type II fibres at rest, as well as identifying several novel proteins as being responsive to exercise training on a fibre type-specific basis (93).

With the declining costs of omics technologies and, therefore their more widespread application, coupled to innovations in the detection of posttranslational modifications (105), the same unbiased approaches can be applied to understanding the molecular networks that are acutely responsive to exercise, and therefore promise to reveal novel regulators of adaptation in skeletal muscle to exercise. One noteworthy example is the exploration of the phosphoproteome in human skeletal muscle after an acute session of aerobic exercise (9–11 minutes at 82–85%W_{max} in untrained men), which identified 1,004 phosphosites on 562 proteins (~12% of the skeletal muscle phosphoproteome) as being exercise-responsive (106). While some of these phosphosites were targets of known exercise-responsive protein kinases, including AMPK, CaMK and mTOR, the majority of kinases and substrates phosphosites had not previously been associated with exercise-induced signal transduction. An analogous investigation of resistance exercise was performed using electrically evoked maximal intensity contractions in mouse tibialis anterior muscle consisting of 10 sets of 6 contractions lasting 3 seconds with 10 seconds between contractions and a 1-minute rest period between sets (107). The analyses identified 5,983 unique phosphorylation sites of which 663 were found to be regulated by the exercise session. Again, known phosphosites of exercise-responsive protein kinases, including p38 MAPK, CaMK and mTOR, were identified, but in contrast to aerobic exercise, a high proportion of the regulated phosphorylation sites were found on proteins that are associated with the Z-disc, with ~75% of Z-disc proteins experiencing robust changes in phosphorylation. The phosphorylation state of two Z-disc kinases, namely striated muscle-specific serine/threonine protein kinase (SPEG) and obscurin, was dramatically altered by the maximal intensity contractions and was proposed as novel kinases potentially playing a role in mechanotransduction in skeletal muscle (107).

While the functional relevance of these novel kinases and substrates in the work of Hoffman et al. (2015) (106) and Potts et al. (2017) (107) remains to be established, especially in the context of having a permissive role in the adaptive response to exercise, these findings highlight the possibility that much remains to be discovered about exercise-induced signal transduction in skeletal muscle. To this end, omics approaches allow the unbiased, large-scale identification of proteins that change their concentration or become modified in response to exercise. This allows researchers to develop new hypotheses and gain new insights into ways that are much faster than with one-target-at-a-time approaches. One recent example is an extension of the work of Potts et al. (2017) (107) by the same research group that resulted in the novel identification of TRIM28 and its Ser473 phosphorylation site as being activated by exercise and being a potential regulator of muscle size and function (albeit in non-physiological expression models in rodents) (108). Another notable finding was that the majority of contraction-induced phosphorylation events were rapamycin-insensitive (108), which is interpreted as further evidence for the emerging concept described above of mTORC-**in**dependent (i.e not dependant on mTORC) signal transduction pathways also being important for the regulation of skeletal muscle hypertrophy after loading (42).

PRACTICAL SIGNIFICANCE OF KNOWLEDGE AND INVESTIGATION OF SIGNAL TRANSDUCTION IN SKELETAL MUSCLE

Sport scientists and practitioners whose interests reside more on the applied side of research and practice could certainly be forgiven for questioning the practical significance of all this endeavour in molecular exercise physiology. In other words, can the knowledge of the molecular responses to exercise ultimately inform practice? The answer to this question at present would be a qualified 'yes'; while molecular responses cannot yet fully explain the specificity of adaptation to divergent types of exercise, there is ample evidence that by using these methods, existing exercise training and co-intervention strategies can be fine-tuned, and new ideas can be explored. Examples of co-intervention strategies, i.e. strategies that enhance the adaptive response to exercise with a view to optimising sports performance, include reduced carbohydrate availability (**Chapter 10**), altitude training and heat exposure (**Chapter 11**), timing of exercise (**Chapter 11**), blood flow restriction and essential aminoacidemia (**Chapter 10**) (35, 109, 110).

The central tenet to the first four strategies listed is that these strategies are undertaken in the belief that imposing greater stress, and therefore larger perturbations to homeostasis, will maximise molecular responses in skeletal muscle (and other tissues and organs), promote superior training adaptation and enhance one (or more) of the factors underpinning fitness and performance (109). In practice, many of these strategies begin with the observation of an effect in the real world, which inspires scientists to investigate the mechanistic bases for the observed effects in subsequent laboratory-based studies. When such mechanisms are better known, an investigation of the acute molecular response under 'normal' conditions compared to 'modified' conditions is then undertaken, and the independent variable is often investigated in a dose-response manner, e.g. gradation of carbohydrate availability, height of simulated altitude and so on. Ultimately, the aim is to determine what might be considered the optimal response, at least acutely, and translate this into applied practice. Similarly, there is increasing interest in using molecular methods to inform optimal exercise prescription, be that in the context of concurrent exercise training (81) as described above or understanding dose-response relationships from MICT to SIT (23).

An important premise that underpins the study of acute responses in these contexts is that there is continuity between acute molecular responses and adaptive effects of long-term exercise training. However, like the discordance between an acute change in MPS to long-term changes in muscle mass, or discordance between predicted interference between signal transduction pathways and concurrent training effect, another example of such discordance is that exercise training with reduced carbohydrate availability does not always translate into a performance benefit despite molecular adaptations suggesting that this perhaps should be the case (e.g. greater mitochondrial adaptation) – discussed in Chapter **10**. Therefore, some caution is needed when interpreting the outcomes of acute studies before translating such strategies into applied practice.

Outside of the sporting domain, there is much interest in the fundamental science of characterising the human response to exercise depending on age, biological sex, body composition, fitness level and previous exposure to exercise training. One of the questions that might be addressed with such approaches is whether resting (basal) expression patterns and/or the acute molecular response to exercise can explain some of the contributors to the vast heterogeneity in response to exercise (111, 112). Recent work examining DNA methylation and gene expression responses to resistance exercise training, detraining and retraining identified several genes that were hypomethylated after a single session of acute resistance exercise, which were then maintained as hypomethylated during training and detraining, but importantly also demonstrated an enhanced gene expression after later retraining (32). One interpretation is that in addition to skeletal muscle possessing an epigenetic memory of anabolic stimuli, it may be possible to identify biomarkers of the acute response that predict later adaptation to exercise. Therefore, an attractive hypothesis is that the exercise response (e.g. aerobic fitness, strength and insulin sensitivity) can be predicted and understood through combined molecular and physiological classifications as a function of the individual and the exercise model (in combination with an individuals genetic potential to respond to exercise- covered in Chapters four and 5). In essence, this would represent the advent of personalised exercise medicine in which the exercise prescription is matched to the individual and the condition or desired outcome, as opposed to the currently broad public health guidelines.

On a similar note, the recently established Molecular Transducers of Physical Activity Consortium (MoTrPAC), funded to the tune of $170 million by multiple agencies at the National Institutes of Health (USA), aims to elucidate how exercise improves health and ameliorates diseases by building a map of the molecular responses to acute and chronic exercise in preclinical rodent and clinical human studies using multi-omic and bioinformatic analyses (113). The overarching themes echo the personalised exercise medicine concept in that a better understanding of these biological processes and pathways would allow for the development of targeted exercise interventions and prescriptions. In their recent overview of the programme (113), the MoTrPAC investigators also stated an aim of providing a foundation for developing pharmacologic interventions, which are broadly termed 'exercise mimetics', and this concept forms the final part of this chapter.

EXERCISE MIMETICS

The observations of exercise training-like phenotypes in transgenic and/or knockout mice, or similar effects produced pharmacologically, have ignited interest over the past two decades in developing so-called **exercise mimetics**. For example, chronic activation of AMPK in the skeletal muscle of mice via overexpression of the $\gamma 3$-subunit of AMPK with an R225Q polymorphism (114), or via consumption by rats of a diet enriched with 1% β-guanadinopropionic acid (β-GPA) (115), that induces mitochondrial biogenesis and an endurance-like phenotype. Similarly, disruption of myostatin activity (a member of the transforming growth factor β (TGF-β) superfamily of growth and differentiation factors and an established inhibitor of muscle growth), either through gene knockout (116) or through pharmacological inhibition (117), produces a muscle hypertrophy phenotype (see **Chapter 8**).

Early interest in exercise mimetics was largely around pharmacotherapy to replicate the metabolic effects of aerobic exercise (118, 119) from which the '**exercise-in-a-pill**' concept emerged. More recently, there has also been considerable interest in the potential of pharmacotherapy to address age-related declines in skeletal muscle mass and function (120). However, leaders with decades of experience in the field of molecular exercise physiology have been largely sceptical of the promise of exercise mimetics from when they were first proposed right up to the present day (121, 122). A somewhat semantic point, but one worth making, is that exercise mimetic is a misnomer because a single pill is unlikely to produce the myriad of physiological and metabolic responses to, and consequent benefits of, acute exercise, while the complexities of the molecular response and systemic multi-organ effects of exercise training are well-beyond the faculty of currently available monotherapeutic approaches. The inability to reproduce the multi-organ effects of exercise has been a major criticism of the exercise mimetic concept, especially due to current efforts being largely skeletal muscle-centric and therefore not addressing impacts of exercise on, for example, atherosclerosis and the cardiovascular system (122).

While exercise-in-a-pill is often represented in the public domain as a shortcut for people who simply do not want to exercise – and no doubt that would be a large and lucrative market – the truest spirit of the exercise mimetic concept is pharmacotherapy to mimic the effects of exercise in persons unable to do so for a variety of reasons, such as physical disability, coma or paralysis. To date, perhaps the only alternative to obtain the benefits of exercise for such individuals is neuromuscular electrical stimulation, which can bring about metabolic benefits and adaptive changes through involuntary exercise (123). However, the value of pharmacotherapy in support of modified or bespoke exercise interventions should not be discounted, i.e. compounds serving as adjunct treatments to exercise that potentiate or augment the acute benefit or adaptive response to exercise.

For example, central to the therapeutic effects of exercise in type 2 diabetes mellitus (T2DM) is the amelioration of skeletal muscle insulin resistance. Ectopic lipid deposition, as one consequence of elevated circulating FFAs, results in increased intramuscular lipid (IMCL) content, and is a major contributor to skeletal muscle insulin resistance. To make matters worse, T2DM patients can suffer from exercise intolerance (124), and are less able to oxidise IMCL during exercise, especially when plasma FFAs are elevated to >300 μM (125). However, when T2DM patients exercise under conditions of pharmacologically reduced plasma FFAs via inhibition of adipose tissue lipolysis, the glucoregulatory and insulin-sensitising action of acute exercise is enhanced (126), even at low levels of exercise intensity (127). Therefore, using pharmacotherapy, albeit not an exercise mimetic, may target fatty acid handling during exercise as a key strategy to ameliorating skeletal muscle insulin resistance, or may allow patients to derive more benefit than would be otherwise expected from a low dose of exercise.

A cautionary point on exercise mimetics has also emerged from the frailty and sarcopenia literature in recent years. Because the molecular regulation of muscle mass has been reasonably well-described, and a variety of the molecular candidates are 'druggable' targets, there has been a raft of pharmacotherapies developed to target sarcopenia that have proceeded to phase 1 and 2 clinical trials (120). These compounds have included myostatin inhibitors, antagonists of the activin receptor and selective androgen receptor modulators (SARMs). This first generation of drugs in this field are designed to specifically address the original defining characteristic of sarcopenia – the loss of muscle mass – with the expectation that a resulting muscle hypertrophy would translate to an increase in muscle strength and improved patient function. To date, results from many trials have shown impressive effects of measurable muscle hypertrophy, but in most cases have shown limited success for improving muscle strength or patient physical function (120). In other words, producing a larger muscle in the absence of an exercise stimulus does not perhaps result in a better functioning muscle. This example highlights the abovementioned challenge of translating the myriad of effects of exercise into a pill, i.e. the translation of increased muscle mass to improved patient function remains the major challenge for current experimental drugs that target skeletal muscle anabolism. However, a provocative thought is whether combining such therapies with even very small quantities of exercise could provide marked benefits to patients compared to either approach alone. This, perhaps, is where exercise mimetics will eventually show their utility.

Another viewpoint is that the complex, multi-organ nature of the response to exercise would require many different types of drugs to even approximate, never mind mimic, this response. In other words, a 'polypill' would be needed rather than a single exercise mimetic. Hawley et al. (2021) (122) have recently made the case that, in theory, such an approach may be possible, but this effectively represents current practice of individuals already taking multiple medications for disease treatment and/or risk reduction. Even though these medications can be somewhat effective, our natural history and evolutionary development involved daily physical activity, and these authors argue that in the absence of physical activity, there is little chance that any pharmacological approach will fully succeed. For readers especially interested in the exercise mimetic concept, the paper by Hawley et al. (2021) (122) is highly recommended for its broad and thought-provoking approach to the topic.

CONCLUSION

As the corpus of this Second Edition suggests, molecular exercise physiology has transitioned from being a new vista in the sport and exercise sciences into a pillar of the field, with an increasing acknowledgement that discoveries made on the bench-top have wide-ranging implications in the efficacy and application of exercise and training in performance, health and disease. The present chapter has been primarily concerned with skeletal muscle and the acute molecular responses that potentially underpin the adaptive response to exercise. While much insight has been garnered over the past few decades, there remains much still to be learned, and in particular a more complete picture of proteins and pathways that confer the specificity of training effects. Another important question that has largely been ignored to date is whether biological sex has a modifying effect on the molecular response to exercise. Overall, the hypothesised role for exercise-induced signal transduction in adaptation to exercise described in this chapter provides a framework for studies of the molecular regulation of skeletal muscle adaptation to exercise training, which may be applied in continuing studies of different populations (e.g. healthy/diseased/athletic), and/or exercise types (e.g. aerobic/intermittent/resistance/concurrent), and/or combined with co-interventions (e.g. nutrition/hypoxia/heat). Such investigations, when combined with the more widespread application of omics approaches, should result in many more novel insights in the coming years.

REVIEW QUESTIONS

- Summarise the molecular response to exercise and associated exercise-induced signal transduction as a means to explain the mechanistic basis for adaptation to exercise in skeletal muscle.

- Explain the six steps that encompass the exercise stimulus through exercise-induced signal transduction through to changes in protein abundance/activity, and do so with specific reference to molecular regulators relevant to each step.

- Create a concept map to illustrate the integrative nature of the exercise-induced signal transduction by detailing an example of continuity between changes in cellular substrates, AMPK and downstream targets such as transcriptional regulators. You may wish to download specialised software for concept mapping such as Visual Understanding Environment (VUE) from Tufts University to assist you in this task.

- Compare and contrast the stimulus provided by aerobic exercise compared to resistance exercise, and include examples of how signal transduction pathways might exhibit differential responses to the divergent exercise types.

- Critically appraise the exercise mimetic concept as it pertains to molecular exercise physiology and likelihood of delivering meaningful health impacts to patients with lifestyle-related chronic disease.

FURTHER READING

Booth FW & Neufer PD (2012). Exercise genomics and proteomics. In: Farrrell PA, Joyner MJ, Caiozzo VJ (eds) *ACSM's Advanced Exercise Physiology*. 2nd edn. Lippincott Williams & Wilkins, Baltimore, MD, pp 669–98.

Egan B & Zierath JR (2013). Exercise metabolism and the molecular regulation of skeletal muscle adaptation. *Cell Metab* 17 (2), 162–84. doi:10.1016/j.cmet.2012.12.012

Hawley JA & Joyner MJ & Green DJ (2021). Mimicking exercise: what matters most and where to next? *J Physiol* 599 (3), 791–802. doi:10.1113/jp278761

Saltin B & Gollnick PD (1983). Skeletal muscle adaptability: significance for metabolism and performance. In: Peachy LD, Adrian RH, Geiger SR (eds) *Handbook of Physiology, Section 10: Skeletal Muscle*. American Physiological Society, Bethesda, MD, pp 555–631.

Seaborne RA & Sharples AP (2020). The interplay between exercise metabolism, epigenetics, and skeletal muscle remodeling. *Exerc Sport Sci Rev* 48 (4), 188–200. doi:10.1249/jes.0000000000000227

REFERENCES

1. Holloszy JO. *J Biol Chem*. 1967. 242(9):2278–82.
2. Booth FW, et al. *Physiol Rev*. 1991. 71(2):541–85.
3. Pedersen BK, et al. *Scand J Med Sci Sports*. 2015. 25(Suppl 3):1–72.
4. Maquet P, et al. *The Law of Bone Remodelling*. Berlin: Springer-Verlag, 1986.
5. Winter EM, et al. *J Sports Sci*. 2009. 27(5):447–60.
6. Coyle EF. *Am J Clin Nutr*. 2000. 72(2 Suppl):512S–20S.
7. Hawley JA, et al. *Cell*. 2014. 159(4):738–49.
8. Egan B, et al. *Cell Metab*. 2013. 17(2):162–84.
9. Hoppeler H, et al. *Compr Physiol*. 2011. 1(3):1383–412.
10. Brook MS, et al. *Eur J Sport Sci*. 2019. 19(7):952–63.
11. Hood DA, et al. *Biochem J*. 2016. 473(15):2295–314.
12. Wackerhage H, et al. *J Appl Physiol*. (1985). 2019. 126(1):30–43.
13. Holloszy JO, et al. *J Appl Physiol*. 1984. 56(4):831–8.
14. Burniston JG, et al. Signal Transduction and Adaptation to Exercise: Background and Methods. In: Wackerhage H, editor. *Molecular Exercise Physiology: An Introduction*. 1st ed. Oxon, UK: Routledge; 2014. pp. 52–78.
15. Perry CG, et al. *J Physiol*. 2010. 588(Pt 23):4795–810.
16. Egan B, et al. *PLoS One*. 2013. 8(9):e74098.
17. Wackerhage H, et al. *J Sports Sci Med*. 2002. 1(4):103–14.
18. Fluck M, et al. *Rev Physiol Biochem Pharmacol*. 2003. 146:159–216.
19. McGee SL, et al. *Nat Rev Endocrinol*. 2020. 16(9):495–505.
20. Seaborne RA, et al. *Exerc Sport Sci Rev*. 2020. 48(4):188–200.
21. Egan B, et al. *Cell Metab*. 2016. 24(2):342–.e1.
22. Ross A, et al. *Sports Med*. 2001. 31(15):1063–82.
23. MacInnis MJ, et al. *J Physiol*. 2017. 595(9):2915–30.
24. Gibala M. *Appl Physiol Nutr Metab*. 2009. 34(3):428–32.
25. Saltin B, et al. *Acta Physiol Scand*. 1976. 96(3):289–305.
26. Coffey VG, et al. *FASEB J*. 2006. 20(1):190–2.
27. McConell GK, et al. *J Physiol*. 2020. 598(18):3859–70.

28. Steenberg DE, et al. *J Physiol.* 2019. 597(1):89–103.

29. Benziane B, et al. *Am J Physiol Endocrinol Metab.* 2008. 295(6):E1427–E38.

30. Fernandez-Gonzalo R, et al. *Acta Physiol (Oxf).* 2013. 209(4):283–94.

31. Mallinson JE, et al. *Scand J Med Sci Sports.* 2020. 30(11):2101–15.

32. Seaborne RA, et al. *Sci Rep.* 2018. 8(1):1898.

33. Turner DC, et al. *Sci Rep.* 2019. 9(1):4251.

34. Hearris MA, et al. *Nutrients.* 2018. 10(3).

35. McGlory C, et al. *J Physiol.* 2019. 597(5):1251–8.

36. Kjøbsted R, et al. *FASEB J.* 2018. 32(4):1741–77.

37. Chin ER. *Exerc Sport Sci Rev.* 2010. 38(2):76–85.

38. Philp A, et al. *Exerc Sport Sci Rev.* 2013. 41(3):174–81.

39. Lindholm ME, et al. *Exp Physiol.* 2016. 101(1):28–32.

40. Kramer HF, et al. *J Appl Physiol (Bethesda, MD: 1985).* 2007. 103(1):388–95.

41. Boppart MD, et al. *Am J Physiol Cell Physiol.* 2019. 317(4):C629–c41.

42. Goodman CA. *J Appl Physiol (Bethesda, MD: 1985).* 2019. 127(2):581–90.

43. Cairns SP, et al. *J Physiol.* 2015. 593(21):4713–27.

44. Rossetti ML, et al. *Mol Cell Endocrinol.* 2017. 447:35–44.

45. Dehkhoda F, et al. *Front Endocrinol (Lausanne).* 2018. 9:35.

46. Schiaffino S, et al. *Skelet Muscle.* 2011. 1(1):4.

47. Martínez-Redondo V, et al. *Diabetologia.* 2015. 58(9):1969–77.

48. Ameln H, et al. *FASEB J.* 2005. 19(8):1009–11.

49. Tintignac LA, et al. *J Biol Chem.* 2005. 280(4):2847–56.

50. Pillon NJ, et al. *Nat Commun.* 2020. 11(1):470.

51. Domańska-Senderowska D, et al. *Int J Sports Med.* 2019. 40(4):227–35.

52. Callahan MJ, et al. *Sports Med.* 2021. 51(3):405–21.

53. ENCODE Project Consortium. *Nature.* 2012. 489(7414):57–74.

54. Cao Y, et al. *Dev Cell.* 2010. 18(4):662–74.

55. Vaquerizas JM, et al. *Nat Rev Genet.* 2009. 10(4):252–63.

56. McGee SL, et al. *FASEB J.* 2006. 20(2):348–9.

57. Durham WJ, et al. *J Appl Physiol (Bethesda, MD: 1985).* 2004. 97(5):1740–5.

58. Baar K, et al. *FASEB J.* 2002. 16(14):1879–86.

59. McGee SL, et al. *J Physiol.* 2009. 587(Pt 24):5951–8.

60. Barres R, et al. *Cell Metab.* 2012. 15(3):405–11.

61. Maasar MF, et al. *Front Physiol.* 2021. 12:619447.

62. Yang S, et al. *J Muscle Res Cell Motil.* 1996. 17(4):487–95.

63. Bartel DP. *Cell.* 2018. 173(1):20–51.

64. Russell AP, et al. *J Physiol.* 2013. 591(18):4637–53.

65. Davidsen PK, et al. *J Appl Physiol.* 2011. 110(2):309–17.

66. Booth FW, et al. Exercise genomics and proteomics. In: Farrrell PA, Joyner MJ, Caiozzo VJ, editors. *ACSM's Advanced Exercise Physiology.* 2nd ed. Baltimore, MD: Lippincott Williams & Wilkins; 2012. pp. 669–98.

67. Hojlund K, et al. *Mol Cell Proteomics.* 2008. 7(2):257–67.

68. Orphanides G, et al. *Cell.* 2002. 108(4):439–51.

69. Gygi SP, et al. *Mol Cell Biol.* 1999. 19(3):1720–30.

70. Handschin C, et al. *Endocr Rev.* 2006. 27(7):728–35.

71. Hughes DC, et al. *Am J Physiol Cell Physiol.* 2021. 320(1):C45–c56.

72. Seaborne RA, et al. *J Physiol.* 2019. 597(14):3727–49.

73. Sugden MC, et al. *Arch Physiol Biochem.* 2006. 112(3):139–49.

74. Baar K, et al. *Am J Physiol.* 1999. 276(1 Pt 1):C120–C7.

75. Bodine SC, et al. *Science.* 2001. 294(5547):1704–8.

76. Rom O, et al. *Free Radic Biol Med.* 2016. 98:218–30.

77. Martin-Rincon M, et al. *Scand J Med Sci Sports.* 2018. 28(3):772–81.

78. Lira VA, et al. *FASEB J.* 2013. 27(10):4184–93.

79. Atherton PJ, et al. *FASEB J.* 2005. 19(7):786–8.
80. Baar K. *Med Sci Sports Exerc.* 2006. 38(11):1939–44.
81. Baar K. *Sports Med.* 2014. 44 (Suppl 2):S117–25.
82. Hickson RC. *Eur J Appl Physiol Occup Physiol.* 1980. 45(2–3):255–63.
83. van Damme R, et al. *Nature.* 2002. 415(6873):755–6.
84. Murach KA, et al. *Sports Med.* 2016. 46(8):1029–39.
85. Coffey VG, et al. *J Physiol.* 2017. 595(9):2883–96.
86. Gibala MJ, et al. *J Appl Physiol (Bethesda, MD: 1985).* 2009. 106(3):929–34.
87. Granata C, et al. *Sci Rep.* 2017. 7:44227.
88. Little JP, et al. *Am J Physiol Regul Integr Comp Physiol.* 2011. 300(6):R1303–10.
89. Konopka AR, et al. *Exerc Sport Sci Rev.* 2014. 42(2):53–61.
90. Porter C, et al. *Med Sci Sports Exerc.* 2015. 47(9):1922–31.
91. Ozaki H, et al. *Eur Rev Aging Phys Act.* 2013. 10(2):107–16.
92. Hokken R, et al. *Acta Physiol (Oxf).* 2020.e13561.
93. Deshmukh AS, et al. *Nat Commun.* 2021. 12(1):304.
94. Leick L, et al. *Am J Physiol Endocrinol Metab.* 2008. 294(2):E463–E74.
95. Rowe GC, et al. *PLoS ONE.* 2012. 7(7):e41817.
96. Islam H, et al. *Metabolism.* 2018. 79:42–51.
97. Schiaffino S, et al. *J Neuromuscul Dis.* 2021. 8(2):169–83.
98. Raue U, et al. *J Appl Physiol.* 2012. 112(10):1625–36.
99. Mitchell CJ, et al. *PLoS One.* 2014. 9(2):e89431.
100. Damas F, et al. *J Physiol.* 2016. 594(18):5209–22.
101. Wilkinson SB, et al. *J Physiol.* 2008. 586(Pt 15):3701–17.
102. Reggiani C, et al. *Eur J Transl Myol.* 2020. 30(3):9311.
103. Goode JM, et al. *Mol Endocrinol.* 2016. 30(6):660–76.
104. Parker KC, et al. *J Proteome Res.* 2009. 8(7):3265–77.
105. Wilson GM, et al. *Exerc Sport Sci Rev.* 2018. 46(2):76–85.
106. Hoffman NJ, et al. *Cell Metab.* 2015. 22(5):922–35.
107. Potts GK, et al. *J Physiol.* 2017. 595(15):5209–26.
108. Steinert ND, et al. *Cell Rep.* 2021. 34(9):108796.
109. Hawley JA, et al. *Cell Metab.* 2018. 27(5):962–76.
110. Preobrazenski N, et al. *Eur J Appl Physiol.* 2021. 121:1835–1847.
111. Timmons JA. *J Appl Physiol (Bethesda, MD: 1985).* 2011. 110(3):846–53.
112. Timmons JA, et al. *F1000Res.* 2016. 5:1087.
113. Sanford JA, et al. *Cell.* 2020. 181(7):1464–74.
114. Garcia-Roves PM, et al. *J Biol Chem.* 2008. 283(51):35724–34.
115. Bergeron R, et al. *Am J Physiol Endocrinol Metab.* 2001. 281(6):E1340–E6.
116. Amthor H, et al. *Proc Natl Acad Sci U S A.* 2007. 104(6):1835–40.
117. Wang Q, et al. *J Physiol.* 2012. 590(Pt 9):2151–65.
118. Lagouge M, et al. *Cell.* 2006. 127(6):1109–22.
119. Narkar VA, et al. *Cell.* 2008. 134(3):405–15.
120. Rooks D, et al. *J Frailty Aging.* 2019. 8(3):120–30.
121. Booth FW, et al. *J Physiol.* 2009. 587(Pt 23):5527–39.
122. Hawley JA, et al. *J Physiol.* 2021. 599(3):791–802.
123. Guo Y, et al. *Mech Ageing Dev.* 2021. 193:111402.
124. Wahl MP, et al. *Front Endocrinol (Lausanne).* 2018. 9:181.
125. van Loon LJ. *J Appl Physiol.* 2004. 97(4):1170–87.
126. van Loon LJ, et al. *Diabetologia.* 2005. 48(10):2097–107.
127. Hansen D, et al. *Eur J Sport Sci.* 2018. 18(9):1245–54.

CHAPTER EIGHT

Molecular adaptation to resistance exercise

Keith Baar

LEARNING OBJECTIVES

By the end of this chapter, you should be able to:

1. Differentiate training strategies to increase muscle mass versus strength and the evidence that support these.
2. Explain how the mTORC1 pathway increases protein synthesis and drives hypertrophy in response to resistance exercise.
3. Describe mTORC1-independent mechanisms for controlling protein synthesis and hypertrophy in response to resistance exercise.
4. Explain the effect of the myostatin-Smad pathway on muscle mass.
5. Describe the role of E3 ubiquitin ligases in the anabolic response to loading.
6. Define satellite cells and explain their function during muscle adaptation to loading versus regeneration after injury.
7. Explain the molecular mechanisms that regulate the identity of satellite cells and their fusion into muscle fibres.

INTRODUCTION

Resistance exercise or strength training is when skeletal muscles work against high loads over a short time. When resistance exercise is repeated at a sufficient frequency, intensity and duration, this type of training improves neural activation and increases muscle size, strength and/or power. We start this chapter with practical resistance training principles and review the ideas of hyperplasia, an increase in fibre number, and hypertrophy, an increase in fibre cross-sectional area (CSA), in response to resistance exercise. After discussing the classical exercise physiology, we will then review the molecular mechanisms that drive the adaptation to resistance exercise.

Resistance exercise activates signal transduction pathways that regulate transcription, muscle protein synthesis, protein degradation, and satellite cell behaviour. This chapter first reviews research aimed at elucidating how our muscles turn the mechanical signal of resistance exercise into a chemical signal that triggers the growth response, mainly in the form of increased myofibrillar protein synthesis. This discussion will focus on the mechanistic target of rapamycin (mTOR) complex 1 pathway, the central

DOI: 10.4324/9781315110752-8

node within our muscles that integrates multiple inputs and determines the degree to which myofibrillar protein synthesis and thus muscle hypertrophy occur. The mTORC1 pathway is more difficult to activate in the presence of metabolic stress, such that muscles do not grow as well when caloric balance is negative, or glycogen is low. An opponent of mTORC1 and muscle growth is the myostatin/Smad signalling pathway. Genetic defects that decrease functional myostatin or Smad signalling (discussed in detail in chapter 4) have a large effect on muscle size through the regulation of phosphoinositide 3,4,5 trisphosphate (PIP3) in the membrane. Next, the focus will shift towards how strength is signalled differently to the increase in muscle mass. Once we have taken a look at signalling within existing myofibres, we will discuss muscle stem cells and show how the key muscle stem cell, the satellite cell, can both self-renew and differentiate to assist in muscle repair. The function of satellite cells in response to resistance training and injury will be reviewed. Satellite cells will also be used to introduce some basic cell physiology concepts such as stem cells and muscle development. To prepare for this chapter, you should first read **Chapter 7** as it covers the theory of signal transduction and molecular responses leading to adaptation. Additionally, we cover the genetics of the muscle growth-regulating mTORC1 and myostatin-Smad pathways in **Chapter 4**. Finally, we refer readers to **Chapter 13** to further extend their knowledge on the role of satellite cells across all modes of exercise (endurance and resistance) as well as provide insight into the role of satellite cell communication with other cell types and the effect of ageing on satellite cell function.

RESISTANCE EXERCISE: CURRENT RECOMMENDATIONS, TRAINABILITY AND SCIENTIFIC EVIDENCE

At the core of a resistance exercise or strength training programme is the overload principle, which states that strength gains occur because of systematic and progressive exercise of sufficient frequency, intensity and duration to cause adaptation. Therefore, planning a resistance training programme requires decisions about: **Load** (% of 1 repetition maximum; abbreviated as RM), **volume** (repetitions, sets, sessions per week), **velocity** (heavy/slow, light/fast, isometric), **rest** (time between sets) and **progression** (how to change variables during a training programme). This plan should also consider **training goals** (strength, power, hypertrophy), the type of **equipment available** (machines, free weights, body weight) and the **muscle action** (eccentric, concentric, isometric) to be used. Moreover, the resistance training programme needs to be specific for the individual as different programmes will be needed for patients rehabilitating following trauma, those living with cancer, children, untrained individuals, older participants or athletes.

Even when a programme is designed and executed, the magnitude of the adaptation in muscle mass and strength will vary dramatically between individuals. This was beautifully shown in a study by Hubal et al. who found that on the same training programme some individuals did not increase muscle size (CSA) or strength (1 repetition maximum (1RM)), whereas others increased size and strength by >40% and >100%, respectively (**Figure 8.1**). You will also note that the figure (8.1) showing the change in strength looks different from that showing the change in muscle size, suggesting that even though the two parameters are associated, the link between size and strength is not absolute. This could reflect the fact that it is often difficult to accurately measure strength, but the data clearly show that there is a large variation in resistance exercise trainability in the human population and thus even a very good training programme will not work for all subjects.

Keeping this high variation of trainability in mind, what training variables can be recommended for effective resistance training? The reality is that there is no single, scientifically proven programme to increase muscle size and strength. Having said that, in the last decade, the important variables for increasing muscle size and strength have been defined. Interestingly, the programme to build size is different than that required to increase strength. This fact was elegantly demonstrated in a series of studies from Professor Stuart Phillips' laboratory. In these studies, Mitchell et al. had participants perform leg extensions with either 30 or 80% of 1RM for either 1 or 3 sets (2). For the last set, the participants lifted

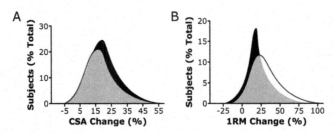

Figure 8.1 **(A)** The increase in the cross-sectional area (CSA) and **(B)** 1 repetition maximum (1RM) in response to a resistance training programme differs dramatically in both females (grey/white curve) and males (black curve). Redrawn from Hubal MJ, et al. (1).

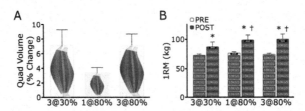

Figure 8.2 **(A)** The increase in the cross-sectional area (CSA) and **(B)** 1 repetition maximum (1RM) in response to a resistance training programme differs according to volume and load, respectively. Redrawn from Mitchell et al. (2).

the weight to **positive failure**, the point at which they could no longer complete a full repetition. Following ten weeks of training, the high-volume strength training routines, those performing more sets, resulted in a greater increase in muscle volume (size) as measured using magnetic resonance imaging (MRI) than the participants performing less volume regardless of the weight lifted (**Figure 8.2**). In contrast, those lifting heavier loads (80% 1RM) increased strength more than when the lighter load was lifted regardless of the number of sets performed (2). These data suggest that, with training the increase in **muscle mass is dependent on the volume of training done to failure**, whereas the increase in **muscle strength is dependent on the load lifted**. The work from Mitchell et al. showed a different association between load and muscle growth than the seminal work of Wong and Booth (performed in rats), who showed that the greater the load, the greater the increase in muscle mass (3). The difference between the studies is that Wong and Booth used electrical stimulation to activate the muscles in the rat, whereas Mitchell relied on their human participants to work as hard as possible. This means that when Wong and Booth exercised their rats, all the motor units within the muscle were activated by the electrical stimulus, whereas in the Mitchell's study, the participants recruited only the motor units required to lift the weight. In fact, the only times when all human motor units are recruited are during a 1RM effort or at failure. Therefore, load activates a signalling cascade that results in an increase in muscle fibre CSA; however, only at failure (when all motor units are active) will all fibres within the muscle sense the load. In other words, when increasing muscle mass is the goal, in the weight room, failure is the goal!

The initial increase of strength in response to resistance training is due to an increase in neuromuscular activation and increases in muscle size only contribute to strength gains at a later stage (4). Specifically, resistance training increases the maximal firing frequency of motor units and action potential doublets can be detected post-training (5). However, even though neural adaptations predominate over the initial period of adaptation, muscle protein synthesis increases immediately after a bout of resistance exercise (6). This apparent discrepancy, that an increase in muscle protein synthesis occurs in untrained individuals even though it contributes little to strength, is because the neural adaptations allow us to recruit more muscle fibres to perform a given movement making us appear stronger even though our individual muscle fibres have not yet grown very much. At this early stage, increased muscle mass cannot be detected even with sensitive techniques such as MRI.

HUMAN MUSCLE FIBRES

Human muscles contain thousands to hundreds of thousands of muscle fibres. For example, the vastus lateralis (outer quadriceps muscle) of young accident victims with a mean age of 19 ± 3 years contained between 393,000 and 903,000 muscle fibres (7). This shows not only the high number, but also the large variability of fibre numbers within a given muscle. This variation in muscle fibre number may explain why some untrained individuals have large muscles, whereas others have small muscles (see **Chapter 4** that discusses the genetic heritability of fibre number). Human muscle fibres can be up to 20 cm long (8) and up to 10,000 μm^2. Questions therefore arise from the fact that human muscle fibres are so large and some of these issues will be addressed below.

Hyperplasia

The first question is: does resistance exercise increase the number of muscle fibres within a muscle, a process known as hyperplasia? As discussed in **Chapter 2**, generally, hyperplasia is thought to contribute little towards the overall increase in muscle mass after resistance training in adult human muscle. However, in response to extreme growth stimuli, muscle fibre number can increase. A meta-analysis of animal studies suggested that in models where muscle fibre mass increases dramatically (by over 50%) such as following synergist ablation (where the gastrocnemius and soleus muscles are removed from the hind limb and the plantaris muscle must take on the extra load, and as a result undergoes significant muscle hypertrophy), approximately ~7% of that increase in muscle mass was due to an increase in fibre number as a result of fibre splitting. This number was later confirmed in a study by Goodman et al., where they demonstrated that synergist ablation in rodents resulted in a ~60% increase in fibre number, which accounted for only ~6% of the increase in muscle mass (9). The use of synergist ablation in contributing to our understanding in molecular exercise physiology was covered in **Chapter 1** and the role of hyperplasia in muscle regeneration is also covered in more detail in **Chapter 13.**

Thus, hyperplasia can occur in instances of extreme muscle growth in rodents and birds, but does not seem to occur after resistance exercise in adult humans. It is also unclear whether long-term muscle growth in humans either from resistance training or from steroid use results in significant hyperplasia in humans. Human fibre number is thought to increase at specific stages of development. For example, in a longitudinal biopsy study, Glenmark et al. found that between 16 and 27 years of age, men increased their body mass by 17% at a time when their muscle fibre CSA remained unchanged (10). This suggests that the increase in muscle mass during puberty could partly reflect an increase in muscle fibre number.

Hypertrophy

Before discussing muscle hypertrophy, which depends on adding sarcomeres in parallel, we will first review several facts related to skeletal muscle and the proteins it contains. In men, ~38% and in women, ~31% of the body mass is skeletal muscle tissue (11). However, this depends on the individual. Lean subjects have a higher percentage, whereas obese subjects have a lower percentage. Muscle comprises, depending on hydration status, 70% water (700 ml per kg muscle) and 30% solids (300 g per kg of muscle). Of the solids, ~70% (215 g per kg muscle) is myofibrillar protein (12). In muscle, thousands of different proteins are expressed but the most abundant proteins are the myofibrillar proteins of the sarcomere. It has been estimated that between 20 and 40% of overall protein (in absolute terms 45–85 g per kg of muscle) is myosin heavy chain (the 'motor') and that ~15% (30 g per kg muscle) is actin (the 'track'; (14)).

Muscle growth from resistance exercise occurs mainly by muscle fibre hypertrophy, an increase in the size of individual fibres without an increase in fibre number. Muscle hypertrophy occurs when myofibrillar protein balance is positive. Myofibrillar protein balance is defined as the difference between myofibrillar protein synthesis and breakdown. Skeletal muscle protein synthesis, and even more so protein breakdown, are challenging to measure in humans; however, by using stabile isotopes, several laboratories across the world can readily determine the turnover of myofibrillar, cytoplasmic, extracellular matrix (ECM) and mitochondrial fraction of proteins within the muscle.

Measuring human muscle protein synthesis

Human muscle protein synthesis is measured by providing stable isotope-labelled amino acids or water to an individual and measuring their accumulation in muscle. For example, ~99% of carbon occurs naturally as 12C, whereas the other ~1% is the heavier 13C stable (i.e. non-radioactive) isotope. For stable isotope protein synthesis measurements, these low percentage isotopes, 13C or 2H, are added to amino acids or water to generate a tracer for the experiment. The tracer is then infused into the body or provided in food or liquids. From there, the tracer is incorporated into muscle protein because of protein synthesis in the same manner as an unlabelled amino acid. To measure muscle protein synthesis, muscle samples are taken and the enrichment of the amino acid tracer into muscle protein is measured by separating the 13C/2H from the naturally occurring 12C/1H using mass spectroscopy. The greater the increase in the stable isotope within muscle proteins over time, the greater the rate of muscle protein synthesis. Classically, these experiments have been performed by labelling amino acids and measuring their incorporation over short time periods (1–6 hours). However, this technique has been modified to measure protein turnover across much longer time frames by providing deuterium water, where the 1H has been replaced by 2H. Over time, alanine and other amino acids become labelled with the 2H and these amino acids are incorporated into muscle. In this way, free living protein turnover across weeks can be directly measured to get a much better sense of the long-term effects of changes in lifestyle (diet and resistance exercise).

Protein synthesis after resistance exercise

It is clear from studies where protein turnover is measured following a single bout of resistance exercise that protein synthesis increases for up to 72 hours in individuals unaccustomed to exercise (**Figure 8.3**; (15)). This phase of elevated protein synthesis is much longer than the one that occurs after a meal. Interestingly, in unaccustomed strength athletes, the increase in protein synthesis following the first bout of resistance exercise does not relate to muscle hypertrophy. It is only after a period of acclimatization that the increase in myofibrillar protein synthesis after a single bout of training correlates with the increase in muscle fibre hypertrophy following training (15, 16). The rate of protein synthesis after the first training bout is significantly higher because of the damage to the fibre as a result of lifting heavy weights for the first time. It is only when injury is minimized that protein synthesis can be focused less on repair and more on growth.

Even though myofibrillar protein synthesis increases following a single bout of resistance exercise, if training is performed in the fasted state, the increase in muscle protein degradation will be greater than the increase in protein synthesis, resulting in a negative protein balance (**Figure 8.3**). Protein balance only becomes positive, and thus muscle fibre hypertrophy only occurs, in the presence of sufficient essential amino acids, such as following a meal (17). Amino acid intake and the molecular regulation of protein synthesis after exercise are covered in detail in **Chapter 10**. Briefly, consuming essential amino acids not only increases muscle protein synthesis, but also decreases protein breakdown following resistance exercise, suggesting that the increase in protein breakdown following resistance exercise in the fasted state occurs in order to supply essential amino acids for protein synthesis. Since essential amino acids only come from the diet or stored protein, they must be supplied by breaking down existing protein if insufficient protein has been ingested. Therefore, following resistance exercise in a fasted state, more protein must be degraded to produce the precise amino acid mix necessary to synthesize novel

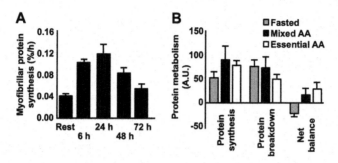

Figure 8.3 (A) Resistance exercise increases myofibrillar protein synthesis for up to 72 hours, redrawn from Damas F, et al. (15). **(B)** Only resistance exercise and amino acids together result in a positive protein balance. Effect of resistance exercise only (fasted, grey bars), additional mixed amino acids (Mixed AA, dark bars) and essential amino acids (Essential AA, open bars) on protein synthesis, protein breakdown and the net balance (where positive values suggest fibre growth). Protein synthesis is only greater than protein breakdown after resistance exercise if mixed or essential amino acids are given. Redrawn from Tipton KD, et al. (17).

proteins. With training, muscles become better at recycling amino acids resulting in better overall protein balance in the fasted state (18).

In summary, resistance exercise to failure in the presence of sufficient essential amino acids leads to an increase in protein balance, largely due to an increase in myofibrillar protein synthesis. Over time, the addition of myofibrillar proteins in parallel results in an increase in muscle fibre CSA and whole muscle hypertrophy. In small mammals, pronounced muscle growth is accompanied by an increase in the number of muscle fibres, termed hyperplasia. During short-term resistance exercise in humans, there is a negligible increase in the number of muscle fibres, and thus the increase in muscle mass is almost solely due to muscle fibre hypertrophy. It is unclear, however, whether muscle fibre numbers increase in response to many years of resistance training in humans.

MOLECULAR RESPONSES TO RESISTANCE TRAINING AND ADAPTATION

Molecular exercise physiologists have made numerous advances in understanding how a mechanical signal, such as the load across a muscle, is converted into chemical signals that promote increases in muscle mass and strength. From this work, there appear to be three/four major regulators of muscle mass:

1. Anabolic mTORC1-dependent or -independent signal transduction pathways that increase protein synthesis resulting in hypertrophy.
2. The catabolic myostatin-Smad2/3 signal transduction pathway that inhibits muscle growth.
3. Satellite cells that aid in the repair of damaged fibres, decrease fibrosis and may provide more ribosomes to enable long-term hypertrophy.

Strength adaptations combine the increase in muscle mass with the production of the better functioning extracellular matrix (ECM) needed to transmit the force produced by the increased amount of muscle protein to the tendons and bones.

In the text below, these signal transduction pathways will be described in detail as will the mechanisms believed to drive the increase in muscle size and strength.

The mTORC1 pathway and protein synthesis

The mechanistic target of rapamycin complex 1, abbreviated mTORC1, is a complex that contains a serine/threonine protein kinase (mTOR), an intermediary protein (raptor) and two kinase inhibitors (DEPTOR and PRAS40). In brief, mTOR phosphorylates proteins, that are identified by raptor, following activation by the small G-protein Rheb (Ras homologue enriched in brain). mTOR can also form a second complex (mTORC2), in which raptor is replaced by rictor (rapamycin insensitive companion of TOR) resulting in the targeting of a different subset of proteins for phosphorylation by mTOR.

Once activated, mTORC1 increases protein synthesis (i.e. the translation of mRNA into protein by the ribosome) and ribosome biogenesis (i.e. the capacity of a cell for protein synthesis) and inhibits multiple forms of protein breakdown, including autophagy (the breakdown of proteins and large protein complexes in the lysosome) and mitophagy (the breakdown of mitochondria). In some cell types, mTORC1 promotes cell division and the transcription of genes involved in anabolism. mTORC1 got its name from the fact that it is selectively inhibited by the macrolide antibiotic, called rapamycin (19). As a result, rapamycin has become a powerful research tool to identify what function mTORC1 has in a cell.

The mTORC1 pathway and resistance exercise

The mTORC1 pathway was first identified as being activated by resistance exercise in rats (20). In this experiment, the researchers demonstrated that the amount of muscle hypertrophy following six weeks of training was directly related to the degree of phosphorylation of p70 S6K1 (also known as S6K1), a known downstream target of mTORC1, 6 hours after a single bout of resistance exercise. Even though this study demonstrated a strong correlation between mTORC1 activity and muscle hypertrophy, it took several other experiments to show that mTORC1 was required for load-induced muscle growth. In the first, Bodine et al. (21) showed in mice that rapamycin (an inhibitor of mTORC1) could prevent load-induced skeletal muscle hypertrophy. In the second, Drummond et al. showed that human volunteers receiving 12 mg of rapamycin prior to performing 11 sets of 10 repetitions of leg extension at 70% of their 1RM (22) did not increase myofibrillar protein synthesis after exercise, whereas those who got a saline injection (no rapamycin) did increase myofibrillar protein synthesis (**Figure 8.4**). In the third and most conclusive experiment, Goodman et al. (9) used mice in which they had mutated mTOR so that it could no longer be inhibited by rapamycin. In this experiment, wild-type (control) and rapamycin-resistant mice underwent synergist ablation that evokes rapid hypertrophy. In the wild-type mice, synergist ablation resulted in an average increase of muscle fibre CSA of 42% over 14 days. Wild-type

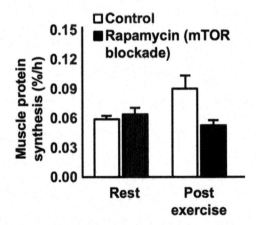

Figure 8.4 The mTOR inhibitor rapamycin blocks resistance exercise-induced increase of protein synthesis in human muscle. Redrawn from Drummond MJ, et al. (22).

mice that underwent synergist ablation, but received daily injections of rapamycin, showed only a 6% increase in fibre CSA. However, mice expressing a rapamycin-resistant version of mTOR showed a similar increase in muscle fibre CSA both in the presence and in the absence of rapamycin. Overall, suggesting inhibiting mTOR with rapamycin after muscle loading blunts hypertrophy; however when mTOR no longer responds to rapamycin (mutated mTOR), hypertrophy can occur as normal, confirming that mTOR is required for load-induced muscle growth. To further demonstrate that it was mTORC1 that was required for muscle hypertrophy, You et al. knocked raptor out of muscles prior to synergist ablation (23). As would be expected from the data above, muscles without functional raptor had negligible S6K1 phosphorylation and showed no increase in fibre CSA following 14 days of overload. Taken together, these data demonstrate that mTORC1 is required for load-induced skeletal muscle hypertrophy.

Even though the above data suggest that mTORC1 is required for muscle hypertrophy, this does not mean that it is sufficient on its own to cause it. To demonstrate that activating mTORC1 was sufficient to induce muscle hypertrophy, Goodman et al. electroporated (used electricity to insert plasmid DNA into muscle fibres) Rheb, the small G-protein activator of mTORC1 (discussed below), into mouse muscles. Seven days after electroporation, they removed the muscles and found that mTORC1 activity was higher in the muscles electroporated with Rheb, and the muscle fibres where they had activated mTORC1 had grown ~40% larger, whereas control plasmid DNA (no Rheb gene) had no effect on muscle fibre size. Thereby, demonstrating active mTORC1 was sufficient and required for muscle hypertrophy.

Activation of mTORC1

As described above, mTORC1 is sufficient and required for load-induced skeletal muscle hypertrophy (**Figure 8.4**). If instead, we were interested in insulin signalling, mTORC2 would be far more important since it regulates insulin-stimulated glucose uptake by phosphorylating Akt/PKB. However, for muscle growth, mTORC1 is central. mTORC1 can be activated in two different ways, with each of these upstream pathways resulting in the GTP loading and activation of the small G-protein Rheb. GTP-bound Rheb can directly bind to the N-terminal region of mTOR's protein sequence and, once bound, Rheb activates mTORC1 by realigning the amino acid residues within the active-site of the mTOR kinase, bringing the amino acids into the correct register for the transfer of the gamma phosphate of ATP to a serine or threonine residue on the target protein (24). Even though the last step in the activation of mTOR is the same, there are several different ways that this activation can occur:

1. **Growth factors such as IGF-I and insulin** activate mTORC1 by binding to the IGF-I receptor and the resulting movement of the insulin receptor substrates (IRS) and phosphoinositide 3-kinase (PI3K) to the plasma membrane. Once at the plasma membrane, PI3K finds its substrate phosphoinositide 4,5 bisphosphate (PIP2) within the lipid bilayer and converts it to phosphoinositide 3,4,5 trisphosphate (PIP3). PIP3 is a very important signalling molecule that is normally very low within the plasma membrane, but increases more than 15-fold within seconds (e.g. after the addition of insulin) (25). PIP3 then functions to recruit more proteins to the membrane. Specifically, proteins that contain a pleckstrin homology (PH) domain, such as PDK1 and PKB/Akt, bind to PIP3 and are therefore moved from the cytoplasm to the membrane. Once at the membrane, PKB/Akt is sequentially phosphorylated by PDK1, and the membrane-associated mTORC2, and this results in PKB/Akt activation (26). Active PKB/Akt can then phosphorylate PRAS40 (27) and TSC2 (28) moving them away from mTOR and Rheb, respectively. Since TSC2 is a GTPase-activating protein (GAP) towards Rheb (29), moving TSC2 away from Rheb allows the GTP binding and activation of Rheb. Rheb can then activate any mTOR located close to it, but without other stimuli, the amount of mTOR located close enough to Rheb is small resulting in a marginal increase in protein synthesis (**Figure 8.5**). The hypertrophic effect of this pathway has been demonstrated by overexpressing IGF-I (30) or constitutively active PKB/Akt in skeletal muscle (21). However, physiological activation of this pathway by insulin or IGF-I is short-lived (~60 minutes) and therefore does not drive hypertrophy on its own.

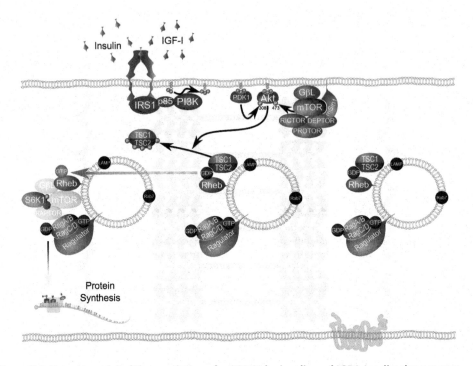

Figure 8.5 Current model of the regulation of mTORC1 by insulin and IGF-I. Insulin alone moves
TSC1/2 away from Rheb but fails to move mTOR to Rheb, resulting in only a partial
activation of mTORC1 and protein synthesis.

2. **Amino acids, specifically leucine,** activate mTORC1 by moving mTOR to Rheb (**Figure 8.6**). The
   movement of mTOR to Rheb is facilitated by another subset of small G-proteins called the Rag
   proteins. The Rag proteins form a heterodimer (a two-protein complex where the two proteins are
   different) with either RagA or B binding to either RagC or D. To simplify the discussion, we will only
   look at the RagA/RagD dimer. When RagA is bound to GTP and RagD is GDP-bound, the dimer is
   able to bind to raptor (31). This is important because the Rag proteins are tethered to the membrane
   in close proximity to Rheb, meaning that when the RagA is GTP-bound, mTORC1 is brought to its
   activator, Rheb. As with Rheb, GTP loading of RagA is tightly regulated. For RagA, the GAP is a
   complex known as GATOR1 (32) and the guanine nucleotide exchange factor (GEF; a protein that
   removes GDP from a G-protein) complex is known as the Ragulator (33). Leucine regulates Rag GTP
   loading by binding to a protein called sestrin. When bound to leucine, sestrin changes its shape in
   such a way as to move away from the protein complex known as GATOR2 (34). GATOR2, when not
   bound to sestrin, functions as an inhibitor of GATOR1. When GATOR1 is inhibited, the Ragulator
   removes the GDP from RagA, and because GTP is more than 10-fold higher than GDP in the cell,
   RagA becomes GTP-loaded and can bind raptor and bring mTOR close to Rheb. Leucine also
   activates the vacuolar protein sorting (Vps) 34. Vps34 is a class III PI3K that promotes the movement
   of membrane-bound organelles around the cell. In this case, Vps34 moves membrane-bound RagA
   around the cell to increase the likelihood of interacting with raptor, bringing more mTOR to its
   activator Rheb. However, without insulin/IGF-I or another stimulus to remove TSC2 from Rheb,
   Rheb will remain inactive resulting in a marginal increase in protein synthesis (**Figure 8.6**).

3. **Resistance exercise** activates mTORC1 when a yet, unidentified mechanoreceptor activates the
   Really **I**mportant **K**inase (RIK), a kinase that targets an R-X-R-X-X-Ser/Thr (where R indicates
   arginine and X indicates any amino acid) protein sequence motif on TSC2 (**Figure 8.7**). Once TSC2

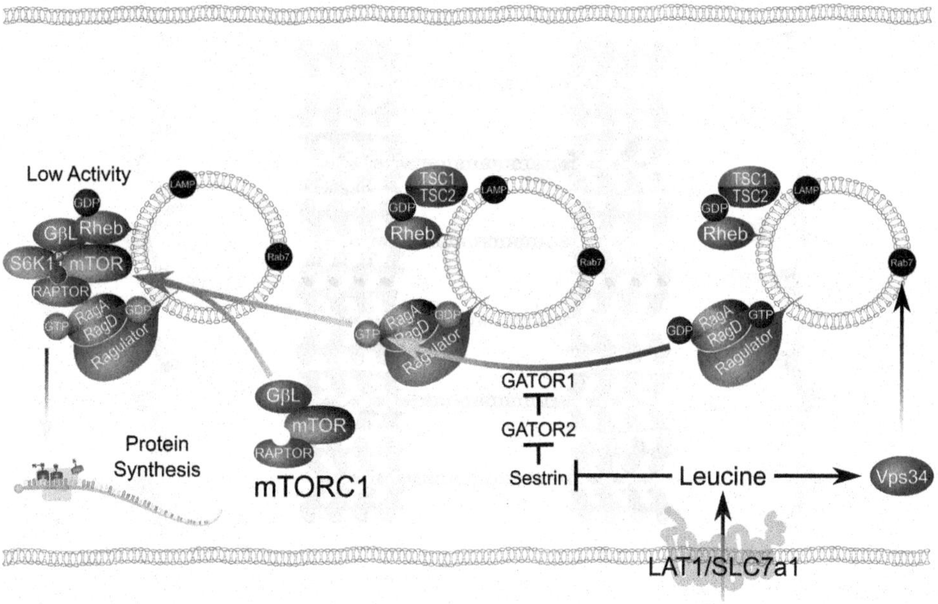

Figure 8.6 Current model of the regulation of mTORC1 by amino acids. Amino acids alone move mTOR to Rheb but fail to activate Rheb, resulting in only a partial activation of mTORC1 and protein synthesis.

is phosphorylated by RIK, it moves away from Rheb (35), allowing Rheb to become GTP-loaded and activated. In this way, resistance exercise activates mTORC1 in a PI3K- (36) and PKB/Akt-independent (37) manner. Further, since both growth factors and resistance exercise do the same thing (move TSC2 away from Rheb), loading can overcome the need for growth factors such as IGF-I for muscle hypertrophy (38).

In summary, growth factors or resistance exercise moves the GAP protein, TSC2 away from Rheb allowing it to become GTP-loaded. Leucine binds to sestrin and removes it from GATOR2 which can now bind to and inactivate the Rag GAP protein complex, GATOR1. Once GATOR1 is inhibited, RagA can remain GTP-loaded and active and therefore can bind raptor as it is moved around the cell by active Vps34, bringing mTORC1 to its activator Rheb. When they are close to one another, active Rheb binds to the N-terminal domain of mTOR and this change in the shape of mTOR is such that it activates the kinase. This is why combining resistance exercise (moving TSC2 away from Rheb) and leucine-rich protein (moving mTOR to Rheb) results in greater protein synthesis and muscle hypertrophy than either stimulus alone (39).

Catabolic control of mTORC1 signalling

The activation of mTORC1 by anabolic signals (growth factors, resistance exercise and dietary protein) is counterbalanced by the negative effects of catabolic signals (caloric deficit, inactivity or disease). For example, in some cell types, the AMP-activated protein kinase can phosphorylate and activate TSC2 (40), and phosphorylate raptor and move it away from mTOR (41). Since AMPK is activated by endurance exercise (see **Chapter 9**), this has long been thought of as the mechanism underlying the concurrent training effect (discussed in Chapter 7), where a high volume of high-intensity endurance exercise limits muscle mass and strength gains resulting from resistance training (42). In fact, chemical activation of AMPK using 5-aminoimidazole-4-carboxamide riboside (AICAR) can block the activation of mTORC1

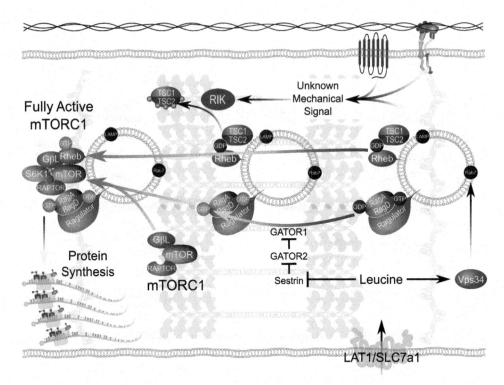

Figure 8.7 Current model of the regulation of mTORC1 by resistance exercise and amino acids. A yet
unknown mechanosensory molecule activates the really important kinase (RIK) and RIK
phosphorylates TSC2 and moves it away from Rheb. In the presence of leucine-rich
protein, RagA becomes GTP-loaded after inhibition of GATOR1, and Vps34 moves RagA
around the cell so that it can bind all of the raptor and bring mTORC1 to Rheb. Active
mTORC1 then phosphorylates S6K1 and increases protein synthesis (translation). In
reality, many more proteins are involved and mTORC1 regulates functions other than
protein synthesis.

in rats (43) and knockout of α1AMPK results in more activation of mTORC1 and muscle hypertrophy in
response to synergist ablation than is observed in wild-type mice (44). Together, these data suggest that
metabolic stress can limit mTORC1 activity and muscle growth in response to resistance exercise.
However, whether this is observed in human muscle after a single bout of moderate endurance exercise is
still an open question (45–47).

mTORC1 and translational activity

The activity of mTORC1 depends on the combined effects of the aforementioned four inputs: (1)
resistance exercise, (2) amino acids, (3) growth factors and (4) negative or catabolic signals. Once
activated, mTORC1 regulates numerous cellular functions. After resistance exercise, the most important of
these is the increase of protein synthesis or more technically, **translational activity** (protein synthesized
per molecule of RNA – see **Chapter 7**). In all cells, proteins are synthesized by the ribosome, an organelle
composed of a total of 79 proteins and 4 ribosomal RNAs. In 2009, Venkatraman Ramakrishnan, Thomas
A Steitz and Ada E. Yonath won the Nobel Prize in chemistry for their studies of 'the structure and
function of the ribosome' which demonstrated how the ribosome translated the information held within
a messenger RNA into a protein. mTORC1-signalling drives protein synthesis by

increasing the activity of existing ribosomes and speeding up initiation and elongation. In part, the activation occurs via the phosphorylation of the ribosome-activating proteins 4E-BP1 and the 70 kilodalton ribosomal S6 protein kinase (S6K1). Phosphorylation of 4E-BP detaches it from the 5'-RNA cap-binding protein eIF4E (48). Once 4E-BP is removed from eIF4E, eIF4G can bind to the site vacated by 4E-BP (49) and this brings the messenger RNA to the ribosome in order to initiate translation. How S6K1 regulates initiation is less clear but no less important. The importance of S6K1 in the regulation of cell size is best seen in the smaller body size of animals that lack S6K1 protein (22, 50; see below). mTORC1-dependent signalling also regulates the movement of the ribosome along the mRNA (translation elongation) through the regulation of elongation factor 2. The result is that when mTORC1 becomes activated, the rate of translation initiation and elongation increases in muscle (51). Therefore, it should be no surprise that protein synthesis increases in an mTORC1-dependent manner in the period after resistance exercise (22, 52).

mTORC1 and translational capacity

Beyond the acute regulation of the activity of the ribosome (the amount of protein synthesized per mRNA), mTORC1 can also regulate the number of ribosomes, or the **translational capacity** of the cell. mTORC1 regulates ribosome mass through the phosphorylation of a key ribosomal RNA transcription factor (upstream binding factor; UBF) and the preferential translation of ribosomal proteins. The transcription of ribosomal DNA is so essential to the function of cells that it has a specialized RNA polymerase (POL I) whose only role is to translate rDNA into the 47S pre-ribosomal RNA. The polycistronic 47S rRNA is then cleaved into the mature 5, 5.8, 18 and 28S rRNAs that make up ~80% of all the RNA in a mammalian cell (53). The transcription of rDNA is thought to be the first step in ribosome biogenesis (the production of new ribosomes) and this process is regulated by the protooncogene myc, and the transcription factors SL1 and UBF. Important for the regulation of ribosome biogenesis in response to resistance exercise, myc transcription increases, and the transcriptional activity of UBF is increased in response to phosphorylation by S6K1. In fact, activation of S6K1 is enough to drive ribosome biogenesis (54). So, following resistance exercise, UBF phosphorylation increases in relation to increased ribosomal RNA synthesis (52).

As stated above, beyond the 4 catalytic ribosomal RNAs, 79 proteins are required to form the large and small ribosomal subunits. All the mRNAs that encode the ribosomal proteins contain a long sequence of pyrimidines in their 5' untranslated region (UTR). This so-called 5' terminal oligopyrimidine (TOP) tract motif normally limits translation of this special subset of proteins by binding LARP1 (La-related protein 1). However, when LARP1 becomes phosphorylated by mTORC1, PKB/Akt and/or S6K1, it dissociates from the 5'UTR of the ribosomal protein mRNAs and this increases the translation of ribosomal proteins (55). Therefore, mTORC1 controls the production of ribosomes by regulating the transcriptional activity of UBF and the translation of ribosomal protein mRNAs through the phosphorylation of LARP1.

To summarize, mTORC1 is activated by growth factors, resistance exercise and amino acids, and inhibited by catabolic stimuli. Active mTORC1 increases ribosomal activity by promoting translation initiation and elongation, resulting in an acute increase in protein synthesis. Over time, active mTORC1 can increase the transcription of ribosomal DNA through phosphorylation of UBF and the translation of ribosomal protein mRNA, through the phosphorylation of LARP1. Together, the acute increase in transcriptional activity combined with the prolonged increase in translational capacity, results in long periods of time when myofibrillar protein synthesis remains higher than protein breakdown, resulting in the addition of sarcomeres in parallel and muscle fibre hypertrophy.

THE HUNT FOR THE ELUSIVE MECHANOSENSOR

Initially, researchers assumed that resistance exercise and other forms of muscle overload activated the mTORC1 pathway by changing the expression of the growth factor IGF-I. However, hypertrophy can occur in response to overload in transgenic mice where the IGF-I receptor is mutated (38) or where the signalling in between IGF-I and PKB/Akt was otherwise inhibited (36). Moreover, the time course of IGF-I expression and mTOR activity does not match. Other results against a requirement of IGF-I followed, leading to a point-counterpoint debate on whether IGF-I is required for overload-induced mTOR activation and hypertrophy in the *Journal of Applied Physiology* (56). What is clear from this work is that tension directly stimulates mTORC1-dependent protein synthesis via a growth factor independent mechanism (57). The mechanosensors studied to date include stretch-activated channels (58), diacylglycerol kinase ζ (59), titin (60) or costameres (61). One of these, or another mechanoreceptor, will couple the load across the muscle fibre to the activation of RIK (**Figure 8.8**) and the phosphorylation and movement of TSC2 away from Rheb (35). However, exactly what senses mechanical stress is still unknown.

MYOSTATIN AS A NEGATIVE REGULATOR OF MUSCLE MASS

Diametrically opposed to the resistance exercise-mTORC1 pathway is the myostatin-Smad2/3 pathway. In contrast to mTORC1, activating the myostatin-Smad2/3 signalling pathway inhibits skeletal muscle growth. This is most obvious in animal models where myostatin activity has been removed. In mice (62), dogs (63) or cows (64) where myostatin has been knocked out or mutated, muscles can be twice the size of animals with normal myostatin. Combining myostatin knockout with overexpression of follistatin, an inhibitor of myostatin and other members of the transforming growth factor beta (TGFβ)/activin/ growth and differentiation factor (GDF)/bone morphogenetic protein (BMP) family of cytokines, can increase muscle mass even more (65). **Figure 8.8** gives an introduction into the signalling of the myostatin-Smad2/3 signalling pathway. Inhibiting myostatin-Smad2/3 signalling further by blocking Smad transcriptional activity with the transcriptional repressor ski also results in huge increases in muscle mass (66). Therefore, the myostatin-Smad pathway clearly regulates skeletal muscle size, but does the increase in muscle mass lead to an increase in muscle strength? The answer to this question is: it depends on how you measure it. In the myostatin knockout mouse, there is an increase in the absolute amount of force the muscle produces, but the specific force (also known as muscle quality), which is the force in relation to muscle size, is lower than control wild-type muscle (67). However, reports of a boy who has no myostatin and his family (68), and the heterozygous myostatin racing dogs (63), suggest that the increase in muscle mass that occurs as a result of inhibiting myostatin-Smad2/3 signalling together with training can improve strength and speed, whereas simply decreasing myostatin-Smad2/3 signalling alone may increase mass, but may actually decrease function. One of the potential explanations for this difference is that myostatin is a member of the TGFβ family of cytokines. These cytokines play a central role in the expression of collagen and other ECM proteins. Interestingly, muscles from myostatin knockout mice that are loaded less show lower collagen content and are more prone to contraction-induced muscle damage than either the same muscle in a wild-type animal or a more frequently loaded muscle in a myostatin knockout (67). These data suggest that blocking myostatin-Smad2/3 signalling alone increases muscle size but not collagen content, whereas combining myostatin-Smad2/3 inhibition with exercise results in both an increase in muscle fibre size and collagen content and this is translated into an increase in muscle strength. In support of this hypothesis, blocking the increase in collagen content that occurs following synergist ablation has no effect on the increase in fibre diameter, whereas maximal force goes up only half as much as control animals (69). Together, these data suggest that in the absence of training, inhibiting myostatin-Smad2/3 activity increases muscle mass but decreases collagen content resulting in a decrease in specific force.

Myostatin-Smad2/3 activity and resistance exercise

Even though genetically inhibiting myostatin-Smad2/3 activity clearly results in muscle growth, how/ whether resistance exercise affects this pathway remains controversial. Louis et al. (70) measured myostatin mRNA using RT-PCR from 0 to 24 hours after resistance and endurance exercise. They observed

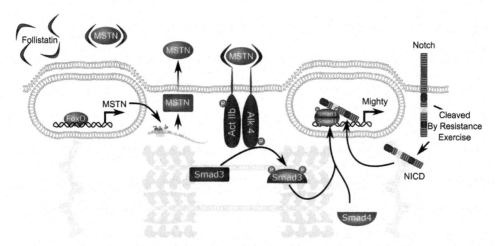

Figure 8.8 Glucocorticoids and other stimuli that drive muscle atrophy commonly increase myostatin mRNA. Transcribed and translated myostatin is then cleaved by proteases and is secreted. Extracellular myostatin can alternatively form inactive complexes with inhibitor proteins, such as follistatin, but free myostatin binds to the activin-type IIb/Alk 4 receptor. When myostatin binds, Alk4 phosphorylates Smad2 or 3. Phosphorylated Smad 2 or 3 together with Smad4 enter the nucleus where they regulate genes that control muscle mass. Following resistance exercise, the transmembrane protein Notch can be cleaved, releasing the notch intracellular domain (NICD). The NICD can move to the nucleus and inhibit Smad transcriptional activity, thus leading to muscle-specific inhibition of myostatin signalling.

a significant decrease in myostatin mRNA after resistance exercise peaking 8 hours after exercise. However, they also observed a decrease in myostatin mRNA after endurance exercise, which does not normally increase muscle mass. Kim et al. also saw a decrease in myostatin mRNA following resistance exercise and by separating their participants into three groups: (1) non-responders, whose mean fibre CSA did not change; (2) moderate responders, whose mean fibre CSA increased ~1,100 μm^2; and (3) extreme responders, whose mean fibre area increased ~2,500 μm^2 over 16 weeks of training; they were able to show that the decrease in myostatin mRNA was not related to the change in muscle mass that occurred with training (71). Therefore, there is no clear relationship between myostatin mRNA after resistance exercise and changes in muscle mass.

When you consider the pathway shown in **Figure 8.8**, it is perhaps not surprising that myostatin mRNA does not relate to muscle growth after exercise. After myostatin mRNA is made in a muscle, it must be translated and shipped out of the muscle where it must be cleaved and activated, but it can also become bound up by inhibitors. When myostatin protein does bind to its receptor, it has to activate Smad2 or 3 and promote their movement into the nucleus with Smad4 leading to the activation of transcription, which can be blocked by inhibitors of Smads such as ski, sno or notch. Further, we all have seen individuals who only lift heavy weights with their upper body and the result is large upper body muscle mass and small leg muscles. If globally decreasing myostatin (a circulating factor) was how resistance exercise inhibited myostatin-Smad2/3 activity, then why would our lazy weightlifter's legs not get bigger too? Some suggestion as to the mechanism was presented by MacKenzie et al. who used the expression of mighty (akirin1), a known myostatin-Smad2/3 target (72), to study how resistance exercise affected myostatin-Smad2/3 activity (73). The authors found that myostatin-Smad2/3 activity, 6 hours after exercise, was inversely related to the increase in muscle mass following 6 weeks of training. Interestingly, the decrease in myostatin-Smad2/3 activity was not related to myostatin mRNA, Smad2 phosphorylation, or ski levels in the muscle. Instead, the Smad2/3 inhibitor notch was activated by resistance exercise, and this was related to myostatin-Smad2/3 inactivation. Notch is normally a large transmembrane protein, and during resistance exercise notch is cleaved, and the notch intracellular domain is released and moves to the nucleus where it inhibits Smad2/3 transcriptional activity. In this way, myostatin-Smad2/3 activity can be decreased specifically in the muscle fibres that have been loaded during the bout of exercise, allowing our lazy weightlifter to maintain their large upper body at the same time their leg muscles remain small and weak.

Mechanism of myostatin-Smad2/3 activity

The inhibition of myostatin-Smad2/3 activity following resistance exercise is different than the genetic models of myostatin inhibition (**Chapter 4**). When myostatin is knocked out, mutated or blocked using follistatin or ski, muscle cells lack myostatin-Smad2/3 signalling from the moment they are created during development. When this happens, muscle cells divide more during development, and this results in more muscle fibres in the adult. Once formed, or if myostatin-Smad2/3 activity is reduced in the adult, using myostatin antibodies, follistatin or resistance exercise, then fibre hypertrophy is the main effect. In both cases, skeletal muscle hypertrophy occurs once again because myofibrillar protein synthesis is greater than degradation resulting in more sarcomeres in parallel. Inhibition of myostatin-Smad2/3 signalling increases protein synthesis in part through mTORC1. The activation of mTORC1 is modulated through the ability of Smad2/3 to control PIP3 levels. When Smad3 is inactivated (myostatin is absent), PIP3 levels in the membrane increase, resulting in the activation of PKB/Akt. By working at the level of PKB/Akt, inhibition of myostatin-Smad2/3 activity can both increase myofibrillar protein synthesis and decrease degradation. The increase in protein synthesis happens through the phosphorylation and movement of TSC2 away from Rheb, and the Rheb-induced activation of mTORC1 described above (74). However, it is important to note that rapamycin can only decrease 50% of the increase in muscle mass that occurs because of inhibition of myostatin-Smad2/3 activity. At least part of the rapamycin-independent effect of inhibiting myostatin-Smad2/3 activity is mediated through a PKB/Akt-dependent decrease in protein degradation. To slow protein degradation, PKB/Akt phosphorylates the FoxO (Forkhead Box Protein, Subclass O) transcription factors and this keeps FoxO out of the nucleus. By excluding FoxO from the nucleus, the production of FoxO target genes, including the so-called 'atrogenes' (or atrophy genes) MuRF1 and MAFbx (E3 Ubiquitin Ligases – introduced in **Chapter 7**), decreases and this slows protein degradation. Therefore, activating myostatin-Smad2/3 decreases myofibrillar protein synthesis by inhibiting PIP3/Akt/mTORC1 signalling and increases muscle protein degradation by inhibiting PIP3/Akt resulting in the activation of FoxO/atrogenes.

E3 ligases and muscle mass

Even though the E3 ligases MuRF1 and MAFbx (introduced in Chapter **7**) were first identified as being upregulated during muscle atrophy and their knockdown could prevent muscle wasting (75), E3 ligases have a far more complex relationship with muscle growth than is generally recognized. First, MuRF1 and MAFbx levels and protein degradation through the proteosome are highest at periods of significant muscle growth (76). The increase in protein breakdown during muscle growth is likely the result of the need for remodelling and potentially the need for amino acids as sarcomeres are added in parallel. Second, novel E3 ubiquitin ligases (UBR4 and UBR5) have been identified as playing a role in anabolism, hypertrophy and recovery from atrophy (77, 78). UBR5 was identified in a methylation screen as containing a hypomethylated DNA signature and increased expression of UBR5 after resistance training in humans (79). Interestingly, single nucleotide polymorphisms (SNPs) of UBR5 have been identified that lead to increased expression of the gene. In a study of 357 athletes and controls, the A alleles of two SNPs (rs10505025 and rs4734621) were strongly associated with larger CSA of fast-twitch muscle fibres and were more frequently observed in strength/power versus endurance/untrained phenotypes. Lastly, knockdown of UBR5 causes reduced total RNA, reduced muscle protein synthesis and ultimately muscle atrophy (80). At the moment, three roles for E3 ligases in anabolism and hypertrophy have been proposed:

1. E3 ligases might be involved in degrading negative regulators of anabolic signalling.
2. E3 ligases might be important in maintaining RNA translation efficiency by regulating translation initiation factor levels.
3. E3 ligases might maintain protein quality control by clearing out damaged proteins before new protein can be synthesized to increase muscle mass.

However, identifying the substrates of these E3 ligases is essential before the role of ubiquitination and protein degradation in the regulation of muscle mass can be fully understood.

Satellite cells, repair after muscle injury and adaptation to resistance exercise

Human muscle fibres can be up to 20 cm long and 10,000 μm² in area (8). They are therefore the second largest cells in the human body after α motor neurones. Each muscle fibre is a single cell that contains many nuclei, termed myonuclei. In rats, there are between 44 and 116 myonuclei per millimetre of fibre with more nuclei in slow, type I fibres than in type II fibres (81). Applying these nuclei-per-millimetre of fibre values to human muscle fibres suggests that the largest human muscle fibres could contain between 9,700 and 23,000 nuclei. This would make muscle fibres by far the most nucleated cells of the human body. A unique feature of a muscle fibre with multiple myonuclei is that they cannot divide. The scientific term for their inability to divide is to say that they are post-mitotic or terminally differentiated. Therefore, if additional myonuclei are required during post-natal growth, to regenerate damaged or injured fibres, or to avoid the dilution of nuclei during muscle fibre hypertrophy after resistance exercise, where do the additional myonuclei come from?

In skeletal muscle, the production of new myonuclei is outsourced to the so-called **satellite cells**. Satellite cells are the resident stem cells of skeletal muscle (as described in **Chapters 2** and **13**). In human muscle, it has been estimated that between 1.4 and 7.3% of all nuclei are satellite cells, with higher frequencies in muscles or around muscle fibres with a slower phenotype (82). Satellite cells were discovered by Alexandra Mauro using an electron microscope to study frog muscle fibres (**Figure 8.9**; 61). In his paper, Mauro described satellite cells as a subset of cells 'wedged' between the plasma membrane, also known as the sarcolemma, and the basement membrane or basal lamina. Because of their position, he named these cells 'satellite cells', noted that satellite cells were almost all

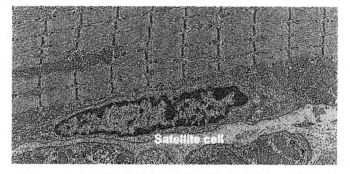

Figure 8.9 A satellite cell from a rat sartorius anterior muscle. Above the satellite cell are the sarcomeres and mitochondria of a muscle fibre. Note the membrane surrounding the satellite cell (sp: satellite cell plasma membrane) and the other around the muscle fibre (mp: muscle plasma membrane). Figure taken from the 1st edition of this book.

nucleus with very little cytoplasm and hypothesized that their function might be to repair skeletal muscle (83).

The regenerative capacity of skeletal muscle has been studied for well over a century, and this early work showed that muscles from small animals such as chick, pigeon, rat and mouse could completely regenerate after catastrophic injury (84). However, much of this work was unappreciated because it was performed in the former Soviet Union and published predominantly in Russian (85). Bruce Carlson visited the USSR as part of an exchange programme between the Academies of Sciences of the US and the USSR, and upon his return to the US, Carlson brought attention to the work and repeated many studies in his own laboratory. The most surprising finding from this early work was that muscles could be removed, minced or completely homogenized, the pieces (or slurry) placed into the site of injury, and within 14 or 28 days (rat and frog, respectively) the muscles would be completely regenerated. These experiments highlighted that muscle fibres can regenerate extremely well, and the discovery of satellite cells suggested that these tissue-specific stem cells underlie this regeneration.

Satellite cells as stem cells

To conclusively demonstrate that satellite cells were necessary and sufficient for muscle repair, a London (UK)-based group transplanted muscle fibres with their resident satellite cells that contained genetic self-renewal or differentiation reporters into a non-transgenic recipient mouse (86). To understand the process of self-renewal and differentiation of muscle cells, see **Chapter 13** 'Satellite Cells and Exercise'. The self-renewal reporter generated a blue dot for each self-renewed satellite cell and the differentiation reporter produced a blue dot whenever a satellite cell differentiated and fused with a muscle fibre or by forming a new muscle fibre. The team found that the blue self-renewal and differentiation reporters were both switched on in regenerating recipient muscle. Moreover, the recipient muscle showed large numbers of self-renewed and differentiated satellite cells. This experiment confirmed that satellite cells are true muscle stem cells and are capable of extensive self-renewal and differentiation/regeneration (86). Perhaps the most impressive experiment on satellite cell self-renewal involved transplanting a single satellite cell (termed muscle stem cell in that study due to the isolation method) expressing a firefly reporter gene into a regenerating muscle. Following several rounds of muscle injury and repair, the investigators estimated that the single satellite cells gave rise to between 20,000 and 80,000 daughter cells (progeny) (87). To conclude, satellite cells are the primary adult stem cells within skeletal muscle. There are some other stem cells within muscle, such as vascular-based pericytes or mesoangioblasts, but their capacity to repair muscle is limited.

Myogenesis

Before reviewing how satellite cells respond to resistance exercise and injury and whether their function is essential for regeneration and hypertrophy, we will first cover the molecular mechanisms that regulate the identity of satellite cells and the development of skeletal muscle, termed myogenesis. Myogenesis occurs not only during embryonic development but also when satellite cells respond to injury or growth stimuli.

In the first major molecular myogenesis experiment, 5-azacytidine was used to convert non-muscle cells (fibroblasts) into muscle cells. The muscle-making agent used 5-azacytidine, a drug that removes the methylation marks from DNA. As discussed in **Chapter 6**, DNA methylation is the primary epigenetic tool to stably alter the structure of the DNA, opening or closing the promoter regions of different genes and thereby making them more or less likely to be transcribed, respectively. Treating the fibroblasts with 5-azacytidine must have 'opened up' crucial myogenic (muscle-making) genes allowing the cells to swap their lineage from fibroblasts to muscle cells. The authors then identified 26 different sequences that were only expressed after 5-azacytidine treatment (labelled A-Z). The fourth of these myogenic genes, MyoD (Myo for myogenic and D from the fourth sequence isolated), was a transcription factor that was sufficient to convert fibroblasts into myoblasts or muscle cells (88, 89). Later, the team demonstrated that

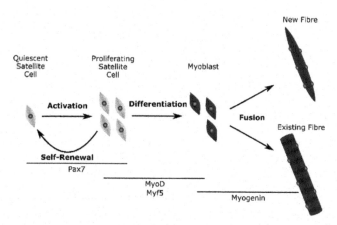

Figure 8.10 Myogenesis. Satellite cells become activated after muscle injury and begin to proliferate. Once sufficient cell mass is produced, the proliferating cells begin to differentiate and fuse to either regenerate existing fibres or generate new fibres.

MyoD could also convert other cells – pigment, nerve, fat and liver cells – into myoblasts, making MyoD the first determination gene, a single gene that can determine the fate of a muscle cell. Soon after, several other proteins that shared the basic helix-loop-helix (bHLH) structure with MyoD were identified and shown to also regulate myogenesis. The genes within this family, MyoD, Myf5, MRF4 (Myf6) and myogenin, are known collectively as myogenic regulatory factors (MRFs). The in vivo function of these MRFs was characterized in several studies by using knockout mice. So, did MyoD knockout prevent the formation of skeletal muscle as one might predict? No, to the surprise of the investigators, MyoD knockout mice developed skeletal muscle. However, soon after it was shown that when MyoD and another MRF, Myf5, were knocked out together, myogenesis was completely prevented (there were no myoblasts (muscle cells) in the mouse). These data show that MyoD and Myf5 are what is known as redundant proteins; they are similar enough to each other that one can compensate when the other is missing (90). Where MyoD and Myf5 are essential to making myoblasts, another MRF, myogenin, is necessary for the fusion of myoblasts into multinucleated myotubes, the precursors of muscle fibres (90). Therefore, MyoD and Myf5 are needed to make myoblasts, help them proliferate and potentially begin differentiation, whereas myogenin is needed to drive the differentiation and fusion of myoblasts into muscle fibres (**Figure 8.10**).

How do myogenic regulatory factors function?

MyoD functions as a transcription factor that binds to ~60,000 MyoD DNA binding sites located all over the genome (91). Once bound to DNA, MyoD recruits acetyl- and methyltransferases that serve to open the DNA around muscle enhancer regions and then MyoD recruits more transcription factors to these enhancers driving the transcription of muscle-specific genes (92).

What is the relevance of myogenic regulatory factors for satellite cells?

MRFs help satellite cells proliferate and then differentiate into a muscle fibre. In inactive, or quiescent, satellite cells, the MRFs are generally expressed at low levels. During the quiescent stage, the paired box transcription factor Pax7 is essential to maintain satellite cells (**Figure 8.10**). In fact, mice where Pax7 has been knocked out, satellite cell numbers are very low indicating that Pax7 is key for the development of satellite cells. In the absence of Pax7$^+$ satellite cells, muscle does not grow as big and is not able to regenerate after injury, demonstrating the importance of satellite cells in developmental muscle growth and regeneration after injury (93). Today, Pax7 is used as a marker to identify satellite cells immunohistochemically. However, Pax7 is only important for generating satellite cells and is not required for their function in adult muscle. This was demonstrated in an experiment where Pax7 was only knocked out in adult muscle, after satellite cells had formed, and muscle regeneration occurred normally (94).

Taken together, in response to muscle damage, satellite cells move through the following stages:

1. **Quiescent satellite cell** – Most satellite cells in an adult muscle are quiescent and express Pax7 and not MyoD.
2. **Activated satellite cell** (also described as satellite cell-derived myoblast) – In response to trauma, satellite cells become activated, break through the basal lamina (**Figure 8.9**), migrate to the injured area and proliferate to form clusters of activated satellite cells (**Figure 8.10**). Activated satellite cells continue to express Pax7 but also express high levels of MyoD.
3. **Differentiating satellite cell** – Activated satellite cells then make a cell fate decision, either they continue to proliferate, return to quiescence (self-renew), or differentiate and fuse either with an existing muscle fibre or to other myoblasts to form a new fibre (hyperplasia). Differentiating satellite cells lose Pax7 expression, still express MyoD and begin to express myogenin, which is required for fusion into existing fibres or to generate new fibres (**Figure 8.10**).
4. **Satellite cells return to quiescence** (also known as **self-renewal**) – Self-renewing satellite cells that exit the cell cycle and lose the expression of MyoD return to a quiescent state to preserve satellite cell number (stem cell pool) in preparation for the next injury.

Satellite cell regulation

With this basic understanding of Pax7 and the MRFs, the next obvious question is 'what signals activate satellite cells?' Unfortunately, there is not a consensus answer to this question. This is also discussed in more detail in **Chapter 13** in respect to aerobic and resistance exercise, ageing and communication of satellite cells with other cell types. Since satellite cells are specialized to respond to muscle damage, it should surprise no one that the immune system is important in the activation of satellite cells after injury. The acute inflammatory reaction in response to injury results in an increase in neutrophils followed by infiltration of macrophages. The first macrophage cells to enter muscle after injury are the pro-inflammatory (or M1) marcophages and these cells predominate in the early phase of repair (88). Depleting the M1 macrophages from muscle prior to injection with the myotoxic drug notexin completely prevents muscle regeneration, suggesting that M1 marcophages are essential to satellite cell activation (95). By the second day following injury, the number of M1 macrophages declines and a second population of macrophages enter the injured muscle. The second subset of marcophages express anti-inflammatory cytokines and are referred to as (M2) macrophages. The transition from M1 to M2 macrophages is essential for the regeneration of skeletal muscle after injury. The M1 macrophages activate and stimulate the proliferation of satellite cells, whereas the M2 macrophages stimulate myoblast differentiation and fusion (95). Even though the immune system can regulate the activation and differentiation of satellite cells, other factors contribute to determining the cell mass generated in response to injury through the regulation of the rate of proliferation.

Numerous factors affect the proliferation rate of satellite cells after injury. As with the regulation of muscle mass, satellite cell proliferation is negatively regulated by myostatin-Smad2/3 signalling (96) and positively regulated by the Notch pathway (97). As would be expected from the discussion above, myostatin-Smad2/3 activity has been shown to inhibit satellite cell proliferation (98), and this could be part of the mechanism by which myostatin inhibits muscle growth especially after birth. Consistent with the negative effects of myostatin-Smad2/3 signalling, activation of Smad1, 3 and 5, transcription factors that compete with Smad2/3 for the co-Smad (Smad4) and entry into the nucleus, has been shown to drive the proliferation of activated satellite cells whilst preventing differentiation (99). Other pathways that regulate satellite cell proliferation include the Wnt/β-catenin pathway (100) and Hippo-Taz/Yap signalling that both promote the proliferation of satellite cells (101, 102). The role of Notch, Smad and Wnt and their role in satellite cells with exercise are discussed in more detail in **Chapter 13**. Together, these data suggest that the activation of satellite cells is stimulated by M1 macrophages but the extent of proliferation and therefore the number of cells generated following injury are regulated by numerous growth factors, including myostatin-Smad2/3.

Satellite cells and muscle mass

Now that we understand satellite cells, the question arises: 'are satellite cells essential for hypertrophy in response to resistance exercise?' This question has been the subject of an intense debate that has taken place within the field for over 30 years. Closely related to this question is the much-discussed myonuclear domain hypothesis, which states that there are mechanisms that maintain a fixed nuclear-to-cytoplasmic volume ratio. If this were true, then new nuclei (presumably derived from satellite cells) would need to be added for fibres to hypertrophy, whereas atrophying fibres should lose nuclei. The debate surrounding whether satellite cells are required for muscle hypertrophy was the focus of a point-counterpoint debate in the *Journal of Applied Physiology* (103). However, like most of biology, there is no clear-cut answer, but this review (104) provides a wonderful and thorough review of all of the literature. This section will summarize a small portion of this work to answer the question: 'is an addition of myonuclei by satellite cells required for hypertrophy and is the myonuclear domain hypothesis valid?'. This is intriguing, given that no answer to the former part of the question leads to the rejection of the myonuclear domain hypothesis.

The most compelling argument against the myonuclear domain hypothesis is from genetic studies that either knocked out myostatin (105) or overexpressed constitutively active PKB/Akt (106). Both genetic manipulations significantly increase muscle mass without the addition of myonuclei and therefore in both cases the myonuclear domain increased significantly, suggesting that myonuclear addition from satellite cells is not essential for hypertrophy and that the myonuclear domain hypothesis is not a strict rule.

Satellite cells and load-induced muscle hypertrophy

To conclusively establish the role of satellite cells in load-induced muscle hypertrophy, one must eliminate satellite cells from an adult skeletal muscle and then test whether the muscle can still hypertrophy in response to overload. Rosenblatt and Parry attempted this using γ-irradiation to eliminate satellite cells and synergist ablation (overload) to induce hypertrophy. Following overload, they observed that muscle mass decreased ≈25% after γ-irradiation and overload, whereas the same overload without γ-irradiation increased muscle mass by ≈20% (107). They interpreted these data to mean that functioning satellite cells were essential for muscle hypertrophy in response to overload. However, a major limitation of this experiment was that γ-irradiation has profound effects on muscle, beyond the loss of satellite cells, that might underlie this detrimental effect. Nearly 20 years later, McCarthy et al. created a transgenic mouse that allowed them to knock out > 90% of the satellite cells in adult skeletal muscle. Removing the satellite cells had no impact on the hypertrophic response to synergist ablation (108), the fibres from the knockout mice after overload demonstrated fewer myonuclei, and therefore showed a greater increase in myonuclear domain than muscles where satellite cells were intact. This experiment demonstrated that satellite cells are not required for hypertrophy induced by acute synergist ablation in rodents. Follow-up work with these mice has shown that muscles from still growing mice (two-month-old) do not hypertrophy as a result of synergist ablation in the absence of satellite cells (109) and that following a more physiological load-induced increase in muscle mass, recovery following unloading is completely normal in mice lacking satellite cells (110). Together, these data suggest that satellite cells are required for developmental growth, but that load-induced muscle hypertrophy after synergistic ablation in mature rodents occurs independently of satellite cells. However, more physiological models of hypertrophy in rodents are now available and satellite cell depletion studies in these models will be discussed in more detail in chapter **13**.

Satellite cells in human hypertrophy in response to resistance exercise

Even though satellite cells may play a limited role in load-induced skeletal muscle hypertrophy after synergist ablation in rodents, this does not mean that they are not important in the response to resistance exercise in humans. Using the non, moderate and extreme responders to resistance exercise described above, Petrella et al. showed that the increase in muscle fibre CSA best correlated with the increase in myonuclear number following 16 weeks of training (111). The relative satellite cell number at baseline predicted the increase in muscle mass with training. Interestingly, work from Luc van Loon's laboratory demonstrated that, in response to 12 weeks of strength training, hypertrophy of small muscle fibres (2,-000–4,000 μm^2) best correlated with an increase in myonuclear domain, whereas the hypertrophy of the largest muscle fibres (8,000–10,000 μm^2) best correlated with the increase in myonuclear number (112). These data suggest that small muscle fibres can hypertrophy without the addition of myonuclei derived from satellite cells, whereas larger fibres are more dependent on satellite cells to grow. This observation could help integrate the rodent data, where satellite cells are not necessary for hypertrophy and the human experiments, where satellite cells/myonuclear number predict hypertrophy, because mean fibre areas in mouse muscles are <1,500 μm^2, whereas in humans mean fibre areas are more on the order of 4,500 μm^2. In this way, the observations that in rodents (small fibres) satellite cells are not required for growth, whereas in humans (large fibres) satellite cells are important in hypertrophy are consistent.

SIGNALLING INCREASES IN STRENGTH

As described above, mTORC1, myostatin-Smad2/3 and satellite cells are important for determining the size of a muscle. However, as we have also seen, this does not always translate into an increase in strength. To make a muscle bigger, more myosin and actin protein need to be added in parallel. To make the muscle stronger, the force produced by the additional myosin and actin needs to be transferred to the bones. Therefore, force transfer is an important determinant of strength. As stated above, when Stantzou et al. performed synergist ablation and then treated animals with halofuninone to prevent an increase in collagen content, they observed a normal, or slightly greater, increase in muscle fibre diameter, but the increase in force was half that of the saline treated animals (69). This suggests that increased collagen may be required to maximally increase force. If this is true, how could load increase collagen production? One possible mechanism is through the activation of a gene called Egr (early growth response)1. Egr1 is known to regulate collagen and ECM production (113) and is one of the most upregulated genes in muscle following resistance exercise (114). Unlike the pathways above that are most related to muscle mass, Egr1 is regulated by ERK1/2 (extracellular regulated kinase) and ERK1/2 activity following resistance exercise is directly related to the load across the muscle (115). This then could explain the observations that began this chapter: to increase strength, you must lift a heavy weight. The heavier the weight, the greater the activation of ERK1/2; the greater the activation of ERK1/2, the more Egr1 is produced; the more Egr1 is produced, the greater the increase in force transfer proteins (collagen and other proteins in the ECM); and the greater the increase in force transfer proteins, the more force is transferred from the motor proteins (actin and myosin) to the tendon and bone. However, there is still alot more research to be done in order to evaluate more deeply the signal transduction pathways involved in increases in strength after resistance exercise.

SUMMARY

Resistance training involves exercising against high loads for short periods of time. There is no 'optimal' training regime. Any exercise performed to failure will increase muscle size, whereas strength gains are proportional to the weight lifted. Therefore, lifting a heavy weight to failure will increase muscle size and strength. The major gross adaptation to strength training is muscle hypertrophy, where myofibres increase in size because of the addition of sarcomeres in parallel. In the fasted state, resistance exercise acutely increases myofibrillar protein synthesis and breakdown. Myofibrillar protein balance will only become positive when sufficient essential amino acids are ingested. Resistance exercise and leucine-rich amino acids lead to the activation of the mTORC1 pathway, which then increases the translational activity of the cell by targeting proteins that increase the rate of initiation and elongation. The translational capacity is also increased by prolonged activation of mTORC1 through the production of new ribosomes. The mTORC1 pathway integrates the input of several positive signals, including hormones, such as IGF-I and insulin, resistance exercise and essential amino acids. Additionally, mTORC1 can be inhibited by catabolic pathways, including AMPK, which is activated by acute endurance exercise, a caloric deficit and low glycogen. Myostatin inhibits muscle growth by decreasing myofibrillar protein synthesis and increasing degradation. Both of these effects could be mediated through the Smad2/3-dependent inhibition of PIP3/Akt signalling. A decrease in Akt signalling would decrease hormonal activation of mTORC1 and increase the activity of FoxO resulting in a decrease in protein synthesis and an increase in degradation, respectively. Following resistance exercise, myostatin-Smad2/3 activity is decreased in the exercised muscle because of the cleavage and activation of notch, a transmembrane protein that can block Smad2/3 activity. Satellite cells are the resident stem cells of skeletal muscle. They express Pax7 and are normally quiescent in uninjured skeletal muscle. With a growth- or injury-stimulus, satellite cells become activated, express MyoD, proliferate until sufficient cell mass is created, and then either turn on myogenin and differentiate or turn off MyoD and return to quiescence, a process called self-renewal. Satellite cells are not essential for hypertrophy of small muscle fibres (like those found in rodents) but are perhaps more important in the hypertrophy of larger fibres (like those found in humans), and for regenerating a skeletal muscle after injury. The force produced by the extra motor proteins added during muscle hypertrophy needs to be transferred to the bone through matrix proteins such as collagen and these proteins are required for the increase in strength with resistance exercise.

REVIEW QUESTIONS

- What is known about the effects of resistance exercise on myofibrillar protein synthesis?

- Discuss the evidence for the hypothesis that mTORC1 is a key mediator of the protein synthesis response to resistance exercise and other forms of skeletal muscle overload.

- Explain how mTORC1 regulates protein synthesis.

- Discuss the evidence that myostatin is a key regulator of the muscle growth adaptation to resistance exercise.

- Describe how myostatin-Smad2/3 activity is inhibited in muscles following resistance exercise.

- Explain what satellite cells are.

- Compare and contrast the function of Pax7 and MyoD in relation to satellite cells.

- Describe the 'myonuclear domain hypothesis' and discuss whether it has been experimentally confirmed.

- Discuss whether satellite cells are required for skeletal muscle hypertrophy and regeneration after muscle injury.

REFERENCES

1. Hubal MJ, et al. Med Sci Sports Exerc. 2005. 37(6):964–72.
2. Mitchell CJ, et al. J Appl Physiol. 2012. 113(1):71–7.
3. Wong TS, et al. 1988. 65(2):950–4. doi:101152/jappl1988652950

4. Sale DG. *Med Sci Sports Exerc.* 1988. 20(5):S135–45.
5. Van Cutsem M, et al. *J Physiol.* 1998. 513(1):295–305.
6. Chesley A, et al. *J Appl Physiol.* 1992. 73(4):1383–8.
7. Lexell J, et al. *J Neurol Sci.* 1988. 84(2–3):275–94.
8. Heron MI, et al. *J Morphol.* 1993. 216(1):35–45.
9. Goodman CA, et al. *J Physiol.* 2011. 589(22):5485–501.
10. Glenmark B, et al. *Acta Physiol Scand.* 1992. 146(2):251–9.
11. Janssen I, et al. *J Appl Physiol.* 2000. 89(1):81–8.
12. Forsberg AM, et al. *Clin Sci.* 1991. 81(2):249–56.
13. Carroll CC, et al. *J Muscle Res Cell Motil.* 2004. 25(1):55–9.
14. Miller BF, et al. *J Physiol.* 2005. 567(3):1021–33.
15. Damas F, et al. *J Physiol.* 2016. 594(18):5209–22.
16. Phillips SM, et al. *Am J Physiol – Endocrinol Metab.* 1997. 273(1 36–1).
17. Tipton KD, et al. *Am J Physiol – Endocrinol Metab.* 1999. 276(4 39-4):628–34.
18. Phillips SM, et al. *Am J Physiol – Endocrinol Metab.* 1999. 276(1 39-1).
19. Sabatini DM, Proc Natl Acad Sci U S A; 2017. 114:11818–25.
20. Baar K, et al. *Am J Physiol – Cell Physiol.* 1999. 276(1):C120–7.
21. Bodine SC, et al. 2001. 3(11):1014–9.
22. Drummond MJ, et al. *J Physiol.* 2009. 587(7):1535–46.
23. You JS, et al. *FASEB J.* 2019. 33(3):4021–34.
24. Yang H, et al. *Nature.* 2017. 552(7685):368–73.
25. Ruderman NB, et al. *Proc Natl Acad Sci U S A.* 1990. 87(4):1411–5.
26. Sarbassov DD, et al. *Science (80-).* 2005. 307(5712):1098–101.
27. Nascimento EBM, et al. *Cell Signal.* 2010. 22(6):961–7.
28. Tee AR, et al. *J Biol Chem.* 2003. 278(39):37288–96.
29. Tee AR, et al. *Curr Biol.* 2003. 13(15):1259–68.
30. Coleman ME, et al. *J Biol Chem.* 1995. 270(20):12109–16.
31. Sancak Y, et al. *Science (80-).* 2008. 320(5882):1496–501.
32. Bar-Peled L, et al. *Science (80-).* 2013. 340(6136):1100–6.
33. Bar-Peled L, et al. *Cell.* 2012. 150(6):1196–208.
34. Wolfson RL, et al. *Science (80-).* 2016. 351(6268):43–8.
35. Jacobs BL, et al. *J Physiol.* 2013. 591(18):4611–20.
36. Hamilton DL, et al. *PLoS One.* 2010. 5(7).
37. Hornberger TA, et al. *Biochem J.* 2004. 380(3):795–804.
38. Spangenburg EE, et al. *J Physiol.* 2008. 586(1):283–91.
39. Cermak NM, et al. *Am J Clin Nutr.* 2012. 96:1454–64.
40. Inoki K, et al. *Cell.* 2003. 115(5):577–90.
41. Gwinn DM, et al. *Mol Cell.* 2008. 30(2):214–26.
42. Hickson RC. *Eur J Appl Physiol Occup Physiol.* 1980. 45(2–3):255–63.
43. Thomson DM, et al. *J Appl Physiol.* 2008. 104(3):625–32.
44. Mounier R, et al. *FASEB J.* 2009. 23(7):2264–73.
45. Moberg M, et al. *Sci Rep.* 2021. 11(1):6453.
46. Apró W, et al. *Am J Physiol – Endocrinol Metab.* 2015. 308(6):E470–81.
47. Apró W, et al. *Am J Physiol – Endocrinol Metab.* 2013. 305(1).
48. Gingras AC, et al. *Genes Dev.* 2001. 15(21):2852–64.
49. Haghighat A, et al. *EMBO J.* 1995. 14(22):5701–9.
50. Montagne J, et al. *Science.* 1999. 285(5436):2126–9.
51. Baar K, et al. *Am J Physiol.* 1999. 276:C120–7.
52. West DWD, et al. *J Physiol.* 2016. 594(2).
53. Henras AK, et al. *Wiley Interdiscip Rev RNA.* 2015. 6(2):225.
54. Hannan K, et al. *Mol Cell Biol.* 2003. 23(23):8862–77.
55. Hong S, et al. *Elife.* 2017. 6: e25237.

56. Stewart CE, et al. *J Appl Physiol.* 2010. 108(6):1820–1.
57. Philp A, et al. *J Appl Physiol.* 2011. 110(2):561–8.
58. Spangenburg E, et al. *J Appl Physiol.* 2006. 100(1):129–35.
59. You J, et al. *J Biol Chem.* 2014. 289(3):1551–63.
60. van der Pijl R, et al. *J Cachexia Sarcopenia Muscle.* 2018. 9(5):947–61.
61. Mathes S, et al. *Cell Mol Life Sci.* 2019. 76(15):2987–3004.
62. AC M, et al. *Nature.* 1997. 387(6628):83–90.
63. Mosher DS, et al. *PLOS Genet.* 2007. 3(5):e79.
64. AC M, et al. *Proc Natl Acad Sci U S A.* 1997. 94(23):12457–61.
65. Lee SJ, et al. *Proc Natl Acad Sci U S A.* 2001. 98(16):9306–11.
66. P S, et al. *Genes Dev.* 1990. 4(9):1462–72.
67. Mendias CL, et al. *J Appl Physiol.* 2006. 101(3):898–905.
68. Schuelke M, et al. *N Engl J Med.* 2004. 350(26):2682–8.
69. Stantzou A, et al. *Neuropathol Appl Neurobiol.* 2021. 47(2):218–35.
70. Louis E, et al. *J Appl Physiol.* 2007. 103(5):1744–51.
71. Kim JS, et al. *J Appl Physiol.* 2007. 103(5):1488–95.
72. Marshall A, et al. *Exp Cell Res.* 2008. 314(5):1013–29.
73. MacKenzie MG, et al. *PLoS One.* 2013. 8(7).
74. Winbanks CE, et al. *J Cell Biol.* 2012. 197(7):997–1008.
75. Bodine SC, et al. *Science (80-).* 2001. 294(5547):1704–8.
76. Baehr LM, et al. *Front Physiol.* 2014. 5 FEB.
77. Hunt LC, et al. *Cell Rep.* 2019. 28(5):1268–1281.e6.
78. Seaborne RA, et al. *J Physiol.* 2019. 597(14):3727–49.
79. Seaborne RA, et al. *Sci Rep.* 2018. 8(1):1–17.
80. Hughes DC, et al. *Am J Physiol – Cell Physiol.* 2021. 320(1):C45–56.
81. Tseng BS, et al. *Cell Tissue Res.* 1994. 275(1):39–49.
82. Kadi F, et al. *Pflugers Archiv Eur J Physiol. Pflugers Arch;* 2005. 451:319–27.
83. Mauro AJ. *Biophys Biochem Cytol.* 1961. 9(2):493–5.
84. Studitsky AN. *Ann N Y Acad Sci.* 1964. 120(1):789–801.
85. Carlson BM. *Anat Rec.* 1968. 160(4):665–74.
86. Collins CA, et al. *Cell.* 2005. 122(2):289–301.
87. Sacco A, et al. *Nature.* 2008. 456(7221):502–6.
88. Lassar AB, et al. *Cell.* 1986. 47(5):649–56.
89. Davis RL, et al. *Cell.* 1987. 51(6):987–1000.
90. Arnold HH, et al. *Int J Dev Biol.* 1996. 40(1):345–53.
91. Cao Y, et al. *Dev Cell.* 2010. 18(4):662–74.
92. Blum R, et al. *Epigenetics.* 2013. 8(8):778–84.
93. Sambasivan R, et al. *Development.* 2011. 138(17):3647–56.
94. Lepper C, et al. *Nature.* 2009. 460(7255):627–31.
95. Welc SS, et al. *J Immunol.* 2020. 205(6):1664–77.
96. McKay BR, et al. *FASEB J.* 2012. 26(6):2509–21.
97. Conboy IH, et al. *Science (80-).* 2003. 302(5650):1575–7.
98. McCroskery S, et al. *J Cell Biol.* 2003. 162(6):1135–47.
99. Ono Y, et al. *Cell Death Differ.* 2011. 18(2):222–34.
100. Otto A, et al. *J Cell Sci.* 2008. 121(17):2939–50.
101. Judson RN, et al. *J Cell Sci.* 2012. 125(24):6009–19.
102. Sun C, et al. *Stem Cells.* 2017. 35(8):1958–72.
103. O'Connor RS, et al. *J Appl Physiol.* 2007. 103:1099–102.
104. Murach KA, et al. *Physiology.* 2018. 33: 26–38.
105. Amthor H, et al. *Proc Natl Acad Sci U S A.* 2007. 104(6):1835–40.
106. Blaauw B, et al. *FASEB J.* 2009. 23(11):3896–905.
107. Rosenblatt JD, et al. *J Appl Physiol.* 1992. 73(6):2538–43.

108. Mccarthy JJ, et al. *Development*. 2011. 138(17):3657–66.
109. Murach KA, et al. *Skelet Muscle*. 2017. 7(1):14.
110. Jackson JR, et al. *Am J Physiol – Cell Physiol*. 2012. 303(8).
111. Petrella JK, et al. *J Appl Physiol*. 2008. 104(6):1736–42.
112. Snijders T, et al. *Acta Physiol*. 2021. 231(4):e13599.
113. Havis E, et al. 2020. 21(5):1664.
114. Chen Y-W, et al. *J Physiol*. 2002. 545(1):27–41.
115. Martineau LC, et al. *J Appl Physiol*. 2001. 91(2):693–702.

CHAPTER NINE

Molecular adaptations to endurance exercise and skeletal muscle fibre plasticity

Keith Baar

LEARNING OBJECTIVES

By the end of the chapter, you should be able to:

1. Prescribe endurance exercise for the general public and/or athletes based on solid scientific evidence.
2. Describe individual differences in response to an endurance training programme.
3. Describe the difference between physiological cardiac hypertrophy, the athlete's heart and pathological cardiac hypertrophy, hypertrophic cardiomyopathy. Describe the signal transduction events that regulate both forms of hypertrophy.
4. Describe the characteristics of slow type I, intermediate type IIa and fast type IIx and IIb muscle fibres. Discuss how fibres respond to endurance exercise, supra-physiological stimuli such as chronic electrical low-frequency stimulation and a lack of innervation.
5. Explain the regulation and function of calcineurin-NFAT signalling and the effect on the expression of 'fast' and 'slow' genes.
6. Describe the genomic organisation of myosin heavy chain genes. Explain how MyoMir-Sox6 and chromatin remodelling affect their expression.
7. Explain how AMPK/SIRT/CaMK/p38 MAPK-PGC-1α signalling stimulates mitochondrial biogenesis in response to endurance exercise.
8. Explain how PGC-1α and hypoxia signalling regulate the expression of angiogenic growth factors such as VEGF and how this promotes exercise-induced angiogenesis.

INTRODUCTION

Endurance exercise is not a panacea; however, to many this form of exercise can help prevent and treat lifestyle-related diseases. Endurance exercise is also the main tool used to condition endurance athletes for events from the 1,500 m run to the Ironman Triathlon in Hawaii, the Tour de France or the Raid Gauloises. This chapter will first explore whether common endurance training recommendations (for example, how hard and how long one should train) can be backed up by

DOI: 10.4324/9781315110752-9

sound scientific evidence. The reality is that there is little reliable scientific evidence for many training recommendations and the training response to a given programme varies considerably among individuals. With this background, the rest of the chapter centres on the question 'what are the molecular mechanisms that mediate the adaptation of cardiac and skeletal muscle to endurance exercise?' First, the chapter discusses the mechanisms that are responsible for the development of an athlete's heart and compares the molecular events that drive the genesis of the athlete's heart with those associated with the disease state we know as hypertrophic cardiomyopathy. Second, the chapter will describe type I, IIa and IIx (and IIb in rodents) muscle fibres and describe the effect of short-term endurance exercise on muscle fibre type: a small reduction in type IIx fibres and increase in IIa fibres. We then describe how calcineurin-NFAT signalling differs between fast and slow muscle fibres, and that it has an effect on fibre-type-specific gene expression. Third, the chapter will look at the evolutionary conserved genomic organisation of myosin heavy chain (MyHC) isoform genes in a fast and slow gene cluster. Discuss how MyoMir-Sox6 signalling contributes to the expression of these genes in such a way as to ensure that most of the time only one MyHC isoform is expressed in a given muscle fibre. The chapter will also describe how the DNA of genes is opened up or closed by epigenetic mechanisms such as acetylation and methylation following endurance exercise. Fourth, the chapter discusses how an increase in energy turnover, intracellular calcium, NAD^+, and other signals are sensed by AMPK/SIRT/CaMK/p38 MAPK/PKA and lead to the activation of the transcriptional co-factor PGC-1α via multiple mechanisms. PGC-1α then regulates the expression of mitochondrial genes encoded in the nucleus and the activity of transcription factors that increase the expression of genes which are encoded in mitochondrial DNA (mtDNA). Finally, the chapter discusses how PGC-1α, ERRα and HIF-1α regulate exercise-induced angiogenesis by affecting the expression of growth factors such as VEGF and of metalloproteinases which allow vascular cells to tunnel through the extracellular matrix to form new capillaries.

ENDURANCE EXERCISE: LIMITATIONS

Endurance performance depends on many limiting factors. One important factor that will not be discussed in this chapter but is central to endurance performance is neurocognitive drive, or the psychological determinants of endurance. Please see this review article (1) for more information on this topic. The primary musculoskeletal factors that limit endurance are as follows:

1. **Maximal oxygen uptake ($\dot{V}O_{2max}$):** defined as the maximum rate of oxygen consumption measured during an incremental exercise test, or the highest rate at which oxygen can be utilised to power endurance exercise.
2. **Lactate threshold:** defined as the exercise intensity at which lactate accumulates exponentially in the blood during endurance exercise. This indicates the percentage of $\dot{V}O_{2max}$ that can be maintained without resulting in a drop in power/velocity.
3. **Mechanical efficiency:** defined as the oxygen uptake or energy utilised to maintain a given power output or velocity.

Each of these limiting factors can be further broken down into a plethora of smaller limiting factors which include cardiac output, stroke volume, capillary density, fibre-type proportions, mitochondrial mass, and tendon and ligament stiffness; many of these adaptations to exercise will be discussed in the molecular part of this chapter. The factors that limit endurance performance all depend on both genetic variation (**Chapter 5**) and environmental factors, of which endurance training is the most important, with diet/nutrition to support training being of secondary importance. Only those individuals with genetic variations that match the requirements of their endurance sport who also engage in the most effective training and support that training with optimal nutrition (**Chapter 10**) will become elite endurance athletes.

ENDURANCE EXERCISE: CURRENT RECOMMENDATIONS, TRAINABILITY AND SCIENTIFIC EVIDENCE

How should one train to increase endurance? As for resistance exercise (see **Chapter 8**), adaptation to endurance exercise training is dictated by the overload principle: '*endurance adaptations occur as a result of systematic and progressive exercise at a sufficient frequency, intensity and duration to cause adaptation*.'As discussed in detail in **Chapter 7**, the key variables to control for an endurance training programme include **training intensity** (set as % $\dot{V}O_{2max}$, %HR_{max}, %HR reserve, % lactate threshold, rate of perceived exertion or velocity/pace), **volume** (time, distance or step count per session, day or week) and **type of session** (continuous exercise, high-intensity interval training or sprint interval training). Other important factors include **training goals** (aerobic capacity, anaerobic capacity, general fitness, caloric expenditure, risk factor reduction, health improvement), **progression/periodisation** (how to change variables during a training programme; also tapering to prepare for competitions) and **exercise mode** (running, cycling, swimming, etc.). Finally, the endurance training programme must be specific for the individual (elite athlete, para-athlete, untrained individual, child, patient participating in a cardiac rehabilitation programme, etc.) as individuals differ in their trainability and may respond better to certain types of endurance training.

Before talking exercise prescription, it is very important to understand that there are two different types of fatigue that affect the musculoskeletal system as a result of endurance exercise. The first is metabolic fatigue that relates to the energy used to perform the exercise. Metabolic fatigue represents how much substrate (CHO, fat or protein) is broken down to perform the exercise and that will need to be replensihed before the next training bout can be performed. The second type of fatigue is the mechanical fatigue that results from repeated impact loading of the musculoskeletal system. To better understand mechanical fatigue, compare rowing and running. Both sports are considered endurance exercise. However, in one (running), the athlete hits the ground ~180 times a minute with an impact force more than twice their body weight (2), whereas in the other (rowing), athletes perform ~30 strokes a minute with a maximal impact force in the catch phase of 1.4 times body weight (3). This means that in each minute, the runner will experience loads on their musculoskeletal system in excess of 360 times their body weight, whereas the rower will absorb only ~45 times their body weight over a minutes exercise. This means that the mechanical load on a runner is 8 times higher than that of a rower and this fact has a significant effect on the size of the athlete in the sport and the volume of training that each athlete can perform. Specifically, elite male 2-man rowers weigh 90–100 kg and can row two and a half hours a day over an Olympic quad (4), whereas in the build-up to an Olympic marathon elite male runners will weigh 55–65 kg and limit themselves to an average of an hour and 30 minutes of training a day (5). Put another way, the perfect endurance training plan (volume and intensity) for a swimmer (low musculoskeletal impact) would cause significant injury to a soccer (high musculoskeletal impact) player.

As a result of the multitude of variables that surround what we generalise as endurance exercise, it is impossible to recommend an 'ideal' endurance training programme based on sound scientific evidence. This is compounded by the fact that the individual response to a single endurance training programme varies greatly in the human population. As discussed in **Chapter 5**, in the first large-scale study to determine individual variability in response to endurance exercise, Bouchard et al. showed that following 20 weeks of an identical supervised cycling programme, approximately 24 of the 481 (5%) participants did not improve their $\dot{V}O_{2max}$, whereas a similar number increased $\dot{V}O_{2max}$ by more than 1 L/min (6). If instead of the change in $\dot{V}O_{2max}$, the change in blood pressure, fasting insulin or blood lipids was measured, a similar response curve is observed with ~10% of individuals worsening in response to an exercise programme (7). Most interestingly, $\dot{V}O_{2max}$, % fat, visceral adiposity, cholesterol or glycaemia do not all change together in response to training. Barber et al. demonstrated that only one person out of 564 participants was a high responder, and another single individual was a low responder, in all seven of these traits following training (8). Also, more than half of the study population was a high responder for 1–2 traits and a low responder for the other 5–6 traits. The variability in training response can either be the result of the relative intensity of the training stimulus (9) or a result of our genetic ability to respond to the exercise (10). This means that there is considerable variability in individual responses to a standard dose of exercise, but that almost all of us (99.8%) will be a high responder in at least one trait/category.

Given the variability of endurance trainability and the many variables that go into a training programme, what recommendations can be given for prescribing endurance training? In the 2011 ACSM position stand, the authors recommended the following for aerobic (endurance) exercise for apparently healthy adults (11):

- **Frequency:** ≥5 days per week of moderate exercise or ≥3 days per week of vigorous exercise, or a combination of moderate and vigorous exercise on ≥3–5 days per week is recommended.

- **Intensity:** Moderate and/or vigorous intensity is recommended for most adults.

- **Time:** 30–60 minutes per day (150 minutes per week) of purposeful moderate exercise or 20–60 minutes per day (75 minutes per week) of vigorous, or a combination of moderate and vigorous exercise per day is recommended for most adults.

- **Type:** Regular, purposeful exercise that involves major muscle groups and is continuous and rhythmic in nature is recommended.

What about endurance training recommendations for endurance athletes? Specific endurance training recommendations are given in thousands of sports books, YouTube™ videos and internet articles but almost all of this information is subjective. We know of no specific recommendations for endurance exercise for athletes that are based on a rich body of data coming from randomised controlled trials, which is seen as the highest level of scientific evidence, or from large-scale epidemiological studies. Instead, observational studies looking at the training habits of elite endurance athletes are used. In a review of this literature, Stöggl and Sperlich (12) showed that during the 'base phase' (the time when an athlete is training in the offseason to increase aerobic capacity) most athletes are performing ~90% of their training in Zone 1 (low intensity with steady state lactate <2mM), ~8% of their training in Zone 2 (high-intensity intervals lasting <5 minutes with lactate at ~4 mM), and <5% of their training in Zone 3 (sprint intervals lasting <30 seconds with lactate >4 mM). As the athletes get closer to competition, they shift their training to ~80% Zone 1, 12% Zone 2 and 8% Zone 3. In other words, most elite athletes use a polarised training plan, with most of their training (≥ 80%) performed at a low intensity and the other 20% performed at high-intensity or sprint intervals. As we describe the molecular signals that drive the aerobic adaptations later in this chapter, why these training intensities are used will hopefully become clearer.

CARDIOVASCULAR RESPONSE TO TRAINING

The function of the cardiovascular system is to deliver O_2 and nutrients to muscles and other organs, to remove CO_2 and other waste products from the tissues, and regulate core temperature and pH. The performance of the system depends on cardiac output, which is the volume of blood pumped by the heart per minute, and the O_2 carrying capacity of the blood (red blood cell mass). Since maximal cardiac output (13), and increasing the oxygen carrying capacity (14) correlates with increased $\dot{V}O_{2max}$, it is widely assumed that oxygen delivery to the working muscle is the primary determinant of $\dot{V}O_{2max}$. Endurance exercise training increases the volume and the oxygen carrying capacity (red blood cell mass) of the blood (15) and at sea level this plays a small role in increasing $\dot{V}O_{2max}$; however, this chapter will focus on the adaptations that occur to the heart and skeletal muscle as a result of training.

How do elite endurance athletes achieve a cardiac output that is in some cases twice as high as that measured in sedentary subjects? Maximal cardiac output is the product of maximal stroke volume multiplied by maximal heart rate. **Table 9.1** summarises some results for stroke volume, heart rate and cardiac output for recreationally active college students before and after endurance training as well as for Olympic athletes (16). The table shows that the high cardiac output (> 30 L/min) of the Olympic athletes can be fully attributed to a stroke volume of nearly 170 mL, 60 mL per beat more than that of the

Table 9.1 Maximal oxygen uptake, cardiac output, stroke volume and heart rate before and after endurance training in college students and values for elite endurance athletes

	College students		Olympic athletes
	Before endurance training	After endurance training	
Maximal oxygen uptake (L/min)	3.3	3.91	5.38
Maximal cardiac output (L/min)	20	22.8	30.4
Stroke volume (mL)	104	120	167
Maximal heart rate (beat/min)	192	190	182

Redrawn from Blomqvist CG, et al. (16)

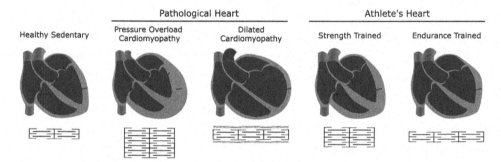

Figure 9.1 Changes in heart structure with exercise and disease. The chamber volume (red line) and wall thickness (blue line) of the heart can change due to high blood pressure (chronic pressure overload cardiomyopathy), disease (dilated cardiomyopathy), strength training (acute pressure overload) or endurance training (volume overload) by either increasing sarcomeres in parallel or series (shown by the sarcomere cartoons beneath each heart).

college students. In contrast, because maximal heart rate decreases with age, maximal heart rate is lower in the Olympians than in college students.

The increased stroke volume with endurance training and that seen in elite endurance athletes is due to exercise-induced physiological cardiac hypertrophy; also known as the athlete's heart (**Figure 9.1**). The athlete's heart has a greater left ventricle volume and a small increase in the thickness of the cardiac muscle (17). In cellular terms, the cardiomyocytes have added sarcomeres in series (increased length) more than in parallel (increased wall thickness). By adding sarcomeres in series, the cardiomyocytes also contract slightly faster allowing the cells to maintain the percentage of the blood pumped out of the heart with each beat (ejection fraction). It is important to remember that large hearts are not always good news: one has to distinguish the beneficial athlete's heart (physiological, left ventricular hypertrophy) from pathological cardiac hypertrophy, also characterised by a greater ventricular volume but more so by thicker ventricular walls. The major functional difference is that pathological hypertrophy of the heart results in a decreased ejection fraction (reduced pumping performance). There are also congenital diseases that lead to bigger hearts. These congenital hypertrophic cardiomyopathies are the leading cause of sudden cardiac death in young athletes (18) but are very difficult to distinguish from athlete's heart. To summarise, the athlete's heart is a physiological adaptation to endurance exercise that increases stroke volume at the same ejection fraction. The endurance athlete's heart is different from pathological conditions such as hypertrophic cardiomyopathy which is a pathological state that can lead to sudden cardiac death.

For the molecular exercise physiologist, there are three key questions related to the athlete's heart:

1. What is the mechanical signal that the heart responds to during exercise?
2. How is this mechanical signal transduced into the chemical signal that drives the athlete's adaptation of the heart in response to endurance exercise?
3. What are the differences in signal transduction between an athlete's heart and hypertrophic cardiomyopathy?

These important questions are difficult to answer in humans because it is difficult to obtain samples from living healthy human hearts. Therefore, virtually all research on cardiac hypertrophy has been performed using cultured cardiomyocytes, cultured hearts (so-called Langendorff-perfused hearts) and transgenic animal models.

Mechanical signal driving the endurance athlete's heart

Endurance exercise is typified by repeated rhythmic contractions of skeletal muscle. Because of the one-way valves in veins, the rhythmic contraction and relaxation of skeletal muscles, this produces a pumping action of the blood back towards the heart. As a greater volume of blood is returned to the heart, the end diastolic volume increases, causing a stretch on the cardiomyocytes. The stretch on the cardiomyocytes during diastole is referred to as the preload on the heart (just like with a slingshot, stretching further – adding more blood to the ventricle – resulting in greater force). The Frank-Starling Law of the heart states that as venous return (preload) increases, there is a proportional increase in stroke volume. Therefore, from the first rhythmic contractions of endurance exercise to the last, the heart sees a higher volume of blood during diastole and therefore the adaptation of the heart to endurance exercise is termed a volume overload.

Transducing volume overload into a chemical signal

To understand how a volume overload is transduced into a chemical signal that increases the volume of the left ventricle, this chapter will first look at studies designed to understand the pathways that regulate heart size and then determine how exercise modulates these pathways.

Shioi et al. were the first to show that they could change heart size genetically when they manipulated phosphoinositide 3-kinase (PI3K) activity in mouse hearts (19). Increasing PI3K activity increased heart size, whereas a dominant negative (dnPI3K) version (that blocks PI3K activity) made the heart smaller. Importantly, in these studies increasing PI3K activity did not produce fibrosis or contractile dysfunction, suggesting that this increase in PI3K could produce an athlete's heart. To further investigate the role of PI3K, dnPI3K and control wild type (WT; i.e. a mouse with no mutations), mice swam seven days a week for four weeks to induce physiological cardiac hypertrophy. The investigators found that the WT mice showed ~40% increase in heart mass, whereas the dnPI3K knockout mice showed only ~15% bigger hearts (20), indicating that PI3K is required for the majority of physiological cardiac hypertrophy. Interestingly, the dnPI3K mice showed normal hypertrophy in response to high blood pressure, suggesting that PI3K is not involved in pathological hypertrophy. A volume overload may signal to PI3K through integrin proteins, and an integrin-associated protein named melusin. Melusin binds to the cytoplasmic domain of the β1-integrin and under volume, overload binds to the p85 subunit of PI3K, bringing PI3K to the membrane where it can phosphorylate phosphoinositide 4,5 bisphosphate (PIP2) resulting in the production of phosphoinositide 3,4,5 trisphosphate (PIP3). PIP3 in the plasma membrane can be bound by proteins containing a pleckstrin homology domain such as protein kinase (PK)B/Akt. When Akt moves to the membrane it becomes activated. Consistent with the hypothesis that integrin/PI3K/Akt signalling is important in signalling a volume overload, mice that lack Akt1 who undertake swim training for four weeks show no cardiac hypertrophy (21). Downstream of Akt are the mechanistic target of rapamycin complex (mTORC)1 and the transcription factor FoxO1. Removing the

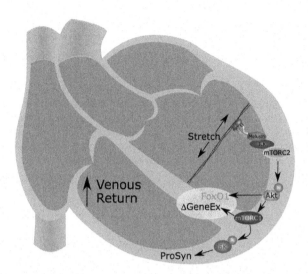

Figure 9.2 Integrin/PI3K/Akt signalling underlying the athlete's heart. The increased venous return due to the rhythmic contraction of muscles during endurance exercise results in an increased end diastolic volume. This stretch is transduced, at least in part, through integrin interaction with the matrix. Integrins signal into the cardiomyocyte through the chaperone proteins integrin-linked kinase and melusin signal to mTORC2, resulting in phosphorylation and activation of Akt. Akt can then increase protein synthesis via the mTORC1 pathway and decrease degradation through FoxO1 resulting in the addition of sarcomeres in series and in parallel. ProSyn = Protein synthesis; GeneEx = Gene Expression.

FoxO1 from the heart also prevents the increase in heart mass as a result of swim training; however, it is not as potent at preventing cardiac hypertrophy as the dnPI3K and Akt1 knockout (22). Taken together, these experiments suggest that the integrin/PI3K/Akt signalling is required for physiological hypertrophy of the heart in response to endurance exercise. Akt can then increase protein synthesis via the mTORC1 pathway and decrease degradation through FoxO1 resulting in the addition of sarcomeres in series and in parallel (**Figure 9.2**).

Mechanical signals driving the strength athlete's heart

Unlike endurance exercise, resistance/strength training does not use rhythmic muscle contractions to pump blood back to the heart. Instead, when lifting heavy weights, muscles contract forcefully for many seconds and this occludes blood vessels within the muscle, resulting in a significant increase in peripheral resistance. For example, during heavy leg press, blood pressure can rise from 120/80 to an average of 320/250 mmHg (23). This means that during resistance exercise the afterload on the heart transiently triples. The result of this 'pressure overload' is that strength athletes have an increase in left ventricular wall thickness with no change in volume (**Figure 9.1**). At the cellular level, this means that the cardiomyocytes have added sarcomeres in parallel, not in series. The stronger muscle with the same chamber size still results in a small increase in ejection fraction and therefore stroke volume.

The molecular mechanism underlying the response to acute pressure overload has been suggested by experiments by Jeffrey Molkentin and his team who showed that the extracellular regulated kinase 1/2 (ERK1/2) pathway could promote physiological cardiac hypertrophy without signs of pathological decompensation or premature death (24). Overexpressing the ERK1/2 kinase MEK1 gene in hearts of mice or rats resulted in thicker walls, an increase in ejection fraction and no change in left ventricle size. Thus, the phenotype of the MEK1 overexpression animals is similar to the cardiac hypertrophy seen in strength athletes. Subsequent research has shown that ERK1/2 alone is not required for chronic pressure overload or volume overload hypertrophy (25). For a more complete review on this topic, we direct readers to (26).

Hypertrophic cardiomyopathy

Unlike in an athlete, the signals for pathological hypertrophy of the heart are derived from chronic changes (e.g. from the elevation of blood pressure, loss of blood flow, genetic mutations) resulting in an increase in wall thickness, chamber volume and collagen that lead to a decrease in ejection fraction (**Figure 9.1**). In arguably the first landmark paper on the molecular mechanism underlying hypertrophic cardiomyopathy, Eric Olson and his team identified two transcription factors, NFAT (nuclear factors of activated T cells) and GATA-4 (GATA binding protein 4), that could bind to the promoter of genes that were upregulated in pathologic cardiac hypertrophy (27). NFATs were interesting because they are regulated by the Ca^{2+}-activated phosphatase, calcineurin. When calcineurin is active, it dephosphorylates the NFATs, they move to the nucleus and together with GATA-4 change gene expression and potentially drive cardiac hypertrophy.

To test the hypothesis that calcineurin-NFAT signalling could promote cardiac hypertrophy, the group created two transgenic mouse strains where either calcineurin or NFAT were overexpressed only in cardiomyocytes. Both transgenic lines developed pathological cardiac hypertrophy and many mice suffered either serious cardiac problems or died prematurely (27). Treating the mice with the calcineurin inhibitor drug cyclosporine A completely reversed the hypertrophy. These data demonstrate that increased calcineurin-NFAT signalling induces pathological cardiac hypertrophy that is reversed by cyclosporine A. Since the initial discovery, other groups have shown that calcineurin can be activated by hypertrophic signals like angiotensin II, endothelin I and chronic stress (26) and that hypertrophic cardiomyopathy induced by pressure overload can be reversed by cyclosporine A (28). Even with the central role of calcineurin, there are many other molecular signals that are important in the development of hypertrophic cardiomyopathy (26).

In humans, high blood pressure is thought to result in a chronic increase in the force (stress) that the heart must produce. The resulting increase in wall stress is sensed by the endothelial cells that line the ventricle (**Figure 9.3**). Under stress, the endothelial cells release endothelin-1 and angiotensin-II, which bind to a G-protein coupled receptor on the surface of cardiomyocytes. Within cardiomyocytes, the endothelin-1 receptor activates $G\alpha q$, which activates phospholipase C, an enzyme that cleaves phosphatidylinositol 4,5-bisphosphate (PIP2) to produce inositol 1,4,5-trisphosphate (IP3) and diacylglycerol. The IP3 induces calcium release that can activate calcineurin, leading to the dephosphorylation of NFAT, resulting in a change in increased hypertrophy marker genes such as atrial natriuretic peptide and B-type natriuretic peptide that are associated with pathological cardiomyopathy.

To summarise, there are three major forms of cardiac hypertrophy. The first is volume-overload hypertrophy, which occurs as a result of increased venous return-induced cardiomyocyte stretch, is physiologically beneficial and results in an increase in sarcomeres in series, increasing left ventricular volume while maintaining ejection fraction. These hearts generate a higher cardiac output during maximal exercise which is a key adaptation to endurance training. The second form of cardiac hypertrophy is physiological pressure overload hypertrophy. This form of hypertrophy occurs as a result of increased pressure (up to 3-fold) induced by occluding blood vessels while lifting heavy weights, is physiologically beneficial and results in an increase in sarcomeres in parallel and therefore a thicker ventricle wall. These hearts generate more force, increasing ejection fraction while maintaining left ventricular volume, resulting in a higher cardiac output. The third form of cardiac hypertrophy is pathological hypertrophy. Pathological pressure overload occurs because of a chronic increase in blood pressure. Chronic pressure overload is pathological and leads to degenerative hypertrophic cardiomyopathy where ejection fraction decreases as the disease progresses. Transgenic mouse studies suggest that physiological and pathological cardiac hypertrophy are regulated by different signal

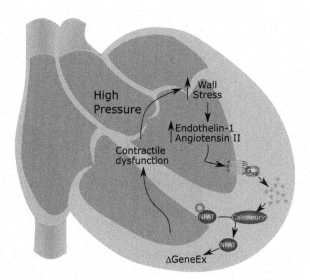

Figure 9.3 Calcineurin signalling in response to chronic pressure overload. As a result of chronic high blood pressure, the force (stress) that the heart has to produce is elevated. Greater wall stress can increase the production of endothelin and Angiotensin II from the endothelial cells that line the heart. These can then increase calcium release through a G-protein-coupled receptor. The chronically high calcium activates the calcium-dependent phosphatase calcineurin resulting in dephosphorylation of its target NFAT. When NFAT is dephosphorylated, it moves to the nucleus and increases cardiac hypertrophy genes such as atrial natriuretic peptide and B-type natriuretic peptide that are associated with pathological cardiomyopathy. GeneEx = Gene Expression.

transduction pathways: endurance exercise increases integrin/PKB/Akt signalling; strength training may signal through ERK1/2; and chronic pressure overload-induced pathological hypertrophy is mediated, in part, by the calcineurin-NFAT pathway which drives hypertrophy together with a change in gene expression that results in a progressive decline in function known as decompensation.

REGULATION OF SKELETAL MUSCLE FIBRE PLASTICITY

A motor unit is defined as a motor neurone and all of the muscle fibres it innervates (**Figure 9.4**). There are different types of motor units which are characterised by their α-motor neurone and muscle fibre type:

- S (slow) motor units comprise small α-motor neurones with a low excitation threshold and small axons that innervate a small number of slow type I fibres.

- FR (fatigue-resistant) motor units comprise moderate-sized α-motor neurones that innervate intermediate type IIa fibres. The α-motor neurone properties are in between those of Fast Fatiguing (FF – see below) and S α-motor neurones.

- FF (fast fatiguing) motor units comprise large, hard to excite but fast-conducting α-motor neurons that innervate many large fibres that contract rapidly but fatigue easily (type IIb and IIx fibres);

The aim in this section of the chapter is to further discuss the different types of muscle fibres and their response to increased or decreased contractile activity.

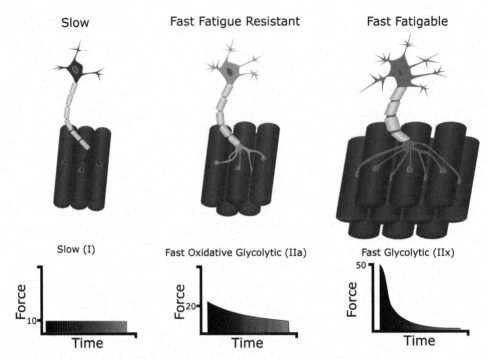

Figure 9.4 Motorneurons and their fibres. Slow motor units are small, easily excited neurons that innervate a small number of slow type I fibres producing low force without much fatigue. By contrast, fast fatigue-resistant motor units comprise moderate-sized α-motor neurones that innervate more type IIa fibres that produce a moderate force and fatigue slowly over time. Lastly, fast fatiguing motor units comprise large α-motor neurons that are difficult to excite and innervate large numbers of big fibres that contract rapidly but fatigue quickly (type IIb and IIx fibres).

Muscle fibre properties

The muscle fibres within a muscle differ greatly in their contraction speed, force output and fatigability. This is a result of differences in both the motor and regulatory proteins within the fibres. The best described difference is that of the motor protein, myosin heavy chain (MyHC), whose ATPase activity is directly proportional to the rate of muscle contraction (29). This relationship between myosin ATPase and contractile velocity (**Figure 9.5**) makes MyHC a key marker of muscle fibre type. Other reasons for making MyHC the protein that is used to classify different types of muscle fibres are as follows:

- High-quality antibodies against the different MyHC isoforms are available (30). These antibodies allow a reliable determination of fibre type.

- MyHC isoforms determine both the maximal shortening velocity (29) and force per cross-sectional area (CSA) of a muscle fibre (31). Thus, the expression of different MyHC isoforms determines the function of muscle fibres.

- Myosin is the most abundant protein within muscle fibres (45–85 mg per g of muscle). This makes detection easier than proteins whose concentrations are lower.

MyHCs are divided into slow (type I) and fast (type II) isoforms. The type II MyHC isoforms are further subdivided into type IIa (intermediate), type IIx (fast) and type IIb (fast, highly fatigable). Note, the type

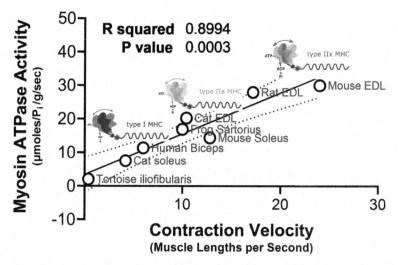

Figure 9.5 **Relationship between muscle contraction velocity and MyHC ATPase activity.** Across all species, there is a direct linear relationship between the rate that a myosin protein breaks down ATP and the rate that the muscle shortens, demonstrating the importance of MyHC in muscle function. Drawn from data from **Bárány M (29).**

Table 9.2 Generalised fibre-type characteristics of the major human and rodent fibre types

Fibre type (major myosin heavy chain expressed)	Type I	Type IIa	Type IIx/IIb*
Description	slow, red, fatigue-resistant	intermediate, red, fatigue-resistant	fast, white, readily fatigued
Nuclei per mm of fibre	high	high	low
Maximal shortening speed when fully activated	slow	fast	slightly higher still
Force development and relaxation rate	slow	medium	fast
ATPase activity under 'physiological' conditions	low	medium	high
Glycolytic enzyme activity	low-medium	medium-very high	high
Myoglobin content	high	medium-high	low
Oxidative capacity or mitochondrial enzyme activity	high	high	low
Capillaries per fibre or mm²	high	medium	low

*Type IIb myosin heavy chain only in rodent skeletal muscle but not in most human muscles. Type IIx is also known as IId.

IIb MyHC gene exists in the human genome but the protein is not expressed in any of the major human muscles, it is only expressed in rodent skeletal muscle (32). An overview over the functional characteristics of different muscle fibre types is given in **Table 9.2**.

Be aware that fibre-type tables such as **Table 9.2** are a gross over-simplification for several reasons. First, fibres can change from one fibre type into another and during the transition phase fibres express more than one myosin heavy chain isoform. Such mixed fibres are termed hybrid fibres. Second, while there are large differences in the concentrations of MyHC isoform proteins, the concentrations of other proteins vary more and may overlap between fibres. Some researchers describe this as a continuum of fibre types to highlight the fact that having three fibre-type classifications is an over-simplification. Third, there are more MyHC genes and proteins than those shown in **Table 9.2**. In muscles that move the eye, extraocular MyHC is expressed; in the jaw, some muscles express MyHC-16, a superfast variety. Embryonic and developmental MyHC are expressed during development and when muscle fibres are regenerating after injury.

Once researchers were able to determine the different muscle fibre types, they started to compare the fibre-type composition between different species, in different muscles of the human body and in the human population. This research revealed the following:

1. **Inter-individual variability:** Due to genetic variation and environmental factors, humans vary in their fibre-type composition. For example, North American Caucasians have a mean of ~50% type I fibres in their vastus lateralis muscle; however, the range extends from 15 to 85% (See Chapter 4). Thus, the fibre-type composition in any given skeletal muscle varies significantly in the human population.
2. **Fibre-type composition of athletes:** Sprint and endurance athletes commonly have a high percentage of fast type II and slow type I fibres in their sport-specific muscles, respectively (33).
3. **Intra-individual variation:** Within a given individual, the fibre-type composition of muscles varies dramatically. For example, the human soleus has ~70% type I fibres, whereas the vastus lateralis has ~70% type II fibres. Generally, postural muscles have more type I fibres, whereas those that contract less frequently and with more power have more type II fibres.
4. **Inter-species variation:** Muscle fibre-type proportions within a muscle differ between species and even between different strains. For example, the soleus muscle of a guinea pig is 100% type I, in a human this muscle is 70% type I, and in a mouse, the soleus has 35% type I fibres.

Fibre type and exercise

Since the discovery of fibre types and ways to distinguish them biochemically, exercise physiologists have studied whether exercise can be used to change the fibre-type composition of an individual (34). The first team to address this question comprehensively included the American biochemist Phil Gollnick and the Scandinavian Exercise Physiologist Bengt Saltin (reviewed in 33). They first compared the fibre-type composition of trained and untrained men and then conducted an endurance training study where six subjects exercised intensely for five months on a cycle ergometer for 1 hour per day for four days per week (35). The subjects were biopsied before and after training. Histochemical assays were then used to determine the percentage of slow twitch (type I) and fast twitch (type II) (the assays used at the time were not able to distinguish between type II subgroups of IIa and IIx) and the area of the slow and fast fibres. They found that endurance training caused a non-significant increase of the percentage of slow twitch fibres from 32 to 36%. Even though the relative number of slow fibres did not change significantly, the slow fibres did get bigger (increased slow twitch fibre CSA). A limitation of the study was that only six subjects were investigated but their data suggest that a small fast-to-slow change in fibre and more importantly an increase in the CSA of slow fibres with endurance training.

The most interesting study in this area looked at a pair of monozygotic twins who had very different activity levels for ~30 years. One of the twins ran 2 marathons, completed an Ironman and two half Ironman triathlons, and recorded >34,000 km of running, whereas the other remained relatively inactive. A biopsy of the vastus lateralis followed by single fibre MyHC analysis on 213 fibres showed that the trained twin had >90% slow fibres, whereas the untrained twin had only 38% slow fibres (36). These data suggest that given sufficient time and training, the fibre-type proportions within a muscle can be changed with persistent endurance exercise. The general rule of thumb for MyHC shifts is that increases in contractile activity results in a slower phenotype, whereas decreased contractile activity will drive fibres towards a faster phenotype on the following scale: **I (slow) ↔ IIa (intermediate) ↔ IIx (fast) ↔ (IIb;**

fast; not expressed in humans). Some pathological, physiological and supra-physiological interventions that change the fibre composition of muscles are as follows:

1. **Endurance exercise (physiological)**: short-term exercise promotes a limited fibre-type shift that involves a decrease of type IIx and increase of type IIa fibres (37). Over decades, it seems that a significant type II-to-type I fibre-type shift can occur (36).
2. **Chronic electrical low-frequency stimulation (supra-physiological)**: a model for continuous low-intensity endurance exercise, can induce complete type II-to-type I fibre-type transitions in animal models (38).
3. **Cross-reinnervation (supra-physiological)**: i.e. connecting a fast muscle nerve to a slow muscle changes the muscle phenotype from fast to slow and vice versa (39). It demonstrates the dominant effect of the firing pattern of the motor neurones on muscle fibre type.
4. **Denervation (for example, after spinal cord injury; pathological):** decreased contractile activity; induces a slow-to-fast phenotype shift, with predominantly type IIx fibres present years after injury in humans (40).

To summarise, fibre composition of skeletal muscles within the human body and between humans varies greatly. Endurance training regimes over several months generally lower the proportion of type IIx fibres and increase the proportion of IIa fibres, but will not convert a sprinter into a Marathon runner. However, decades of endurance training may trigger type II-to-I conversions. Finally, chronic low-frequency electrical stimulation induces a fast-to-slow fibre transformation whereas a lack of innervation results in a slow-to-fast change.

MyHC plasticity and the calcineurin-NFAT pathway

For molecular exercise physiologists the challenge is always to link changes in skeletal muscle fibre phenotype to the signals that cause the adaptation. One of the potential signals underlying the fibre-type adaptation to prolonged endurance exercise is calcium. Every time that a muscle contracts, there is a 100-fold increase in intracellular calcium levels. Following these bursts of calcium release through the ryanodine receptor, there is a second long-lasting low-amplitude calcium signal that comes from the activation of the inositol 1,4,5 trisphosphate (IP_3) receptor (41, 42, 43). Blocking the IP_3 receptor, using the drug xestospongin, decreases the slow calcium wave and decreases the activity of the Ca^{2+}-activated phosphatase, calcineurin, which as discussed above activates the transcription factor NFAT by dephosphorylation (42). The calcineurin-NFAT pathway has been identified as a regulator of muscle fibre phenotype. To determine whether the calcineurin pathway is activated differently in slow and fast muscle fibres, a team led by Stefano Schiaffino, overexpressed NFAT tagged with green fluorescent protein (GFP) in muscle fibres (**Figure 9.6**). The researchers observed that NFAT was largely in the nucleus in fibres taken from the slower soleus, whereas in the faster tibialis anterior NFAT was predominantly found in the cytoplasm (44). Inactivation of the muscle by denervation reduced NFAT in the nucleus, whereas increasing muscle activity through chronic low-frequency (20 Hz) electrical stimulation increased NFAT in the nucleus (44). Interestingly, short bouts of high-frequency (100 Hz) electrical stimulation, as would be seen in fast fibres, did not increase nuclear localisation. All this suggests that the Ca^{2+} patterns found in slow type I fibres, and that are induced by endurance exercise, activate calcineurin. Calcineurin then dephosphorylates NFAT, and NFAT then moves to the nucleus in type I fibres and in response to low-frequency stimulation. In contrast, the lowering of Ca^{2+} or short high-frequency bursts of Ca^{2+} do not activate calcineurin and consequently NFAT remains in the cytoplasm where it cannot regulate gene expression.

Thus, the activation pattern of the calcineurin pathway is consistent with the hypothesis that calcineurin and NFAT activity contributes to regulate the muscle fibre phenotype (**Figure 9.7**). But what is the evidence that calcineurin regulates the phenotype of muscle fibres? In a landmark paper by Eva Chin, the function of calcineurin in skeletal muscle was directly investigated (45). First, the authors found that

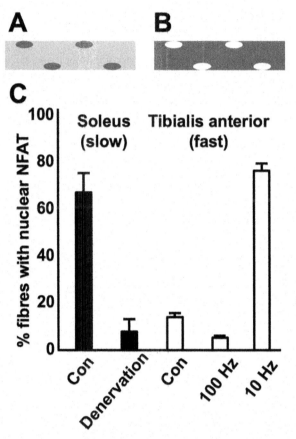

Figure 9.6 Schematic depicting the localisation of the transcription factor NFATc1 in slow and fast
muscle fibres and its response to changes in innervation (**A**). Green fluorescent protein-
tagged NFATc1 is mainly in the nucleus in the slower soleus muscle, whereas (**B**) in the
faster tibialis anterior muscle NFATc1 is predominantly in the cytoplasm. (**C**) Quantification
of NFAT1c localisation shows that the protein is mainly nuclear in the control soleus muscle
(black bars), whereas denervation decreases nuclear NFATc1 in the soleus. By contrast, the
control fast TA muscle (open bars) and stimulation with infrequent high-frequency 100 Hz
electrical pulses prevents NFATc1 nuclear localisation. By contrast, continuous low-frequency
10 Hz stimulation increases NFAT in the nucleus. Redrawn from Tothova J, et al. (44).

when calcineurin was over expressed, muscle phenotype shifted, with slow genes (troponin and
myoglobin) expressed at higher levels. Furthermore, this shift was prevented by inhibiting calcineurin
with cyclosporine A. They then treated rats with the calcineurin inhibitor, cyclosporine A, for six weeks
and found that this doubled the average percentage of fast type II fibres in the soleus muscle. These data
suggest that active calcineurin is central to reducing the number of type II fibres and increasing the
number of type I fibres. In subsequent experiments, researchers overexpressed or knocked out calcineurin
in transgenic mice and showed that high calcineurin activity increased the percentage of slow type I
fibres in vivo (46, 47). However, no intervention that targets the calcineurin-NFAT pathway has converted
all type II fibres into type I fibres. To summarise, the unique Ca^{2+} patterns in slow type I and fast type II
muscle fibres, specifically the IP_3-activated slow current, differentially activates calcineurin in type I and
type II fibres. Calcineurin then activates the transcription factor NFAT in type I fibres which drives the

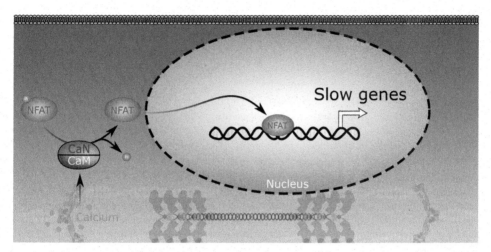

Figure 9.7 Model for calcineurin signalling in response to long-term changes in muscle activity. The pattern of calcium release in slow type I muscle fibres activates an IP$_3$-dependent slow calcium current resulting in calcineurin activation in type I fibres. Calcineurin then dephosphorylates the transcription factor NFAT which drives the expression of 'slow' genes such as slow MyHC, troponin and myoglobin.

expression of 'slow' genes such as slow MyHC, troponin and myoglobin. Endurance exercise can also activate this pathway and over time has a similar effect.

Is calcineurin the sole regulator of skeletal muscle fibre-type identity?

No. Many other pathways have been suggested to regulate fibre type, including PGC-1α (48), MEF2 (49), members of the MAPK pathway (50), Tead1 (51) and the myogenic regulatory factor myogenin (52). To summarise, the calcineurin-NFAT pathway is a major regulator of fibre-type regulation, but many other signalling proteins are also involved in the regulation of skeletal muscle fibre phenotype.

Myosin heavy chain gene clusters and chromatin remodelling

In mammalian genomes, the genes encoding the MyHC isoforms are located close together in one of two genomic locations: the slow/cardiac cluster or in the fast/developmental cluster. The slow/cardiac cluster harbours MyHC types Iα (cardiac-specific) and Iβ (slow/cardiac), whereas the fast/developmental cluster encodes the type IIa/x/b subtypes as well as embryonic and perinatal isoforms (**Table 9.3**).

Within the nucleus, our genome is condensed through tightly wrapping the long stranded DNA around histones. This tertiary structure of DNA needs to be remodelled (opened) before any gene can be expressed. The opening and closing of DNA is termed chromatin remodelling or epigenetic regulation and it has been shown that such regulation is key for myogenesis and probably also plays an important role in the regulation of MyHC isoform expression and fibre-type specification. Support for the important role of chromatin remodelling driving myogenesis comes from early experiments where fibroblasts were treated with 5-azacytidine, a drug that opens up DNA. Within days, the fibroblasts had turned into myoblasts because regions of the genome where muscle genes are located had become more accessible (53). Chromatin remodelling is tightly regulated by DNA methylation and by histone acetylation and methylation on lysine residues as described by the 'histone code' hypothesis (54).

Table 9.3 Genomic locations (chromosome number: location in mega bases (Mb)) of skeletal and cardiac muscle human myosin heavy chain (MyHC) genes and their orthologues in mouse

Isoform	MyHC Iα	MyHC Iβ	MyHC IIa	MyHC IIx/d	MyHC IIb
Gene	MYH6	MYH7	MYH2	MYH1	MYH4
Cluster	Slow/cardiac	Slow/cardiac	Fast/developmental	Fast/developmental	Fast/developmental
Human	14; 22.92Mb	14; 22.95Mb	17; 10.37Mb	17; 10.34Mb	17; 10.29Mb
Mouse	14; 46.91Mb	14; 46.91Mb	11; 66.78Mb	11; 66.83Mb	11; 66.87Mb

As described in **Chapter 6**, histone acetyltransferases (HATs) and deacetylases (HDACs), which regulate the acetylation of histones, and methyltransferases and demethylases, that add or remove methyl groups, are involved in the differentiation of skeletal muscle. Acetylation and methylation describe the addition of an uncharged acetyl or methyl group onto a charged amino acid (lysine), respectively. Therefore, acetylation or methylation removes a positive charge that may have been used to facilitate protein-protein interactions, bind a nucleotide (ATP or GTP), or interact with a negatively charged molecule like DNA. During the differentiation of C2C12 myoblasts (mouse muscle cell line) into myotubes (see **Chapter 2**), which involves an upregulation of MyHC expression, Asp et al. (55) found that trimethylation of lysine 27 in histone 3 (H3K27me3), which is associated with gene silencing, decreased around the promoters of muscle genes, whereas trimethylation of lysine 36 (H3K36me3), which is found on actively transcribed genes, increased during C2C12 differentiation into myotubes. Another marker of actively transcribed genes, trimethylation of lysine 4 of histone 3 (H3K4me3), correlated with the expression of MyHC genes. In the rat plantaris (faster phenotype), H3K4me3 was higher in the promoters of the fast MyHC genes, whereas this mark was highest around the promoter of type I MyHC in the soleus (slower phenotype) muscle (56). Interestingly, when the investigators hind limb suspended rats (used the rats tail to suspend their feet above the ground), a stimulus that results in a slow-to-fast phenotype shift in the soleus muscle, H3K4me3 decreased around the type I promoter and increased around the IIx promoter. These data are consistent with the hypothesis that histone modifications contribute to the opening up of muscle-specific genes when myoblasts differentiate into myotubes and may play a role in the expression of specific MyHC genes.

Myosin heavy chain on/off regulation, MyoMirs and Sox6

One unique feature of MyHC isoform regulation is that their regulation is on/off or binary in contrast to many other genes where expression is regulated along a continuum. In other words, individual muscle fibres, apart from hybrid fibres, express only one MyHC isoform, while the expression of all other MyHC isoforms is almost fully switched off. This is in stark contrast to, for example, glycolytic or oxidative enzymes, which are always expressed but increase or decrease in response to endurance exercise or inactivity, respectively. A team led by Eric Olson has identified one mechanism by which the on/off regulation of MyHC isoforms is achieved. The research team discovered that the slow MyHC 1β gene harbours two regulatory microRNAs (miRNA), miR-208b and miR-499 (termed MyoMirs) in its introns (57). This means that when the MyHC 1β gene is transcribed, the two miRNAs are made at the same time. Therefore, a muscle that expresses MyHC 1β mRNA would have high levels of miR-208b and miR-499. miRNAs are post-transcriptional epigenetic regulators as described in detail in **Chapter 6**. When miRNAs match a target sequence on an mRNA perfectly, the bound mRNA is rapidly degraded. When the sequence does not match exactly, miR binding prevents the translation of the associated mRNA into protein. Eva van Rooij et al. discovered that miR-208b and miR-499 bind imperfectly to Sox6 mRNA. Sox6 is a transcription factor that prevents the expression of slow genes and upregulates fast genes (58). When miR-208b and miR-499 bind the Sox6 mRNA, they prevent the production of Sox6 protein and in this way increase the

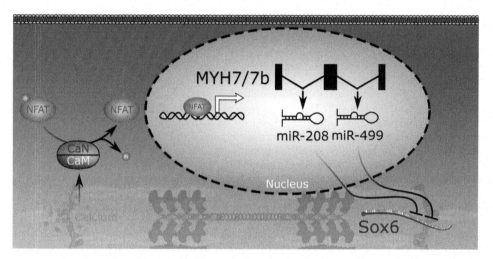

Figure 9.8 MyoMirs (miR-208b and miR-499) and muscle phenotype. Fibre-type signalling pathways such as calcineurin upregulate the expression of slow type I myosin heavy chain (MYH7/7b). Within the introns of MYH7 are two miRNAs, miR-208b and miR-499. Thus, whenever the MYH7 is expressed, these MyoMirs are also expressed. The MyoMirs then prevent the translation of the transcription factor Sox6 mRNA into protein. Since Sox6 normally inhibits the expression of type I MHC and increases the expression of fast myosin heavy chains, this ensures that fast myosin heavy chains are not expressed in fibres where the type I myosin heavy chain is expressed. In this way muscle fibres maintain a single fibre type phenotype through the coordinated expression of genes and miRNAs.

expression of slow muscle genes. In support of the role of decreasing Sox6 on slow gene expression, in mice where Sox6 has been knocked out, there is a 7- to 80-fold increase in the expression of slow muscle genes. Together, these data suggest that muscle fibres maintain a single phenotype through the coordinated expression of genes and miRNAs. Evolutionarily, incorporating miRNAs into the mRNA of a MyHC gene that can then decrease the production of the opposing phenotype allows binary control of muscle phenotype (**Figure 9.8**).

PGC-1α AND THE ENDURANCE PHENOTYPE

Even though shifts in MyHC can occur over a long period of continuous training, other changes occur within skeletal muscle over a much shorter time frame, including increases in mitochondrial mass, capillary density, fat oxidation enzymes and insulin sensitivity. Many of these muscular responses to endurance exercise lie down stream of a single transcriptional co-activator, PGC (peroxisome proliferator-activated receptor-gamma co-activator) -1α. Most importantly, in response to exercise PGC-1α mRNA is increased, with much of the increase the result of activation in the alternative promoter (59) discussed in **Chapter 6**, with this form of the gene known as PGC-1α2/3. PGC-1α acts as an effector by increasing the transcriptional activity of transcription factors. In other words, transcription factors, like the nuclear respiratory factor 1 (NRF1), the estrogen-related receptor α (ERRα) and the peroxisome proliferator-activated receptor-delta (PPARδ), bind DNA in a sequence-specific manner to increase the expression of mitochondrial, angiogenic and fat oxidation genes, respectively. By contrast, a co-activator like PGC-1α does not bind DNA. Instead, it binds to transcription factors and helps them open DNA around, and recruit polymerase II to, the promoter of the gene. In this section, the regulation and downstream effects of PGC-1α will be discussed in detail.

SIGNAL TRANSDUCTION IN RESPONSE TO ENDURANCE EXERCISE

Skeletal muscle produces force in response to repeated release of calcium, which binds to troponin C and opens the myosin binding sites on actin, activates cross-bridge cycling, and while calcium remains high the cycle continues. For each step in the process, the energy needed for contraction, calcium reuptake, $Na^+/K^+/ATPase$ activity and almost all other active processes within the cell are derived from the hydrolysis of ATP (ATP $\rightarrow$ ADP + Pi). The concentration of ATP in muscle is ~3.5–8 mmol/kg in resting mammalian muscle and declines only during maximally fatiguing exercise (**Figure 9.9**). ATP is kept constant, and even though ATP hydrolysis can increase by more than 200-fold during maximal sustained muscle contractions. The stability of the ATP levels in the muscle is achieved through the rapid resynthesis of ATP. ATP can be resynthesised using one of four different systems: the myokinase reaction (ADP + ADP $\rightarrow$ ATP + AMP); the Lohmann reaction (ADP + phosphocreatine $\rightarrow$ ATP + creatine), glycolysis and/or oxidative phosphorylation (**Figure 9.9**). Mitochondria are the main site for ATP resynthesis at rest and during long duration (>2 minutes) endurance exercise, whereas during high-intensity exercise (15 seconds to 4 minutes) ATP is regenerated using glycolysis, and during sprints (<30 seconds) ATP is regenerated using the myokinase and Lohmann reactions. There is also overlap between these times at the points of transition between metabolic processes. In the process of regenerating ATP, hydrogen ions are transferred throughout the cell using nicotinamide adenine dinucleotide (NAD^+). NAD^+ bioavailability increases during exercise. Since NAD^+ plays an important role in the regulation of metabolism and is increased in muscle by exercise training, this metabolic co-factor could be an important exercise signal (also see **Chapter 7**). As exercise intensity increases, muscle relies more on its internal store of glycogen, and this results in a decrease in glycogen content of the muscle. Further, increasing exercise intensity produces a whole-body stress response that activates the sympathetic nervous system and therefore epinephrine levels rise. In the following section, how Ca^{+2}, ATP turnover, glycogen depletion and

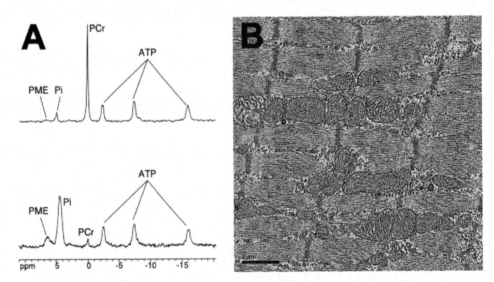

Figure 9.9 Signals that lead to the endurance exercise phenotype. Long duration exercise: (i) increases calcium release activating CaMKII; (ii) increases NAD+ resulting in the activation of SIRT1; and (iii) depletes glycogen activating p38MAPK. High-intensity exercise: (i) increases epinephrine release that signals through cAMP to PKA; (ii) uses ATP faster than it can be produced increasing ADP and activating AMPK; and (iii) depletes glycogen resulting in the activation of p38MAPK. All the enzymes listed above can alter PGC-1α activity or transcription resulting in a stimulus to increase mitochondrial mass, angiogenesis and fat oxidation enzymes. From Greenhorn and Wackerhage (unpublished) and taken from the 1st edition of this book.

epinephrine levels are sensed by the muscle and how these signals are sensed by **sensors** that, in turn, activate **signal transduction** and **effectors** that control **gene transcription** ('signal transduction hypothesis' described in **Chapter 7**), to stimulate **mitochondrial biogenesis** (production of additional mitochondria) will be explained.

How is exercise duration sensed?

As described above, each time a muscle contracts, intracellular calcium levels increase >100-fold. This means that whether a cardiac rehab patient exercises for 3 minutes, or 3 hours, their muscle will see a high Ca^{+2} concentration for the whole time. Therefore, Ca^{+2} is a likely signalling molecule to encode the duration of exercise (**Figure 9.10**). Calcium can influence the endurance phenotype through the 'sensors' called calcium-calmodulin-activated protein kinases (CaMKs). The primary CaMK in skeletal muscle is CaMKII. CaMKII is activated by exercise (60) and when over expressed in skeletal muscle CaMK can increase PGC-1α expression, and PGC-1α acts as the effector by increasing mitochondrial mass, enzymes for fat oxidation and fatigue resistance (61). These data suggest that exercise duration can be sensed through the repeated increase in Ca^{+2} activating CaMKII leading to an increase in PGC-1α2/3 and muscle endurance (**Figure 9.10**).

Another metabolic change that occurs throughout exercise to evoke a molecular signal is the shuttling of hydrogen ions through the NAD^+/NADH to harness energy from the breakdown of glucose, fatty acids and amino acids in glycolysis, β-oxidation and the citric acid cycle. As NAD^+ levels rise within the cell, a family of NAD^+-dependent enzymes act as sensors for NAD^+ and become activated. Best known of these enzymes are the sirtuins (SIRT), a family of NAD^+-dependent deactetylases (enzymes that remove acetyl groups from lysine within proteins). Within the cytoplasm and nucleus, SIRT1 is the predominant sirtuin, whereas in the mitochondria SIRT3 predominates (62). During exercise, the activity of SIRT1 increases (63), suggesting that NAD^+ levels in the cytoplasm and/or nucleus go up during exercise (**Figure 9.10**). The increase in SIRT1 levels is potentially important since one of its molecular targets is PGC-1α and the deacetylation of PGC-1α increases its activity as a transcriptional co-activator (64).

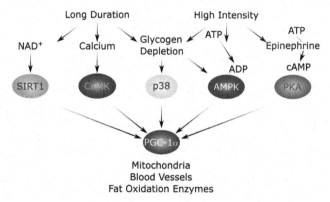

Figure 9.10 Signals that lead to the endurance exercise phenotype. Long duration exercise: **(i)** increases calcium release activating CaMKII; **(ii)** increases NAD+ resulting in the activation of SIRT1; and **(iii)** depletes glycogen activating p38MAPK. High-intensity exercise: **(i)** increases epinephrine release that signals through cAMP to PKA; **(ii)** uses ATP faster than it can be produced increasing ADP and activating AMPK; and **(iii)** depletes glycogen resulting in the activation of p38MAPK. All the enzymes listed above can alter PGC-1α activity or transcription resulting in a stimulus to increase mitochondrial mass, angiogenesis and fat oxidation enzymes.

How is exercise intensity sensed?

Whenever ATP is used, its millimolar concentration decreases only by a few micromoles because the Lohmann and myokinase reactions rapidly stabilise the ATP concentration. The small decrease of ATP translates into a large relative increase in the concentration of ADP and AMP as these substances lie in the micromolar range at rest. Thus, a small relative drop in ATP (less than 1%) can translate into a doubling of ADP and AMP concentrations. ADP and AMP are sensed and regulate the activity of AMP-activated protein kinase (AMPK). AMPK is a protein kinase that functions as a heterotrimer (a complex of three different proteins):

- the α catalytic subunit (the actual kinase whose activity is regulated by Thr172 phosphorylation);

- the β subunit that holds the enzyme together and binds glycogen;

- the γ subunit which binds ATP, ADP or AMP.

AMPK is maximally activated when the γ subunit is ADP- or AMP-bound and when glycogen is low. Thus, the rise of ADP and AMP at the beginning of endurance exercise or when performing high-intensity or sprint intervals leads to the activation of AMPK (65). AMPK activity is prolonged by phosphorylation at Thr172. This phosphorylation accumulates because binding of ADP or AMP prevents the enzyme from being dephosphorylated. As a result, AMPK activity increases during high-intensity endurance exercise and AMPK starts to phosphorylate its downstream signal transduction targets (**Figure 9.10**).

How is the stress of exercise sensed?

The stress of intense exercise is sensed locally through the depletion of glycogen and globally through the release of the sympathetic hormone epinephrine. As mentioned above, low glycogen increases the activity of ATP/ADP turnover sensor, AMPK (66). Another kinase that senses the signal of low glycogen is the mitogen-activated kinase (MAPK) p38. p38 MAPK activity is therefore higher when exercise is started with low glycogen (67), and the γ isoform is necessary for the acute increase in PGC-1α activity after exercise. One way that p38 MAPK acts as an effector is through phosphorylation of the regulatory domain of PGC-1α. Phosphorylation of PGC-1α at Thr262, S265, and T298 by p38 MAPK increases PGC-1α activity and can drive mitochondrial biogenesis, angiogenesis and increases in fat oxidation enzymes (68).

The stress of exercise is signalled globally by an increase in circulating epinephrine. The rise in epinephrine is one of the driving forces of glycogen depletion through its regulation of the activity of the enzyme glycogen phosphorylase, which drives the breakdown of glycogen. Phosphorylase is activated by epinephrine through the first signalling cascade ever discovered. In the 1930s and 40s, Gerty Cori discovered that the enzyme phosphorylase catalyses the breakdown of glycogen and that this process could be accelerated by epinephrine. She crystallised two forms of the enzyme, one phosphorylated and the other in the dephosphorylated form and showed that a protein that required adenylic acid (now called cAMP) catalysed the conversion of phosphorylase 'B' to 'A' and facilitated the breakdown of glycogen. What she had discovered was a part of a kinase cascade that would be flushed out in subsequent decades and her discovery made her the first American (though she had been born in Prague) female to win the Nobel Prize. Today, we know that when epinephrine binds to the β-adrenergic receptor, it activates a G-protein that stimulates adenylyl kinase and increases the production of cAMP. Cyclic AMP activates protein kinase (PK)A, which phosphorylates and activates phosphorylase kinase, the kinase that eventually activates glycogen phosphorylase allowing the breakdown of glycogen. Along with the breakdown of glycogen, the activation of PKA also increases the transcription of PGC-1α2/3 (69), suggesting that it might play a role in the response and adaptation to endurance exercise.

Therefore, exercise duration can be encoded via calcium-CaMKII and NAD^+-SIRT1 signalling to increase PGC-1α2/3; high-intensity exercise can be signalled through the drop in glycogen and the accelerated use of ATP to AMPK and p38 MAPK; and the whole-body stress of exercise can signal through epinephrine-PKA (**Figure 9.10**). The following section will discuss how these signals drive the adaptation to endurance exercise.

Increasing PGC-1α through post-translational modifications

As stated above, PGC-1α protein can be modified by phosphorylation and acetylation. When PGC-1α is phosphorylated at amino acids Thr262, S265, and T298 by p38 MAPK and when PGC-1α is deacetylated, it has the greatest activity (64, 70). These post-translational modifications would be expected to happen rapidly during a bout of exercise and may underlie the observation that PGC-1α activity increases before there is an increase in either PGC-1α2/3 mRNA or protein (71). The phosphorylation of PGC-1α by p38 MAPK occurs in the regulatory domain where a repressor known as RIP140 can bind to PGC-1α and inhibits its ability to co-activate transcription (72). This suggests that exercise that depletes glycogen activates existing PGC-1α within the cell as a result of p38 MAPK-induced phosphorylation and uncoupling from its inhibitor RIP140.

The regulation of PGC-1α acetylation by exercise is more complex. PGC-1α is acetylated by the acetyltransferases p300 and GCN5 (64, 73) and deacetylated by SIRT1 (64). PGC-1α becomes deacetylated following exercise, even when SIRT1 is knocked out, as a result of a decrease in binding with GCN5. Therefore, during exercise PGC-1α becomes deacetylated through the inactivation of acetyltransferases and the activation of deacetylases. When PGC-1α is deacetylated, it is more active (64) because of the increase in positively charged amino acids (lysine) with the removal of the neutral acetyl groups. This change in protein charge could provide sites for the protein-protein interactions needed to increase transcription or allow PGC-1α to bind to a protein that pulls it into the nucleus. However, even though acetylation of PGC-1α has the potential to alter the activity of the protein, either there is too much redundancy in the system or this post-translational modification is secondary to transcriptional control since overexpressing SIRT1 either with (74) or without knockout of GCN5 (63) has no effect on mitochondrial biogenesis in response to exercise.

Increasing PGC-1α transcription

In one of the most elegant molecular exercise experiments, Zen Yan and his group inserted the PGC-1α2/3 promoter upstream of the firefly luciferase gene into the muscles of mice (75). They then either rested or exercised the muscles and used a high-power camera to measure the activation of the PGC-1α2/3 promoter (luciferase). Next, they mutated the promoter to discover which regions were necessary for exercise-induced activation of PGC-1α2/3 transcription. In this way, Akimoto et al. showed that in response to exercise PGC-1α2/3 transcriptional activity was dependent on two myocyte enhancing factor (MEF)2 sites and a cAMP-response element (CRE). Interestingly, the promoter of the insulin-stimulated glucose transporter (GLUT4) uses the same sites (76), suggesting that PGC-1α and GLUT4 transcription are regulated simultaneously by exercise.

MEF2

MEF2 transcriptional activity is regulated through its interaction with class II histone deacetylases (HDACs). When bound to HDACs, MEF2 is unable to stimulate transcription. The interaction between MEF2 and HDACs is regulated by phosphorylation (**Figure 9.11**). When HDACs become phosphorylated, they release MEF2, bind to the chaperone protein 14-3-3, and are removed from the nucleus (77). Therefore, the phosphorylation of HDACs is central to the regulation of MEF2 activity. The two primary kinases thought to regulate HDAC phosphorylation in response to exercise are CaMKII and AMPK (60, 78). CaMK phosphorylates HDAC4 at Ser246 and Ser467 and HDAC5 at the homologous sites Ser259 and Ser498, resulting in the export of the class II HDACs from the nucleus (77). AMPK phosphorylates these same residues (79) and as a result, after exercise HDAC4 and 5 are decreased in the nucleus (78), suggesting that MEF2 is activated by either continuous (through CaMKII) or high-intensity (through AMPK) endurance exercise.

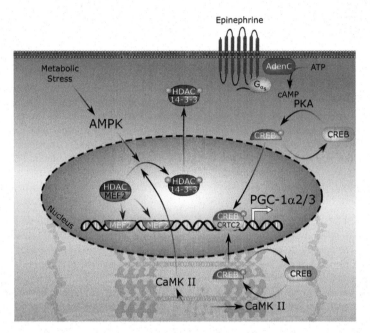

Figure 9.11 Transducing endurance exercise. (TOP) High-intensity exercise increases AMPK activity, and this results in the phosphorylation of class II HDACs and export from the nucleus by chaperone protein14-3-3. The global stress of the high-intensity exercise can also signal through β-adrenergic receptors to adenylyl kinase resulting in an increase in cAMP and activation of protein kinase A. PKA can phosphorylate and move CREB into the nucleus where together with its co-factor CRTC2, it binds the cAMP response element (CRE). Together, MEF2 and CREB increase the expression of PGC-1α2/3. (BOTTOM) Exercise duration is encoded by calcium release from contraction increasing the activity of CaMK resulting in the phosphorylation of CREB and the class II HDACs. Once phosphorylated, CREB moves to the nucleus and binds the CRE, whereas the class II HDACs release MEF2 and bind to the chaperone protein 14-3-3. 14-3-3 moves the HDACs out of the nucleus allowing MEF2 to bind to the PGC-1α2/3 promoter and together CREB and MEF2 increase PGC-1α2/3 transcription.

CRE site activation

The cyclic AMP response element in the PGC-1α2/3 promoter binds the members of the cyclic AMP response element binding protein (CREB) family: CREB, cAMP response element modulator (CREM) and activating transcription factor-1 (ATF-1). CREB was initially identified as a protein kinase A activated transcription factor, therefore epinephrine-induced activation of PKA by exercise is the obvious mechanism for activating the CRE element. Consistent with this hypothesis, the β-agonist clenbuterol increases PGC-1α2/3 transcription in a CRE dependent manner (69). Beyond PKA, CREB can be phosphorylated by both CaMKII (80) and AMPK (81). Other CREB family members can also be phosphorylated by exercise-regulated kinases. For example, ATF-1 can be phosphorylated by p38 MAPK, AMPK and CaMKII.

Signalling endurance exercise to PGC-1α2/3 transcription

Together, these data suggest that during long bouts of endurance exercise the calcium released during each contraction activates CaMKII and CaMKII can phosphorylate HDACs and CREB family members

(**Figure 9.11**). Phosphorylation of HDACs move them away from MEF2 and phosphorylation of CREB increases its transcriptional activity resulting in increased transcription of PGC-1α2/3. Similarly, high-intensity exercise increases ATP breakdown and the resulting rise in ADP increases AMPK activity. Like CaMKII, AMPK can phosphorylate HDACs and CREB family members resulting in increased transcription of PGC-1α2/3. All exercise where glycogen is depleted can activate p38 MAPK resulting in the phosphorylation of ATF-1 which can contribute to the activation of the CRE. Lastly, high-intensity exercise causes a whole-body stress, which through epinephrine can increase PKA activity resulting in greater CREB activity. The ability of long slow endurance and high-intensity exercise to increase PGC-1α2/3 through these different mechanisms likely explains why the two types of exercise increase endurance equivalently (82).

PGC-1α FUNCTION

PGC-1α is a transcriptional co-factor. A transcriptional co-factor is a protein that increases transcription without binding DNA directly. Instead, co-factors activate other transcription factors that bind DNA in a sequence-specific manner. In this way, a co-factor can increase the transcription of genes that regulate multiple pathways simultaneously. The next few sections will provide an overview of how PGC-1α or PGC-1α2/3 regulate mitochondrial biogenesis, angiogenesis and fat oxidation enzymes.

Mitochondria are unusual organelles since they have their own DNA, abbreviated as mtDNA, which encodes for some of the mitochondrial proteins. **Table 9.4** shows the proteins of the electron transfer chain complexes and how many of the proteins are encoded in the mtDNA versus nuclear DNA.

WHY DO MITOCHONDRIA HAVE THEIR OWN DNA?

Mitochondria have their own DNA because they evolved from prokaryotic organisms that infected eukaryotic cells hundreds of millions of years ago. The endosymbiotic hypothesis states that free-living aerobic bacteria were taken inside a eukaryotic cell and their ability to use oxygen to provide energy meant that the symbiotic pair could combine glycolytic and oxidative metabolism; a huge selective advantage.

How does PGC-1α increase the expression of the proteins encoded in both the nuclear and mtDNA?

Evolution then led to the development of regulatory systems where nuclear DNA was used to increase the number of proteins within the mitochondria and regulate the replication of the mtDNA such that in cases of increased energy demand the number of mitochondria could increase.

Table 9.4 Numbers of subunits of electron transfer chain complexes encoded in either mtDNA or nuclear DNA

Electron transfer chain complex	mtDNA	nuclear DNA
I NADH dehydrogenase complex	7	>25
II Succinate dehydrogenase	-	4
III Cytochrome bc1 complex	1	10
IV Cytochrone c oxidase	3	10
V F0F1 ATP synthase	2	11

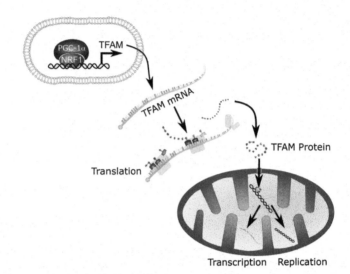

Figure 9.12 Regulation of mitochondrial biogenesis by PGC-1α and TFAM. PGC-1α controls nuclear and mitochondrial genomes through the co-activation of nuclear respiratory factor (NRF)1. NRF1 binds to the promoter of many nuclear-encoded mitochondrial genes and importantly to the promoter of the mitochondrial transcription factor A (TFAM). After translation in the cytosol, TFAM moves to the mitochondrial and initiates mtDNA transcription and replication.

Coordination of nucDNA and mtDNA

The coordination of the production of mitochondrial proteins by the nucleus and mitochondrial genomes is mediated by the mitochondria transcription factors (mtTFA/TFAM and TFBM). TFAM and TFBM are encoded in nuclear DNA, translated in the cytosol and then imported into mitochondria where they activate mtDNA gene expression and mtDNA replication (**Figure 9.12**). The regulation of TFAM by PGC-1α was first described by work from Bruce Spiegelman's lab. In this seminal article, Wu et al. identified PGC-1α and showed that PGC-1α increased TFAM transcription by binding to nuclear respiratory factor (NRF)1. In the wild-type promoter, PGC-1α increased TFAM transcription 4-fold. When the NRF1 binding site was mutated, PGC-1α was unable to increase TFAM and when the binding of PGC-1α and NRF1 was prevented, PGC-1α could not increase mitochondrial mass (83). These data suggest that PGC-1α drives mitochondrial transcription by co-activating NRF1 and producing more TFAM mRNA. TFAM is then translated and moves to the mitochondria and drives mtDNA replication. NRF1 can also bind to the promoters of many of the genes encoding proteins within the electron transport chain. Overexpressing NRF1 therefore results in partial mitochondrial biogenesis (84), suggesting that the increase in mitochondrial mass is regulated primarily by the co-activation of NRF1 by PGC-1α in response to exercise.

Regulation of fat oxidation enzymes

As with mitochondrial regulation by PGC-1α co-activating NRF1, fatty acid oxidation enzymes are regulated by PGC-1α co-activating the PPAR transcription factors. The primary PPAR in muscle is PPARβ/δ. In situations where there is low muscle glycogen, PPARδ activity is increased (85) and over time this can lead to an increase in the capacity to use fat as a fuel (86). As the name (PPARγ co-activator) suggests, PGC-1α was discovered as a co-activator of the PPARs. Therefore, the increase in PGC-1α activity after exercise, through its co-activation of PPARδ, likely underlies the increase in fat oxidation that occurs with endurance training (87).

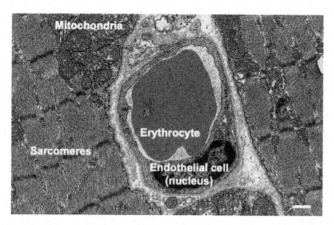

Figure 9.13 Electron microscopy image of a capillary within a mouse soleus muscle. An endothelial cell and its nucleus are visible. The capillary is surrounded by sarcomeres and mitochondria. Inside the capillary, an erythrocyte or red blood cell can be seen. Scale bar = 0.5 μm. Image taken from 1st edition of this book.

Exercise-induced angiogenesis

Endurance training not only increases stroke volume and therefore maximal blood flow but also the number of capillaries in the trained muscles through a process termed angiogenesis (**Figure 9.13**). Classical exercise physiologists demonstrated that slow type I fibres have a greater capillary density than faster type II fibres, that endurance athletes have a greater capillary-to-muscle fibre ratio in their muscles than untrained individuals (88), and that endurance training increases capillary density in skeletal muscle (89). This increase in capillary density can be replicated by over expressing PGC-1α2/3 in skeletal muscle (90), suggesting that PGC-1α underlies angiogenesis as well.

PGC-1α increases capillaries by co-activating the estrogen related receptor (ERR)α. Together, PGC-1α and ERRα upregulate the expression of vascular endothelial growth factor (VEGF) and other angiogenic growth factors (69). When ERRα is knocked out, PGC-1α is unable to increase VEGF expression and mice lacking ERRα who exercise for 14 days see no difference in the capillary:fibre ratio, whereas wild-type mice who run for 14 days increase capillaries by ~75% (69).

A second pathway that is involved in exercise-induced angiogenesis is the hypoxia-inducible factor (HIF) 1α pathway (discussed in detail in **Chapter 11**). Muscle oxygen tension or PO_2 has been estimated to shift from 30 mmHg at rest to 3–4 mmHg during moderate to heavy exercise. Given that oxygen is the key molecule for aerobic metabolism it is no surprise that it is sensed by cells and that adaptations such as angiogenesis are triggered when PO_2 is reduced. When oxygen is high, HIF-1α undergoes hydroxylation by the prolyl hydroxylase domain protein (PHD) and then is rapidly ubiquitinated and degraded by the proteasome. During intense exercise or at high altitude the PO_2 in muscle can drop and can cause an increase in HIF-1α (91). Similar to PGC-1α-ERRα complexes, HIF-1α also increases the expression of VEGF by binding to a DNA binding site which is termed hypoxia response element (HRE). To test the role of HIF-1α in determining capillary density in mice, Nunomiya et al. knocked down PHD2 as a way to more than double the HIF-1α in resting muscle (92). The result was an ~40% increase in capillaries at rest; however, more interesting was the fact that the PHD2 mice still increased capillary density by the same amount (~20%) as wild-type animals in response to four weeks of treadmill running. These data suggest that HIF-1α can increase capillaries in muscle but may not be involved in angiogenesis in response to exercise (92).

To summarise, PGC-1α increases the expression of VEGF and other growth factors that stimulate angiogenesis following exercise by co-activating ERRα. VEGF appears to be the master regulator of angiogenesis, and as such, regulates the proliferation, migration, elongation, network formation, branching and leakiness of endothelial cells. HIF-1α stabilisation can increase mitochondrial mass in response to low PO_2; however, this does not seem to be necessary for the exercise-induced increase in capillaries. One could hypothesise that HIF-1α plays a role in increasing mitochondrial mass at altitude; however, while there is a transient increase in VEGF at altitude, there is no increase in capillaries in response to eight weeks at 4,100 m, and individuals who are native to 4,100 m altitude show normal VEGF levels and FEWER capillaries per muscle fibre (93). Together, these data suggest that HIF-1α contributes to baseline capillary density in muscle, whereas exercise increases capillaries in a PGC-1α-ERRα-dependent manner.

SUMMARY

The major limiting factors or quantitative traits for endurance performance are maximal oxygen uptake ($\dot{V}O_{2max}$), the percentage of $\dot{V}O_{2max}$ which can be utilised during endurance exercise (lactate threshold), and mechanical efficiency. Prescribing an endurance training programme requires setting variables such as exercise intensity and volume. There is good scientific evidence that endurance training is effective for most, but limited evidence to give specific endurance training recommendations. One general recommendation is to polarise training; perform ≥80% of training at moderate intensity and ≤20% high-intensity and sprint interval training. It is also important to remember, individuals will respond differently to an endurance training programme, and while the vast majority of individuals will improve endurance-related variables such as $\dot{V}O_{2max}$ and disease risk factors, a small minority may not respond and may even worsen some risk factors. A key adaptation to endurance exercise is increased cardiac output which is dependent on an increased stroke volume. The increase in stroke volume depends on the development of physiological cardiac hypertrophy or an athlete's heart. Research suggests that this relies, at least partially, on integrin/PI3K/Akt signalling. In contrast, strength training results in a physiological pressure overload hypertrophy in the heart that may depend on ERK signalling. Lastly, increased calcineurin signalling stimulates pathological cardiac hypertrophy. Muscle fibres can be subdivided into slow type I fibres, intermediate type IIa fibres, fast type IIx fibres and even faster type IIb fibres. The name derives from the MyHC that is predominantly expressed in a muscle fibre. For example, type IIx fibres mainly express type IIx myosin heavy chain. Importantly, while humans have a MyHC IIb gene in their genomes, the MyHC IIb protein is not expressed in major muscles. Calcineurin-NFAT signalling is active in type I fibres and increases in response to electrical stimulation. When active, calcineurin-NFAT signalling promotes the expression of 'slow fibre' genes. Calcineurin-NFAT may contribute to the decrease of type IIx fibres and increase of IIa fibres that occurs in response to endurance training programmes over several months. More pronounced fibre-type conversions may occur in response to years or decades of endurance training, but the evidence is limited. MyHC genes are organised in a slow/cardiac cluster and a fast/developmental cluster and this is evolutionary conserved. Expression of MyHC I also increases the expression of MyoMirs, which via the transcription factor Sox6, prevent the expression of faster MyHCs. This can in part explain the on/off regulation of MyHCs and muscle fibre type specification. Chromatin remodelling also plays a role in which MyHC isoforms are actively transcribed in a muscle fibre. This is also known as an epigenetic regulation. Endurance exercise increases the number of mitochondria which is termed exercise-induced mitochondrial biogenesis. Here, the major players are **CaMKII and SIRT1**, which are the sensors of signals calcium and NAD^+ respectively, produced as a result of exercise duration, **AMPK**, which is the sensor of the rise in AMP and ADP and the fall of glycogen during exercise and **p38 MAPK and PKA**, which are the sensors of signals such as glycogen content, and epinephrine levels in response to the stress of exercise. These signal transduction molecules then activate the transcriptional co-factor PGC-1α by increasing the phosphorylation and by reducing the acetylation of the full-length protein and driving the expression of an alternative form of the gene (PGC-1α2/3), which lacks the repressive domain that normally binds RIP140. Active PGC-1α then acts as an effector, where it co-activates transcription factors to increase the transcription/expression of mitochondrial genes encoded in nuclear DNA and of transcription and replication factors that promote the expression of genes encoded in the mitochondria and therefore the replication of mitochondrial DNA. PGC-1α, together with ERRα, drives the expression of VEGF, which results in the formation of new capillaries after exercise.

REVIEW QUESTIONS

* Describe and explain the strategy you would recommend for the prescription of an endurance training programme which takes the variation in trainability into account.

* Give an example for and explain a signal transduction pathway that regulates physiological (athlete's heart) and pathological cardiac hypertrophy (hypertrophic cardiomyopathy).

* Compare and contrast type I, IIa, IIx and IIb muscle fibres. What is special about IIb fibres? What is the effect of endurance exercise on fibre-type percentages?

* Describe the arguments for and against the hypothesis that the calcineurin pathway is a major regulator of fibre type.

* Explain and discuss mechanisms that may contribute to the on/off regulation of myosin heavy chain isoforms in skeletal muscle.

* Discuss how endurance exercise stimulates mitochondrial biogenesis.

* Explain how an increased energy turnover, changed oxygen levels and increased blood flow may stimulate capillary growth in response to endurance exercise.

FURTHER READING

Egan B & Zierath JR (2013). Exercise metabolism and the molecular regulation of skeletal muscle adaptation. *Cell Metab* 17, 162–84.
Hardie DG (2011). Energy sensing by the AMP-activated protein kinase and its effects on muscle metabolism. *Proc Nutr Soc* 70, 92–99.
Schiaffino S (2010). Fibre types in skeletal muscle: a personal account. *Acta Physiol (Oxf)* 199, 451–63.

REFERENCES

1. McCormick A, et al. *Sports Med.* 2015. 45:997–1015.
2. Gottschall JS, et al. *J Biomech.* 2005. 38(3):445–52.
3. Colloud F, et al. *J Sports Sci.* 2006. 24(5):479–93.
4. Plews DJ, et al. *Int J Sports Physiol Perform.* 2017. 12(5):697–703.
5. Stellingwerff T. *Int J Sport Nutr Exerc Metab.* 2012. 22(5):392–400.
6. Bouchard C, et al. *J Appl Physiol.* 1999. 87(3):1003–8.
7. Bouchard C, et al. *PLoS One.* 2012. 7(5):e37887.
8. Barber JL, et al. *Br J Sports Med.* 2022. 56:95–100.
9. McPhee JS, et al. *Exp Physiol.* 2009. 94:684–94.
10. Timmons JA, et al. *J Appl Physiol.* 2010. 108(6):1487–96.
11. Garber CE, et al. *Med Sci Sports Exerc.* 2011. 43(7):1334–59.
12. Stöggl TL, et al. *Front Physiol.* 2015. 6:295.
13. ASTRAND PO, et al. *J Appl Physiol.* 1964. 19:268–74.
14. Ekblom B, et al. *J Appl Physiol.* 1975. 39(1):71–5.
15. Montero D, et al. *Am J Physiol – Regul Integr Comp Physiol.* 2017. 312(6):R894–902.

16. Blomqvist CG, et al. Annu Rev Physiol. 1983. 45:169–89.

17. Scharhag J, et al. J Am Coll Cardiol. 2002. 40(10):1856–63.

18. Chandra N, et al. J Am Coll Cardiol. 2013. 61:1027–40.

19. Shioi T, et al. EMBO J. 2000. 19(11):2537–48.

20. McMullen JR, et al. Proc Natl Acad Sci U S A. 2003. 100(21):12355–60.

21. DeBosch B, et al. Circulation. 2006. 113(17):2097–104.

22. Weeks KL, et al. Am J Physiol – Heart Circ Physiol. 2021. 320(4):H1470–85.

23. MacDougall JD, et al. J Appl Physiol. 1985. 58(3):785–90.

24. Bueno OF, et al. EMBO J. 2000. 19(23):6341–50.

25. Purcell NH, et al. Proc Natl Acad Sci U S A. 2007. 104(35):14074–9.

26. Nakamura M, et al. 2. Nat Rev Cardiol. 2018. **15:**387–407.

27. Molkentin JD, et al. Cell. 1998. 93(2):215–28.

28. Wilkins BJ, et al. Circ Res. 2004. 94(1):110–8.

29. Bárány M. J Gen Physiol. 1967. 50(6):197–218.

30. Schiaffino S. FEBS J. John Wiley & Sons, Ltd; 2018. 285:3688–94.

31. Harridge SDR, et al. Pflugers Arch Eur J Physiol. 1996. 432(5):913–20.

32. Smerdu V, et al. Am J Physiol – Cell Physiol. 1994. 267(6 Pt 1):C1723–8.

33. Costill DL, et al. J Appl Physiol. 1976. 40(2):149–54.

34. Ingalls CP. J Appl Physiol Am Physiol Soc. 2004. 97:1591–2.

35. Gollnick PD, et al. J Appl Physiol. 1973. 34(1):107–11.

36. Bathgate KE, et al. Eur J Appl Physiol. 2018. 118(10):2097–110.

37. Konopka AR, et al. J Gerontol – Ser A Biol Sci Med Sci. 2011. 66 A(8):835–41.

38. Pette D, et al. Rev Physiol Biochem Pharmacol. 1992. 120:115–202.

39. Weeds AG, et al. Nature. 1974. 247(5437):135–9.

40. Biering-Sørensen B, et al. Muscle Nerve. 2009. 40:499–519.

41. Eltit JM, et al. Biophys J. 2004. 86(5):3042–51.

42. Jordan T, et al. J Cell Sci. 2005. 118(10):2295–302.

43. Hennig R, et al. Nature. 1985. 314(6007):164–64.

44. Tothova J, et al. J Cell Sci. 2006. 119(8):1604–11.

45. Chin ER, et al. Genes Dev. 1998. 12(16):2499–509.

46. Naya FJ, et al. J Biol Chem. 2000. 275(7):4545–8.

47. Parsons SA, et al. Mol Cell Biol. 2003. 23(12):4331–43.

48. Lin J, et al. Nature. 2002. 418(6899):797–801.

49. Wu H, et al. EMBO J. 2000. 19(9):1963–73.

50. Murgia M, et al. Nat Cell Biol. 2000. 2(3):142–7.

51. Tsika RW, et al. J Biol Chem. 2008. 283(52):36154–67.

52. Hughes SM, et al. J Cell Biol. 1999. 145(3):633–42.

53. Lassar AB, et al. Cell. 1986. 47(5):649–56.

54. Jenuwein T, et al., Science. 2001. 293:1074–80.

55. Asp P, et al. Proc Natl Acad Sci U S A. 2011. 108(22):E149-E158.

56. Pandorf CE, et al. Am J Physiol – Cell Physiol. 2009. 297(1).

57. van Rooij E, et al. Dev Cell. 2009. 17(5):662–73.

58. Quiat D, et al. Proc Natl Acad Sci U S A. 2011. 108(25):10196–201.

59. Baar K, et al. FASEB J. 2002. 16(14): 1879–86.

60. Smith JAH, et al. Am J Physiol – Endocrinol Metab. 2008. 295(3):E698–704.

61. Wu H, et al. Science (80-). 2002. 296(5566):349–52.

62. White AT, et al. Am J Physiol – Endocrinol Metab. 2012. 303:308–21.

63. Philp A, et al. J Biol Chem. 2011. 286(35):30561–70.

64. Rodgers JT, et al. Nature. 2005. 434(7029):113–8.

65. Terada S, et al. Acta Physiol Scand. 2005. 184(1):59–65.

66. McBride A, et al. Cell Metab. 2009. 9(1):23–34.

67. Egan B, et al. J Physiol. 2010. 588(10):1779–90.

68. Puigserver P, et al. *Mol Cell*. 2001. 8(5):971–82.
69. Chinsomboon J, et al. *Proc Natl Acad Sci U S A*. 2009. 106(50):21401–6.
70. Pogozelski AR, et al. *PLoS One*. 2009. 4(11):e7934.
71. Wright DC, et al. *J Biol Chem*. 2007. 282(1):194–9.
72. Hallberg M, et al. *Mol Cell Biol*. 2008. 28(22):6785–95.
73. Wallberg AE, et al. *Mol Cell*. 2003. 12(5):1137–49.
74. Svensson K, et al. *Am J Physiol – Endocrinol Metab*. 2020. 318(2):E145–51.
75. Akimoto T, et al. *Am J Physiol – Cell Physiol*. 2004. 287(3):C790–6.
76. Thai M V., et al. *J Biol Chem*. 1998. 273(23):14285–92.
77. McKinsey TA, et al. *Proc Natl Acad Sci U S A*. 2000. 97(26):14400–5.
78. McGee SL, et al. *J Physiol*. 2009. 587(24):5951–8.
79. McGee SL, et al. *Diabetes*. 2008. 57(4):860–7.
80. Shimomura A, et al. *J Biol Chem*. 1996. 271(30):17957–60.
81. Thomson DM, et al. *J Appl Physiol*. 2008. 104(2):429–38.
82. Gibala MJ, et al. *J Physiol*. 2006. 575(3):901–11.
83. Wu Z, et al. *Cell*. 1999. 98(1):115–24.
84. Baar K, et al. *FASEB J*. 2003. 17(12):1666–73.
85. Philp A, et al. *PLoS One*. 2013. 8(10):e77200.
86. Hulston CJ, et al. *Med Sci Sports Exerc*. 2010. 42(11):2046-55.
87. Molé PA, et al. *J Clin Invest*. 1971. 50(11):2323–30.
88. Ingjer F, et al. *Eur J Appl Physiol Occup Physiol*. 1978. 38(4):291–9.
89. Andersen P, et al. *J Physiol*. 1977. 270(3):677–90.
90. Tadaishi M, et al. *PLoS One*. 2011. 6(12):e28290.
91. Ameln H, et al. *FASEB J*. 2005. 19(8):1009–11.
92. Nunomiya A, et al. *Acta Physiol*. 2017. 220(1):99–112.
93. Lundby C, et al. *J Exp Biol*. 2004. 207(22):3865–71.

CHAPTER TEN

Molecular sport nutrition

Mark Hearris, Nathan Hodson, Javier Gonzalez and James P. Morton

LEARNING OUTCOMES:

By the end of this chapter, the reader will be able to:

1. Outline the molecular responses regulating muscle glucose uptake and glycogen storage.
2. Explain the impact of high-fat diets on intramuscular triglyceride storage and the role of free fatty acids as signalling molecules.
3. Outline the molecular response by which protein feeding stimulates muscle protein synthesis.
4. Critically evaluate the role of macronutrient availability on the molecular regulation of skeletal muscle adaptation to exercise training.

INTRODUCTION

In its simplest terms, nutrition provides the basis of all physical performance. What we eat provides the substrates from both intramuscular (i.e. glycogen and intramuscular triglyceride (IMTG)) and extramuscular sources (i.e. plasma glucose and free fatty acids (FFAs)) that are necessary to fuel muscle contraction whilst also promoting recovery from exercise. Although the role of nutrition in promoting exercise performance and recovery has been recognised for almost a century, the last 20 years has led to a rapid advancement in our understanding of how nutrition can regulate adaptations to training in skeletal muscle. In this regard, it is now accepted that the ingestion of macro- and micronutrients before, during and/or after exercise has the capacity to enhance or attenuate key signal transduction pathways with regulatory roles in training adaptation. As such, manipulation of macronutrient intake and timing has the capacity to support skeletal muscle adaptations inherent to both endurance and strength training. For this reason, it is now recognised that an athlete's daily nutritional needs should likely change in accordance with the workload and goal of the upcoming training session as well as the competitive schedule. Accordingly, the concept of "nutritional periodisation" is now a hot topic amongst athletes, coaches, sport scientists and molecular exercise physiologists.

With this in mind, the remit of the modern-day sport nutritionist has expanded towards the goal of:

> Strategic periodisation of energy, macro- and micronutrient availability (alongside targeted use of supplements and ergogenic aids) to improve body composition, training adaptations, performance, health and recovery.

In this chapter, we present a contemporary view of molecular sport nutrition by providing the molecular exercise physiologist or sport nutritionist with an understanding of the molecular processes supporting fundamental principles of performance nutrition. After a brief introduction of how appetite is controlled,

DOI: 10.4324/9781315110752-10

we proceed to discuss the molecular pathways that control muscle glucose uptake and glycogen storage. Such processes support the fundamental principle of carbohydrate (CHO) loading (i.e. the supercompensation of muscle glycogen stores prior to intense and prolonged exercise). In keeping with the theme of "fuels for the fire", we then outline the molecular pathways supporting "fat adaptation" protocols, a nutritional strategy that can be employed to enhance IMTG storage and spare the use of muscle glycogen during exercise. Having considered the basis of energy storage, we subsequently provide a critical overview of how the availability of nutrients supports muscle protein synthesis (MPS) and oxidative adaptations in human skeletal muscle. In relation to the latter, specific attention is given to the role of reduced CHO availability in regulating mitochondrial biogenesis, the so-called "train-low" strategy. Due to space constraints, it is not possible to build upon themes presented in the previous edition of this chapter. For a detailed discussion on principles of nutrient sensing by target tissues such as the brain and muscle, the reader is therefore directed to the previous edition of this textbook, and chapter by Hamilton et al. (2014).

OVERVIEW OF APPETITE REGULATION

Definitions of appetite

Appetite is a general term which covers the motivation, preference and/or selection of food intake. Some more specific definitions include hunger: the urge to eat, satiation: the process that leads to finishing eating, and satiety: the process that leads to preventing further eating [increase in fullness/decline in hunger] (1). Satiation (which determines energy intake at a particular meal) can occur in response to boredom of taste during a meal or by the feeling of fullness after a meal. Therefore, satiation is underpinned by both sensory factors – like stomach stretch – in addition to cognitive factors – the prior beliefs we have developed from eating similar foods in the past. Whilst these cognitive factors can play an important role, the molecular and physiological regulation of appetite will be the focus for the remainder of this chapter.

Convergence of peripheral appetite signals in the hypothalamus

The physiological regulation of appetite involves the brain sensing a variety of signals which either stimulate an increase in appetite or suppress appetite. These signals are directed to an area of the brain known as the arcuate nucleus (ARC), which is found within the hypothalamus, and acts via an area known as the paraventricular nucleus (PVN) (see **Figure 10.1**) (2). Within the ARC, neurons produce four key neuropeptides known as neuropeptide Y (NPY), agouti related peptide (AgRP), pro-opiomelanocortin (POMC) and cocaine- and amphetamine-regulated transcript (CART). NPY and AgRP stimulate food consumption (energy intake), whereas POMC and CART suppress food consumption. The combination of hormones and how active our metabolism is (through increased physical activity or exercise) provide stimuli to these appetite centres in the brain to promote either a reduction in appetite (anorexigenic) or an increase in appetite (orexigenic).

The gut releases the majority of hormones that control appetite. Some are also secreted by fat (adipose) tissue. Hormones which can directly reduce the amount of food consumed include peptide tyrosine tyrosine (PYY), glucagon-like peptide-1 (GLP-1), and oxyntomodulin (all secreted by the L-cells of the intestine); cholecystokinin (secreted by I-cells of the intestine); insulin, glucagon, pancreatic polypeptide and amylin (secreted by the pancreas); and leptin (secreted by fat cells/adipocytes). The only gut hormone currently identified that stimulates (increases) appetite is ghrelin, which is mainly secreted by the gastric cells of the stomach.

Exercise, appetite and energy balance

Long-term changes in body weight are the result of changes in energy balance (the difference between energy intake from food and energy expenditure from metabolism and physical activity). When the

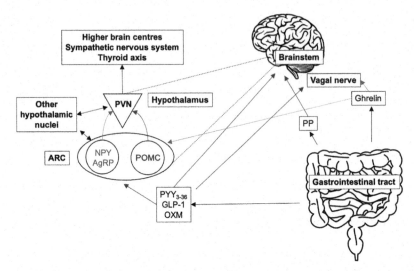

Figure 10.1 Regulation of appetite by gut hormones. Blue lines indicate stimulation. Red lines indicate inhibition. PYY$_{3-36}$, GLP-1 and OXM are peptide hormones secreted by the gastrointestinal tract following the ingestion of food. These hormones can directly reduce/suppress appetite via pathways in the brainstem and hypothalamus and may also influence the vagal nerve that supplies information to and from the brain and digestive system. Pancreatic polypeptide is secreted by the pancreas following food ingestion and may lead to reduced appetite via the brainstem. Ghrelin is released from the stomach in response to periods of no food consumption (fasting) and is thought to stimulate food intake via the hypothalamus and/or the vagal nerve. Many appetite-related signals are integrated in the ARC within the hypothalamus, whereby NPY and AgRP neurons stimulate (increase) appetite and POMC neurons suppress (reduce) appetite by signalling to the PVN. ARC, arcuate nucleus; PVN, paraventricular nucleus; NPY, Neuropeptide Y; AgRP, agouti-related protein; POMC, proopiomelanocortin; PYY, peptide tyrosine tyrosine; GLP-1, glucagon-like peptide-1; OXM, oxyntomodulin; PP, pancreatic polypeptide. Figure redrawn from Murphy and Bloom (2). Some of the images created using smart.servier.com. No permissions required. Attribution 3.0 Unported (CC BY 3.0) licence, https://creativecommons.org/licenses/by/3.0/.

amount of energy the body uses is more than the amount of energy taken in by food (energy deficit), body weight will decrease. In contrast, when the amount of energy the body uses is less than the amount of energy taken in by food (energy surplus), body weight will increase. Observations of army cadets in the 1950s showed that within a single day, the amount of energy used by the body during daily activity plus resting metabolism did not correlate with energy taken in from food (energy intake) (3). This suggested that in the short term, the appetite system is perhaps not as finely in tune with the amount of energy our bodies are using up (energy expenditure) from our daily physical activities. However, over longer-time frames, there is a suggestion that the amount of energy we use with daily activities affects how our appetite system is regulated. When people have a low level of physical activity, they tend to eat more food than is needed to maintain their body weight, and so their body weight increases. However, when people have a high physical activity level, they are more likely to eat the right amount of food to maintain their body weight. In this study, it meant that the body weight of the physically active was around 15 kg less than their counterparts with lower physical activity levels (4). This is also supported by research demonstrating that appetite is regulated more effectively when people are required to perform more physical activity (5). The differences in appetite with high versus low physical activity levels might be due to changes in gut hormones, such as higher GLP-1 and lower ghrelin concentrations when individuals are more physically active. The exact mechanisms that underpin how exercise might change GLP-1 and ghrelin concentrations are unclear but may involve changes in energy sensing by the gut and/or changes in the amounts of specific fuels available in the body, discussed below.

Substrate metabolism during exercise, appetite and energy balance

In addition to hormones, the availability of fuels (metabolic signals) could also contribute to alterations in appetite with exercise. Higher rates of fat use (oxidation) are often associated with fat loss during exercise training. One proposed reason that could explain these responses relates to the ability of humans to store energy as carbohydrate is much more limited than the ability to store energy as fat. Therefore, a more controlled use of carbohydrates should take place, and the body may try to preserve carbohydrate stores more than fat stores. In other words, the rate at which we use carbohydrates during exercise (glycogen and glucose oxidation) should be more tightly related to the amount of energy that is taken in, to ensure that there is an adequate supply of available carbohydrates for later use (6, 7).

Associations between fuel use (substrate utilisation) and self-selected (ad libitum) energy intake have been reported. People who display a higher respiratory exchange ratio [the ratio between the production of carbon dioxide (CO_2) and the uptake of oxygen (O_2), which reflects carbohydrate use] during exercise tend to display a more positive energy balance and/or food intake after exercise when provided with food ad libitum (8, 9). However, the respiratory exchange ratio represents the contribution of all carbohydrate sources to energy expenditure, and is not able to tell us whether specific carbohydrate sources are more important than others. For example, liver glycogen might be the carbohydrate source that is most important, whereas muscle glycogen might be less important for appetite. Studies using tracer methods (labelled molecules to "trace" how the body is using fuels) have shown that the use of carbohydrate from the liver during exercise correlates with energy intake from food after exercise (10), whereas the use of muscle glycogen shows no such correlation, suggesting that liver glycogen stores may be more important for appetite. Further support for this is provided in studies using mice, where overexpression of a gene called protein targeted to glycogen (ptg) in the liver – which increases liver glycogen stores – reduces appetite and food consumed. This suggests an increase in liver glycogen leads to reduced appetite. To help confirm that liver glycogen levels were the important factor in appetite, when these mice had surgery to cut the vagal nerve, which prevented the brain from detecting the levels of glycogen in the liver, the influence of ptg overexpression on reducing appetite was prevented. Therefore, this evidence indicates that liver glycogen is a signal for appetite control that may act via the vagal nerve (11).

Key points

- Appetite control involves the combination of both psychological and physiological components.

- The physiological component of appetite includes the convergence of hormonal and metabolic signals in the hypothalamus of the brain, whereby the balance of appetite-supressing (anorexigenic) or appetite-stimulating (orexigenic) stimuli dictate the drive to eat food.

- Practically, dietary strategies which target increasing the secretion of satiety (feeling full) hormones (e.g. GLP-1, leptin and PYY), whilst suppressing hunger hormones (e.g. ghrelin) and maintaining carbohydrate balance, may facilitate better adherence to an energy deficit and therefore assist with weight loss maintenance.

FUELS FOR THE FIRE

From a practical perspective, a fundamental goal of sport nutritionists is to ensure that athletes commence training and competition with sufficient carbohydrate stored in the muscle and the liver (glycogen) to complete the desired exercise task. To this end, athletes have been long advised to consume diets that contain enough carbohydrate (CHO) to ensure adequate glycogen stores before and during exercise. Following digestion and absorption of CHO sources, glucose is then delivered to the liver and muscle to subsequently increase glycogen storage. The daily CHO requirements of athletes are now thought to range on a sliding scale between 3 and 12 g/kg body mass per day, dependent on the amount of energy required to complete the expected exercise demands.

Muscle glucose uptake and glycogen storage in response to carbohydrate feeding

Glucose enters skeletal muscle through a transport protein called GLUT4 and can only enter the muscle cell when GLUT4 is on the cell membrane (rather than within the cytoplasm). In addition, the concentration gradient from outside-to-inside the muscle cell is also important for glucose transport (12). Three key sites of regulation for muscle glucose uptake are (1) glucose delivery to muscle; (2) glucose transport across the muscle membrane and (3) glucose metabolism within the muscle cell. All these processes can be regulated by the hormone insulin, and by muscle contraction. In resting conditions, it is thought that glucose transport across the muscle membrane is rate-limiting, whereas during high-intensity exercise it is thought that processes within the muscle cell (glucose phosphorylation) could become rate-limiting.

At rest, the primary regulation of the location of GLUT4 is by insulin. In a fasted state when insulin concentrations are low, very little GLUT4 is found on the membrane of skeletal muscle (or adipose tissue) and the rate of glucose uptake into these tissues is low. After eating a meal containing carbohydrate or protein (such as that recommended in the pre-exercise meal), the rise in blood glucose and amino acid (AA) concentrations – combined with the release of some gut hormones – results in the release of insulin by the pancreas. Insulin is transported in the bloodstream and binds to the insulin receptor on muscle cells. Once insulin has bound to the insulin receptor on the muscle membrane, a cascade of insulin signalling begins, which stimulates the movement of GLUT4 from storage within the cell to the plasma membrane. By increasing the amount of GLUT4 at the muscle membrane, this allows more glucose to enter the cell (**Figure 10.2A**) which can then lead to glycogen storage.

Muscle glucose uptake during exercise

At rest, muscle glucose uptake is mainly responding to food intake (and the insulin response to food intake). During exercise, however, muscle contraction is the key stimulus for increasing muscle glucose uptake. During exercise, the source of glucose for muscle comes from the breakdown of liver glycogen and/or from CHO consumed during exercise. Because muscle glycogen stores can become depleted after 2–3 hours of exercise, athletes are often advised to consume CHO during exercise at a rate of 30–90 g/h dependent on the exercise intensity and intended duration. Consumption of CHO during exercise improves the ability to sustain an exercise intensity (exercise capacity) and exercise performance via mechanisms related to preventing liver glycogen depletion, and maintaining blood glucose levels and whole body CHO oxidation.

Rates of muscle glucose uptake during exercise can increase to 100-fold that seen at rest, and this increases with both intensity and duration of exercise. Blood glucose can contribute up to 40% of the energy required to exercise when muscle glycogen is low. Like at rest, the primary regulatory steps to muscle glucose uptake during exercise include glucose delivery to the muscle, glucose transport across the muscle membrane and glucose metabolism within the muscle cell (13).

Muscle glucose delivery is increased during exercise due to increases in blood flow to the muscle, with the increase in blood flow (up to 20× resting) vastly greater than the increase in the difference in glucose concentration between the arteries supplying the muscle and the veins draining the muscle (up to 4× resting) (13). In addition, the recruitment of extra capillaries increases the surface area available for glucose uptake. As glucose uptake is also dependent on the concentration gradient, maintaining a higher

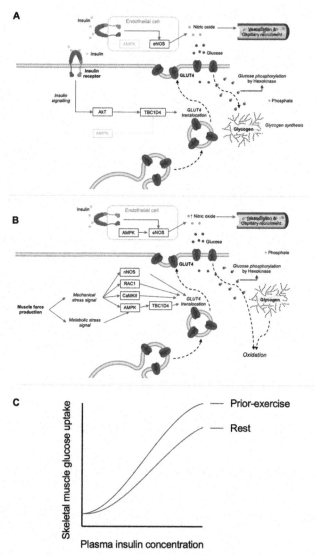

Figure 10.2 (A) In the rested state, insulin is the primary stimulus for muscle glucose uptake by increasing vasodilation, capillary recruitment (via eNOS) and GLUT4 translocation (via intramuscular insulin signalling). Once glucose has entered the muscle cell, further regulation is provided by feedback inhibition of hexokinase activity by glucose-6-phosphate. **(B)** Exercise stimulates muscle glucose uptake during and after exercise by increasing blood flow, GLUT4 translocation and intramuscular glucose flux. This is likely mediated by increases in AMPK activity in both endothelial cells and skeletal muscle cells, at least in the post-exercise period. In response to subsequent insulin stimulation, TBC1D4 phosphorylation is further enhanced by prior muscle contraction, resulting in greater insulin-stimulated muscle glucose uptake **(C)**. Blue lines indicate stimulation. Red lines indicate inhibition. GLUT4, glucose transporter 4; eNOS, endothelial nitric oxide synthase; nNOS, neuronal nitric oxide synthase; RAC1, Ras-related C3 botulinum toxin substrate 1; TBC1D4, TBC1 domain family member 4; CAMKII, Calcium/calmodulin-dependent protein kinase II; AMPK, adenosine monophosphate activated protein kinase; Akt, protein kinase B. Some of the images created using smart.servier.com. No permissions required. Attribution 3.0 Unported (CC BY 3.0) licence, https://creativecommons.org/licenses/by/3.0/.

glucose concentration will also support glucose uptake, which is achieved by increased breakdown of liver glycogen in the early phases of exercise, and can be maintained with CHO ingestion during prolonged exercise.

Glucose transport across the muscle membrane during exercise is still largely regulated by movement of GLUT4 to the muscle membrane (**Figure 10.2B**). The mechanisms underlying exercise-induced GLUT4 movement, however, are different to those at rest as insulin concentrations are low during exercise. The signalling for insulin-stimulated versus exercise-stimulated GLUT4 translocation differ at first, but then likely meet at the level of some key proteins (TBC1D1, TBC1D4 and Rac1) (14). GLUT4 is clearly required for exercise-induced increases in glucose uptake in mice, since mice lacking GLUT4 display little-to-no exercise-induced increase in muscle glucose uptake. However, the increase in GLUT4 movement with exercise is relatively modest (~2-fold) compared with the large increase in muscle glucose uptake. This has led to suggestions that the ability of membrane-bound GLUT4 to transport glucose (i.e. the intrinsic activity) is also increased during exercise, and this may contribute to increased glucose uptake independent of the amount of GLUT4 at the muscle membrane. The idea of increased GLUT4 intrinsic activity, however, is still widely debated (13).

The precise mechanisms underlying exercise-induced GLUT4 translocation and muscle glucose uptake are also unclear. Earlier studies suggested that adenosine monophosphate-activated protein kinase (AMPK) was a key regulator, as demonstrated by increases in AMPK activity with exercise, and increased muscle glucose uptake with drug-based activation of AMPK. Indeed, AMPK activity is increased when the energy status of a cell is low (i.e. increases in ADP and AMP relative to ATP). In addition, low glycogen and high non-esterified fatty acid availability can also increase AMPK activity. However, relatively high-intensity and/or prolonged exercise is required to produce robust increases in AMPK activity in human skeletal muscle, whereas leg muscle glucose uptake increases even during the first hour of low-to-moderate-intensity (~45% VO_{2peak}) exercise, where AMPK activity is not increased (15). This suggests that glucose uptake into muscle during exercise is not always due to increased AMPK activity.

Other potential mechanisms for stimulating GLUT4 movement include calcium (Ca^{2+}) mediated signalling, mitogen activated protein kinases (MAPKs), nitric oxide and reactive oxygen species. Incubating muscle in caffeine (which causes Ca^{2+} release) has been shown to stimulate glucose uptake, which does not require muscle contraction. However, it is possible that these effects could be mediated by the energy stress associated with ion pumping, rather than Ca^{2+} per se. Mitogen activate protein kinases, such as ERK1, ERK2, p38, and JNK are all activated during muscle contraction, although the current evidence suggests they are unlikely to play an important role in stimulating muscle glucose uptake during exercise. Nitric oxide production is increased during contraction, and inhibition of nitric oxide synthase has been shown to decrease contraction-induced muscle glucose uptake, which may be specific to the type of muscle fibre ("fast" or "slow" twitch) (13). In addition, infusing an inhibitor of nitric oxide synthase (L-NMMA) into humans can inhibit nitric oxide synthase without affecting total blood flow, and can reduce leg glucose uptake during exercise. Therefore, nitric oxide may play an important role in muscle glucose uptake during exercise. Finally, other reactive oxygen species have been implicated in contraction-induced muscle glucose uptake. Incubating muscle in hydrogen peroxide (H_2O_2) has been shown to increase glucose uptake, and exercise results in an increase in the production of reactive oxygen species. However, attempts to inhibit reactive oxygen species during exercise in humans do not seem to affect exercise-induced glucose uptake. Therefore, the role of reactive oxygen species in exercise-induced glucose uptake is not fully known. Clearly, muscle glucose uptake during exercise may be regulated by multiple signalling pathways.

When glucose has entered the muscle cytoplasm, a phosphoryl group is rapidly and irreversibly bound to the glucose (phosphorylated) to produce glucose-6-phosphate by the enzyme hexokinase II. During high-intensity exercise, this process could also become limiting to muscle glucose uptake. This has been demonstrated by increased intramuscular glucose concentrations (not bound to a phosphoryl group) during maximal exercise, suggesting that rates of glucose phosphorylation could be limiting. As higher glucose-6-phosphate concentrations inhibit the activity of hexokinase II, it is possible that rapid breakdown of muscle glycogen (resulting in glucose-6-phosphate accumulation), could inhibit hexokinase II activity and limit phosphorylation of glucose entering from the circulation (13, 14).

Post-exercise muscle glucose uptake and glycogen storage

It has been known since the late 1960s that prior exercise increases glycogen storage in the post-exercise period in response to a given dose of CHO. This phenomenon is known as "glycogen supercompensation" and is the theoretical basis underpinning the practical strategy of CHO loading. In this regard, athletes traditionally performed a bout of glycogen-depleting exercise prior to a 1–3-day period of high CHO intake so as to supercompensate and "load" the muscle with glycogen. It is now known that such an aggressive glycogen depletion protocol is not required to facilitate CHO loading, though the fact remains that prior exercise seems to potentiate the muscle to store glycogen.

Muscle glycogen can be found in at least three distinct subcellular locations, subsarcolemmal (located in between the outermost myofibril and the sarcolemma), intermyofibrillar (located between myofibrils) and intramyofibrillar (located within myofibrils) (16). Muscle glycogen is an important fuel source during moderate-to-high exercise intensities. Since the capacity to store muscle glycogen is relatively small, muscle glycogen stores can be depleted to critically low levels relatively quickly (i.e. within 2 hours of high-intensity exercise). The intramyofibrillar glycogen pool appears to be utilised most readily during exercise, and the depletion of this pool seems to be most closely linked to fatigue. The restoration of muscle glycogen stores after exercise can take as much as 24 hours, even with aggressive feeding strategies. Therefore, the use and restoration of muscle glycogen can be a critical factor for success in endurance sports. In addition, the capacity to store carbohydrate as muscle glycogen is a primary defect in people at an increased risk of developing type 2 diabetes, and therefore understanding how muscle glycogen is regulated is important for both sports performance and metabolic health.

Glycogen concentrations reflect the balance of glycogen breakdown (glycogenolysis) and glycogen production (glycogen synthesis). Glycogenolysis is mainly regulated by the enzymes glycogen phosphorylase and debranching enzyme, which act on the terminal α-1,4-glycosidic linked glucose residues, and the α-1,6-branchpoints in the glycogen molecule, respectively. Glycogen phosphorylase is activated by AMP or IMP and inhibited by ATP and glucose-6-phosphate. Finally, muscle glycogen is also subject to autoregulation, which means that a high glycogen concentration stimulates glycogen breakdown, which is likely to be by activating glycogen phosphorylase (17).

Glycogen synthesis is mainly regulated by the enzymes glycogen synthase and branching enzyme, which catalyse the incorporation of UDP-glucose into glycogen via α-1,4-glycosidic linkages, and the formation of α-1,6-branchpoints, respectively (17). Glycogen synthase is stimulated by insulin and by exercise. It is also speculated that glycogen synthase activity is inhibited by glycogen, such that a low muscle glycogen concentration is a stimulus for increased glycogen synthase activity, which may be regulated by protein phosphatase 1. This provides the mechanistic underpinning for glycogen supercompensation.

Post-exercise, muscle displays increased glucose uptake which is likely to be first due to residual increases in blood flow and the amount of GLUT4 at the muscle membrane, but later on in the recovery period, also due to greater insulin-stimulated glucose uptake (known as insulin sensitivity; **Figure 10.2C**) (14). This insulin-sensitising effect of exercise can last for many hours (and even days) after exercise. It is noteworthy, however, that muscle-damaging exercise can, in fact, impair glucose uptake, which is likely due to a reduction in the amount of GLUT4 within the muscle cell, in addition to decreased insulin signalling. It is this insulin-sensitising effect of an acute bout of prior exercise that contributes to the classical supercompensation of muscle glycogen, and it has been suggested that residual AMPK activity from glycogen depletion plays an important role in this response.

From a practical perspective, the acute post-exercise period is therefore a prime opportunity for muscle glycogen restoration due to increased muscle blood flow, residual GLUT4 at the muscle membrane and increased glycogen synthase activity. Therefore, provision of adequate CHO as soon as possible after exercise is a key goal when rapid recovery is required between two exhaustive bouts of endurance exercise. This can be achieved by providing at least 1 g CHO per kilogram body mass per hour, for the first 4 hours post-exercise. Maintaining low muscle glycogen concentrations activates pathways stimulating muscle glucose uptake, and therefore may play a role in the benefits of exercise for glucose control.

Key points

- The primary sites of regulation for muscle glucose uptake are glucose delivery to muscle, glucose transport across the muscle membrane and glucose metabolism within the muscle cell.

- Insulin and muscle contraction are two key (independent) stimuli for all three regulatory sites of muscle glucose uptake.

- Glycogen supercompensation post-exercise is likely due to residual increases in blood flow and GLUT4 at the muscle membrane, in addition to enhanced insulin sensitivity following exercise, which is thought to be largely driven by increased AMPK activity.

FAT LOADING AND FAT ADAPTATION

Whilst glycogen forms the primary fuel source utilised during both prolonged and high-intensity endurance exercises, the capacity to store glycogen within human skeletal muscle is relatively limited (18). In contrast, even in the leanest of individuals, lipid storage within skeletal muscle, and in particular adipose tissue, is highly abundant. In fact, lipid storage within adipose tissue is sufficient for numerous days of continuous exercise. As such, there is increasing interest amongst athletes in strategies that increase the capacity to store and utilise lipids in the hope of sparing muscle glycogen and delaying the onset of fatigue during prolonged exercise (19). Although an increase in the ability to store, transport and oxidise fat forms one of the classical adaptations of endurance training (20), the adoption of a high-fat diet whilst undertaking daily training has the ability to further enhance such adaptation. This dietary strategy is more commonly referred to as fat adaptation or fat loading and is based upon the premise that such adaptations will promote higher rates of fat utilisation during exercise and allow for the sparing of the limited glycogen pool (21). Below, we look at how fatty acids derived from the diet activate the molecular signalling pathways that control the synthesis of proteins involved in the transport and utilisation of fat as a fuel during exercise.

Time-course of fat adaptation

Adaptations to a high-fat diet are rapid, occurring within as little as five days, and result in drastic shifts in metabolism that favour a greater reliance on lipids as fuel for energy production. In fact, rates of fat oxidation can be enhanced by as much as 200% within as little as 5 days (22), and are comparable to rates previously observed within 3–4 weeks (23). The rapid shift towards the metabolism of lipids as fuel is primarily underpinned by an enhanced capacity to store more lipids within skeletal muscle (known as IMTGs), an increased ability of skeletal muscle to uptake fatty acids from the circulation and, once inside the muscle itself, improved transport of fatty acids across the mitochondrial membrane where they undergo oxidation. Finally, high-fat diets also result in a reduction in the activity of enzymes that control glycogenolysis, resulting in a shift away from the metabolism of carbohydrate as fuel for energy production. Together, these adaptations result in a sparing of muscle glycogen during exercise (24).

For circulatory fatty acids, derived from either the ingestion of a high-fat meal or adipose tissue lipolysis, to enter the muscle cell they are transported across the cell membrane by various fatty acid transporters. These include fatty acid translocase (also known as CD36) and fatty acid binding protein (FABP), whereby a greater presence of fatty acid transporters on the muscle membrane results in a greater uptake of fatty acids into the muscle. In this regard, the adoption of a high-fat diet can increase the content of these transporters and allow for a greater delivery of fatty acids into the muscle. Indeed, as little as five days of fat adaptation is sufficient to increase resting gene expression and total protein content of the fatty acid transporter CD36 (25). These responses do, however, occur in the absence of changes in FABPpm, suggesting that an increase in CD36 is the primary mechanism by which fat adaptation occurs (25). Furthermore, the increase in CD36 content that is achieved with fat adaptation is rapidly reversed within just one day of consuming a high CHO diet (26), demonstrating how sensitive this transport protein is to changes in dietary fat intake.

Once inside the muscle, fatty acids are transported to the outer mitochondrial membrane in preparation for oxidation. Whilst short- and medium-chain fatty acids can directly enter the mitochondria without a transporter, long-chain fatty acids require the action of the mitochondrial carnitine palmitoyl transferase 1 (CPT-1). As such, increases in the content of CPT-1 transporters allow for a greater number of long-chain fatty acids to enter the mitochondria where they undergo oxidation to support energy production. Similar to that of CD36, increases in the protein content of CPT-1 have been observed following fifteen days of fat adaptation (27).

Fatty acids and cell signal transduction

The increased availability of fatty acids that occurs in response to fat adaptation provides the primary signal that serves to activate the complex signalling transduction pathway that controls the synthesis of new fatty acid transport proteins, such as CD36 and CPT-1. This signalling pathway is initiated by the binding of fatty acids to a group of nuclear receptor proteins, known as the peroxisome proliferator-activated receptors (PPARs), which function as transcription factors to control the mRNA expression of fatty acid transporters. Of the family of PPARs, PPARδ is the most abundant PPAR within skeletal muscle. Once activated, these transcription factors bind to specific sequences of DNA that code for the transporter proteins of interest (e.g. CD36 and CPT-1) to create new messenger RNA which can be ultimately used to synthesise new fatty acid transport proteins.

In addition to the enhanced ability to transport fatty acids into skeletal muscle and across the mitochondrial membrane, fat adaptation also increases the ability to store lipids within the muscle as IMTGs. In relation to timing, it is noteworthy that as little as two days of a high-fat diet has been shown to increase resting IMTG concentrations by 36% in endurance-trained cyclists (28). The increase in IMTG content appears to provide the main substrate responsible for the elevated rates of fat oxidation observed following fat adaptation, as inhibiting adipose tissue lipolysis (via the administration of a pharmacological lipolysis inhibitor, acipimox) does not significantly impact this response (28). The increase in IMTG concentrations following fat adaptation suggest an increase in lipid synthesis (the process of creating new lipids) above that of lipid degradation (the breakdown of lipids into their fatty acid constituents) during the rest periods between daily workouts. Whilst the increase in fatty acid transporters clearly plays a role in the ability to uptake fatty acids into skeletal muscle, the storage of fatty acids as triglycerides requires the attachment of fatty acids to the glycerol backbone. As this process requires the activity the enzymes glycerol-3-phosphate acyltransferase (GPAT) and diacylglycerol acyltransferase (DGAT) it seems plausible to consider that both enzymes may be upregulated in response to fat adaptation. Unfortunately, human studies to support this theory are currently lacking, although rodent studies have provided preliminary data that demonstrate increases in the mRNA expression of DGAT1 in skeletal muscle and the activity of GPAT within the liver, suggesting that both proteins are sensitive to alterations in fatty acid availability (29, 30). Clearly, more evidence is needed before definitive conclusions can be made in relation to how fat adaptation influences enzymes involved in lipid storage.

Fatty acids and CHO metabolism

Fat adaptation clearly has profound effects on the regulation of proteins involved in fatty acid uptake, storage, transport and subsequent oxidation. In addition to this, the adoption of a high-fat diet also results in distinct alterations to proteins involved in the metabolism of carbohydrate and subsequently results in a reduction in the reliance of skeletal muscle to utilise CHO as a fuel source to support energy production. Although a reduction in the reliance on CHO may appear beneficial, given its limited storage capacity, the suppression of CHO oxidation demonstrates a reduced ability to utilise this important fuel source. This shift in metabolism has important implications for those athletes required to perform at high aerobic work rates (>80% of $\dot{V}O_{2max}$) which are highly dependent on the ability to utilise CHO (31). Furthermore, even in events which necessitate lower aerobic work rates, successful athletes require the ability to sustain intense periods of exercise to counter changes in terrain (e.g. hill climbs), breakaway attempts by other competitors and sprints to the finish line, all of which require the ability to utilise CHO at high rates. In fact, a series of recent studies has clearly demonstrated the performance impairing nature of fat adaptation, reporting an ~8% reduction in performance in a matter of weeks (22, 23, 32). Although the exact mechanisms for such impairments are unknown, a greater reliance on fat as fuel results in a reduction in exercise economy (the volume of oxygen required to move at a given speed or power output), a strong predictor of endurance performance (33, 34).

The reduction in CHO metabolism that occurs in response to fat adaptation primarily occurs due to a rapid reduction in the activation of pyruvate dehydrogenase (PDH), the key regulatory enzyme involved in the oxidation of CHO, both at rest and across a range of exercise intensities (35, 36). This downregulation of PDH occurs via the upregulation of PDH kinase (PDK), a protein kinase that phosphorylates PDH and renders it to its inactive state (**Figure 10.3**). As we have already seen with other proteins, changes in the activity of PDK are rapid and occur within just two days following fat adaptation (37). It should also be noted that fat adaptation also represents a reduction in carbohydrate (CHO) availability which too has a profound effect on the regulation of PDK activity. For example, the restriction of CHO before, during and after exercise augments the expression of PDK4 mRNA, demonstrating PDK is also sensitive to low CHO availability as much as it is too high-fat availability (38–40). Nonetheless, the increased fatty acid availability that occurs in response to fat adaptation, as well as the associated reduction in CHO intake, provides the initial primary signal to activate the PPAR family of transcription factors that are responsible for the increase in PDK activity (41).

Key points

- Adaptations to a high-fat diet are rapid and occur within as little as five days. Such responses are underpinned by an enhanced capacity to store IMTGs, enhanced skeletal muscle fatty acid uptake and transport across the mitochondrial membrane and a concomitant reduction in the activity of enzymes that control CHO oxidation.

- Fatty acids serve as the primary signal which activates the family of peroxisome proliferator-activated receptors (PPARs), a group of nuclear receptor proteins which function as transcription factors to control the transcription of proteins involved in lipid transport and metabolism.

- Fat adaptation results in a reduction in the ability to utilise carbohydrate during exercise through the downregulation of PDH kinase, the key regulatory enzyme involved in the oxidation of carbohydrate. Such adaptations are likely to impair the ability to sustain exercise at high aerobic work rates that are highly dependent on carbohydrate.

PROTEIN NUTRITION AND MUSCLE PROTEIN SYNTHESIS

Defining protein turnover

Skeletal muscle is a protein-rich, highly adaptable tissue which can regularly change in size and function throughout an individual's lifespan, particularly in response to exercise and nutrition. As described in Chapter **8**, for skeletal muscle to change in size, an alteration in the balance between MPS and breakdown

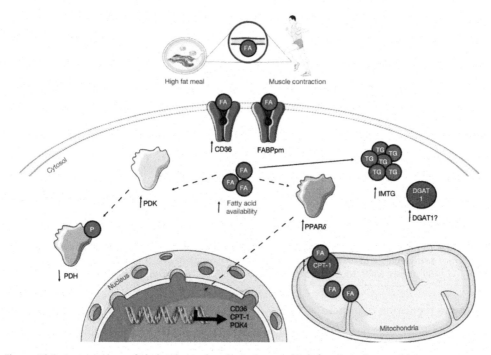

Figure 10.3 An overview of skeletal muscle adaptations to high fat diets. Fatty acids serve as ligands for PPARδ to control the expression of proteins involved in lipid metabolism, including the lipid transport proteins CD36 and CPT-1. Fat adaptation also increases IMTG storage, potentially through enhanced expression of DGAT1 but downregulates the activity of PDH via the upregulation of PDK kinase (PDK). CD36, cluster of differentiation 36; CPT-1, carnitine palmitoyltransferase 1; DGAT1, diacylglycerol acyltransferase; FABPpm, plasma membrane fatty acid binding protein; FA, fatty acid; IMTGs, intramuscular triglycerides; PDH, pyruvate dehydrogenase; PDK, pyruvate dehydrogenase kinase; TG, triglycerides. Runner and high-fat meal are images taken from publicdomainvectors.org that require no permissions Creative Commons Deed CC0, https://creativecommons.org/publicdomain/zero/1.0/.

(MPB) must occur. The difference between these two processes is referred to as net protein balance (NPB), with a positive NPB indicating muscle growth (i.e. hypertrophy) whilst a negative balance suggesting muscle loss (i.e. atrophy). When you have not eaten in more than 6–8 hours (i.e. the fasted state), rates of MPB are elevated, whereas MPS is low which results in a negative NPB. However, when you ingest a meal, particularly one with high amounts of protein, MPS is elevated and MPB is inhibited producing a positive NPB. In most stereotypical individuals who consume a balanced diet and meet physical activity guidelines, these periods of negative and positive NPB are equal across a 24-hour period and allow for muscle mass maintenance (**Figure 10.4A**) (42). In response to resistance exercise, MPS is further elevated, due to direct effects of the exercise and an increased sensitisation of muscle to protein/amino acids (AAs). As a result, NPB becomes even greater causing an overall positive NPB across a day (**Figure 10.4B**) (42). If you repeat this sequence regularly, long-term skeletal muscle protein accretion (i.e. hypertrophy) will then occur. Changes in muscle mass are often sought by athletes, and it is therefore important for practitioners to understand the molecular processes which control muscle protein metabolism to create nutritional programmes more precisely for their athletes. This section of the chapter will focus on these molecular processes and how research regarding them can be utilised to inform nutritional guidelines for athletes.

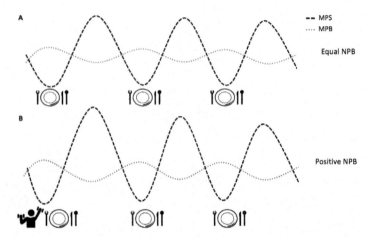

Figure 10.4 Depiction of daily MPS and MPB rates, in response to meals, in (**A**) habitually active individuals and (**B**) individuals undertaking resistance exercise. MPS, Muscle Protein Synthesis; MPB, Muscle Protein Breakdown; NPB, Net Protein Balance.

Molecular regulation of protein turnover

As discussed above, both MPS and MPB contribute to NPB. However, MPS is more sensitive to exercise and nutrition than MPB, especially in the absence of disease states (43, 44). As such, the subsequent sections of this chapter will focus on MPS regulation only.

One particular protein complex, discussed in detail in **Chapter 8**, has been suggested to be an important regulator of changes in muscle protein metabolism and therefore skeletal muscle size. This complex, called mechanistic target of rapamycin complex 1 (mTORC1) (45), is a serine/threonine kinase complex meaning its main role is to phosphorylate other proteins on serine/threonine AAs in that proteins structure. This phosphorylation changes the activity of the proteins which ultimately affects the two main processes involved in MPS, translation initiation and elongation (introduced in Chapter 3, explained in Chapter 7 and discussed with resistance exercise in Chapter 8). Due to this, as well as an important converging role for multiple signals such as growth factors, mechanical load/tension and AA availability, mTORC1 is often referred to as a "master regulator of protein synthesis" (46). However, the issues with terming individual proteins or complexes as 'master regulators' is outlined in Chapter 7. In human skeletal muscle, an inhibitor of mTORC1 has been used to show the importance of this protein complex for MPS. In these studies, MPS responses to both AA ingestion and resistance exercise no longer occurred when the mTORC1 inhibitor (rapamycin) had been ingested by participants before exercise/feeding (47, 48). mTORC1 therefore seems to be essential for MPS changes in response to anabolic stimuli like resistance exercise and protein/AA ingestion. Importantly, resistance exercise and AAs are able to activate mTORC1 by unique mechanisms (introduced in Chapter 8- molecular adaptions to resistance exercise), and it is this which allows these stimuli to have a synergistic effect on MPS (49, 50).

To understand how mTORC1 is activated following AA ingestion, we must first focus on a cellular organelle called the lysosome. This organelle is the place in the cell where autophagy (breakdown of cell components) occurs and means the lysosome contains a large abundance of AAs. Investigations in cells have found that mTORC1 needs to be associated with the lysosome membrane to be activated following AA provision (51). As well as being close to a large amount of AAs, this position also results in mTORC1 being close to its two direct activators, Ras homolog enriched in brain (Rheb) and phosphatidic acid (51, 52). So, when mTORC1 is located on the lysosomal membrane, it can associate with its direct activators and has a supply of AAs, so becomes more active and phosphorylates its downstream target proteins. Specifically, in response to elevated levels of intracellular AAs, increases in AA concentrations inside the lysosome are sensed by the vacuolar ATPase (v-ATPase), which activates the Ragulator-Rag complex. This complex can then bind with proteins within mTORC1 and recruit the kinase complex to the lysosomal membrane (53, 54).

Although this is the established mechanism of mTORC1 activation by AAs, research discovering this was mainly undertaken in non-muscle cells (51, 53, 54). In human skeletal muscle, the mechanism through which mTORC1 is activated following AA ingestion seems to be slightly different. mTORC1 still needs to be localised to the lysosomal membrane to be active; however, it seems that mTORC1 does not dissociate from the lysosome in fasted human skeletal muscle (55, 56). In response to AA feeding, there is then no change in mTORC1 colocalisation with the lysosome; however, mTORC1 activity still increases suggesting that another molecular process is contributing (50, 57). Recently, it has been shown that mTORC1 moves to peripheral regions of muscle fibres following AA feeding which positions mTORC1 in close proximity to upstream activators (Rheb), downstream target proteins and the organelles responsible for protein synthesis, the ribosomes (50, 55, 57, 58). This movement occurs in response to either AA ingestion or resistance exercise; however, it is more prolonged when the stimuli are combined, showing how this mechanism also displays the synergistic effect of these anabolic stimuli (50, 58). Indeed, as discussed in Chapter **8**, this is beacause resistance exercise moves the protein TSC2 away from Rheb allowing it to become GTP-loaded and activate Rheb, whereas and leucine-rich protein moves mTOR to Rheb (described below under 'Why is leucine so special'). Then, when mTORC1 is close to its activator Rheb, Rheb binds to the N-terminal domain of mTOR and this change in the shape of mTOR is such that it activates the kinase. Therefore, both mechanisms combined, resistance exercise (moving TSC2 away from Rheb) and leucine-rich protein (moving mTOR to Rheb) results in greater and prolonged protein synthesis and muscle hypertrophy than either stimulus alone. To summarise then, mTORC1 must be associated with the lysosome to be active and AAs can potentially increase this localisation and cause the movement of mTORC1-lysosomal complexes to peripheral regions of muscle fibres allowing mTORC1 to be activated by Rheb. These mechanisms are depicted in **Figure 10.5**.

Amino acid ingestion and protein synthesis

Although the ingestion of a mixture of all AAs, both essential (EAAs) and non-essential (NEAAs), can activate mTORC1 and elevate MPS, EAAs are the most integral to this (59). This is understandable given the inability of the human body to produce large amounts of these AAs, whereas NEAAs are readily synthesised from TCA cycle intermediates and transamination products. The importance of EAAs for skeletal muscle anabolism was first observed by Tipton et al. (59), who showed that if NEAAs were removed from a mixed AA drink, rates of MPS were not changed. The importance of EAAs for mTORC1 activation has also been shown as EAA ingestion alone can elevate mTORC1 activity substantially (60), whereas the ingestion of an mTORC1 inhibitor removes the effects of EAA ingestion on MPS (47), showing mTORC1 is essential to this. As such, it is apparent that EAAs are the most important AAs to ingest to activate mTORC1 and stimulate MPS in human skeletal muscle.

Of the 9 EAAs, recent research has discovered that the branched chain AAs (BCAAs) (valine, isoleucine and leucine), but in particular leucine, are most able to stimulate mTORC1 activity. For example, the ingestion of 3.4 g leucine on its own is able to elevate mTORC1 activity (61) and mTORC1 activation is significantly reduced if leucine is removed from an EAA drink (2.5–3.4 g removed from 15 to 20 g EAAs) consumed following resistance exercise (60, 62). Although leucine has these effects, the ingestion of a full complement of EAAs, however, is better able to stimulate mTORC1 activity than leucine alone (63) so a complete EAA supplement is most preferable. This is also true in relation to MPS as, even though both leucine and BCAA ingestion alone are able to elevate MPS (61, 64), the only nutritional method which maintains MPS increases for 5h following resistance exercise is the ingestion of a complete EAA beverage (49). Therefore, protein sources rich in leucine and a full compliment of the other EAAs are optimal for elevating MPS.

What is so special about leucine?

As we have mentioned in the previous section, leucine alone can have a pronounced effect on mTORC1 activation. This effect can be explained by examining research on molecular processes, which have identified several potential "leucine sensors". The most well-researched of these "leucine sensors"

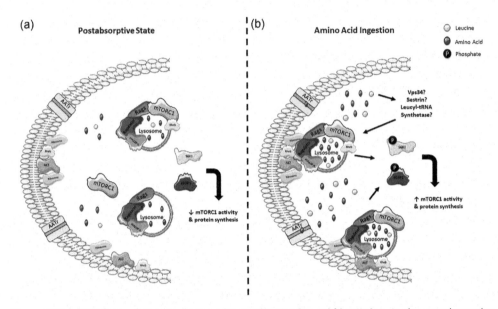

Figure 10.5 mTORC1 regulation in response to nutrient/amino acid ingestion. In the postabsorptive state, low nutrient availability state **(A)**, mTORC1 is located either in the cytosol away from the lysosomal surface (non-muscle cells/rodent skeletal muscle) or associated with the lysosome in central regions of cells (human skeletal muscle). As a result, mTORC1 is not in close proximity to amino acid transporters, ribosomes or upstream activators (AKT & Rheb) and therefore has low kinase activity. Downstream substrates of mTORC1 (S6K1 & 4EBP1) are not phosphorylated and protein synthesis rates are low. In response to elevated nutrient availability/amino acid feeding (B) intracellular amino acid/leucine levels are elevated and sensed by Vps34/Sestrin/Leucyl-tRNA Synthetase. mTORC1 is then either recruited to the lysosomal surface (non-muscle cells/rodent skeletal muscle) or translocated to peripheral regions of fibres (human skeletal muscle). Here, mTORC1 is in close proximity to sites of amino acid entry (AATr), ribosomes and upstream activators (AKT & Rheb) resulting in an elevation in mTORC1 kinase activity. Downstream substrates of mTORC1 are then phosphorylated causing an elevation in rates of protein synthesis. AATr, Amino Acid Transporter; AKT, Protein Kinase B/AKT; Rheb, Ras homolog enriched in brain; Rags, Rag family of GTPases; mTORC1, mechanistic Target of Rapamycin Complex 1; v-ATPase, vacuolar-type ATPase; S6K1, p70 ribosomal protein S6 kinase 1; 4EBP1, eukaryotic translation initiation factor 4E-binding protein 1; Vps34, Vacuolar protein sorting 34/Class III PI 3-kinase. Some of the images created using smart.servier.com. No permissions required. Attribution 3.0 Unported (CC BY 3.0) licence, https://creativecommons.org/licenses/by/3.0/.

are the Sestrin proteins which can bind with leucine when it enters a cell. This interaction between leucine and Sestrin proteins results in Sestrins being removed from their interaction with a protein complex called GATOR2 (GTPase activating proteins towards Rags) (65). GATOR2 is then activated and inhibits another protein complex, GATOR1, which ultimately further increases mTORC1 recruitment to the lysosomal membrane (66). However, as we previously discussed, mTORC1 is almost always localised with the lysosome in human skeletal muscle so this molecular process may not be contributing in this tissue (55). Other leucine "sensors" have been proposed in skeletal muscle, including leucyl t-RNA synthetase, which can activate Rag proteins (67), and Vps34, which translocates to the periphery of the fibres to colocalise with mTOR (57), however, further research is needed to confirm their roles.

PROTEIN SUPPLEMENTATION FOR ATHLETES AND EXERCISERS

Protein supplementation for resistance exercise

The most common exercise modality after which to consume supplemental protein is resistance exercise. This is to provide the muscle with both signals and substrates for MPS with the aim to cause skeletal muscle hypertrophy over time. As such, a lot of researchers have focussed on identifying the ideal protein supplementation strategy to optimally activate MPS. A common question often asked by athletes is: "*How much protein do I need to eat after resistance exercise?*" Initial research identified that 20 g high-quality protein (isolated egg or whey protein) was able to maximally stimulate MPS, and if more protein was consumed, this only resulted in AA oxidation (i.e. AA breakdown) rather than further elevations in MPS (68, 69). However, as athletes from varying sports often have drastically different body weights/compositions, expressing this protein dose as an absolute value is problematic when providing nutritional advice. Indeed, a recent analysis of several previously published studies revealed that a maximal MPS response is achieved when high-quality, EAA/leucine-rich protein at a dose of 0.31 g/kg body weight is consumed following resistance exercise (70). This means that an Olympic weightlifter who weighs 120 kg would need to consume 37 g protein to maximally stimulate MPS where a gymnast weighing 50 kg would only require 15 g protein.

The frequency of protein ingestion following resistance exercise is also important for skeletal muscle anabolism as it is believed that the greater the amount of time spent in a state of muscle protein accretion (i.e. positive NPB), the greater skeletal muscle hypertrophy can be achieved. This was investigated by Areta et al. (71) comparing three different post-exercise protein supplementation strategies, 10 g protein ingested every 1.5 hours, 20 g protein every 3 hours or 40 g protein every 6 hours across a 12-hour recovery period. Here, the 20 g every 3 hours was able to elicit a greater MPS response across the entire recovery period compared to the other two strategies. This therefore suggests that consuming optimal amounts of protein at regular intervals is an ideal strategy. It is often assumed by athletes, and regular gym-goers, that the first of the protein doses should be consumed immediately following an exercise bout in order to capitalise from what is a short-lived "anabolic window" (72). However, this appears to be a misnomer as several studies have failed to find a beneficial effect of immediate protein ingestion on MPS responses. For example, EAA supplements consumed either 1 hour or 3 hours following a resistance exercise bout elicit similar effects on MPS and NPB (73), and the AA sensitising effects of acute resistance exercise on skeletal muscle remain at 24-hour post-exercise albeit not as pronounced as during early recovery (<5 h) (74). Therefore, the previously assumed short "anabolic window" is more prolonged than first thought and although immediate post-exercise protein ingestion may be preferred by athletes, it does not need to be prioritised. When combined with the data reported earlier, this suggests that consuming a high-quality protein dose (high leucine/EAA content) able to maximally stimulate MPS (~0.31 g/kg BW) at regular intervals (~3 h) is the optimal method to achieve ideal skeletal muscle anabolism.

Protein supplementation for endurance exercise

Generally, athletes prioritise other macronutrients than protein during recovery from endurance exercise to maximise energy replenishment and performance recovery. However, it is becoming increasing apparent that protein itself is an important macronutrient for endurance athletes to replenish those AAs oxidised as fuel for energy during endurance exercise (75). AA oxidation contributes to at least 5% of energy production during endurance exercise, with a greater contribution during higher intensities or in glycogen depleted states. These AAs must be replaced following exercise in order to prevent skeletal muscle mass loss in these athletes. Immediately following a bout of endurance exercise, ingestion of high-quality protein sources (milk and whey protein), can elevate rates of MPS when compared to CHO ingestion on its own (76, 77). In addition, protein ingestion has been shown to elevate mTORC1 activity (76, 77) suggesting AAs are also able to activate this kinase complex following endurance-type exercise.

A similar dose-response investigation to those completed following resistance exercise (68, 69) has recently been conducted in response to an acute bout of endurance exercise. Here, maximal MPS was observed following ingestion of 30 g milk protein, with no further stimulation when 45 g protein was consumed (78). The authors also conducted extra analysis to identify the relative protein intake required for maximal MPS, finding a consumption of 0.49 g/kg BW is an optimal dose (78). This is approximately 60% greater than the dose suggested following resistance exercise (0.31 g/kg BW) indicating protein requirements are elevated following endurance-type exercise, probably due to the need to replenish AAs lost to energy production as well as stimulate MPS (75). This research described so far in this section was conducted only on the contractile proteins of skeletal muscle, i.e. proteins needed for force production; however, endurance exercise also requires mitochondria for oxidative fuel production. Interestingly, ingested AAs do not seem to affect mitochondrial protein synthesis (76, 79), even with high protein doses (78). The reason for this could involve mTORC1, as the use of an mTORC1 inhibitor (rapamycin) following endurance exercise in rats did not change post-exercise mitochondrial protein synthesis rates but did reduce contractile protein synthesis (80). Therefore, it seems a yet-to-be identified molecular process, other than mTORC1, that regulates mitochondrial protein synthesis. Nevertheless, post-endurance exercise, protein supplementation is important for proper skeletal muscle recovery, remodelling and adaptation with experts suggesting similar supplementation plans to those proposed following resistance exercise, i.e. EAA/leucine rich protein sources consumed in early recovery (<4 h) and at regular intervals (every 3–4 hours), should be undertaken by endurance athletes albeit with higher relative protein doses (0.49 g/kg BW).

Key points

- Skeletal muscle NPB is the difference between rates of MPS and MPB, of which MPS is the most sensitive to nutritional/exercise stimuli.

- At the molecular level MPS is regulated primarily by mTORC1, although endurance exercise-stimulated elevations in mitochondrial protein synthesis occur in a mTORC1-independent manner.

- Essential amino acids (EAAs), in particular leucine, have the greatest ability to activate mTORC1 and MPS.

- Following exercise, EAA/leucine rich protein sources should be consumed early in recovery (<4 h) and at regular intervals thereafter (every 3–4 h) to maximise recovery/adaptation.

NUTRITIONAL MODULATION OF TRAINING ADAPTATION

As a highly "plastic" tissue, skeletal muscle can undergo major adaptations and alter its phenotype in response to exercise stimuli. This process involves a complex signalling network that ultimately initiates the replication of specific DNA sequences (transcription) and their subsequent translation into an organised sequence of AAs to create new proteins (translation). Chronic endurance training elicits a variety of metabolic and morphological changes, including increased mitochondrial content, enhanced reliance on lipids for energy production and the formation of new blood vessels (known as angiogenesis). Such adaptations are functionally recognised by a rightward shift of the lactate threshold curve (81).

The start of this signalling transduction pathway begins with the accumulation of multiple metabolic signals that occur in response to muscle contraction. For example, a rise in calcium concentrations occur in response to contractile activity whilst an increase in AMP and ADP occurs due to increased ATP usage. These ultimately act as primary messengers which activate specific signalling kinases within the signal transduction pathway. Using the example above, the calcium released in response to muscle contraction binds to a protein called calmodulin and activates the calmodulin-activated protein kinase CaMKII. Once activated, these signalling kinases act as secondary messengers and activate transcription factors and co-activators to initiate the transcription of specific DNA sequences. Whilst these molecular signalling events are relatively short-lived, returning to baseline levels within 24 hours, repeated increases in transcriptional activity and protein synthesis, that occur with regular training, eventually lead to a change in the content of specific proteins (82), discussed in detail in the 'signal transduction hypothesis' **Chapter 7**. Although the modality, intensity and duration of exercise dictate the extent to which these molecular signalling pathways are activated (83–86), such signalling pathways are also sensitive to nutrient availability (87, 88). As such, altering the availability of certain nutrients provides an opportunity to up-regulate the adaptive response to training.

The glycogen granule as a regulatory molecule

Alongside its role as an energy substrate, muscle glycogen also acts as a regulatory molecule that is able to alter the activity of signalling kinases that play an important role in the regulation of mitochondrial-related genes (89). Furthermore, the availability of glycogen within skeletal muscle also acts as a potent modulator of substrate utilisation during exercise which directly influences the accumulation of various metabolic byproducts (e.g. lactate and FFA availability), that act as primary messengers within the signal transduction pathway. In this way, altering muscle glycogen availability provides an opportunity to modulate the adaptive response to training in two ways. Firstly, through direct interactions with protein kinases within the signal transduction network and, secondly, through alterations in whole body metabolism which influence the metabolic environment within the muscle. When you think about this in this way, it is remarkable that the storage of just 500 g of substrate can have such a large impact on the response of skeletal muscle to exercise.

Various metabolic proteins are known to interact and localise within the glycogen granule (89). Of these proteins, the mammalian AMP-activated protein kinase (AMPK), which plays a central role in monitoring cellular energy status, contains a glycogen binding domain on its β-subunit. This binding domain permits glycogen to bind to this kinase and allows AMPK to act as a sensor of endogenous glycogen stores. In this way, under conditions of low glycogen availability, the exercise-induced activity and phosphorylation of AMPK are all enhanced (88, 90, 91). Despite its apparent regularly role, the signal created by low glycogen alone (i.e. without exercise) may be insufficient to activate AMPK (90, 91) given the comparatively large increase in AMPK activation that occurs when exercise is commenced with low muscle glycogen. As such, it appears that the combination of low glycogen and exercise provides the most powerful signal to activate AMPK. In fact, in some cases, the typical increase in AMPK that occurs in response to exercise can be completely supressed under conditions of normal glycogen stores (90, 91). As AMPK plays an important role in the activation of various transcription factors and co-activators within the signal transduction network, it is clear to see the regulatory role that glycogen exerts. For example, AMPK can phosphorylate and histone deacetylase 5 (HDAC5) resulting it its removal from the nucleus, subsequently allowing for the myocyte enhancer factor 2 (MEF2) to bind and activate PGC-1α. As discussed in chapter **9**, the activation of PGC-1α is central to the transcription of many genes associated with endurance training adaptations.

Commencing exercise with low glycogen availability also results in dramatic shifts in whole body metabolism. For example, exercise with low glycogen availability substantially reduces absolute glycogen utilisation (87, 92) whilst concomitantly increasing circulating FFA (88) and catecholamine (93) availability (**Figure 10.6**). As discussed earlier, these metabolites and hormones can alter the activity of various kinases within the signal transduction pathway. For example, FFAs serve to activate the peroxisome proliferator-activated receptor (PPAR), to ultimately control the transcription of enzymes involved in lipid metabolism, including lipid transport proteins carnitine palmitoyltransferase (CPT-1) and fatty acid translocase (CD36). Furthermore, as discussed in chapter **9**, the increase in circulatory catecholamines that occurs with low glycogen exercise may act as an additional primary signal to activate the cAMP response element-binding protein (CREB), a transcription factor that also plays a role in the transcription of the master regulator protein, PGC-1α.

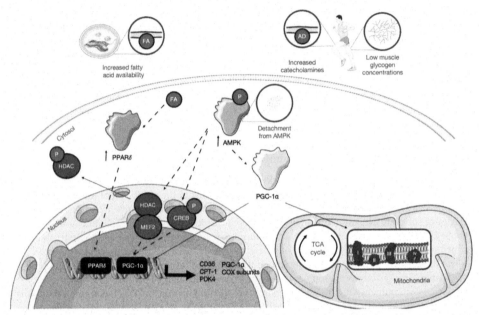

Figure 10.6 An overview of the potential exercise-nutrient-sensitive cell signalling pathways regulating the enhanced skeletal muscle adaptations associated with low glycogen availability. Commencing exercise with low glycogen availability results in increased fatty acid and catecholamine availability as well as less glycogen bound to the β-subunit on AMPK. Reduced glycogen availability enhances the activity and phosphorylation of AMPK which results in the activation and translocation of PGC-1α to both the nucleus and mitochondria. Upon entry into the nucleus, PGC-1α co-activates additional transcription factors (e.g. NRF1/2) to increase the expression of COX subunits as well as autoregulating its own expression. Within the mitochondria, PGC-1α co-activates Tfam to coordinate the regulation of mitochondrial DNA and induces the expression of key mitochondrial proteins of the electron transport chain. AMPK is also able to regulate the expression of PGC-1α through alternative pathways. AMPK induced phosphorylation of HDAC5 disrupts the interaction between MEF2 and HDAC5 which leads to the removal of HDAC5 from the nucleus by the chaperone protein 14-3-3 and allows MEF2 to bind to the PGC-1α promoter. AMPK can also directly phosphorylate CREB which upregulates PGC-1α gene expression. Exercise with low muscle glycogen also enhances fatty acid availability which activates the nuclear transcription factor PPARδ to increase the expression of proteins involved in lipid metabolism. Ad, adrenaline; AMPK, AMP-activated protein kinase; CD36; cluster of differentiation 36, CPT-1, carnitine palmitoyltransferase 1; CREB, cAMP response element binding protein; FA, fatty acid; HDAC, histone deacetylase; MEF2, myocyte enhancer factor 2; PDK, pyruvate dehydrogenase kinase, PGC-1α, peroxisome proliferator-activated receptor gamma co-activator 1-alpha; PPARδ, peroxisome proliferator-activated receptor delta. Runner and high-fat meal are images taken from publicdomainvectors.org that require no permissions Creative Commons Deed CC0, https://creativecommons.org/publicdomain/zero/1.0/.

Practical strategies to train with reduced CHO availability

Whilst ensuring adequate glycogen availability is essential to facilitate high-quality performance output and promote recovery, the deliberate restriction of CHO before and/or during exercise can be used as a strategy to enhance endurance training adaptations within skeletal muscle. This dietary periodisation approach has been termed "train-low, compete-high" (94) and provides a model which promotes carefully scheduled periods of CHO restricted training to enhance training adaptations, but ensures high CHO availability around high-intensity training sessions and competition. Although a wide variety of strategies can be used to achieve periods of CHO restricted training, twice-per-day, fasted training, and the more recent sleep-low approach form the most commonly adopted strategies by athletes.

Twice-per-day training

The twice-per-day model requires the completion of a morning training session to reduce muscle glycogen followed by CHO restriction throughout the post-exercise recovery period prior to the second training session of the day. In this way, the latter exercise session is commenced with a ~50% reduction in muscle glycogen concentrations (91). Commencing exercise under such conditions results in changes to the metabolic environment within the muscle, most notably through an increase in the concentration of FFAs and catecholamines such as adrenaline (93). As discussed above, these metabolites and hormones serve as the primary signals which activate a variety of signalling kinases within the signal transduction network and ultimately regulate the transcription of specific regions of DNA that encode for mitochondrial-related proteins. When performed regularly, twice-per-day training enhances the content of various oxidative enzymes within the TCA cycle and electron transport chain (93, 95, 96).

Fasted training

Performing endurance training in the fasted state represents a simpler train-low model whereby exercise is performed prior to breakfast. Although this approach does not alter muscle glycogen concentrations, liver glycogen is depleted by around 40% following an overnight fast (97). Similar to training with low muscle glycogen, fasted training also results in lowered plasma insulin and circulating glucose, whilst FFA availability and catecholamine concentrations are elevated (98–101). Similar to that of twice-per-day training, regular fasted training increases the activity of key enzymes involved in oxidative metabolism (e.g. citrate synthase and β-HAD), indicating an increase in mitochondrial content (102). Additionally, fasted training also has a profound effect on the regulation of substrate metabolism and results in a shift towards a greater reliance on lipid metabolism which is demonstrated through a reduction in glycogen breakdown (98) and a concomitant increase in intra-myocellular (IMCL) breakdown in both type I and II fibres (103).

Sleep-low train-low

In the sleep-low, train-low model, athletes perform an evening training session, restrict CHO throughout the overnight recovery period and complete a further training session the subsequent morning in the fasted state. Using this approach, the accumulative total time spent with low glycogen could extend to 12–14 hours depending on the timing and duration of the training sessions and sleep period. Akin to both twice-per-day and fasted training, the adoption of this model enhances the activation of various signalling kinases and the subsequent transcription and translation of proteins involved in oxidative metabolism (87, 88, 104). This model is also the first to demonstrate clear performance enhancing benefits amongst elite endurance athletes when compared with traditional high carbohydrate approaches (105, 106).

The glycogen threshold hypothesis

Given that the enhanced training response associated with train-low models is potentially mediated by the availability of glycogen within the muscle, it seems logical to ask the question "how much glycogen is required to facilitate the train-low response?". Although a difficult question to answer, the available train-low studies appeared to suggest the enhanced signalling response associated with low glycogen were particularly evident when exercise was commenced with glycogen concentrations below 300 mmol.kg^{-1} dw, more than 50% less than typical resting values (107). However, more recent studies have begun to suggest that similar activation of the signal transduction network can be achieved with high muscle glycogen stores if the exercise bout is of sufficient intensity to ensure muscle glycogen reaches low values (e.g. below 300 mmol.kg^{-1} dw) by the end of exercise. As such, the focus for the athlete should perhaps be to achieve low post-exercise glycogen concentrations at the end of the training session to enhance such signalling responses. Clearly, attaining low post-exercise muscle glycogen concentrations would require less absolute work if the exercise bout was commenced with already reduced glycogen availability, but nonetheless, can still be achieved through the completion of exercise of sufficient intensity and/or duration.

Key points

- The presence of the glycogen binding domain within AMPK allows the protein kinase to act as a sensor of endogenous glycogen stores whereby its activity, localisation, structure and function can be directly regulated by muscle glycogen availability. AMPK subsequently activates downstream transcriptional factors and co-factors that regulate training adaptations within skeletal muscle.

- Exercise with low muscle glycogen causes distinct shifts in metabolism and enhances the availability of fatty acids and catecholamines which may act as primary signals to alter the cell signalling pathways that regulate training adaptations.

- Exercise with low glycogen availability can be achieved through multiple strategies, including twice-per-day training, fasted training, and the more recent sleep-low, train-low model.

- Similar cell signalling responses can still be achieved when commencing exercise with high muscle glycogen if the exercise bout is of sufficient duration and intensity to reduce absolute muscle glycogen concentrations to low (< 300 mmol.kg dw) values. Clearly, this will require more total work compared with commencing the exercise bout with already reduced muscle glycogen.

FAT AVAILABILITY AND THE MOLECULAR REGULATION OF TRAINING ADAPTATION

As we have already discussed earlier in this chapter, the increased availability of fatty acids that occur in response to training with low CHO availability can serve as primary signals to activate the signal transduction network that controls endurance training adaptations. As such, it seems entirely plausible that increasing fatty acid availability, through the consumption of a high-fat diet, may further enhance such training adaptations through the activation of the PPAR family. The rationale for such a theory mainly comes from rodent studies where artificial elevations in FFA availability enhance the content of various proteins that form the mitochondrial respiratory chain, including cytochrome c, COXI, COXIV and ATP synthase, representing a clear indication of increased mitochondrial content (108). However, despite its theoretical promise, such findings have yet to be replicated within human skeletal muscle. For example, previous studies that have elevated FFA availability through high-fat feeding have failed to observe any increase in the mRNA expression of genes associated with mitochondrial-related training adaptations. Furthermore, studies in which the normal increase in FFA availability is artificially suppressed (via the administration of acipimox, a pharmacological inhibitor of lipolysis) fail to observe any reduction in the expression of such genes. Taken together, it appears that acute increases in fatty acid availability do not further augment the expression of mitochondria-related genes.

Whilst there appears to be no benefit to consuming a high-fat diet to enhance the response to training, of greater concern to the athlete is that such dietary practices may impair the skeletal muscle response to training in as little as three days. For examples, previous studies have reported that the mRNA expression of various mitochondrial proteins involved in oxidative phosphorylation, have all been shown to be decreased following the adoption of a high-fat diet (109). Furthermore, the adoption of a high-fat diet has been shown to reduce skeletal muscle mitochondrial respiration and the expression of proteins COXI and COXIV which play an integral within the mitochondrial respiratory chain (110, 111). These impairments in mitochondrial function may also help to explain the functional reductions in exercise economy (the oxygen cost of a given speed or power output) that occur in response to low-carbohydrate, high-fat diets.

Key points

- Fatty acids have the ability to alter the cell signalling pathways that regulate endurance training adaptations through the activation of the PPAR family.

- Acute increases in fatty acid availability do not appear to further augment mitochondrial-related signalling events that regulate training adaptation.

- Fat adaptation strategies may impair mitochondrial related signalling and attenuate markers of training adaptation in as little as three days and may explain the functional reductions in exercise efficiency that is commonly observed with such strategies.

SUMMARY

The emergence of molecular laboratory techniques in the discipline of sport nutrition has sparked a wealth of research addressing both fundamental (i.e. regulation of substrate storage) and contemporary (i.e. nutritional modulation of muscle plasticity) research areas that are of interest to sport and exercise scientists. In taking a contemporary view of research largely conducted in the last decade, the present chapter has sought to outline the relevant molecular exercise physiology related to pertinent components of practical sports nutrition. In this way, we initially outlined the molecular regulation of muscle glucose uptake so as to provide an understanding of the pathways supporting glycogen storage, glucose uptake during exercise and supercompensation of glycogen storage in the post-exercise period. Additionally, we subsequently reviewed how ingestion of macronutrients (i.e. fat, protein and CHO) can modulate cell signalling pathways that regulate adaptations inherent to both strength (i.e. MPS and hypertrophy) and endurance training (i.e. elevated IMTG storage, lipid oxidation and mitochondrial biogenesis). In the coming decade, it is likely that our understanding of such areas will continue to develop such that training and nutritional strategies for athletes can be further refined and optimised. Indeed, the elite athlete of the coming decade will likely have an individualised and highly personalised plan that is aligned to the core performance priorities of optimising training adaptation, body composition, fuelling and recovery. Additionally, our understanding of the role of nutrition in modulating wider performance priorities (e.g. optimising gut function, cognitive function, sleep and risk of injury) is also likely to develop significantly. It is our view that molecular biology based techniques will continue to be at the forefront of sport nutrition research.

REVIEW QUESTIONS

1. Outline the molecular pathways by which both insulin and muscle contraction regulate muscle glucose uptake.
2. What are the molecular processes supporting supercompensation of muscle glycogen stores in the post-exercise recovery period?

3. Explain the molecular processes regulating mTORC1 activity in response to amino acid ingestion.
4. Critically evaluate the role of macronutrient availability (with specific emphasis on CHO and fat) in regulating training-induced oxidative adaptations in skeletal muscle.

REFERENCES

1. Blundell J, et al. *Obes Rev.* 2010. 11(3):251–70.
2. Murphy KG, et al. *Nature.* 2006. 444(7121):854–9.
3. Edholm OG. *J Hum Nutr.* 1977. 31(6):413–31.
4. Mayer J, et al. *Am J Clin Nutr.* 1956. 4(2):169–75.
5. Hagele FA, et al. *J Clin Endocrinol Metab.* 2019. 104(10):4481–91.
6. Hopkins M, et al. *Sports Med.* 2011. 41(6):507–21.
7. Flatt JP. *Obes Res.* 2001. 9(Suppl 4):256S–62S.
8. Almeras N, et al. *Physiol Behav.* 1995. 57(5):995–1000.
9. Hopkins M, et al. *Br J Sports Med.* 2014. 48(200):1472–6.
10. Edinburgh RM, et al. *Am J Physiol-Endocrinol Metab.* 2018. 315(5):1062–74.
11. López-Soldado I, et al. *Diabetologia.* 2017. 60(6):1076–83.
12. Sylow L, et al. *Curr Opin Physiol.* 2019. 12:12–19.
13. Richter EA, et al. *Physiol Rev.* 2013. 93(3):993–1017.
14. Sylow L, et al. *Cell Metab.* 2021. 33(4):758–80.
15. McConell GK. *Am J Physiol Endocrinol Metab.* 2020. 318(4):E564–E7.
16. Ortenblad N, et al. *Scand J Med Sci Sports.* 2015. 25(Suppl 4):34–40.
17. Jensen TE, et al. *J Physiol.* 2012. 590(5):1069–76.
18. Areta JL, et al. *Sports Med.* 2018. 48(9):2091–102.
19. Burke LM. *J Physiol.* 2021. 599(3):819–43.
20. Hawley JA, et al. *Sports Med.* 2001. 31(7):511–20.
21. Yeo WK, et al. *Appl Physiol Nutr Metab.* 2011. 36(1):12–22.
22. Burke LM, et al. *J Physiol.* 2021. 599(3):771–90.
23. Burke LM, et al. *J Physiol.* 2017. 595(9):2785–807.
24. Burke LM, et al. *J Appl Physiol (1985).* 2000. 89(6):2413–21.
25. Cameron-Smith D, et al. *Am J Clin Nutr.* 2003. 77(2):313–8.
26. Yeo WK, et al. *J Appl Physiol (1985).* 2008. 105(5):1519–26.
27. Goedecke JH, et al. *Metabolism.* 1999. 48(12):1509–17.
28. Zderic TW, et al. *Am J Physiol Endocrinol Metab.* 2004. 286(2):E217–25.
29. Henriksen BS, et al. *Diabetol Metab Syndr.* 2013. 5:29.
30. Kawanishi N, et al. *Am J Physiol Regul Integr Comp Physiol.* 2018. 314(6):R892–901.
31. Hawley JA, et al. *Sports Med.* 2015. 45(Suppl 1):S5–12.
32. Burke LM, et al. *PLoS One.* 2020. 15(6):e0234027.
33. Joyner MJ, et al. *J Physiol.* 2008. 586(1):35–44.
34. Saunders PU, et al. *Sports Med.* 2004. 34(7):465–85.
35. Putman CT, et al. *Am J Physiol.* 1993. 265(5 Pt 1):E752–60.
36. Stellingwerff T, et al. *Am J Physiol Endocrinol Metab.* 2006. 290(2):E380–8.
37. Peters SJ, et al. *Am J Physiol Endocrinol Metab.* 2001. 281(6):E1151–8.
38. Cluberton LJ, et al. *J Appl Physiol (1985).* 2005. 99(4):1359–63.
39. Hammond KM, et al. *J Physiol.* 2019. 597(18):4779–96.
40. Pilegaard H, et al. *Metabolism.* 2005. 54(8):1048–55.
41. Wu Z, et al. *Cell.* 1999. 98(1):115–24.
42. Churchward-Venne TA, et al. *Nutr Metab (Lond).* 2012. 9(1):40.
43. Biolo G, et al. *Am J Physiol.* 1997. 273(1 Pt 1):E122–9.

44. Sepulveda PV, et al. *Clin Exp Pharmacol Physiol.* 2015. 42(1):1–13.
45. Saxton RA, et al. *Cell.* 2017. 169(2):361–71.
46. Laplante M, et al. *J Cell Sci.* 2009. 122(Pt 20):3589–94.
47. Dickinson JM, et al. *J Nutr.* 2011. 141(5):856–62.
48. Drummond MJ, et al. *J Physiol.* 2009. 587(Pt 7):1535–46.
49. Churchward-Venne TA, et al. *J Physiol.* 2012. 590(11):2751–65.
50. Hodson N, et al. *Am J Physiol Cell Physiol.* 2017. 313(6):C604–11.
51. Sancak Y, et al. *Cell.* 2010. 141(2):290–303.
52. You JS, et al. *J Biol Chem.* 2014. 289(3):1551–63.
53. Bar-Peled L, et al. *Cell.* 2012. 150(6):1196–208.
54. Zoncu R, et al. *Science.* 2011. 334(6056):678–83.
55. Hodson N, et al. *Exerc Sport Sci Rev.* 2019. 47(1):46–53.
56. Korolchuk VI, et al. *Nat Cell Biol.* 2011. 13(4):453–60.
57. Hodson N, et al. *Exp Physiol.* 2020. 105(12):2178–89.
58. Song Z, et al. *Sci Rep.* 2017. 7(1):5028.
59. Tipton KD, et al. *J Nutr Biochem.* 1999. 10(2):89–95.
60. Apro W, et al. *FASEB J.* 2015. 29(10):4358–73.
61. Wilkinson DJ, et al. *J Physiol.* 2013. 591(11):2911–23.
62. Moberg M, et al. *Appl Physiol Nutr Metab.* 2014. 39(2):183–94.
63. Moberg M, et al. *Am J Physiol Cell Physiol.* 2016. 310(11):C874–84.
64. Jackman SR, et al. *Front Physiol.* 2017. 8:390.
65. Wolfson RL, et al. *Science.* 2016. 351(6268):43–8.
66. Chantranupong L, et al. *Cell Rep.* 2014. 9(1):1–8.
67. Han JM, et al. *Cell.* 2012. 149(2):410–24.
68. Moore DR, et al. *Am J Clin Nutr.* 2009. 89(1):161–8.
69. Witard OC, et al. *Am J Clin Nutr.* 2014. 99(1):86–95.
70. Moore DR. *Front Nutr.* 2019. 6:147.
71. Areta JL, et al. *J Physiol.* 2013. 591(9):2319–31.
72. Lemon PW, et al. *Curr Sports Med Rep.* 2002. 1(4):214–21.
73. Rasmussen BB, et al. *J Appl Physiol (1985).* 2000. 88(2):386–92.
74. Burd NA, et al. *J Nutr.* 2011. 141(4):568–73.
75. Moore DR, et al. *Appl Physiol Nutr Metab.* 2014. 39(9):987–97.
76. Breen L, et al. *J Physiol.* 2011. 589(Pt 16):4011–25.
77. Lunn WR, et al. *Med Sci Sports Exerc.* 2012. 44(4):682–91.
78. Churchward-Venne TA, et al. *Am J Clin Nutr.* 2020. 112(2):303–17.
79. Abou Sawan S, et al. *Physiol Rep.* 2018. 6(5):e13628.
80. Philp A, et al. *J Physiol.* 2015. 593(18):4275–84.
81. Holloszy JO, et al. *J Appl Physiol Respir Environ Exerc Physiol.* 1984. 56(4):831–8.
82. Perry CG, et al. *J Physiol.* 2010. 588(Pt 23):4795–810.
83. Combes A, et al. *Physiol Rep.* 2015. 3(9):e12462.
84. Egan B, et al. *J Physiol.* 2010. 588(Pt 10):1779–90.
85. Fiorenza M, et al. *J Physiol.* 2018. 596(14):2823–40.
86. Stephens TJ, et al. *Am J Physiol Endocrinol Metab.* 2002. 282(3):E688–94.
87. Bartlett JD, et al. *Am J Physiol Regul Integr Comp Physiol.* 2013. 304(6):R450–8.
88. Wojtaszewski JF, et al. *Am J Physiol Endocrinol Metab.* 2003. 284(4):E813–22.
89. Philp A, et al. *Am J Physiol Endocrinol Metab.* 2012. 302(11):E1343–51.
90. Steinberg GR, et al. *Appl Physiol Nutr Metab.* 2006. 31(3):302–12.
91. Yeo WK, et al. *Exp Physiol.* 2010. 95(2):351–8.
92. Arkinstall MJ, et al. *J Appl Physiol (1985).* 2004. 97(6):2275–83.
93. Hansen AK, et al. *J Appl Physiol (1985).* 2005. 98(1):93–9.
94. Burke LM. *Scand J Med Sci Sports.* 2010. 20(Suppl 2):48–58.
95. Hulston CJ, et al. *Med Sci Sports Exerc.* 2010. 42(11):2046–55.

96. Morton JP, et al. *J Appl Physiol* (1985). 2009. 106(5):1513–21.
97. Iwayama K, et al. *NMR Biomed.* 2020. 33(6):e4289.
98. De Bock K, et al. *J Appl Physiol* (1985). 2008. 104(4):1045–55.
99. Horowitz JF, et al. *Am J Physiol.* 1997. 273(4):E768–75.
100. Montain SJ, et al. *J Appl Physiol* (1985). 1991. 70(2):882–8.
101. Stocks B, et al. *Am J Physiol Endocrinol Metab.* 2019. 316(2):E230–8.
102. Larsen S, et al. *J Physiol.* 2012. 590(14):3349–60.
103. Van Proeyen K, et al. *J Appl Physiol* (1985). 2011. 110(1):236–45.
104. Chan MH, et al. *FASEB J.* 2004. 18(14):1785–7.
105. Marquet LA, et al. *Med Sci Sports Exerc.* 2016. 48(4):663–72.
106. Marquet LA, et al. *Nutrients.* 2016. 8(12):755.
107. Impey SG, et al. *Sports Med.* 2018. 48(5):1031–48.
108. Garcia-Roves P, et al. *Proc Natl Acad Sci U S A.* 2007. 104(25):10709–13.
109. Sparks LM, et al. *Diabetes.* 2005. 54(7):1926–33.
110. Leckey JJ, et al. *FASEB J.* 2018. 32(6):2979–91.
111. Skovbro M, et al. *J Appl Physiol* (1985). 2011. 110(6):1607–14.

CHAPTER ELEVEN

Altitude, temperature, circadian rhythms and exercise

Henning Wackerhage, Kenneth A. Dyar and Martin Schönfelder

LEARNING OBJECTIVES

At the end of the chapter, you should be able to:

1. Describe by what mechanisms our body adapts to acute and chronic high altitude exposure.
2. Explain the genetics of high altitude populations such as Tibetans and Sherpas.
3. Explain how cells sense temperature and the mechanisms by which our body responds to cold or hot environments.
4. Describe the genes and proteins that act as a clock in our body and how they affect circadian exercise and other behaviours.

INTRODUCTION

On the 8th of May 1978 just after 13.00, Reinhold Messner and Peter Habeler reached the top of Mount Everest without supplemental oxygen. Before starting their expedition, they had been warned by brain scientists that the low oxygen pressure above 8,000 m altitude might cause long lasting brain damage (1). So, what enabled Messner and Habeler to climb Everest without bottled oxygen? How did they acclimate? Do they carry DNA variants that help them to acclimate to extreme hypoxia? (low oxygen)

Messner's and Habeler's Everest ascent is one of the most compelling environmental exercise physiology stories and a starting point to explore how the human body behaves at high altitude. In this chapter, we will additionally explore exercising humans in hot and cold environments and how day-night rhythms affect us. First, we will generally introduce the classical exercise physiology and then focus on molecular exercise physiology and the mechanisms of adaptation. We will also review what we know about the genetics of those that live in extreme places such as the Sherpas, Tibetans and Inuits. Finally, we include exciting excerpts from Nobel prize winner, Greg Semenza., who discovered HIF1, an important sensor of low oxygen, and Karyn Esser in the discovery of CLOCK genes controlling day-night rhythms in skeletal muscle with exercise.

ALTITUDE AND EXERCISE

The air we breathe today consists of 20.9% of oxygen (O_2). However, this was not always the case. For almost half of the Earth's history, oxygen was mostly less than 0.001%. This changed during the "great

DOI: 10.4324/9781315110752-11

oxidation event" between 2.4 and 2.1 billion years ago which increased global oxygen levels. Oxygen-producing photosynthesis by plants consumed CO_2 and generated O_2 (2). This dramatic change in the gas concentrations of the atmosphere allowed for the evolution of new species that consumed O_2 and generated CO_2.

While unicellular organisms can exchange gases by diffusion, multicellular organisms, termed metazoans, need respiratory and cardiovascular systems to draw O_2 into the body through the lungs and then to deliver O_2 to mitochondria in the cells. These systems are tightly regulated so that O_2 delivery is linked to O_2 utilisation which varies greatly from ≈ 3.5 mL min^{-1} kg^{-1} at rest (3) to 95 mL min^{-1} kg^{-1} during maximal exercise in some of the best endurance athletes, such as cross-country skiers. This is almost a 30-fold difference in the amount of oxygen that humans can extract and utilise for oxidative phosphorylation.

While the 20.9% oxygen in the atmosphere does not change with altitude, the barometric pressure (P_B) does. P_B varies a little due to weather conditions and declines greatly with increasing altitude. Because the per cent mixture of gases in the atmosphere is almost constant, the so-called partial pressure of oxygen PO_2 declines proportionally with the barometric pressure P_B. Thus, the oxygen partial pressure is always 20.9% of the pressure of dry air. John Dalton's law of partial pressures describes this relationship (PO_2 = Fraction of O_2 (in %) x P_B minus PH_2O). In **Figure 11.1**, we have plotted P_B and PO_2 in relation to altitude, assuming a barometric pressure (P_B) of 760 mmHg and a PO_2 of 149 mmHg at sea level (i.e. 20.9% of P_B −47 mmHg water vapour pressure which is 47 mmHg in the upper airways at 37°C at all altitudes).

Humans can adapt to the lower PO_2 at high altitude through several mechanisms:

1. We **hyperventilate** at higher altitudes, which is a systemic response when blood oxygenation (PaO_2) drops.

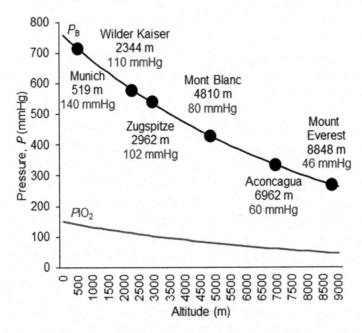

Figure 11.1 Relationship between altitude and the barometric pressure (**P_B**) as well as oxygen partial pressure (**PO_2** or more precisely **PIO_2** where the "I" stand for inspired air). **PO_2** is 20.9% of **P_B** minus the water vapour pressure which is 47 mmHg in the upper airways at 37°C at all altitudes.

2. **Hypoxia-signal transduction** is a cell-specific response to hypoxia that triggers adaptation such as an erythropoietin-mediated stimulation of erythrocyte production (polycythemia) and other adaptations that allow cells and the body to compensate and adapt to the lower oxygen levels.
3. **Genetic inheritance in humans living at higher altitudes** (e.g. people living on the Tibetan plateau, the Andean altiplano and in the East African highlands) because of natural selection of specific genotypes.

We will now discuss these three points.

What mechanisms make us ventilate more at high altitude?

At high altitude, ventilation (V_E or breathing, units L/min) is regulated mainly by the arterial CO_2 tension ($PaCO_2$) but it additionally increases when the arterial O_2 tension (PaO_2) drops substantially. Such a drop of the PaO_2 is sensed by a peripheral chemoreceptor in the vascular system, termed the **carotid body**. The carotid body then stimulates the ventilation-regulating centres in the brain stem. Ventilation at rest is around 5–10 L/min and responds to changes in CO_2 tension (PCO_2) which is sensed by both peripheral (i.e. aortic and carotid) and central hypothalamic chemoreceptors. However, a high $PaCO_2$ is not the only ventilation-driving signal. We also ventilate more when the arterial O_2 tension (PaO_2) drops to 60 mmHg or lower, which is termed **hypoxia** (4). Such hypoxia-driven ventilation is important when we ascend to high altitude as the CO_2 production by oxidative phosphorylation might be unchanged so under these conditions' hypoxia is the key ventilation-increasing signal.

How is the oxygen concentration sensed? The ventilation-regulating signals of O_2, CO_2 and pH are sensed by peripheral chemoreceptors such as the carotid and aortic bodies and by central chemoreceptors in the brain. The ≈20 mm³ large carotid body, located at the bifurcation of the carotid artery that supplies the head and brain with oxygen, is a sensitive oxygen sensor that monitors the arterial O_2 tension (PaO_2) and stimulates ventilation when the PaO_2 drops below ≈60 mmHg. The carotid body contains three main cell types (5, 6):

1. Oxygen-sensing glomus cells (type 1),
2. Subtentacular cells (type 2),
3. Afferent nerve fibres whose axons convey the hypoxia message to the ventilation-regulating parts of the brain.

The exact mechanism of oxygen sensing is incompletely understood; however, the following mechanisms have been proposed by which glomus cells could sense oxygen (5, 6):

1. Metabolites, including ATP, AMP, NADH, reactive-oxygen species, or gases such as CO, NO or H_2S may be involved (5, 6). A key oxygen-sensing protein is the protein encoded by the *Ndufs2* gene. The Ndufs2 protein is part of the electron transport chain and knockout of this gene in mice abolishes the hyperventilatory response to hypoxia, but not to elevated PCO_2 levels (7);
2. Low oxygen sensed by a Ndufs2-related mechanism then inhibits potassium (K^+) channels leading to a depolarisation of the glomus cells;
3. Depolarisation opens calcium (Ca^{2+}) channels which increase the Ca^{2+} concentration inside glomus cells;
4. Increased Ca^{2+} levels trigger a release of poorly characterised mediators (acetylcholine and ATP are candidates) that can activate innervating nerve fibres and carry the hypoxia message to the ventilation-regulating centres of the brain.

As a result of the accelerated ventilation to increase oxygen uptake, we exhale more CO_2 than is produced by metabolism, which is termed hypercapnia. Along with hypercapnia the pH increases and makes the blood more alkaline, termed respiratory alkalosis (8). This respiratory alkalosis is compensated by the kidneys, leading to an increased excretion of bicarbonate and loss of fluids.

How do cells sense hypoxia and how does this change gene expression?

Oxygen is not only sensed by the carotid body but also by many cells in the human body through the activation of transcription factors called hypoxia-inducible factors (HIFs), discovered by Nobel Prize winner Greg Semenza (**Box 11.1**). HIF-1 is a transcriptional complex formed from heterodimers of oxygen-sensitive alpha (HIF-1α, HIF-2α) and beta (HIF-1β) subunits. When the oxygen tension drops inside the cell, cytosolic HIF-1α or HIF-2α proteins are stabilised and can translocate into the nucleus to form heterodimers with nuclear HIF-1β proteins. Accumulating HIF heterodimers then drive expression of genes such as red blood cell development-promoting erythropoietin (EPO) or genes that promote angiogenesis, such as vascular endothelial growth factor (VEGF). Hypoxia induces gene expression through HIFs by three steps (6, 9, 10):

1. **Hypoxia inactivates the prolyl hydroxylases PHD1–3**: PHD1–3 are oxygen-activated enzymes that add an OH-group (hydroxylation) to proline amino acids on other proteins. Thus, under hypoxia, PHD1–3 hydroxylate their targets less.
2. **Hypoxia reduces HIF-1α hydroxylation**: When the oxygen tension drops, PHD1–3 hydroxylate HIF-1α less frequently at Pro402 or Pro564, whereas HIF-2α hydroxylation is reduced on Pro405 and Pro531. Additionally, during hypoxia the asparaginyl hydroxylase FIH-1 reduces the hydroxylation of HIF-1α at Asn803 and HIF-2α at Asn847, which increases the activity of HIF-1. Furthermore, all this is stimulated by iron (Fe^{2+}).
3. **Hypoxia reduces HIF-1α degradation, resulting in a rise of HIF-1α:** Under normoxia, hydroxylated HIF-1α or HIF-2α bind to the von Hippel-Lindau protein (pVHL) E3 ubiquitin ligase complex. This complex adds ubiquitin groups to HIF-1α and HIF-2α, targeting them for degradation by the protein-digesting proteasome. Therefore, HIF-1α levels remain low when there is enough oxygen. In contrast, hypoxia reduces HIF-1α hydroxylation and degradation, resulting in an increase of HIF-1α.
4. **The transcription factor HIF-1 regulates the adaptation to hypoxia**: Under hypoxia, the HIF-1α concentration is relatively high. HIF-1α and HIF-1β then heterodimerise to form the active transcription factor HIF-1. HIF-1 then binds to G/ACGTG DNA motifs known as the hypoxia-response element (HRE) and activates transcription of hypoxia-induced genes such as erythropoietin (EPO). See **Figure 11.2** for an overview of HIF-1's regulation of gene expression.

HIF-1 regulates the expression of many genes in response to hypoxia, but the actual target genes differ among different tissues. These include glycolytic enzymes, genes that manage the growth of the capillary network (angiogenesis) and the production of erythropoietin (EPO).

HIF-1-EPO-mediated regulation of red blood cell numbers

Blood carries oxygen. While 1 litre of water can only carry 0.03 mL O_2 $L^{-1}mmHg^{-1}$ PO_2, one gram of haemoglobin can bind ≈ 1.34 mL of O_2 at body temperature (11). Thus, the haemoglobin-containing erythrocytes are essential to ensure that our blood can transport high amounts of oxygen. A key adaptation to hypoxia is the increased production of erythrocytes, which increases the **haematocrit** and the **total haemoglobin mass.** The haematocrit is defined as the percentage of cellular components in relation to whole blood, represented almost entirely by the mass of red blood cells (erythrocyte). Therefore, the haematocrit determines the total oxygen transport capacity of the blood. The haematocrit ranges in males from ≈ 42–52% and in females ≈ 37–47% but increases slowly after ascending to high altitude. Most studies report a lower haematocrit in athletes than sedentary controls (11). This might be due to the destruction of erythrocytes when blood vessels are mechanically stressed as for example during running (11).

The haematocrit and total haemoglobin mass are not static but can increase due to the production of erythrocytes, termed **erythropoiesis**. A high haematocrit is beneficial for endurance sports as it increases oxygen transport, and therefore VO_2max. This explains the abuse of "blood doping" with erythropoietin

BOX 11.1 Gregg Semenza Nobel Prize winner in the discovery of HIFs

A Wild Ride

When I arrived at Johns Hopkins at the start of my postdoctoral fellowship, I set out to use transgenic mouse technology to identify the genomic sequences required for tissue specific expression of the human EPO gene, encoding erythropoietin, the hormone that controls red blood cell production. It was known that EPO was produced by the foetal liver and, after birth, by the kidneys. In collaboration with the lab of the late John Gearhart, we injected a 4-kilobase human genomic DNA fragment that encompassed the entire gene plus a small amount of flanking sequences. These mice were polycythemic (haematocrit of ~60% as compared to ~45% in non-transgenic mice) due to the production of human EPO in addition to the endogenous mouse Epo. When adult transgenic mice were subjected to phlebotomy, haemolysis or ambient hypoxia, expression of the EPO transgene was induced in the liver but not in the kidney. We subsequently determined that a much larger DNA fragment containing extensive 5′-flanking sequences was required for hypoxia-induced expression in the kidneys. These transgenic mice were highly polycythemic, with haematocrits of over 80%. We fashioned small bicycles for them to ride and found that their endurance was greatly increased (just kidding). These mice were generally healthy but were often done in by the "rigor" of breeding, which is to say they would be found the next morning, with rigor (mortis) and a smile on their face.

Our next challenge was to identify the sequences that mediated the response to hypoxia. We shifted over to use of the Hep3B cell line, in which EPO gene expression is induced by culturing the cells at 1% O_2 rather than standard tissue culture condition of 20% O_2 (this is simple math that, judging from the literature, escapes many a poor soul: the cells are cultured in 95% air and 5% CO_2; $0.95 \times [21\%\ O_2] = 20\%\ O_2$). We identified a short sequence in the 3′-flanking region of the EPO gene that is now known as the HRE. We used the HRE to discover hypoxia-inducible factor 1 (HIF-1) as a DNA-binding activity present in the nucleus of Hep3B cells exposed to 1% O_2 and absent from cells cultured in 20% O_2. We subsequently showed that HIF-1 was induced by hypoxia in all mammalian cell lines tested, which told us that it did much more than just turn on EPO gene transcription. We used binding to the HRE to purify HIF-1 by DNA affinity chromatography using nuclear extract prepared from 120 L of HeLa cells grown in suspension culture.

The discovery and cloning of HIF-1 led to the discovery by several labs of the HIF prolyl hydroxylases, which led to the development by multiple drug companies of small molecule inhibitors of the hydroxylases, which turn on EPO gene expression and erythropoiesis. These drugs are superior to recombinant EPO because they are small molecules that can be taken by mouth rather than injection and alter the expression of many genes that promote hypoxic adaptation. HIF stabilisers even increase erythropoiesis in anephric patients, probably by inducing EPO expression in the liver, as in our transgenic mice. They are now the (illicit) drugs of choice for competitive cyclists – murine or human.

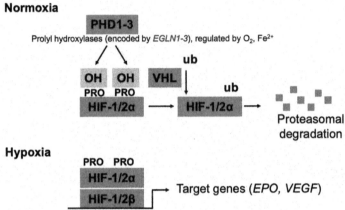

Normoxia

PHD1-3

Prolyl hydroxylases (encoded by *EGLN1-3*), regulated by O_2, Fe^{2+}

OH OH | PRO PRO | HIF-1/2α | VHL | ub ↓ ub | HIF-1/2α → | Proteasomal degradation

Hypoxia

PRO PRO | HIF-1/2α | HIF-1/2β → Target genes (*EPO, VEGF*)

HIF1/2α and HIF1/2β as a heterodimer together form the HIF-1 transcription factor

Figure 11.2 Regulation of hypoxia-dependent gene expression under normoxia and hypoxia. Under normoxia HIF-1/2α are hydroxylated by oxygen-regulated PHD1–3. This leads to the degradation of HIF-1/2α and a low concentration. Under hypoxia, HIF-1/2α hydroxylation and degradation are reduced and so the concentration of HIF-1/2α and of the genes that are regulated by HIF-1/2α increases.

in endurance sports where a high VO_2max is desired. However, while a high haematocrit may increase endurance performance, high haematocrits are also associated with a greater risk of venous thromboembolism in the general population (12). Thus, a too high haematocrit is a health risk, and because of this in 1997, the Union Cycliste International (UCI) decided to set a 50% haematocrit as a limit for health reasons. This seems sensible because while it is possible to identify EPO in a urine sample (13) this does not seem to be a reliable test due to issues in detection at differing times of sampling, which might have been responsible for the negative anti-doping test outcomes for Lance Armstrong.

Moving back to the biology. Knowing what we now know from the above, we might ask ourselves: How does hypoxia increase erythropoiesis to increase the haematocrit, and what is the role of HIF-1 and erythropoietin (EPO)? French researchers first showed that hypoxia increases erythrocyte numbers and suggested that a hormone, later termed erythropoietin, might be responsible for the increased erythropoiesis in response to hypoxia (14). A dissection study then identified the kidney as the main source of systemic EPO synthesis (15), with liver also contributing, but to a lesser extent (16).

Hypoxia in the liver of the foetus and then later in the kidney, specifically increases HIF-2α expression which is similar to HIF-1α but probably more important for the expression of the EPO gene. The evidence for this is that HIF-2α knockout mice suffer from anaemia, suggesting that the HIF-2α is the main transcription factor to regulate EPO expression (17). HIF-2α protein can also bind to HIF-1β to form the HIF-1 transcriptional complex. HIF-1 then increases the expression of EPO up to hundred-fold. EPO is a heavily glycosylated hormone as 40% of its 30 kDa mass is glycosyl groups (i.e. carbohydrate) (16). EPO then enters the circulation. Circulating EPO binds to EPO receptors (EPOR) especially on the so-called colony-forming units-erythroid (CFU-Es) cells in the bone marrow. This promotes their survival, proliferation and differentiation into reticulocytes (immature erythrocytes), which later develop into mature erythrocytes (**Figure 11.3** (10)).

Are some people genetically adapted to high altitude?

Our hominin ancestors evolved East of the East African rift approximately 7 million years ago. While there were earlier waves of migration out of Africa, our direct ancestors were anatomically modern humans

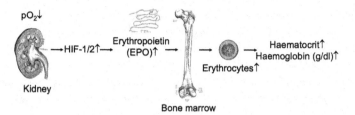

pO₂↓

Kidney →HIF-1/2↑→ Erythropoietin (EPO)↑ →

Bone marrow

→ Erythrocytes↑ → Haematocrit↑ Haemoglobin (g/dl)↑

Figure 11.3 Schematic showing the regulation of erythropoiesis by EPO. Original figure with images taken from Grey's Anatomy that do not require permissions under the Creative Commons CC0 License https://creativecommons.org/publicdomain/zero/1.0/; and/or Creative Commons Attribution-ShareAlike License, https://creativecommons.org/licenses/by-sa/4.0/

that migrated out of Africa ≈55–65,000 years ago and then colonised the rest of the world (18). During their journey, our ancestors met and interbred with archaic Neanderthals and Denisovans and as a consequence, a few per cent of our genome can be traced back to these archaic humans (18).

Within a relatively short space of time, humans then settled in a wide variety of environments all over the Earth encountering different nutrients and pathogens as well as extremes of temperature and altitude (19). Humans also migrated to high altitude. About 25,000 years ago, humans started to settle on the **Qinghai-Tibetan Plateau** (≈4,000 m) and roughly 500 years ago today's Sherpas migrated to Nepal (20). Humans migrated to America 14–15,000 years ago (18) and then started to settle on the **Andean altiplano** which is just under 4,000 m of altitude from ≈11,000 years ago (20). Less is known about the time when people started to populate the **Semien Plateau** (≈2,000–2,500 m) in Ethiopia (21). Thus, there are several populations that have lived for thousands of years at high altitude. Today's highest human settlement is La Rinconada, a mining town of 60,000 people at 5,100 m in the Peruvian Andes.

At high altitude, people are exposed to chronic hypoxia. For example, if the barometric pressure at sea level is 760 mmHg then the inspired partial pressure of oxygen (PIO₂) will be 149 mmHg, 101 mmHg in a settlement at 3,000 m, 89 mmHg at 4,000 m and 78 mmHg at 5,000 m. Such hypoxia is a great challenge to the human body and of the more than 140 million people that live above 250 m, 5–10% are at risk of developing chronic mountain sickness. Chronic mountain sickness is a major health problem which is associated with a high haemoglobin concentration and haematocrit, a reduced oxygen saturation and often a high blood pressure in the lung circuit, termed pulmonary hypertension (22). Thus, the questions are: Are populations that live permanently at high altitudes genetically adapted to chronic hypoxia? Do they carry DNA sequence variants such as those possessed by a Finnish family, where a heterozygous EPO receptor gene (EPOR) mutation increased haematocrit (23)?

Today we know that Tibetans and the inhabitants of the Andes adapted differently to high altitude. On average, Tibetans have a higher mean resting ventilation of 15.0 l/min as compared to 10.5 l/min for those living on the Andean altiplano (20). In contrast, and perhaps surprisingly, Tibetans living at ≈4,000 m have a relatively normal (i.e. comparable to lowlanders) haemoglobin concentration of 14–16 mg/dl, whereas people in the Andeans have a high haemoglobin concentration of 17–18 mg/dl (20). Our knowledge of altitude adaptation of Tibetans, Andeans and Ethiopians is summarised in **Table 11.1**.

The relatively normal average haematocrit and haemoglobin concentrations of Tibetans was already noted in a 1970s study where researchers found that Sherpas living permanently at 4,000 m do not have a higher haematocrit and haemoglobin content when compared to people living at sea level (24). Also, the haemoglobin concentrations of Tibetans are lower than those of Han Chinese living at 4,525 m (**Figure 11.4** (25)), suggesting that Tibetans have a less sensitive hypoxia-EPO system.

Two questions arise: Firstly, why do Tibetans have a relatively normal (i.e. sea level-like) haemoglobin concentration and haematocrit at altitude? Secondly, Do we know the DNA sequence variants that are responsible? The answer to the first question is that a too high haemoglobin concentration of haematocrit is detrimental, as we have mentioned before. For example, a too high haemoglobin concentration

Table 11.1 Summary of the adaptation to populations that live at high altitude (21)

Phenotype	Tibetan	Andean	Ethiopian
Resting ventilation	50% higher	No increase	Not reported
Hypoxic ventilatory response	Similar to sea level	Blunted (low)	Not reported
Arterial oxygen saturation increase	No increase	Elevated	Minimal
Haemoglobin concentration	Lowered	Elevated	Minimal increase
Birth weight	Elevated	Elevated	Not reported

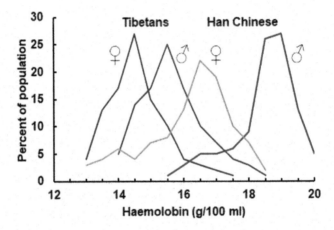

Figure 11.4 Frequency of haemoglobin concentrations in 16–60-year-old Tibetan females and males and Han Chinese females and males living at 4,525 m. Figure redrawn from Wu T, et al. (25).

increases the risk for problems at birth, including stillbirth (26). A high haematocrit is also associated with thromboembolism (12) and haemoglobin concentrations above 17 g/dL are associated with an increased risk of heart disease (27). Moreover, a study of ≈50,000 individuals reported a U-shaped relationship between haematocrit and mortality (28). Thus, when people moved to higher altitude, they were exposed to a strong natural selection for alleles that limit the rise of the haemoglobin concentration or alleles that reduce the negative consequences of a high haemoglobin concentration.

The second question has been answered by comparing the genome of populations that live at high altitude with the genome of people that live at sea level. The first major breakthrough resulted from a study where the researchers compared the DNA sequence of 50 ethnic Tibetans with that of the closely related Han Chinese. This revealed a 78% frequency difference in one SNP near the EPAS1 gene which encodes the hypoxia-related transcription factor HIF-2α (29). Further studies revealed a whole host of genes with evidence for natural selection of alleles in humans that permanently live at high altitude (**Table 11.2**).

Of all these genes, a DNA variant of EPAS1 shows the strongest evidence for natural selection. To identify the origin of these DNA variants, researchers re-sequenced the EPAS1 gene in 40 Tibetans and 40 closely related Han individuals and then searched databases to identify the source of the EPAS1 DNA variants in the Tibetans. They discovered that the EPAS1 DNA variant of Tibetans best matched the DNA of the so-called Denisovans, which are extinct archaic humans similar to Neanderthalers (30). The name

Table 11.2 Some HIF-1-EPO and AMPK-related genes where DNA variants have been identified in people living permanently at high altitude (21). The protein abbreviations and names are also shown

Gene symbol	Protein
EPO	Erythropoeitin
HIF1A	HIF-1α, hypoxia-inducible factor 1, alpha subunit
EPAS1	HIF-2α, hypoxia-inducible factor 2, alpha subunit
ARNT	HIF-1β, aryl hydrocarbon receptor nuclear translocator
ARNT2	HIF-2β, aryl hydrocarbon receptor nuclear translocator
EGLN1	PHD2, prolyl hydroxylase domain-containing protein 2
EGLN2	PHD1, prolyl hydroxylase domain-containing protein 1
EGLN3	PHD1, prolyl hydroxylase domain-containing protein 3
PRKAA1	AMPK, AMP-activated protein kinase, α1 Catalytic Subunit
PRKAA2	AMPK AMP-activated protein kinase, α2 Catalytic Subunit
VHL	Von Hippel-Lindau Tumour Suppressor, E3 Ubiquitin Protein Ligase

"Denisovan" is derived from the Denisovan cave in the Altai Mountains, where researchers found a finger bone fragment of a female and sequenced the DNA. Thus, the archaic Denisovans might have already been adapted to high altitude and Denisovan-human interbreeding allowed the EPAS1 DNA variant to first enter the human genome. Migration to the Tibetan plateau then promoted a strong positive selection of the Denisovan EPAS1 allele.

Also, EGLN1 DNA variants have been selected in Tibetans, Andeans and Ethiopians that live at high altitude (21). Recently, researchers have identified a DNA variant in exon 1 of the EGLN1 gene of Tibetans with strong evidence for natural selection (31). To explain, this DNA variant changes the 4th amino acid of the prolyl hydroxylase PHD2 from asparagine to glutamine and amino acid 127 a cysteine to serine. The mutant PHD2 protein impairs the proliferation of erythroid progenitor cells under hypoxia, suggesting that the mutant PHD2 promotes greater HIF degradation, resulting in less proliferation or erythrocyte precursor cells and a reduced production of erythrocytes (31).

Another adaptation is linked to the number of ATP molecules that is produced for each oxygen atom as this depends on whether glucose or fatty acids such as palmitate are oxidised (32). This has been termed the P/O ratio and the values are as follows:

Glucose : $C_6H_{12}O_6 + 6\ O_2 \rightarrow 6\ CO_2 + 6\ H_2O_2$; P/O ratio : 2.41

Palmitate $C_6H_{32}O_2 + 23\ O_2 \rightarrow 16\ O_2 + 16\ H_2O$; P/O ratio: 2.1

Thus, for each oxygen atom, 2.41 ATPs are produced when glucose is oxidised but only 2.1 ATPs when palmitate is oxidised. This is a ≈15% difference and one can imagine that this can become critical at high altitude, as glucose and carbohydrate are more oxygen efficient. So are high altitude dwellers carbohydrate burners? Researchers compared two strains of wild mice living at 4,000–4,500 m in the Peruvian Andes with related strains caught at 100–300 m altitude. The researchers found that the mice living at higher altitudes oxidised more carbohydrates than their low level counterparts during exercise and while exposed to hypoxia (33). This suggests that increased carbohydrate oxidation was one evolutionary adaptation in these mice. In another study, Horscroft et al. investigated the metabolism of the Himalayan Sherpas and found that Sherpas had a lower capacity for fatty acid oxidation and thus presumably higher carbohydrate oxidation and increased lactate dehydrogenase activities than lowlanders (34). The genetic or other mechanisms that explain this difference in carbohydrate metabolism at altitude are unknown.

In summary, Tibetans that live permanently at high altitude have experienced a natural selection of DNA variants of hypoxia-sensing genes such as *EPAS1* which encodes HIF-2α, and EGLN1 which encodes PHD2. These mutations limit the increased haematocrit and haemoglobin concentrations at high altitude, allowing the carriers to avoid associated health problems. This is the opposite from the EPOR mutation carriers in Finland who have a higher haematocrit but live at sea level (23). In contrast, people that live in the Andean Altiplano have elevated haemoglobin concentrations, but perhaps selection of additional unidentified alleles that limit the negative consequences of high haemoglobin concentrations. Additionally, there is some evidence that high altitude mice and humans may favour carbohydrate oxidation which is associated with a greater production of ATP per oxygen atom (i.e. P/O ratio).

TEMPERATURE AND EXERCISE

Temperature is a key factor of our environment and fluctuations of temperature are a challenge to our bodies as we need to maintain our core body temperature within a narrow range. The current hottest and coldest temperatures on Earth are 56.7°C measured in Death Valley, USA, and −89.2°C at Vostok station, Antarctica, respectively, which is a difference of nearly 150°C. Humans have permanently settled in hot and cold climates, ranging from hot environments near the equator to Greenland and the Arctic. Humans are able to do so due to genetic adaptation to the local environment, our ability to regulate our core body temperature and because of technology and culture in the form of clothing, heating and air conditioning. Here, we will first discuss the physiology and molecular mechanisms of thermoregulation and will then discuss the evolution and genetics of populations that live in places with extreme temperatures.

Humans have evolved a regulatory system that maintains the temperature of our body within a narrow range. In relation to temperature, our body has two parts. First, our shell includes our skin and extremities, and here the temperature can vary considerably, for example when we have cold hands or feet. Second, our core comprises the vital inner organs and here the temperature is on average close to ≈37°C. Maintaining our core body temperature is important, as temperature has profound effects on the molecules of life. For example, the activity of enzymes increases with temperature before a too hot temperature causes enzymes and other proteins to denature. If we become too hot then we can suffer a **heat stroke** which is **hyperthermia** (core body temperature of >40°C) accompanied by systemic inflammation which can cause defects of several organs and eventually death (35). In contrast, when our body cools down, then enzymes are less active, membranes are more rigid and ion fluxes are reduced. **Hypothermia** is a medical syndrome when the core body temperature falls to below 35°C. When we are hypothermic, we start to shiver and become lethargic and this can progress further to confusion, coma and death.

PHYSIOLOGY AND MOLECULAR MECHANISMS OF THERMOREGULATION

So how do we maintain our core body temperature? Several processes either increase or decrease our body temperature. Body temperature is increased mainly by two processes (**Figure 11.5** (36)):

1. **Non-shivering thermogenesis**: While most of our adipose tissue is white adipose tissue, we have two types that can generate heat via the "short cut" protein UCP1. These are brown (37) and so-called beige (or brite) adipose tissue (38). Hormones such as catecholamines, secretin (39) as well as metabolites, myokines and miRNAs regulate the expression of the UCP1 gene or the activity of the heat-generating UCP1 protein which generates heat, raises energy expenditure and has beneficial metabolic effects (39, 40).
2. **Muscle contraction** in the form of exercise or shivering: Skeletal muscle is an organ that converts the chemical energy of nutrients into work (i.e. muscle contraction and exercise) and heat. For example, during cycling only approximately 20% of the chemical energy is converted into work,

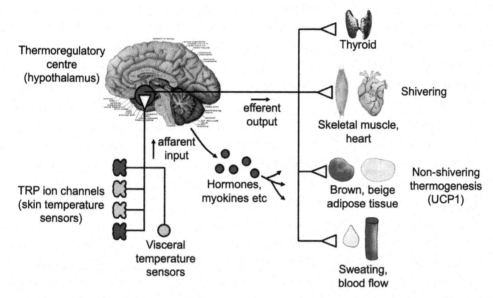

Figure 11.5 Overview over the thermoregulation system. TRP ion channels and visceral temperature sensors sense skin and body temperature, and this information is conveyed to the thermoregulatory centre of the brain in the hypothalamus. This centre then regulates the body temperatures via the thyroid, muscles, adipose tissue as well as sweating and blood flow. Moreover, stimuli such as infections (fevers), food uptake and exercise can alter thermoregulation either via the thermoregulatory centre of the brain or via hormones, myokines, metabolites and other factors that circulate in the blood. Original figure with some images taken from Grey's Anatomy and do not require permissions under the Creative Commons CC0 License, https://creativecommons.org/publicdomain/zero/1.0/; and/or Creative Commons Attribution-ShareAlike License, https://creativecommons.org/licenses/by-sa/4.0/.

whereas 80% of it is typically converted into heat. This is why we become warm when we exercise. Moreover, involuntary muscle contractions during shivering also generate heat via the same mechanism.

We lose heat and reduce our body temperature through the following processes (**Figure 11.5**):

1. **Sweat evaporation** cools our body as water vapourises on our skin. The heat of evaporation has been estimated at 2.5 kJ to transfer 1 g of water to water vapour at 35°C (41).
2. **Radiation** is heat transfer by objects that are not in contact. A good example is the sun heating us, but similarly we can radiate heat.
3. **Conduction** is heat transfer to objects that directly touch each other. An example is heat loss when you lie on a cold floor.
4. **Convection** is heat transfer by a moving gas or liquid. For example, you lose heat through convection when it is windy (it additionally increases heat loss through evaporation) and when you are swimming.

Additionally, humans are unique as our intelligence allows us to use technical solutions for dealing with temperature fluctuation. A key function of housing, heating, air conditioning and clothing is to help us maintain our temperature. This has allowed humans to live in extreme environments such as the arctic or regions near the equator.

So how does the thermoregulatory system function? Like all adaptation systems it has three main parts which are as follows: the sensing of signals, computation and the regulation of an output by effectors (42):

1. **Temperature sensors** such as the temperature-sensitive (TRP) ion channels (42) sense skin, visceral and deep brain temperature and convey these information through thermoafferent neurons to the brain.
2. In the brain, the **hypothalamus** receives and computes this information.
3. Efferent neurons then control organs that can change body temperature. These include skeletal muscle, the heart, adipose tissue and the thyroid gland.

We will now discuss these three parts in detail.

How is temperature sensed?

A key requisite for thermoregulation is that some proteins must act as biological thermometers. In our body, the TRP (transient receptor potential) ion channels have that function. In contrast to a technical thermometer, however, there is not just one TRP ion channel but a family of TRP's that are either warm-sensitive (TRPV1–4, TRP2–5) or cold-sensitive (TRPM8, TRPA1, TRPC5) and sense everything from cold pain, to cold, warmth and heat pain. Typically, once a certain temperature is reached, the TRP ion channels open and Na^+ and Ca^{2+} ions flow into and depolarise any neuron that expresses these TRPs. Several knockout mouse models for the TRP ion channels have been generated and suggest that together, the TRPs sense temperature (42).

The response to temperature is regulated in the thermoregulatory centre of the hypothalamus

Temperature is sensed by TRP ion channels in the skin, viscera, spinal cord and by the brain. But where does this information go to be processed and to trigger responses to a "too hot" or "too cold" sensation? First, researchers noted that animals rapidly increased their body temperature when they damaged the base of the brain, suggesting that this houses the thermoregulatory centre (43, 44). Today we know that thermoregulatory centre is located in the **hypothalamus**, specifically in the preoptic area and in the anterior part.

The thermoregulatory centre receives temperature information from the skin, viscera and the hypothalamus itself, computes this information and triggers either "just right", "too hot" or "too cold" thermoregulatory responses. But what neurons and neural circuits compute the sensed temperature information and trigger countermeasures such as sweating or shivering? Here, studies utilised measurements of Ca^{2+} (45), of the active neuron marker gene cFOS (46) or of the active neuron marker phosphorylated S6 (47) to identify neurons that are activated by warm or cold stimuli in the hypothalamus. The researchers then either activated or inhibited these neurons and discovered that this triggered thermoregulatory responses such as hypothermia or fever. Together, these experiments identified warm-sensitive and cold-sensitive neurons that not only respond to temperature but additionally regulate thermoregulatory responses (45–47). In these neurons, glutamine and GABA are neurotransmitters, but their exact role is still incompletely understood (44).

Importantly, temperature is not the only signal that is computed by the thermoregulatory system and elicits a response. For example, prostaglandin E2 (PGE2) is a metabolite that is produced as a result of inflammation or infection that triggers a heating response by binding to prostaglandin receptors in the hypothalamus. Also, when we eat spicy peppers, the capsaicin binds a specific TRP ion channel and will trigger a thermoregulatory response. This is why some of us sweat when we eat a hot Vindaloo curry.

Regulation of the effectors by the thermoregulatory centre: sweating and skin blood flow

The aforementioned thermoregulatory centre in the hypothalamus then regulates the thermoregulatory adaptations to a "too hot" or "too cold" signal. The main responses are sweating when it is too hot and the control of skin blood flow, specifically vasoconstriction in the cold and vasodilation in the heat.

In relation to exercise in the heat, sweating and the resultant fluid loss and dehydration are key reactions and issues. Humans have ≈4 million exocrine sweat glands. Sweat generates heat loss as 2.5 kJ are required to transfer 1 g of water to water vapour at 35°C (41). Of these sweat glands, 90% are the smaller eccrine glands and the remainder are larger apocrine glands (48). Why do we sweat and how does the resultant loss of water (dehydration) and salts/ions affect our performance? During exercise, the whole body sweat rate differs greatly in between humans. In a hot environment, we typically lose between 0.5 and 2 L of sweat per hour but in ≈2% of athletes this can reach more than 3 L per hour with extremes of nearly 6 L of sweat lost per hour (49). To counteract this, the ACSM recommends drinking enough to prevent a loss of more of 2% body weight loss from water deficit and to avoid excessive changes in the electrolyte balance (50). During sweating, water losses are not the only issue as sweat contains salts and sodium (Na^+) is a key one. The saltiness of our sweat differs between individuals, with Na^+ concentrations ranging from 10 to 90 mmol/L (49). Thus, the harder and longer we exercise in high temperatures, and the more we are genetically predisposed to lose water and Na^+, the more we will dehydrate and become hyponatraemic, and this will affect our exercise performance (50). Exercise-associated hyponatremia refers to blood Na^+ concentrations below 135 mmol/L and is a health hazard that can lead to death. It typically occurs when athletes drink too much hypotonic, Na^+-low fluids such as water so that Na^+-losses through sweating and urine exceed the Na^+-intake. Depending on the severity, hyponatremia can cause major health issues and it is likely that it was responsible for the death of several athletes (51).

So how does the thermoregulatory centre regulate sweating? A key brain area for sweating seems to be the brainstem and here the rostral ventromedial medulla, which is activated when humans sweat (52). This then innervates sweat glands through neurons that use acetylcholine and noradrenaline as neurotransmitters. Acetylcholine is the key neurotransmitter for sweating in response to temperature, as especially the eccrine sweat glands express the acetylcholine-binding M_3-muscarinic receptors and some noradrenaline-binding adrenoceptors (48). The M_3-muscarinic receptors are coupled to G proteins which then via some intermediate steps triggers a Ca^{2+} release, which, in turn, induces a translocation of aquaporin 5 water channels to the membrane. More aquaporin 5 channels in the membrane allow more water to enter the sweat gland, resulting in sweat production and secretion (48). The Aquaporin 5 (gene symbol: Aqp5) water channel seems essential for sweating, because the number of active sweat glands is dramatically reduced in Aqp5 knockout mice (53).

Skin blood circulation is another major site of thermoregulation. The skin is our largest organ and a key barrier in between our warm interior and a colder exterior (as our core body temperature is usually 37°C). Skin blood flow is small in the cold but increases when exercising in the heat to up to ≈8 L/min. These 8 L/min are 40% of the maximal cardiac output of 20 L/min of an untrained individual and 20% of the 40 L/min cardiac output in an elite endurance athlete (54). The neural projections go from the thermoregulatory centre in the hypothalamus via nuclei in the brainstem to the smooth muscles of blood vessels in the skin. Moreover, a inhibition or cooling of the thermoregulatory centre in the hypothalamus results in vasoconstriction, whereas warming or excitation induces vasodilation (44) and this is mediated by warm-sensitive neurons, which trigger vasodilation when stimulated (47). Skin blood flow is regulated by arteriovenous anastomoses, which are muscular, connecting vessels between small arteries and small veins. If these anastomoses are open then blood does not reach the skin and when they are closed then warm blood reaches the skin, allowing heat to dissipate (55).

Genetics of temperature regulation

Living organisms populate extreme "thermal niches", ranging from polar species such as ice bears to thermophilic bacteria such as *Methanopyrus kandleri* that can survive and reproduce in temperatures up to 122°C (56). For the polymerase chain reaction (PCR), a key molecular biology method (see Chapter 2), biologists are using the genetically heat-resistant polymerase of thermophilic *Thermus aquaticus* as the PCR reaction requires temperatures of nearly 100°C.

The first big change in the evolution of human thermoregulation was the evolution of hominins as endurance runners (57). All exercise is associated with a large amount of heat, as muscle often only converts ≈20% of the chemical energy in nutrients into work, whereas the rest is converted into heat that warms the body. For example, if we exercise with an oxygen uptake of 3 L/min, which equates to roughly 60 kJ of chemical energy used per minute, then we would generate 200 Js (equivalent to 200 W) of work and 800 Js (800 W) or 48 kJ/min as heat. This high heat load must be removed from the body to avoid a critical rise in body temperature. Here, the loss of fur and the increase in our capacity to sweat are evolutionary changes of our thermoregulation system that have enabled us to better exercise in the heat (41).

During the peopling of the world, humans have moved from their hot home in the African Savannah to different climates all over the world, including the Arctic (18). This includes the hot settlement of Dallol, where the Afar locals endure average high temperatures that reach ≈45°C from June to August. Conversely, Oymyakon is probably the coldest permanently inhabited place where the average low temperature is an extremely chilly −30°C and below from November to March. From Dallol to Oymyakon, temperature, sun exposure and food availability vary greatly. Is there evidence that this has led to the natural selection of temperature, sun exposure and food availability-related DNA variants?

To test for evidence for natural selection, Mathieson et al. compared the sequenced DNA of ancient Eurasians that lived from 6,500 to 300 years BC with that of over 2,000 modern-day humans (58). This analysis revealed several DNA variants with evidence for selection, suggesting that humans have adapted to their respective environments. They found evidence for selections of alleles in the SLC45A2, SLC45A5 and GRM genes, that encode skin pigmentation and to a DNA variant near HERC2/OCA2, which is linked to light eye colour (58). Generally, human skin colour-influencing alleles were strongly selected when humans migrated to less sunny latitudes, as the skin is a main site of vitamin D production which depends on ultraviolet B (UVB) radiation. Thus, human skin colour is the result of an evolutionary tug of war balancing photoprotection near the equator and sufficient vitamin D production in regions with less sunlight (59).

There is also evidence that DNA variants related to the temperature-sensing TRP ion channels are under positive selection. Key et al. found that the frequency of a DNA variant close to the cold-sensing ion channel TRPM8 varied with latitude, ranging from 5% in Nigeria to 88% in Finland (60), suggesting positive selection. Thus, this or related DNA variants may explain why Finns perceive cold differently than Nigerians.

Finally, Fumagalli et al. compared DNA variants of Greenlandic Inuit with DNA variants of individuals of European ancestry and Han Chinese. Here, the strongest evidence for selection was linked to fatty acid desaturases (FADS1–3) and this is probably related to the fish and meat-rich diet of the Inuit. TBX15 was another gene with evidence for selection which might be linked to the role of TBX15 in the development of adipose tissue and perhaps thermogenesis (61).

Collectively, these studies suggest again that modern humans have adapted, at least partially, to the specific challenges of their local environments. This includes skin colour adaptation to balance sun exposure with vitamin D synthesis, diet-related DNA variants and DNA variants directly linked to thermoregulation.

TIME, CIRCADIAN CLOCKS AND EXERCISE

Life on Earth evolved under 24 h oscillating rhythms of light and temperature. Because Earth rotates about its axis, each dawn brings a new wave of solar energy, warming the land and providing crucial electromagnetic energy needed for plants to produce oxygen and carbohydrates via photosynthesis. At dusk, as Earth turns away from the sun, this energy wanes. This rhythmic oscillation drove the evolution of homeostatic circadian timing systems across all kingdoms of life. These molecular clocks allow organisms to anticipate the cycling environmental conditions of life on Earth, and to coordinate their physiology and behaviour accordingly.

Because of their roles in metabolism and energy homeostasis, day-night cycles are also relevant for sport and exercise. Imagine exercising at 11 h in the morning, at 17 h in the evening and at 2 h at night. Athletic performance, core body temperature and the concentrations of hormones and nutrients all differ depending on the time of day (62, 63). To give just one example, Kline et al (64) demonstrated how the time of day affected swimming performance while trying to keep constant many of the other factors that might fluctuate with time and might affect performance. They observed a circadian variation in performance, with peak performance from around midday to late evening, and lower performance during the night and early morning (63, 64) (**Figure 11.6**).

Interindividual differences in athletic performance are based on relationships between genes and the environment. These interactions may also determine relative time-of-day differences in athletic performance. For example, a person's preferred time of sleep and activity, termed chronotype, is a behavioural manifestation of complex interactions between environmental time and their personal biological time, which is determined by their genetically coded circadian clock (65). Human chronotypes show large variation, with phase differences between extreme early and extreme late chronotypes spanning across 12 h. Earlier chronotypes perform better in the morning, whereas later

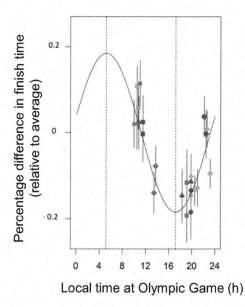

Figure 11.6 Swim performance and time of day at Olympic games. Different colours indicate different competitions. Figure Taken from Lok et al., (63) published in *Scientific Reports* (2020, Springer Nature) that allows reuse in any medium under Creative Commons Attribution 4.0 International License, https://creativecommons.org/licenses/by/4.0/.

chronotypes perform better in the evening (62). While each age group shows normal distribution among early and late chronotypes, average chronotype changes according to sex, and throughout one's lifetime according to age (66). Finally, athletes travelling across time zones for competitions may experience jet lag, which is an abrupt mismatch between their internal biological time and their external environmental time. This can result in sleep disturbances, increased lethargy, digestive problems, and impaired cognitive and physical performance (67). These issues tend to be worse than the more time zones that are crossed, and particularly when travelling East rather than West.

The aim of this section is to discuss the molecular clocks of our bodies. How are they regulated? How do they control the body's functions? How are they influenced by exercise and nutrients? Do they affect the association between performance and time of day?

Circadian (Latin *circa diem*, around a day) clocks are metabolic/transcriptional/translational/post-translational oscillators present in all nucleated cells of our body, and they maintain cellular time according to the rotation of the Earth (68). While each cell has its own robust 24 h molecular clock, in order for tissues, organ systems or multicellular organisms to operate efficiently as a functional unit, the individual clocks among cells must be synchronised by periodic stimuli called *zeitgebers* (German, time-giver). Earth's 24 h light-dark cycle is the main zeitgeber for the **suprachiasmatic nucleus (SCN)**, the **master pacemaker** in the hypothalamus. Photic information about the time of day is transmitted directly from the eye to the SCN by specialised photosensitive retinal ganglion cells. In turn, the SCN activates the sympathetic nervous system and regulates feeding patterns, locomotor activity and body temperature rhythms. In this way, the master clock in the brain coordinates production and distribution of a diverse array of metabolites. These include different steroid hormones, lipids, carbohydrates, vitamins and cofactors that can collectively modulate circadian gene expression and metabolism in tissues throughout the body. Daily exposure to sunlight is therefore important to maintain strong phase coherence between central and peripheral clocks, and to ensure strong coupling between our tissue clocks (**Figure 11.7**).

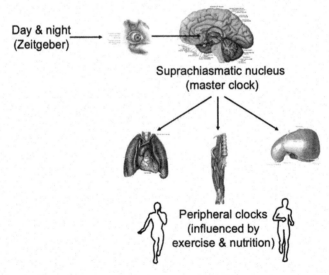

Figure 11.7 Circadian rhythms and the master and peripheral clocks of the body. Original figure with the organ images taken from Grey's Anatomy that do not require permissions under the Creative Commons CC0 License, https://creativecommons.org/publicdomain/zero/1.0/; and/or Creative Commons Attribution-ShareAlike License, https://creativecommons.org/licenses/by-sa/4.0/.

1. The **master circadian clock** located in the **suprachiasmatic nucleus**, a small bundle of neurons in the hypothalamus of the brain. This master clock drives the ~24 h rhythm and is driven by ambient light sensed by specialised retinal ganglion cells in the back of the eye (i.e. light is the *Zeitgeber*, "time setter" of this clock).

Peripheral circadian clocks for example in heart, skeletal muscle or the liver. The peripheral clocks also tend to follow the rhythm set by the master circadian clock, but their rhythm can be perturbed by exercise or food uptake.

Clock genes and circadian rhythms

The circadian clock works as a transcriptional regulator that switches genes on and off. At the centre of the circadian rhythm generator are the so-called basic helix-loop-helix (bHLH) PER-ARNT-SIM (PAS) domain-containing transcription factors CLOCK (circadian locomotor output cycles kaput) and BMAL1 (brain and muscle ARNT-like 1). CLOCK and BMAL1 proteins bind to each other, forming heterodimers, and then bind to the so-called E-boxes ("enhancer boxes" with CACGTG DNA motifs) in the promoters or enhancers of genes throughout the genome. The CLOCK:BMAL1 transcription factor then drives the expression of thousands of target genes in a circadian fashion.

As described in **Box 11.2** by Professor Karyn Esser, the *Clock* gene was discovered by the group of Joseph Takahashi using large scale screens for overt changes in circadian locomotor activity in mice after exposure to the mutagen ENU (N-ethyl-N-nitrosourea). His team monitored 24 h activity patterns in mice after placing them in constant darkness. Under these conditions the master clock in the SCN is free-running (i.e. without entrainment from the light-dark cycle) and the endogenously generated circadian period for each individual can be calculated based on their locomotor activity rhythm. The research team noted one mouse whose circadian period lengthened under constant darkness. After breeding this mouse, they found the offspring also inherited disturbed circadian rhythms suggesting that genetic mutations were responsible (69). Later, they pinpointed the mutated locus to the *Clock* gene (68) demonstrating that circadian rhythms are under genetic control.

What mechanism allows CLOCK:BMAL1 transcription factors to switch between being "active" and "inactive", and synchronised with the light-dark cycle? The core intracellular mechanism is a transcriptional/translational/post-translational feedback loop. In the positive arm of the loop, CLOCK:BMAL1 heterodimers drive expression for thousands of target genes, including their own repressors. These include the period genes (PER1/2) and the cryptochrome genes (CRY1/2) (**Figure 11.8** (68)). Over several hours, PER1/2 and CRY1/2 proteins accumulate and translocate into the nucleus to inhibit CLOCK:BMAL1 transcriptional activity (68). As a consequence, CLOCK:BMAL1 are inactive when PER1/2 and CRY1/2 are active, and vice versa. PER1/2 and CRY1/2 are time-limited, as they are phosphorylated by kinases such as casein kinase I, ubiquitinated and degraded in a circadian fashion (68). This allows CLOCK:BMAL1 to become transcriptionally active again and takes ~24 h to complete one cycle. Additional targets of CLOCK:BMAL1 include the highly abundant, yet transient nuclear hormone receptors REV-ERBα/β. These transcriptional repressors turn off genes by binding to specific "AGGTCA" DNA motifs near target genes, and recruiting corepressor proteins, including NCOR1 and the histone deacetylase HDAC3. Together, these proteins form a repressor complex that can modify chromatin structure and block accessibility of the transcriptional machinery, leading to repression of target genes, including BMAL1. This secondary 24 h loop adds additional stability to the circadian cycle (70), along with additional signalling events affecting phosphorylation, ubiquitination and localisation of clock transcription factors (71). In summary, CLOCK:BMAL1 transcription factors regulate thousands of genes that oscillate in a circadian fashion, including PER1/2, CRY1/2, and REV-ERBα/β. Once PER1/2, CRY1/2, and REV-ERBα/β are transcribed and translated, the CLOCK:BMAL1 activity is repressed as PER1/2 and CRY1/2 and REV-ERBα/β directly and indirectly inhibit CLOCK:BMAL1 transcriptional activity.

BOX 11.2 Karyn Esser on the discovery of clock genes in relation to exercise

My entry into circadian rhythms field is a story of curiosity, new technology, good timing, good luck and good people.

In 2001, I was newly minted Associate Professor at the University of Illinois, Chicago. My lab was studying mechanisms that link high force muscle contractions with hypertrophy. In particular we were interested in both transcriptional and translational mechanisms that linked a specific set of contractions with growth. I was fortunate at that time to have an opportunity to talk with Eric Hoffman and Yiwen Chen who were actively using gene array technology at Children's National Medical Center and George Washington University. These discussions lead to our collaboration in we used the high force contraction model to generate control and experimental tissues for gene arrays. I remember being very excited for our first venture into unbiased approaches using the Affymetrix Rat U34A gene chips with ~8,000 genes analysed. The results of this project were published in 2002, Chen et al., **J. Physiology** and we outlined changes in gene expression due to transcription and/or translation following contractions known to induce hypertrophy. What is not included in that publication was our identification of the core clock gene, Bmal1, as a gene that was transcriptionally changed in response to contractions. I cannot tell you why I was so fascinated by this observation, but I just was.

I spent the next 1–2 weeks reading about **Bmal1** and learning more and more about the circadian clock mechanism. Timing is another critical part of this story because the publication detailing the discovery of the first mammalian circadian factor, Clock, was in 1994 and so the cloning of the other components of the core clock mechanism came following this set of papers. This resulted in circadian clock research being very active in the late 1990s and early 2000s. These papers included the discovery of Bmal1 and its role in the clock as well as the observations that the core clock factors were functional in peripheral tissues as well as the CNS. The more I read about Bmal1 and the clock mechanism, the more it made sense to me as a way to maintain cell homeostasis.

Now to good people and their importance in science. As I became more fascinated with the circadian clock and its potential role in skeletal muscle, I reached out to Dr. Joseph Takahashi. Joe's lab had done the original cloning of Clock in mice and he was a leader in the field of the molecular regulation of circadian function. Joe was quick to respond to my initial email asking to talk to him about our finding of Bmal1 gene expression changes in muscle following contractions. I drove up to his lab at Northwestern University and we talked for over an hour about our work and his enthusiasm helped fuel my own enthusiasm that the concept of clocks in muscle was likely something biologically and physiologically important.

So back to another aspect of good timing. I had just been promoted to Associate Professor with tenure and had now been at UIC long enough to be eligible for a sabbatical. I had decided I would take a sabbatical to step away from departmental responsibilities and enhance the lab's approaches. I was considering two very different labs and am grateful for my longtime friend and colleague, Susan Kandarian, as her input helped me talk through this decision. Ultimately, I decided to take advantage of Joe Takahashi's willingness to open his lab to me and officially went on sabbatical in 2002.

It was in Joe's lab that I really learned the terminology of the circadian field, fundamentals of circadian biology and the core clock mechanism. My goal on sabbatical was to build a floxed Bmal1 gene construct to use for homologous recombination to generate a mouse for conditional knockout use. Needless to say, I was unsuccessful in my technical goal but the amount I learned from the students, postdocs and staff in the lab made up for my lack of success at the bench. In particular, I was fortunate to learn from Erin McDearmon, Hee-kyung Hong, Seung-Hee Yoo and Ethan Buhr and hopefully I was able to teach them a few things about skeletal muscle.

During this time, I participated in my first 48 h circadian collection and the tissues from this collection contributed to the first publication in my CV on circadian gene expression, Miller et al., 2007. The other outstanding aspect of Joe's lab was access to genetic mouse models, including the Bmal1 KO and Clock mutant mice. Analysis of both genetic circadian mutants demonstrated that skeletal muscles exhibit several pathological changes, including weakness, altered myofilament structure and diminished mitochondrial content and function. These findings were ultimately published in 2010, Andrews et al. and helped contribute to our understanding of the importance of circadian clock function in skeletal muscles.

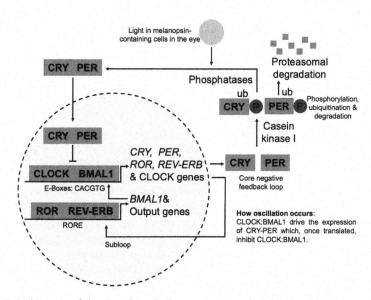

Figure 11.8 Key elements of the circadian rhythm-generating clock genes.

How is the master pacemaker in the brain synchronised to the environmental day-night cycle? The cryptochrome-encoding genes CRY1/2 are key candidates for linking the master clock to ambient light as cryptochromes are blue light-sensing proteins. However, Cry1/2 knockout mice still increase the expression of Per1 and Per2 in response to light (72), suggesting CRY1 and CRY2 proteins are not needed to synchronise the master clock with the day/night cycle. So where are the light sensors that help to entrain the master clock in the suprachiasmic nucleus? Researchers found that if they knocked out a protein in mice called melanopsin, containing intrinsically photosensitive retinal ganglion cells of the eye, the animals demonstrated normal pattern vision, yet struggled to link their circadian rhythms to the day-night cycle (73). As melanopsin responds mostly to blue light, this wavelength of visible light keeps melanopsin more active during the day and less active at night.

Circadian rhythms and exercise

How are circadian rhythms relevant for skeletal muscle and exercise? Many of the benefits attributed to exercise may actually stem from its maintenance of circadian health. Because exercise elicits such dramatic shifts in temperature and blood flow, along with metabolic and hormonal alterations, exercise can be harnessed as a powerful zeitgeber. Accordingly, clinicians often prescribe exercise as a way to realign circadian disruption in shift workers and patients with sleep disturbances, and to mitigate complications associated with jet leg.

General exercise performance and responses to exercise are both known to show time-of-day-dependent variations, but circadian researchers are only beginning to understand the precise molecular underpinnings. Thousands of genes oscillate over 24 h in skeletal muscles from mice (74, 75) and from humans (76). These oscillations can be driven directly by feeding and activity-dependent rhythms controlled by the SCN, or indirectly through the regulation of the muscle circadian clock, which is likewise coupled to the SCN master pacemaker (77). Accordingly, most 24 h oscillating muscle genes depend on muscle type, and fluctuating metabolic and mechanical signals resulting from feeding and activity patterns.

A direct effect of exercise on muscle clock function was first demonstrated in a human resistance training study (78). Healthy male subjects performed a single acute bout of one-legged isotonic knee extensions at 14:00, early in the afternoon. Researchers then collected vastus lateralis muscle biopsies from both exercised and non-exercised legs 6 h later in the evening (~20 h), and the next morning (~08 h), 18 h after exercise. Comparing global gene expression patterns, they identified hundreds of genes regulated by resistance exercise in a time-dependent manner, including clock genes. In particular, they found that resistance exercise shifted the phase of BMAL1, PER2, and CRY1, demonstrating that resistance exercise can modulate circadian clock phases.

Circadian researchers have also investigated how clock function impacts performance and endurance capacity. Jordan et al. (2017) found increased endurance capacity in Cry1 and Cry2 double knock mice. CRY1/2 inhibits transcriptional activity of the nuclear receptor PPARδ, a known mediator of increased endurance capacity (79). Together with additional studies in mice and humans (80–84) these data demonstrate that circadian clock proteins can impact time-dependent differences in exercise performance by directly controlling skeletal muscle metabolism.

These relationships are especially relevant when considering exercise type and intensity. In mice, high-intensity exercise capacity is greater during the early part of the active phase (at night, because mice are nocturnal), whereas moderate- and low-intensity exercise capacity is greater during the latter part of the active phase (85, 86). Various clock-related metabolic sensors are known to play a role here, including the sirtuins and AMP-activated protein kinase (AMPK) (87–89). These data suggest time-of-day differences in low and moderate intensity exercise capacity reflect interactions between circadian clocks and nutritional state.

On the other hand, day-night differences in high-intensity exercise capacity are linked to exercise-induced hypoxemia and a glycolytic gene programme driven by hypoxia-inducible factor 1α (HIF-1α)

(86). Like CLOCK and BMAL1, HIF-1α is a helix-loop-helix (bHLH) PER-ARNT-SIM (PAS) domain-containing transcription factor that can bind to the same regulatory sites in the genome. Interestingly, HIF-1α shows a bidirectional relationship with the circadian clock (90, 91). CLOCK and BMAL1 can drive 24 h *Hif1a* expression, and HIF-1α can synergise with BMAL and drive expression of common circadian clock and HIF target genes, including *Per2* and *Cry1*. Accordingly, hypoxia can phase-shift clocks in a tissue-dependent manner (92). In mouse skeletal muscles this was shown to occur by a time-of-day-dependent interaction between HIF-1α and clock transcription factors (93). Further studies are needed to fully elucidate these relationships, but this data collectively suggests that moderate reductions in oxygen levels during high-intensity training, or even training under hypoxic conditions may be a more potent zeitgeber than lower or more moderate exercise bouts.

In addition to a wide variety of metabolic genes and energy sensors, many muscle-specific genes cycle over 24 h, including *Myod1*, coding for the myogenic regulatory factor MyoD. During development, MyoD acts as a myocyte pioneer factor by opening closed chromatin at regulatory DNA sites in muscle-specific genes throughout the genome. After directing progenitor cells into a muscle cell-type lineage, MyoD maintains their differentiated state by keeping these sites open. Normally expressed at low levels in quiescent muscle stem cells (See **Chapter 8 and 13**), MyoD expression can become activated by exercise or damage, promoting muscle tissue remodelling and regeneration. Just like BMAL1, CLOCK and HIF-1α, MyoD is another basic helix-loop-helix transcription factor that can activate target genes after binding to similar "E-Box" DNA sequences in promoters and enhancers. Professor Karyn Esser's group (Box 11.2) determined that *Myod1* is also a direct CLOCK:BMAL1 target (94) and vice versa (83). This bidirectional regulation is thought to strengthen circadian alignment, and to amplify 24 h oscillation of other circadian clock-controlled muscle genes, including *Telethonin* (*Tcap*). This 24 h oscillating CLOCK:BMAL1 target gene is a Z-disc protein that regulates sarcomere assembly and T-tubule function. Intriguingly, muscles from *Clock*[Δ19] mice (mutant mice have a point mutation that causes a deficiency in the 19th exon of the Clock gene) which are completely arrhythmic, show severely blunted 24 h *Myod1* transcript and protein levels. This was associated with impaired muscle function, as specific force of extensor digitorum longus muscles was ~30% lower in *Clock*[Δ19], *Bmal1* knockout and *Myod1* knockout mice compared to wildtype controls (94). However, 24 h *Myod1* oscillation was reported to be increased in a muscle-specific *Bmal1* knockout mouse model (77). These mice had normal 24 h activity and feeding patterns, yet also showed reduced force production in gastrocnemius muscles, and blunted oscillation of circadian genes, including *Telethonin* (95). These contrasting data suggest additional work is needed to clarify the roles of MyoD in driving circadian gene regulation in differentiated muscle tissue.

Muscle protein degradation genes, like *Atrogin1* (*MAFbx*/*Fbxo32*) and MuRF1 (*Trim63*) also oscillate over 24 h (74), and show increased expression upon muscle-specific *Bmal1* ablation (81). This suggests that in addition to regulation by rhythmic hormonal (glucocorticoids) and metabolic cues (fasting), muscle breakdown may be locally regulated by the muscle circadian clock. A direct connection between MuRF1 expression and muscle clock transcription factors was verified by combining chromatin immunoprecipitation assays performed in adult mouse skeletal muscle with 24 h gene expression profiling of muscle-specific *Bmal1* knockout mice (81). REV-ERBα was shown to represses MuRF1 expression near a genomic locus normally activated by the glucocorticoid receptor. Loss of repression was associated with a transient increase in glucocorticoid receptor transcriptional activity during the feeding phase, along with increased cycles of protein turnover.

Muscle protein synthesis was also shown to be under circadian clock control. Expression of a dominant negative CLOCK mutant in developing zebrafish abolished day-night differences in muscle protein synthesis and impaired overall muscle growth (96). In addition to altered protein metabolism, circadian clock-disrupted mice also showed reduced insulin-dependent glucose uptake and oxidation, and important changes in lipid and amino acid metabolism (82, 83, 97). Collectively, these alterations highlight a close connection between muscle circadian clock function, protein turnover and general metabolic homeostasis.

The 24 h sleep-wake cycle is another circadian clock-regulated process that can impact exercise performance, as well as adaptations and recovery after training (98). Intriguingly, sleep quality has also been linked to skeletal muscle clock function. In constant darkness Bmal1 knockout mice are completely arrhythmic, which also impacts their sleep quality and recovery after sleep deprivation (99). To identify the organ responsible for these changes, researchers re-introduced constitutively high levels of BMAL1 in different organs of Bmal1 knockout mice (100). This kind of supraphysiological BMAL1 rescue in the brain did not restore normal sleep, suggesting that Bmal1 expression outside the brain may be responsible. They next expressed a constitutively high BMAL1 gene selectively in skeletal muscles of the otherwise Bmal1-deficient mice. This was sufficient to restore the amount of NREM (Non rapid eye movement) sleep and improve the ability to recover from sleep loss. While these data showed that skeletal muscle-specific BMAL1 can affect sleep, the precise mechanisms by which muscle communicates with the brain, the key organ for sleep, remain unidentified.

In summary, circadian clocks are major regulators of muscle metabolism and exercise capacity. Here we have highlighted only a few key studies to emphasise the many complex reciprocal interactions that exist between circadian rhythms and exercise (type, intensity, duration, time of day), relative nutritional state and sleep. In a bidirectional relationship, exercise can impact entrainment and function of peripheral circadian clocks, especially in skeletal muscle, but exercise performance and muscle metabolism also depend on relative circadian clock alignment.

SUMMARY

A 30 min jog at an altitude of 4,500 m, at $-20°C$ at 3 am is much more challenging than the same jog at sea level, at 15°C at lunchtime, highlighting the effects of high altitude, temperature and circadian rhythms on our ability to exercise.

At higher altitudes, low arterial O_2 tension (PaO_2) is detected by Ndufs2-related sensing molecules within the carotid body which then increases ventilation by stimulating neurons that control ventilation in the brainstem. This mechanism increases ventilation and aids oxygen delivery at high altitude. Moreover, low O_2 tension in tissues such as skeletal muscle induces the hypoxia-induced factors (HIFs) which then trigger longer term adaptations such as gene expression in muscle and the production of erythropoietin, which stimulates the formation of red blood cells (erythrocytes). Sherpas and Tibetans as well as inhabitants of other altitude regions have DNA sequence variants of genes such as EPAS1 which affect their haematocrit and ventilation and helps life at high altitude.

Hot and cold temperatures are sensed by TRP (transient receptor potential) ion channels which sense cold and cold pain, warmth and heat pain. Once a critically high or low temperature is reached, TRP ion channels open and Na^+ and Ca^{2+} ions flow into and depolarise neurons that express these TRPs. This information is then relayed to the central nervous system which regulates sweating, thermogenesis by brown adipose tissue or muscle shivering and blood flow, to respond to hot or cold temperatures.

Circadian rhythms are controlled by cellular clocks that are synchronised by feeding, activity and temperature rhythms. These clock proteins are transcription factors that regulate day-night oscillations in gene expression and metabolism. Since exercise changes hormones, metabolism, temperature and blood flow, it can be used to synchronise internal clocks. Conversely, because circadian clocks direct metabolic gene expression, they can also impact athletic performance throughout the day.

REVIEW QUESTIONS

1. What molecular "sensors" sense high altitude and how do our bodies respond to it?
2. What advantages do Sherpas and Tibetans have at high altitude when compared to humans that live near sea level?

3. How do our bodies sense warmer and colder temperatures and how do we adapt to hot and cold temperatures?
4. How do circadian rhythms affect our ability to exercise and how does exercise affect our circadian rhythms?
5. How might altitude affect the circadian clock?

FURTHER READING

Gabriel BM & Zierath JR (2019). Circadian rhythms and exercise — re-setting the clock in metabolic disease. *Nat Rev Endocrinol* 15, 197–206.

West JB, Schoene RB, & Millede JS (2012). *High Altitude Medicine and Physiology*, Taylor & Francis Ltd.

REFERENCES

1. Habeler P. "Ich bin kein Eroberer" Alpin Interview mit Peter Habeler 2008 [Available from: http://www.alpin.de/home/news/4873/artikel_das_ausfuehrliche_alpin-interview_mit_peter_habeler.html.
2. Lyons TW, et al. *Nature*. 2014. 506(7488):307–15.
3. Ainsworth BE, et al. *Med Sci Sports Exerc*. 1993. 25(1):71–80.
4. Dempsey JA, et al. *Physiol Rev*. 1982. 62(1):262–346.
5. Lopez-Barneo J, et al. *Am J Physiol Cell Physiol*. 2016. 310(8):C629–42.
6. Samanta D, et al. *Wiley Interdiscip Rev Syst Biol Med*. 2017. 9(4).
7. Fernandez-Aguera MC, et al. *Cell Metab*. 2015. 22(5):825–37.
8. West JB. *Am J Respir Crit Care Med*. 2012. 186(12):1229–37.
9. Schofield CJ, et al. *Nat Rev Mol Cell Biol*. 2004. 5(5):343–54.
10. Jelkmann W. *J Physiol*. 2011. 589(Pt 6):1251–8.
11. Mairbaurl H. *Front Physiol*. 2013. 4:332.
12. Braekkan SK, et al. *Haematologica*. 2010. 95(2):270–5.
13. Lasne F, et al. *Nature*. 2000. 405(6787):635.
14. Jelkmann W. *Eur J Haematol*. 2007. 78(3):183–205.
15. Jacobson LO, et al. *Nature*. 1957. 179(4560):633–4.
16. Haase VH. *Blood Rev*. 2013. 27(1):41–53.
17. Gruber M, et al. *Proc Natl Acad Sci U S A*. 2007. 104(7):2301–6.
18. Nielsen R, et al. *Nature*. 2017. 541(7637):302–10.
19. Lopez S, et al. *Evol Bioinform Online*. 2015. 11(Suppl 2):57–68.
20. Beall CM. *Proc Natl Acad Sci U S A*. 2007. 104(Suppl 1):8655–60.
21. Bigham AW. *Curr Opin Genet Dev*. 2016. 41:8–13.
22. Villafuerte FC, et al. *High Alt Med Biol*. 2016. 17(2):61–9.
23. de la Chapelle A, et al. *Proc Natl Acad Sci U S A*. 1993. 90(10):4495–9.
24. Morpurgo G, et al. *Proc Natl Aca Sci U S A*. 1976. 73(3):747–51.
25. Wu T, et al. *J Appl Physiol (Bethesda, MD: 1985)*. 2005. 98(2):598–604.
26. Gonzales GF, et al. *J Matern Fetal Neona*. 2012. 25(7):1105–10.
27. Chonchol M, et al. *Am Heart J*. 2008. 155(3):494–8.
28. Boffetta P, et al. *Int J Epidemiol*. 2013. 42(2):601–15.
29. Yi X, et al. *Science*. 2010. 329(5987):75–8.
30. Huerta-Sanchez E, et al. *Nature*. 2014. 512(7513):194–7.
31. Lorenzo FR, et al. *Nat Genet*. 2014. 46(9):951–6.
32. Brand MD. *Biochem Soc Trans*. 2005. 33(Pt 5):897–904.
33. Schippers MP, et al. *Curr Biol*. 2012. 22(24):2350–4.
34. Horscroft JA, et al. *Proc Natl Acad Sci U S A*. 2017. 114(24):6382–7.

35. Porter RS, et al. *The Merck Manual*. 19 ed. Whitehouse Station, NJ: Merck Sharp & Dohme Corp, 2010.
36. Tansey EA, et al. *Adv Physiol Educ*. 2015. 39(3):139–48.
37. Betz MJ, et al. *Diabetes*. 2015. 64(7):2352–60.
38. Rosen Evan D, et al. *Cell*. 2014. 156(1):20–44.
39. Li Y, et al. *Cell*. 2018. 175(6):1561–74.e12.
40. Bartelt A, et al. *Nat Rev Endocrinol*. 2014. 10(1):24–36.
41. Lieberman DE. *Compr Physiol*. 2015. 5(1):99–117.
42. Wang H, et al. *Temperature (Austin)*. 2015. 2(2):178–87.
43. Siemens J, et al. *Pflugers Archiv: Eur J Physiol*. 2018. 470(5):809–22.
44. Tan CL, et al. *Neuron*. 2018. 98(1):31–48.
45. Song K, et al. *Science (New York, NY)*. 2016. 353(6306):1393–8.
46. Zhao ZD, et al. *Proc Natl Acad Sci U S A*. 2017. 114(8):2042–7.
47. Tan CL, et al. *Cell*. 2016. 167(1):47–59.e15.
48. Hu Y, et al. *Br J Dermatol*. 2018. 178(6):1246–56.
49. Baker LB. *Sports Med (Auckland, N Z)*. 2017. 47(Suppl 1):111–28.
50. American College of Sports M, et al. *Med Sci Sports Exerc*. 2007. 39(2):377–90.
51. Hew-Butler T, et al. *Front Med (Lausanne)*. 2017. 4:21.
52. Farrell MJ, et al. *Am J Physiol Regul Integr Comp Physiol*. 2013. 304(10):R810–7.
53. Nejsum LN, et al. *Proc Natl Acad Sci U S A*. 2002. 99(1):511–6.
54. Gonzalez-Alonso J, et al. *J Physiol*. 2008. 586(1):45–53.
55. Walloe L. *Temperature (Austin)*. 2016. 3(1):92–103.
56. Takai K, et al. *Proc Natl Acad Sci U S A*. 2008. 105(31):10949–54.
57. Bramble DM, et al. *Nature*. 2004. 432:345–52.
58. Mathieson I, et al. *Nature*. 2015. 528(7583):499–503.
59. Jablonski NG, et al. *Int J Paleopathol*. 2018. 23:54–9.
60. Key FM, et al. *PLoS Genet*. 2018. 14(5):e1007298.
61. Fumagalli M, et al. *Science*. 2015. 349(6254):1343–7.
62. Teo W, et al. *J Sports Sci Med*. 2011. 10(4):600–6.
63. Lok R, et al. *Sci Rep*. 2020. 10(1):16088.
64. Kline CE, et al. *J Appl Physiol (Bethesda, MD: 1985)*. 2007. 102(2):641–9.
65. Wittmann M, et al. *Chronobiol Int*. 2006. 23(1–2):497–509.
66. Fischer D, et al. *PLoS One*. 2017. 12(6):e0178782.
67. Leatherwood WE, et al. *Br J Sports Med*. 2013. 47(9):561–7.
68. Partch CL, et al. *Trends Cell Biol*. 2014. 24(2):90–9.
69. Vitaterna MH, et al. *Science (New York, NY)*. 1994. 264(5159):719–25.
70. Preitner N, et al. *Cell*. 2002. 110(2):251–60.
71. Hirano A, et al. *Nat Struct Mol Biol*. 2016. 23(12):1053–60.
72. Okamura H, et al. *Science (New York, NY)*. 1999. 286(5449):2531–4.
73. Guler AD, et al. *Nature*. 2008. 453(7191):102–5.
74. McCarthy JJ, et al. *Physiol Genomics*. 2007. 31(1):86–95.
75. Dyar KA, et al. *Mol Metab*. 2015. 4(11):823–33.
76. Perrin L, et al. *eLife*. 2018. 7:e34114.
77. Schiaffino S, et al. *Skeletal Muscle*. 2016. 6:33.
78. Zambon AC, et al. *Genome Biol*. 2003. 4(10):R61.
79. Wang YX, et al. *PLoS Biol*. 2004. 2(10):e294.
80. Loizides-Mangold U, et al. *Proc Natl Acad Sci U S A*. 2017. 114(41):E8565–e74.
81. Dyar KA, et al. *PLoS Biol*. 2018. 16(8):e2005886.
82. Harfmann BD, et al. *Skeletal Muscle*. 2016. 6:12.
83. Hodge BA, et al. *eLife*. 2019. 8:e43017.
84. van Moorsel D, et al. *Mol Metab*. 2016. 5(8):635–45.
85. Ezagouri S, et al. *Cell Metab*. 2019. 30(1):78–91.e4.
86. Sato S, et al. *Cell Metab*. 2019. 30(1):92–110.e4.

87. Nakahata Y, et al. *Cell*. 2008. 134(2):329–40.
88. Masri S, et al. *Sci Signal*. 2014. 7(342):re6.
89. Lamia KA, et al. *Science*. 2009. 326(5951):437–40.
90. Wu Y, et al. *Cell Metab*. 2017. 25(1):73–85.
91. Peek CB, et al. *Cell Metab*. 2017. 25(1):86–92.
92. Manella G, et al. *Proc Natl Acad Sci U S A*. 2020. 117(1):779–86.
93. Adamovich Y, et al. *Cell Metab*. 2017. 25(1):93–101.
94. Andrews JL, et al. *Proc Natl Acad Sci U S A*. 2010. 107(44):19090–5.
95. Dyar KA, et al. *Mol Metab*. 2014. 3(1):29–41.
96. Kelu JJ, et al. *Proc Natl Acad Sci U S A*. 2020. 117(49):31208–18.
97. Dyar KA, et al. *Mol Metab*. 2014. 3(1):29–41.
98. Watson AM. *Curr Sports Med Rep*. 2017. 16(6):413–8.
99. Laposky A, et al. *Sleep*. 2005. 28(4):395–409.
100. Ehlen JC, et al. *eLife*. 2017. 6:e26557.

CHAPTER TWELVE

Cancer and exercise

Tormod S. Nilsen, Pernille Hojman and Henning Wackerhage

Pernille Hojman contributed to this chapter but then sadly passed away.

LEARNING OBJECTIVES

At the end of the chapter, you should be able to:

1. Explain how mutations of cancer genes transform cells-of-origin to become cancer cells with different hallmarks of cancer.
2. Describe the standard treatments for cancer.
3. Explain the relationship between physical activity and the risk of cancer as well as the effects of exercise in cancer patients.
4. Explain some established and potential mechanisms that explain the anti-cancer effects of exercise.

INTRODUCTION

Cancer is a relatively common disease with improving but still imperfect treatment. According to the World Health Organization, 18.1 million new cancer cases were diagnosed worldwide in 2018, and an estimated 9.6 million people died from cancer. Roughly 50% of patients survive cancer for ten years, but this differs markedly between cancer types, with testicular cancer having the highest and pancreatic cancer having the lowest survival rates. Breast, prostate, lung and colon (bowel) cancers are the most common cancers in the UK and most Western countries as they account for roughly half of all cancers (1).

Meeting physical activity recommendations has been shown to lower the risk of several cancers (2) and is also an effective treatment for many other diseases (3). In a study where 12 prospective cohort studies were merged, allowing researchers to investigate the influence of leisure-time physical activity on cancer risk in 1.44 million responders, higher activity levels were associated with a lower risk in 13 of the 26 types of cancer studied (4). Additionally, physical activity before and after cancer diagnosis reduces mortality (5, 6) and has additional beneficial effects such as improving the quality of life (QoL) for those with cancer (7). Thus, physical activity is an important tool to reduce the number of people who will develop cancer, and to help to treat those who do have cancer. Based on this evidence, associations such as The American College of Sports Medicine (ACSM) issue guidelines on exercise programmes with cancer patients and exercise is increasingly offered to cancer patients (8).

In this chapter, we first cover the basics of what cancer is, how it happens, and briefly comment on common treatment options. Thereafter we discuss the potential "anti-cancer" mechanisms of exercise. We

DOI: 10.4324/9781315110752-12

start by explaining how cancers develop and explain why cancer is still such a complex disease to treat. Cancer is a genetic disease, so we also explain how mutations change the so-called driver genes that turn normal cells into cancer cells and then exhibit the so-called "hallmarks of cancer". We then introduce the standard treatments of cancer, including the idea of precision medicine. After that, we primarily discuss observational studies that link physical activity to lower cancer risk and identify exercise as a cancer treatment. Finally, we will come back to molecular exercise physiology. Here we discuss the potential molecular anti-cancer mechanisms of exercise. We also discuss the potential of exercise to counteract cancer-induced skeletal muscle wasting, termed cachexia, at the molecular level.

What is cancer, and why does it happen?

Cancer occurs when mutations change the DNA sequence of so-called cancer genes in a cell-of-origin. Not all mutations, however, cause cancer. This is because many mutations do not alter the expression or function of cancer genes. Mutations that are present in a cancer cell but do not contribute to a cancer are termed **passenger mutations**. In contrast, some mutations hit cancer genes and contribute to turning a normal cell into a cancer cell. These mutations are termed **driver mutations** (9).

Cancer genes typically encode proteins that regulate the cell division, survival, growth or metabolism, termed the hallmarks of cancer, which we describe below. There are two main types of cancer genes:

- **Oncogenes** are genes whose gain-of-function by mutation promotes cancer.

- **Tumour suppressor genes** are genes whose loss of function by mutation or other factors promotes cancer.

So how does a cancer cell differ from a normal cell? Cancer cells generally carry several driver mutations that are usually acquired step-by-step and that begin to turn a normal cell into a cancer cell. The unique properties of cancer cells have been termed the **hallmarks of cancer** (10).

The hallmarks of cancer cells are as follows:

1. Proliferation (undergo cell division) - in situations where normal cells would not proliferate.
2. Unlimited cell division - whilst normal cells have a so-called Hayflick cell division limit (which is where a typical normal human cell can divide 40–60 times in its lifespan before it can no longer divide).
3. Resistance to cell death - in situations where normal cells would die, e.g. through apoptosis.
4. More growth (anabolism) - more than normal cells.
5. Metabolic reprogramming - many cancer cells have a more glycolytic-anabolic metabolism, termed the Warburg effect, than normal cells.
6. Immune evasion.
7. Metastasis - as some cancer cells can leave their original site and can undergo metastases (secondary malignant growth from their primary site) elsewhere in the body.
8. Angiogenesis - which is the formation of blood vessels.
9. Inflammation - produced by a tumour (mass of cancer cells).
10. Genomic instability - which is the unfaithful division of chromosomes to daughter cells.

Not every cancer cell displays all the above hallmarks of cancer but usually, the cells-of-origin accumulate several of these hallmarks whilst progressing from a pre-cancerous benign lesion to a full-blown, malignant tumour.

Given that driver DNA mutations are the main cause of cancer, where do these mutations come from? Mutations can be either inherited, occur due to exposure to environmental factors (i.e. carcinogens) or can occur during cell replication/division, i.e. where the DNA is copied and passed on.

Most mutations occur at random when cells divide, as the whole DNA needs to be duplicated during replication. In the adult, it is primarily the tissue-specific stem cells that divide. Such stem cells are often the so-called cells-of-origin or cells that turn into cancer cells. In a simple but insightful analysis, the mathematician Christian Tomasetti and the cancer geneticist Bert Vogelstein plotted the estimated number of divisions of self-renewing cells in tissue versus the lifetime risk of developing cancer in that tissue. They found a strong, significant correlation of r=0.81. This suggests that most cancers occur because of "bad luck" and that the frequency of mutations is strongly linked to the number of cell divisions (11). In other words, if there are many dividing cells in a tissue, then the cancer risk in that tissue is high.

The tissue-specific mutation rate can be increased further by exposure to environmental factors such as tobacco smoke or sunlight. Environmental factors (e.g. carcinogens and mutagens), inheritance or cell division, i.e. the doubling of DNA, can cause mutations that include cancer driver gene mutations. In a follow-up analysis, the same research group tried to quantify the contribution of environmental factors, inheritance and replication to mutations and cancer frequency. They concluded that mutations that occur during cell division account for two-thirds of the mutations in human cancer (12). Different mutations leave different mutational signatures, which are patterns of how the cancer-causing agent is changing the DNA (13). For example, the mutagens in tobacco smoke typically cause C-to-A mutations which are detectable in tissues exposed to tobacco smoke such as the lung.

Inherited mutations are termed germline mutations. One example of inherited **germline mutations** are mutations of the Breast cancer type 1 and type 2 susceptibility genes (BRCA1/2) or tumour protein p53 (TP53) gene, commonly known as simply p53 (Li Fraumeni syndrome) (14). Germline mutations will be present in all cells of the body.

What are the cancer genes, and why do cancers often progress from a benign overgrowth to an aggressive, metastatic cancer?

Let us get specific about mutated cancer genes. In 2007 researchers published the results of sequencing the DNA of the 514 kinases in cancers and then compared the cancer DNA to normal DNA (15). Surprisingly, the gene with most mutations was TTN which encodes titin, a muscle protein. Titin is only expressed in skeletal muscle and the heart and probably almost never in cancer cells. Thus, even if the titin gene is mutated in cancer cells it will not have an effect because the titin protein will not be present. TTN mutations are thus typical **passenger gene mutations**.

So how did researchers identify driver mutations? Cancer genes were mainly discovered by studying the function of genes and through improved bioinformatical analyses of cancer genomes (16). The most frequently mutated cancer gene is the tumour suppressor also known as the "Guardian of the Genome" p53 (17) as its normal function is to prevents cells with DNA damage from expanding. If p53 is lost, cells with DNA damage will proliferate. A common cancer oncogene is KRAS, where gain-of-function mutations generate a highly active KRAS-protein that drives cancer development and growth. Gain-of-function mutations of KRAS often change the regulatory glycines at position 12 and 13 as well as the glutamine at position 61 within the protein which makes the KRAS-protein more active (18). The loss of p53 through a disabling mutation or the gain of KRAS through an activating mutation in many cell types progresses these cells towards cancer (18). More formally, tumour suppressor genes and oncogenes are defined as genes whose inactivation or activation by mutation increases the selective growth advantage of the cell in which is resides (19). However, with few exceptions, one gene's function changing mutation is not enough to drive cancer. In most cases, several mutations are required (19) to trigger the hallmarks that contribute to a full-blown cancer (10).

There are also many parallels between cancer and Darwinian selection. Each mutated cell is subjected to Darwinian selection in the "ecosystem" of the tissue where it resides. If a cell's mutations promote more proliferation or other hallmarks of cancer, then that cell or clone will expand more within its organ ecosystem than other cells (9). Additionally, not all the cells within a tumour are the same with respect to genetics, cell types or properties, and different cancer cell subpopulations may "work" together. This phenomenon is a result of the often increased mutation rate in cancer (genetic instability) together with Darwinian selection and ultimately results in cancer heterogeneity (20).

FROM CANCER CELLS TO BULK TUMOUR

Above, we discussed how mutations could transform healthy cells into cancer cells. However, cancer cells alone do not make a tumour. To sustain growth and metastasise, cancer cells make a symbiotic relationship with several "normal" cells, which constitute the tumour microenvironment (21). These normal cells include predominantly fibroblasts, endothelial cells, and immune cells, as well as many others.

Fibroblasts help build the scaffolding of the tumour. They secrete large amounts of collagens and other matrix proteins, which helps maintain the extracellular matrix. Many types of tumours have a stiff extracellular matrix, making them hard and difficult to penetrate, and it also makes them palpable if the tumour is not located too deep in the tissue. Delivery of nutrients and oxygen is also highly warranted in the fast-growing and metabolic active cancer cells. The capillary network of tumours, however, is often very irregular, comprising sprouting vessels, which are poor at blood perfusion. This immature capillary network thus leads to insufficient nutrients, oxygen and immune cell infiltration into the tumour, as well as difficulty in the delivery of anti-cancer drugs, especially if the treatment is administered intravenously. In particular, low delivery of oxygen will lead to hypoxic areas within the tumours. These areas are known to have a higher rate of mutations, and thus a stronger drive towards making the cancer cells metastasise.

Lastly, infiltrates of immune cells play an important role in the tumour environment. Cytotoxic immune cells, i.e. NK and CD8 T cells acts to control tumour growth through eradication of cancer cells, and high levels of these immune cells within a tumour is associated with a better prognosis. In contrast, other immune cell subtypes may add to the pro-inflammatory profile of the intratumoral microenvironment. If left uncontrolled, this pro-inflammatory milieu might promote tumour growth through the release of various growth factors. This process mimics wound healing, and tumours have been described as "wounds that do not heal" (22).

WHAT ARE THE CURRENT TREATMENTS FOR CANCER?

The main objective of cancer therapy is to remove all malignant cancer cells whilst sparing normal cells. Despite recent improved anti-cancer treatments, radical resection, which is the complete surgical removal of the tumour, is still considered the main option for curing cancer. However, the oncologists' toolbox also contains several other approaches, which, for the most part, are efficient in controlling advanced stage cancer. Thus, many patients can live with cancer for years.

The most common anti-cancer treatments include:

1. **Surgery** to cut out the cancer and a layer of surrounding tissue after its (ideally) early detection. Surgery is often combined with adjuvant/additional chemotherapy or other therapy before or after surgery not only to remove cancer cells but also to remove cancer cells located outside of the "resection margin" or in the circulation (23).
2. **Chemotherapy** is an umbrella term covering various drugs that may be used alone or in combination to target key features of cancer cell development. The history of chemotherapy started with nitrogen mustard compounds during WWII that caused remissions in lymphoma patients. The compounds were developed, and chemotherapy also became an adjuvant therapy to surgery and radiotherapy (24).
3. **Radiotherapy** causes cellular stress, directly and indirectly, through the generation of reactive oxygen species in all cells within the radiation fields, which leads to DNA damage, organelle damage and autophagy (25). Since normal cells have more robust antioxidant capacity compared to cancerous cells, radiation therapy aims to destroy the cancer tissue with radiation whilst sparing normal tissues (26).
4. **Hormone therapy** lowers sex hormones, such as oestrogens and androgens. It is used to treat hormone-sensitive tumours, such as some forms of prostate and breast cancers depend on the presence of hormone receptors (27).

5. **Immune therapy** involves strategies aimed at turning immune cells (e.g. T cells) against cancer cells (28). For example, several cancer cells express the membrane protein PD-L1, which is the ligand of programmed cell death-protein 1 that regulates the cytotoxic effect of T-Cells. Thus, in, e.g., malign melanoma cancers, cancer cells can "turn off" the immune response and evade detection by the immune system. By using a class of medication termed checkpoint inhibitors, the PD-L1 on the cancer cells is blocked, and the tumour is more easily detected by the immune cells.

6. **Stem cell transplantation** to replace blood cells, either from pre-harvested own stem cells (autologous) or from donor stem cells (allogeneic), following high-dosage chemotherapy in, e.g., blood and bone marrow cancers.

In summary, the most common treatments of cancer aim to remove all cancer cells by cutting the tumour out, target the cancer cells themselves, or to turn the microenvironment against the tumour to suffocate it by anti-angiogenic (i.e. blood vessel formation-stopping) treatment or by making the immune system attack the cancer more aggressively using immune therapy. The problem is that normal cells are also affected by some of these treatments. For example, cytotoxic chemotherapy not only kills dividing cancer cells but also other dividing cells such as gut and hair follicle cells. In turn, this leads to acute side-effects often seen in patients who receive these treatments, like loss of hair and gastrointestinal issues with chemotherapy.

In the targeting of cancer cells, one recent strategy is **precision, or personalised medicine** (29). The idea of precision medicine is to use "prevention and treatment strategies that take individual variability into account". Cancer is a good example of this approach because, as described earlier in the chapter, the driver mutations vary (cancer heterogeneity) even in the same type of cancer (16). Next-generation sequencing allows researchers to determine the driver mutations within an individual patient with cancer. An example of personalised medicine in cancer can be found in patients diagnosed with chronic myelogenous leukaemia, and the discovery of the Philadelphia chromosome. In 1959, researchers discovered that chromosome 9 was unusually long and that chromosome 22 was unusually short in patients diagnosed with myelogenous leukaemia. This deviation from "normal" chromosomes, was due to reciprocal translocation of genes between the two chromosomes. The *ABL1* gene, which encodes a tyrosine kinase, was translocated from chromosome 22 to chromosome 9 and formed the *BCR-ABL1* gene, which encodes for a specific tyrosine kinase that is always "on". In turn, this leads to impaired DNA binding and unregulated cellular division. Based on this knowledge, imatinib, a specific BCR-ABL1 tyrosine kinase inhibitor, which induces apoptosis in the cancer cells, was developed. This is an example of genetic precision medicine and has dramatically improved prognosis in patients diagnosed with Philadelphia chromosome-positive chronic myelogenous leukaemia (30).

However, it is not clear that precision medicine is the strategy that will cure other cancers. Mutated cancer genes are also active in other cells and targeted treatments will change the function of normal cells, which might cause side-effects. In addition, the rate of actual treatment progress is slow, testing drugs in patients with multiple driver mutations, and the costs of health care are important issues that might limit the success of precision medicine (31).

THE CANCER CONTINUUM AND CONSIDERATIONS FOR EXERCISE

Cancer arises from mutations in individual cells, which then start to grow, interact with surrounding tissue to form a microenvironment, and finally spread via the process of metastasis. From the patients' point of view, once they are diagnosed with cancer, the tumour has already formed, and in some cases also started to spread. From this point of verified cancer diagnosis, the patients move through a treatment trajectory, described as the cancer continuum, comprising primary curative-intended treatment (often surgery and/or radiotherapy), adjuvant therapy (chemotherapy, anti-hormonal therapy, targeted therapy), following which, most patients can move into their post-cancer survivorship phase (outlined in

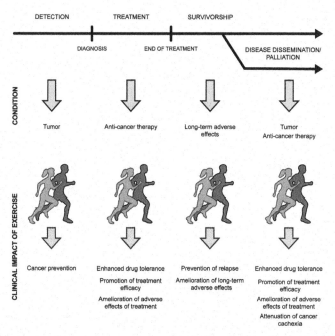

Figure 12.1 The cancer continuum comprises detection, primary treatment and cancer survivorship with cancer-free survival, or disease dissemination and palliation. Exercise has different roles through the cancer continuum, ranging for lowering the risk of certain cancers pre-diagnosis, help prepare patients for the treatment (i.e. prehabilitation), help reduce treatment side-effects such as cancer-related fatigue, and help lower the risk of relapse and other treatment late effects. Thus, each phase offers opportunities for exercise, but also presents unique treatment-specific boundaries. Reprinted from Hojman et al., (32), Molecular Mechanisms Linking Exercise to Cancer Prevention and Treatment, Cell Metabolism, 2018 Jan 9;27(1): pages 10-21. Hojman P, Gehl J, Christensen JF, Pedersen BK Copyright (2017 / 2018), with permission from Elsevier.

Figure 12.1). For patients with advanced stage cancer, there may not be treatments that can still cure the patient from cancer. Instead, they will receive palliative treatment, which intends to prolong life and (often) reduce pain. For some cancer patients, palliative treatment can go on for several years if the patient tolerates the treatment. This trajectory is characterised by several features, including hospitalisation, systemic therapy, treatment toxicities, mental distress, bodily pain and exhaustion (often called cancer-related fatigue), all of which can influence the patient' ability to exercise. It is therefore imperative to consider the different cancers, anti-cancer treatments, as well as the different phases of the cancer trajectory when delivering targeted exercise interventions in cancer settings.

In light of the cancer continuum, any exercise interventions should consider the timing of the intervention in relation to the phases of treatment. Several preventive trials have been conducted in healthy people at high risk of developing breast cancer. Here, the aim was to reduce breast cancer incidence using exercise training through a reduction in systemic risk factors, in particular sex hormones levels (e.g. oestrogens). Thus, these trials are considered cancer preventive trials. These trials demonstrated that modest reductions in sex hormone levels can be obtained with exercise training, especially if this is associated with weight loss. However, if this targeting of breast cancer risk factors using exercise translates into lower breast cancer incidence remains to be determined through longer-term definitive trials.

Numerous exercise intervention trials have been conducted during adjuvant chemotherapy. Again, breast cancer patients comprise the major population of patients participating in these trials. Here, the focus has been on symptom control. However, next-generation studies will and should focus on decreasing treatment toxicity (e.g. cancer-related fatigue, impaired quality of life or QoL, lymphedema, neuropathies and pain) and dose reductions (i.e. increasing the number of patients receiving the indented drug dosage). For some of these endpoints (e.g. reduce cancer-related fatigue and prevention of QoL impairments), it is already well established that exercise can make a difference. However, in some other endpoints (e.g. treatment tolerance and prevention of treatment dose reductions) the level of scientific evidence is less strong. In fact, preclinical data indicate that aerobic exercise may increase tumour angiogenesis (33, 34), which in theory might increase the efficacy of anti-cancer drugs administered to the circulation. Evidence from clinical trials is, however, still lacking. At the other end of the cancer continuum, exercise training might help cancer patients where cancer has already spread, to reduce treatment toxicity, ameliorate side-effects and improve survival. However, this end of the trajectory is still poorly studied, although several studies show that patients with advanced stage cancer are capable of undertaking exercise training.

EXERCISE FOR CANCER PREVENTION AND TREATMENT: INTERVENTION AND OBSERVATIONAL STUDIES

Can cancer patients exercise, and does it affect their symptoms?

Thirty years ago, the dogma was that cancer patients should rest and avoid physical exertion during their treatment. This dogma was first challenged in the work by Winningham and MacVicar (1988), who conducted the first clinical exercise study in women undergoing adjuvant chemotherapy. The results showed that exercise was safe, feasible, and also associated with improvements in physical function and patient-reported symptomology (e.g. lower feeling of nausea amongst exercisers than control subjects). Since this pioneering study, a plethora of reports has demonstrated the safety and feasibility of exercise in cancer patients, and its capacity to improve treatment-related adverse effects during and following primary anti-cancer therapy. To exemplify this broad approach, the range of endpoints in the conducted exercise intervention trials in cancer patients is enormous. It includes direct physiological effects of the exercise training, such as; fitness levels, oxygen consumption, muscle mass and strength, across exercise-related outcomes such as functional capacity and body composition, to biological and psychosocial outcomes, including QoL and levels of fatigue, depression, anxiety, empowerment and self-esteem. No completed studies have yet been undertaken to address significant clinical outcomes like disease recurrence and cancer-specific mortality. However, studies are currently being conducted to address these clinically important endpoints.

For the vast majority of these findings, cancer patients, and cancer survivors may respond as any healthy individual engaging in exercise training. The underlying mechanisms for these adaptations are likely to follow what is known from the general exercise physiology literature. Yet, in particular for adaptations in muscle mass, anti-cancer therapy has been shown to affect the adaptive training response. For instance, in prostate cancer patients undertaking androgen deprivation therapy, which is associated with loss of lean mass, participating in resistance or combined resistance and endurance training, has highly variable muscular responses and shows large individual differences. Where some patients respond as expected, whilst other patients do not respond at all (i.e. with no changes in muscle mass) despite large training volumes. Whilst the terms "responder" and "non-responder" (or "less-responder") are well described in the exercise literature and illustrate the heterogenous response to similar exercise regimens within a cohort of study participants, one might speculate that there is similar heterogeneity in the response to anti-cancer treatments between patients. In turn, this could help explain the highly heterogenous responses to exercise observed in cancer patients undertaking treatment, however, it requires more evidence to confirm this. As for a more direct effect of exercise on cancer, research is still in the early phases of understanding this interaction, and the role that exercise could play as a cancer treatment.

Exercise and cancer incidence and progression – data from observational studies

Currently, observational studies suggest that regular physical activity and exercise training lowers both the risk of cancer and the risk of disease recurrence and overall mortality. Conservative estimates state that an individual's physical activity level may lower the risk of colorectal cancer, postmenopausal breast cancer and endometrial cancer (35). However, in the largest observational study so far, involving up to 1.44 million people, Moore et al. showed that higher levels of physical activity were associated with a lower risk for several types of cancer, including cancers of the oesophagus, liver, stomach, kidney, blood, head and neck, rectum and bladder (4). There are trends linking higher physical activity to a lower risk for other cancers, except for melanoma (skin cancer), where physical activity was associated with an increased risk (4). This might be explained by the fact that physically active individuals are more frequently outside, where intermittent sun exposure is associated with a higher risk of melanoma (36).

Along similar lines, Fridenreich et al. reported reduced risk of recurrence following cancer diagnoses across 26 studies in breast, colorectal and prostate cancer by comparing incidence rates amongst the most versus the least active patients (37).

Therefore, the evidence suggests that increased physical activity can help lower the risk of cancer incidence. Further, exercise after the time of diagnosis seems to help cancer patient's survival. The first study demonstrating such an inverse relationship between self-reported physical activity level, and cancer-specific and overall mortality risk after a diagnosis of breast cancer was identified in a study by Holmes et al. (2005). Similar inverse relationships have now been documented in epidemiological studies across several malignancies (i.e. types of cancer), with the strongest evidence for exercise reducing cancer associated mortality in colorectal, breast, endometrial and prostate cancer (38). A systematic review suggests fairly consistent evidence that physical activity before or after diagnosis is associated with lower cancer-specific and overall mortality in breast and colorectal cancer (5). In a large prospective study of 293,511 cancer patients, researchers found that those with higher physical activity levels pre-diagnosis had a lower risk of death from colon, liver, lung cancer and Hodginks lymphomas (6). Finally, in a recent meta-analysis amongst 23,041 breast cancer survivors, higher levels of self-reported physical activity were associated with a reduced the risk of all-cause mortality by 42% (random effects HR=0.58, 95% CIs: 0.45–0.75) compared to women that reported low levels (39). Although, any causal relationship cannot be established from cross-sectional studies (e.g., some survivors may be more physically active simply because they are healthier or less affected by their cancers and/or treatments), this indicates that increased physical activity levels may improve survival amongst cancer survivors.

One caveat is that almost all the observational evidence for the anti-cancer effects of physical activity or exercise is based on associations, and associations do not prove causation. Thus, the key question is: does exercise directly affect the development and growth of a tumour? Or is the anti-cancer effect of exercise indirect, where we see improvements in the function and performance of other body tissues with exercise? For example, the exercise-induced reduction in body fat, which is a risk factor for cancer (40), or are there other mechanisms at play? Lastly, is exercise always good or can it be detrimental under certain circumstances, for example, because it may interfere with an acute drug treatment?

What are the anti-cancer mechanisms of exercise?

Experimental evidence from animal studies does demonstrate that exercise training can control tumour growth. One of the first studies is from 1943 (Rusch and Kline, Cancer Res 1943), where Rusch and Kline placed mice in rotating cylindrical cages for 16 hours a day and found that the growth of transplantable fibrosarcomas was significantly slower in the trained mice. Nine years later, Rashkis (Rashkis, Science 1952) showed that swimming in glass jars between 1.5 and 4.25 hours per day slowed tumour growth of Ehrlich ascites tumours in mice. In recent years, animal studies into the effect of exercise on cancer have gained momentum, as reviewed comprehensively by Ashcraft et al, (81).

Indeed, the effects of exercise on tumours can be examined in the three fundamental developmental events of cancer: (1) mutagenesis and establishment; (2) tumour growth and (3) metastasis. These will be discussed below.

Exercise effect on mutation rate and tumour establishment

In line with the reduced risk of most cancers, as seen in observational studies, many animal studies have shown that exercise training reduces the incidence of cancers. Here, the best models include genetic tumour models, where mice carry oncogenic driver mutations, which across time results in tumour formation within the animals. Yet, chemically induced tumours may also lead to various oncogenic mutations, which further will lead to tumour formation over time. Using different modes of exercise, including voluntary wheel running, forced treadmill running and swimming, most studies demonstrate the exercise training decreased the incidence and multiplicity of tumours (Colbert et al, 2009 and McClellan et al, 2014).

Currently, there is little evidence to suggest whether this is caused by an effect of exercise on mutagenesis. However, as almost 2/3 of all cancers are caused by mutations whose frequency is determined primarily by the rate of cell proliferation, any exercise-induced change of cell proliferation may affect how likely it is that a cancer develops. Another cause of mutations is genomic instability which might be linked to telomere length. Telomeres are the caps at the ends of our chromosomes, and these become shorter each time a cell divides ultimately meaning cells become senescent and stop dividing. There are a few studies that suggest exercise training affects telomere length. For example, Puterman et al (2010) demonstrated that exercise might inhibit telomere shortening (41), and later Sjögren et al demonstrated that less sedentary behaviour was associated with telomere lengthening (42). However, causality between lower incidence of cancer and increased telomere length in exercisers remains to be established (43).

In contrast, exercise training may be able to affect the rate of mutations caused by external factors. Although criticised by some researchers, it is estimated that 29% of cancer-causing mutations are linked to external factors, other than bad luck (44). In cancer prevention, less exposure to environmental carcinogens and smoking cessation are often highlighted as external factors leading to cancer. But lifestyle factors such as a healthy diet and sufficient amounts of physical activity are important, modifiable external factors. For example, in a seminal paper published in 2008, McTiernan and co-workers showed that physical activity was linked to lower risk of cancer, based on positive changes in circulatory cancer risk factors, such as sex hormones, insulin and IGFs, as well as inflammatory markers (45).

However, the chance is that cancer-causing mutations occur, and minuscule nascent tumours develop. Here, our immune system plays a major role in clearing nascent tumours before they establish into macroscopic detectable tumours. In fact, the levels of cytotoxic T cells and natural killer cells (NK cells) within tumours are associated with a better prognosis (46, 47). However, cancer cells often express immune cell inhibiting ligands, such as PD-L1 that regulate the cytotoxic activity of immune cells or may secrete high levels of TGF-β and other cytokines that suppresses the function of infiltrating immune cells (48). Exercise training has a paramount impact on our immune function, priming the innate immune defence to act quickly and efficiently and with a fast resolution of the induced inflammation. Voluntary wheel running in rodents has been shown to induce immune cell redistribution from circulation to tumours and to increase the cytotoxic activity of infiltrating NK-cells (49). Furthermore, in a later paper Rundqvist and co-workers showed that wheel running was associated with reduced tumour burden in mice that were injected with I3TC cancer cells, and that the anti-cancer effect of running was abolished upon depletion of CD8+ T (immune) cells (50), suggesting that the anti-cancer effect of exercise requires functioning immune cells. Overall, some of the cancer risk-reducing effect of exercise may be attributed to more favourable blood profiles and to immune cell redistribution, mobilisation, and activation, leading to reduced growth stimuli for cancer cells and improved detection of nascent tumours.

Exercise effect on tumour growth

The vast majority of preclinical studies have investigated the effect of exercise on tumour growth. Research in this area has been conducted since the 1940s when Rusch and Kline investigated the effects of forced running on tumour growth in mice (51). It was already known that energy restriction (i.e.

reduced food intake) had inhibitory effects on tumour development, and Rusch and Kline hypothesised that increased energy expenditure would have similar effects. The researchers installed rotating floors in the cages for exercising mice and not in the control cages and observed tumour growth in subcutaneously injected cancer cells. Following the exercise protocol, the researchers observed slower growth rates in the exercising mice. Since then, numerous studies have confirmed that various exercise modalities inhibit tumour growth. In these studies, the macroscopic measurement of tumour growth is a balance between cell death and cell proliferation. Therefore, a range of events leading to reduced tumour growth may occur.

Slower cellular proliferation and induction of apoptosis/cell death

By incubating cultured cancer cells from two different prostate cancer cell lines (LNCaP and LN-56) with serum from trained and untrained men, Leung et al. investigated the effects of the systemic/secreted factors as a result of exercise on cellular division (i.e. tumour growth) and apoptosis (52). The LN-56 prostate cancer cell line (derived from LNCaP cells) was rendered non-functional for the gene p53 by the expression of a dominant negative fragment of the gene. Interestingly, when LNCaP cancer cells were incubated with exercise serum, there was a 30% reduction in growth and proliferation cell nuclear antigen protein (a marker of cell cycle activity and thus cell replication), compared to cells incubated with serum from the untrained men. Suggesting secreted factors in the serum following exercise could slow cancer cell growth rates. In addition, apoptosis (programmed cell death) was increased by 371%, which was accompanied by a doubling of p53 abundance, meaning that secreted factors from exercise not only slowed cancer cell growth but also helped kill some of the cancer cells by committing a type of cell suicide (apoptosis). On the contrary, when LN-56 cells (non-functional for p53) were incubated, no significant changes occurred, perhaps suggesting that functional p53 was crucial for the exercised serum effect on reducing cancer cell growth and inducing apoptosis. Later, the same group established that this effect was mediated by the higher circulating levels of IGF-I found in the untrained men (53). In the context of this book, it is worth mentioning that the acute elevation of IGF-I levels following exercise, which facilitates muscular adaptation, likely serves no cancer risk. However, chronically elevated IGF-I levels have been linked to increased risk of several cancers (54), as IGF-I stimulates cancer cell proliferation and inhibits apoptosis (55).

Increased p53 was also observed in exercised mice by Higgins et al, when mice were injected with A549 lung adenocarcinoma cells, and randomly allocated wheel running or no wheel running (56). Exercised mice showed a slower growth rate, initiated by an upregulation of tumour suppressor, p53 and a three-fold increase in cells expressing Caspase 3 activity. Increased Caspase activity, which is involved in apoptosis itself, were also suggested as a mechanism for tumour growth inhibition when breast cancer cells (MCF-7 cells) were exposed to serum from exercised mice (57).

In summary, exercise has been shown to influence growth in tumours both in vivo and in vitro. Reasons for the slower growth rates seem to be attributed to reduced cell cycle activity, but in particular, increased apoptosis. Mechanistically, exercise seems to improve cell cycle control through upregulation of p53, which is known to maintain genomic integrity by responding to cellular stress and DNA damage through cell cycle arrest and DNA repair or induction of apoptosis.

Immune destruction

For cancer cells to survive and proliferate, they must evade the host immune system (58). In fact, tumour cells use different strategies to diminish immune cell recognition and infiltration (48, 59), including inhibiting receptor ligands and the secretion of TGF-ß and other immune-suppressing factors. However, it is now well known that both an acute bout of exercise and chronic exercise training alters the quality and quantity of several immune cell types in the circulation (60) and in various bodily tissue compartments (61, 62). Research has suggested that exercise might also inhibit tumour development through improved immune cell infiltration and increased cytotoxic activity. For example, Pedersen and co-workers showed that wheel running decreased tumour growth in mice through an exercise-dependent mobilisation of cytotoxic natural killer cells (NK-cells) (49). Furthermore, Abdalla and co-workers demonstrated that exercise promoted a shift towards an anti-tumour response pattern of immune cells (63). Specifically, exercise lowered the number of leukocytes and macrophages expressing pro-tumour activity (i.e. cells expressing IL-4, IL-10 and TGF-β), and increased the frequency of cells showing anti-tumour capabilities (i.e. cells expressing interferon-γ, IL-2, IL-12 and TNF-α). Although there is a potential, and a strong rationale, for exercise to show cancer suppressive effects through immune modulation, mechanisms and any clinical implications are currently unclear.

Altered metabolism

As described earlier in this chapter, it has long been known that tumour cells have an altered metabolism compared to normal cells. Tumours reprogramme their metabolism so that the rate of glycolysis is comparatively high even under aerobic conditions. A key function of this reprogramming is that cancer cells can use glycolytic intermediates and other metabolites as substrates necessary for rapid cell proliferation (64). In addition, due to the chaotic and immature tumour vasculature of solid tumours, many cancer cells survive in a hypoxic milieu (65), which could also explain the energy substrate preference. A consequence of anaerobic energy turnover is the accumulation of lactate, which, in turn, causes a number of pro-tumorigenic adaptations, such as promotion of an immunosuppressive environment, increased tumour cell invasion with the formation of metastasis, and subsequently this correlates with worse clinical outcomes (64, 66).

Exercise, often being a high energy-consuming activity, induces dramatic changes in the body's metabolism. Realising that tumours are not isolated entities in the body, such activity also alters intratumoral metabolism. For example, seven weeks of treadmill running inhibited tumour growth in mice injected with MC4-L2 breast cancer cells (67). The slower growth rate was accompanied by decreased lactate levels, monocarboxylate transporter 1 and lactate dehydrogenase A (LDH-A). The later has been demonstrated to be upregulated in breast cancer cells as a result of hypoxia (68, 69). Thus, it can be speculated whether the lactate reduction observed following seven weeks of running in mice, was a consequence of lower glucose availability, forcing the cancer cells to shift to slower energy substates (e.g. fatty acids), or was a result of improved vascularisation and increased intratumoral blood flow as discussed below.

There are several preclinical studies in the literature that suggest tumours are susceptible to an exercise-induced shift in metabolism. In particular, tumour cells with an inherent high metabolism seem to be affected by exercise. In simple terms, this could be interpreted as a reprogramming of the cancer cells due to low energy substrate availability. But how this affects tumour growth and the metastatic rate is not fully understood (32).

Angiogenesis and normalisation of the tumour microenvironment

Tumour blood vessels have been described as "abnormal", meaning that they are both structurally and functionally different from healthy tissues. Consequently, tumours are inadequately oxygenated and consist of regions with clinically relevant areas of hypoxia (70), which is associated with aggressive tumour cell phenotypes and increases the risk of metastases (71). Large hypoxic areas are associated with poor response to radiotherapy (72). In addition, several intravenously administered cancer treatment modalities are dependent upon sufficient blood perfusion in order to deliver its anti-cancer effects to the tumours (71, 73).

Interestingly, preclinical studies have shown that aerobic exercise promotes blood perfusion in tumours. By applying the radionuclide-tagged microsphere technique, McCullough et al. evaluated acute effects of aerobic exercise on prostate tumour blood flow and vascular resistance in vivo in male prostate cancer-bearing rats (74). The results showed increased tumour blood perfusion during exercise, which was contrary to the general blood flow of the prostate tissue. The authors explain the increase in tumoral blood perfusion by increased vasoconstriction in the healthy tissue, leading to increased blood flow through the abnormal tumoral blood vessels. The increased blood flow also diminished tumour hypoxia, which may be clinically relevant as it may induce a shift towards less aggressive phenotypes. Jones and co-workers explored this hypothesis. Using an orthotopic mouse model of prostate cancer, they investigated the effects of voluntary wheel running on tumour growth and metastasis. They also explored underlying mechanisms for increased growth control by exercise (33). Here, tumour blood perfusion was assessed by MRI. Although the primary tumour growth rates were comparable between exercising and sedentary mice, increased blood perfusion was associated with lower expression of prometastatic genes. Also, the exercise-induced stimulation of HIF-1 and increased VEGF expression (see Chapter 11) was associated with enhanced tumour perfusion and increased maturation status of tumour blood vessels. Here it is worth mentioning that tumour hypoxia activation of HIF-1 normally results in pathological angiogenesis, abnormal vessel formation and impaired blood perfusion (75, 76). Thus, exercise-induced activation of HIF-1 may have contrary effects, leading to increased blood perfusion and a shift to less aggressive cancer cell phenotypes.

In addition, it is well known that exercise alters the blood content of several potential signalling molecules, including metabolites (77) and a variety of proteins (78) (**Figure 12.2**). If tumour cells express receptors for these exercise-altered molecules, the increased tumour blood perfusion might lead to increased impact on tumour cells.

Likewise, increased tumour blood perfusion may also have therapeutic potential in drug delivery to tumours, as several cancer therapies are intravenously administered. For example, Betof et al. investigated the effects of voluntary wheel running on tumour growth, blood perfusion, hypoxia, and other factors related to angiogenic and apoptotic processes in mice implanted with syngeneic 4t1 murine breast cancer cells orthotopically into the dorsal mammary fat pad (34). Mice randomised to wheel running showed slower growth rates compared to sedentary mice, with an increased frequency of tumour cells expressing markers of apoptosis. Furthermore, tumours collected from exercising mice showed increased microvessel density and reduced areas of hypoxia. When exercising and sedentary mice were treated with chemotherapy (cyclophosphamide), there was an increased treatment effect. The authors speculated that

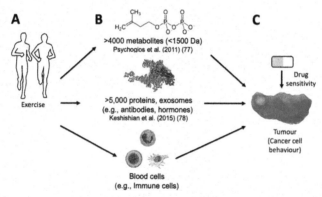

Figure 12.2 Blood could be a link by which (A) exercise can affect a tumour, a mix of **(B)** metabolites, proteins and blood cells. **(C)** Exercise-conditioned blood then perfuses a tumour where it might affect cancer cell behaviour, drug sensitivity and immune checkpoints. Original figure with blood cells images taken from Grey's Anatomy that do not require permissions under the Creative Commons CC0 License, https:// creativecommons.org/publicdomain/zero/1.0/; and/or Creative Commons Attribution-ShareAlike License, https://creativecommons.org/licenses/by-sa/4.0/. Protein structure and metabolite images taken from Wikipedia that does not require permissions under a Attribution-ShareAlike 3.0 Unported licence, https://creativecommons.org/licenses/ by-sa/3.0/, and the public domain copyrights, respectively.

this resulted from increased tumour blood perfusion, which led to an increased drug uptake into the cancer cells. Similar results were also reported by Schadler et al., who also included a mechanistic explanation for why exercise could help normalise the intratumoral vasculature (79). Shear stress, the mechanical stimuli exerted on the blood vessel walls, initiated remodelling of the non-functional tumour vessels into more normal vessels through calcineurin-NFAT-TSP1 signalling in the endothelial cells. Consequently, moderate exercise in combination with chemotherapy showed greater anti-tumour effects than chemotherapy alone.

In summary, aerobic exercise may lead to increased tumoral blood perfusion, which may initiate blood vessel remodelling leading to normalisation of the tumour vasculature. This may result in decreased area of hypoxia, which is associated with more aggressive cancer cell phenotypes and resistance to anti-cancer treatment modalities. Exercise mediated increases in tumoral blood perfusion have been shown to induce a shift to less aggressive cancer cell phenotypes and increase the efficacy of chemotherapy in different mouse models. It is, however, worth mentioning that any clinical evidence for such mechanisms is scarce (80).

Exercise effect on rate of metastasis

Several mouse studies have attempted to investigate the effect of exercise on metastasis. Yet, the animal models for this process are not very advanced and there are several concerns with these studies, which have been nicely discussed by Ashcraft et al. (81). In short, the preferred model for studying effects of exercise on metastasis seem to be by intravenously injecting cancer cells, which evaluates the ability of cancer cells to survive in the circulation and to colonise bodily tissues. However, it does not take into account the earlier steps in the metastatic cascade. Furthermore, the exercise modality may also influence the outcomes, as forced treadmill running and swimming may induce significant stress in the animals, which may, in turn, affect tumour biology.

Nevertheless, for metastasis, exercise training indicates a protective effect, and some mechanisms have been suggested. First, the whole metastatic process could be promoted by intratumoral hypoxia, creating a hostile environment, which promotes mutagenesis and cancer cell migration. Given that exercise training can normalise the tumour microenvironment and reduce hypoxia, as discussed above, this drive to metastasise might decline. Second, Ju et al. showed that exercise training regulated the adhesion molecules Cadherins in ApcMin mice, which may prevent circulating cancer cells to colonise distant tissues (82).

Thus, there is a theoretical rationale for exercise to reduce the risk of metastasis, but there are major methodological limitations to be able to conclude this with absolute certainty.

EXERCISE AND THE CONTROL OF CANCER TREATMENT SIDE-EFFECTS

So far, we have been discussing the effects of exercise on tumour characteristics in preclinical studies. However, results obtained in animal models may not always reflect the effects seen in human cancer patients. So, what are the benefits of exercise in cancer patients? Does exercise influence common treatment-induced side-effects?

First and foremost, it is important to realise that the side-effect types and burden vary significantly between different cancer types, cancer sites and with different treatments. Furthermore, the side-effect burden may show huge variation from one patient to the next even when treated similarly. Thus, this section is limited to a few well-known side-effects that range across different cancer sites and treatment options.

Cancer patients are at risk of early mortality, which is primarily attributed to second malignancy and cardiovascular diseases (83, 84). It has long been stated that exercise may improve survival in cancer patients. Historically, there has been limited evidence to support this claim. However, recent data from prospective cohort studies and randomised controlled trials in cancer populations with a high risk of mortality have provided new insight into this area. For example, amongst 5,689 adult survivors of childhood cancer, exercise was associated with a 40% reduced risk of all-cause mortality in survivors that maintained high-levels of exercise (self-reported >9 MET-hours per week), or that increased their exercise behaviour (risk ratio=0.60, 95% CI: 0.44–0.82) across eight years (85). However, these results are limited by the self-reported nature of the physical activity data, and it is also well known that exercisers may also make other favourable lifestyle choices. Therefore, the effects of the exercise itself might therefore be inflated. To begin to address this issue, a 2020 meta-analysis included eight studies in breast cancer, lung cancer, patients undergoing allogeneic hematopoietic stem cell transplant and mixed cancer groups, which used exercise as the study intervention, and assessed mortality following the intervention as an endpoint (86). Across the studies, exercise was shown to lower the risk of mortality by 24% (risk ratio=0.76, 95% CI: 0.40–0.93). Any causal relationship between exercise and survival outcomes is, however, limited given retrospective study design of these studies. It is also worth mentioning that none of the studies were designed to investigate the effect on mortality. In summary, the effects of exercise on mortality following cancer diagnoses are still unknown. However, larger-scale randomised control trials with sufficient statistical power to explore such endpoints and improved activity monitoring may provide more accurate knowledge in this area in the coming years.

Health-related QoL (quality of life), which is defined as an individual's self-perceived physical, mental, social, and functional health, and is often summarised by a common score across the different domains in validated questionnaires, is critical in cancer care. A cancer diagnosis, and also going through cancer treatment often affect patient's QoL. According to an extensive review of exercise and cancer literature, health-related QoL is the most frequent outcome reported in exercise trials in cancer settings (80). A meta-analysis of pooled data sets from 34 individual studies, including 4,519 patients participating in different exercise trials in cancer settings, found statistically significant benefits of exercise on QoL, although the observed effects were small ($\beta = 0.15$, 95% CI = 0.10; 0.20) (87). The meta-analysis further explored potential moderators of the response, including; demographic (e.g. age, sex, marital status and education), clinical (e.g. body mass index, cancer type and presence of metastasis), intervention-related (e.g. intervention timing, delivery mode and duration, and type of control group) and exercise-related variables (exercise frequency, intensity, type, time). The only significant moderator was whether sessions were performed under supervision or not. Thus, there might be other factors than those strictly related to improved fitness that may induce improvements in QoL. The relationship with the trainer might be just as important for this specific outcome.

One of the most prevalent and troublesome side-effects from cancer, and different cancer treatment options, is cancer-related fatigue. This is described as a distressing, persistent, subjective sense of physical, emotional and/or cognitive tiredness or exhaustion. Although it is a transient phase for most cancer patients, it may persist for years post-treatment in some survivors and approximately 30% of breast cancer survivors (88), and lymphoma survivors (89) still report symptoms ten years post-diagnosis. Exercise has been shown to lessen the burden of cancer-related fatigue to some extent, as participants randomised to exercise groups typically report less symptoms and less distress compared to participants in the non-exercising control groups in randomised controlled trials undertaken during or shortly after treatment (90). It is, however, worth mentioning that only small to moderate effects sizes are reported in the meta-analysis. Furthermore, it is still unknown if exercise influences more the chronic phases of fatigue (i.e. if the condition persists for 6 months or longer following diagnosis) as very few trials have investigated this to date. In summary, it seems that exercise is one of few strategies that may help cancer patients cope with cancer-related fatigue, but whether these effects of exercise apply at later stages of cancer survivorship is largely unknown.

A patient's cardiorespiratory fitness, measured as VO_{2max}/VO_{2peak}, has been reported to decrease between 5% and 26% during various systemic combinational treatment regimens in different cancer settings (91–93). This may recover shortly after treatment cessation (e.g. if the lower VO_{2peak} is caused by reduced hemoglobin /Hb levels during treatment and the bone marrow is still intact). In other cases, the VO_{2peak} may not recover (91, 94–96). It is well known that aerobic exercise training holds the potential to improve cardiorespiratory fitness. In a 2018 meta-analysis, where 48 individual randomised controlled trials, comparing the change in cardiorespiratory fitness between cancer patients in an exercise group and a control group were included. The analysis showed a difference in mean change in VO_{2peak} of 2.13 mL $O_2 \times kg^{-1} \times min^{-1}$ (95% CI, 1.58 to 2.67). Such improvements may be of clinical importance, as impaired cardiorespiratory fitness correlates with higher symptom burden (97) and poorer clinical outcomes (91, 98).

Similar to cardiorespiratory fitness, muscle mass and – function is pivotal for locomotion and physical function. In more cancer-specific settings, patients muscle mass has received more attention over the last decade. Low muscle mass has been shown to correlate with a higher incidence of postoperative complication, poor prognosis and increased length of hospital stay post-surgery across different cancers (99, 100). Low muscle mass also increases the risk of drug toxicity during chemotherapy (101), and loss of muscle mass during treatment has been associated with increased frequency of serious adverse events and mortality in patients with unresectable colorectal cancer (102).

As discussed earlier, there are numerous treatment modalities for different cancers. Some of these treatment options have the potential to interfere with muscle mass and adaptation to exercise training. For instance, cisplatin, a chemotherapy drug used in different cancers (e.g. testicular cancers and lung cancer), may increase muscle protein breakdown through increased activity in the ubiquitin-proteasome system (103). Radiation therapy may interfere (locally) with the available pool of satellite cells (104). Moreover, androgen deprivation therapy may lead to decreased muscle protein synthesis at rest (105) and is known to induce loss of muscle mass (106). Glucocorticoids are also used under different settings in cancer care, for example in managing side-effects such as pain and nausea and as a part of treatment for lymphoid cancers (107, 108). Glucocorticoids increases the expression of a protein regulated in development and DNA responses-1 (REDD1), which is a potent inhibitor of mTOR (109), and activates the UPS and results in increases in MAFbx and MuRF-1 mRNA, leading to reduced protein synthesis, increased protein breakdown and ultimately muscle atrophy with concurrent skeletal muscle atrophy (110, 111) (see Chapter 7 and 8). As a result, dose-dependent rates of muscle mass loss have been reported during glucocorticoid treatment (112).

Although there are likely to be direct effects of such treatments on muscle cells as outlined in **Figure 12.3**, this is still poorly described in the literature to date. In clinical practice, both loss of appetite (leading to anorexia-like conditions) and fatigue may contribute to reduced energy intake and reduced activity levels. In turn, this may confound research into mechanisms behind treatment-induced loss of muscle mass in patients.

Does exercise during concurrent cancer treatment intervene with the normal effects of exercise? Currently, such questions are difficult to answer, as there are only three studies available in the literature that included muscle biopsies pre- and post-intervention during anti-cancer treatment. First, Christensen and co-workers explored the effects of resistance training during cisplatin-based chemotherapy treatment for germ cell cancer on muscle outcomes. Interestingly, participants in the non-exercise control group (i.e. those who received standard care) showed reduced mean muscle fibre cross-sectional area and demonstrated a trend towards a shift in muscle phenotype to a larger proportion of type II fibre types (113). The high-intensity resistance training performed throughout treatment by the intervention group did not induce significantly different changes when compared to the control group. Second, Mijwel and co-workers compared two different exercise programmes on biochemical, histological, and protein level outcomes in skeletal muscle tissue and satellite cells. This included, aerobic exercise alone or in combination with resistance exercise, in women diagnosed with breast cancer undergoing chemotherapy (i.e. anthracyclines (epirubicin) or taxanes, or a combination of the two) (114). During the 16-week chemotherapy period, citrate synthase activity, muscle fibre cross-sectional area, the number of capillaries per fibre, and the proportion of type I fibres measured in resting muscle biopsies were reduced in the group undergoing chemotherapy with no exercise, measures that were preserved by performing exercise training (both aerobic and aerobic plus resistance groups). Also, in accordance with the training principle of specificity, aerobic exercise alone led to increases in electron transport chain protein levels, and an increased number of satellite cells were observed in the group that performed resistance training in combination with aerobic exercise training.

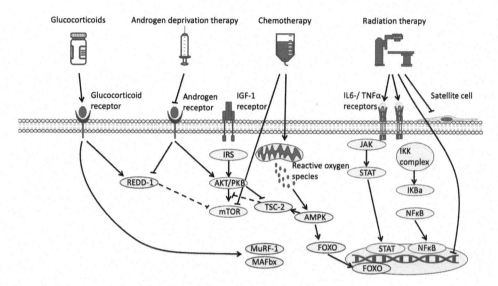

Figure 12.3 Suggested pathways by which some anti-cancer therapies may hamper anabolic signalling (i.e. reduced mTOR activity) and induce catabolic signalling (i.e. increased expression of MuRF-1/MAFbx and translocation of FOXO).

Third, Nilsen and co-workers explored the effects of high-volume resistance training for 16 weeks compared to usual care in men with prostate cancer treated with androgen deprivation therapy (ADT) in skeletal muscle. Here, exercise led to an increased cross-sectional area of type II muscle fibres but no change in the number of satellite cells. Furthermore, the results may also indicate a fibre-type-specific effect of ADT. There was a trend towards a reduced cross-sectional area of type I fibres within the control group and an increased number of myonuclei in the type I fibres (that did not increase in size) within the exercise group. However, this remains speculative, provided the limited number of studies available.

In summary, cancer treatment may induce detrimental side-effects in skeletal muscle. Although exercise holds the potential to counteract such side-effects, the current literature is limited in the number of studies and also suffers from methodological challenges such as small sample sizes. Furthermore, given the complexity of different treatment modalities in cancer care, chances are that the knowledge base in this area will remain limited for the overseeing future.

WHAT IS CANCER CACHEXIA AND CAN EXERCISE BE USED TO TREAT IT?

Cancer cachexia is a condition where cancer patients lose body weight as a consequence of a specific loss in muscle mass, with or without simultaneous loss of fat mass (115). It has been estimated that cachexia impacts as much as 60% of all US cancer patients but is likely more prevalent in advanced stage cancers (116). Although the condition may be partially mended by nutritional support, such strategies do not seem to reverse or prevent such losses. The term "cachexia" descends from Greek and can be translated into "poor physical state", which is suited for the gradual loss of physical function reported in these populations. In addition to poor physical function, impaired QoL, increased resistance towards cancer therapies, and increased risk of mortality has been reported (115).

The mechanisms behind cachexia-induced muscle loss are likely multifactorial and are still under investigation However, at least two leading hypotheses seem plausible: first, malignant tumours take up circulating nutrients and amino acids, resulting in energy and amino acid deficits for other tissues such as the skeletal muscle mass (117). Mathematical modelling of tumour energy costs in metastatic settings suggests that the high glucose turnover resulting from anaerobic metabolism can result in cachexia. Second, tumour cells secrete numerous catabolic factors. These may lower muscle protein synthesis and activate proteolysis in skeletal muscle, both through the ubiquitin-proteasome system and through autophagy (118). Different cytokines and pro-inflammatory molecules, generated through tumour-immune system interactions, leads to an increase in circulating stress hormones (e.g. adrenalin, cortisol and glucagon), which results in increased resistance towards insulin and other growth factors in muscle and therefore impaired anabolism (119, 120). Furthermore, transcription of autophagy- and ubiquitin-proteasome system related genes are directly activated by several pro-inflammatory factors originating from the tumour, immune cells or both (118). Also, the low circulating levels of amino acids, as a result of the high tumour uptake, may further activate muscle protein breakdown (121, 122). In summary, immune cell-tumour cell interactions leads to lowered muscle protein synthesis and increased muscle protein breakdown (**see Chapters 7 and 8**), resulting in body weight loss specifically caused by loss of skeletal muscle mass.

From a clinical perspective, it is still unknown whether exercise may counteract or prevent cancer cachexia in patients. A 2021 Cochrane review by Grande and co-workers only identified four clinical trials evaluating the effect of exercise in cachexic settings (123). The authors conclude that together these four studies offer minimal information on the effectiveness, acceptability and safety of exercise in these settings and that the body of evidence is shallow with a high risk of bias. There are, however, studies underway at the moment that may help shed more light on this area in the future.

Summary

Cancer is one of the leading causes of death in the Western world. Due to improvements in early detection and treatment, relapse-free survival is improving for several cancer forms. Nevertheless, cancer treatment is still associated with troublesome side-effects and late effects (i.e. "side-effects" occurring or persisting beyond a year after completing treatment).

Exercise may reduce the risk of developing several cancers. Emerging evidence from preclinical studies have improved our understanding of the potential role of exercise in risk of cancer incidence. These mechanisms include improved genomic control and increased cancer cell apoptosis, improved tumour vascularisation and reduced tumour hypoxia, altered cancer cell metabolism, and improved immune detection. Exercise may also improve the circulatory environment, in terms of lower levels of stimulating hormones and growth factors. Furthermore, exercise is a valuable strategy during cancer treatment, as it may help relieve the symptom burden (i.e. less cancer-related fatigue and improved QoL) and help patients maintain their physical function. In addition, exercise prior to cancer treatment (i.e. prehabilitation) may help prepare patients for major cancer surgery. Thus, exercise plays an important role in all phases of the cancer continuum.

Review questions

1. What is cancer, and how does it occur?
2. What are the most common ways of treating cancer?
3. In which way may exercise reduce the risk of developing cancer?
4. In which way may exercise improve the effects of intravenously administered anti-cancer therapies?
5. How can exercise help cancer patients going through treatment?
6. What is the potential effect of exercise in cancer cachexia, and what do we know from clinical research in this area?

FURTHER READING

Ashcraft and Betof et al. (2019). *Exercise as adjunct therapy in cancer.* https://pubmed.ncbi.nlm.nih.gov/30573180/
Campbell et al. (2019). Exercise guidelines for cancer survivors: Consensus statement from international multidisciplinary roundtable.
Christensen et al. (2018) *Exercise training in cancer control and treatment.* Compr Physiol. 2018 Dec 13;9(1):165–205

REFERENCES

1. CRUK. Worldwide cancer statistics 2017 [Available from: http://www.cancerresearchuk. org/health-professional/cancer-statistics/worldwide-cancer.
2. Physical-Activity-Guidelines-Advisory-Committee. Physical Activity Guidelines Advisory Committee report 2008. Washington, DC: U.S. Department of Health and Human Services; 2008.
3. Pedersen BK, et al. *Scand J Med Sci Sports.* 2015. 25(Suppl 3):1–72.
4. Moore SC, et al. *JAMA Intern Med.* 2016. 176(6):816–25.
5. Ballard-Barbash R, et al. *J Natl Cancer Inst.* 2012. 104(11):815–40.
6. Arem H, et al. *Int J Cancer.* 2014. 135(2):423–31.
7. Fong DYT, et al. *BMJ (Clin Res ed).* 2012. 344:e70.
8. Campbell KL, et al. *Med Sci Sports Exerc.* 2019. 51(11):2375–90.
9. Greaves M, et al. *Nature.* 2012. 481(7381):306–13.
10. Hanahan D, et al. *Cell.* 2011. 144(5):646–74.
11. Tomasetti C, et al. *Science.* 2015. 347(6217):78–81.
12. Tomasetti C, et al. *Science (New York, NY).* 2017. 355(6331):1330–4.
13. Alexandrov LB, et al. *Nature.* 2013. 500(7463):415–21.
14. Rahman N. *Nature.* 2014. 505(7483):302–8.
15. Greenman C, et al. *Nature.* 2007. 446(7132):153–8.
16. Lawrence MS, et al. *Nature.* 2014. 505:495–501.
17. Lane DP. *Nature.* 1992. 358(6381):15–6.
18. Marcus K, et al. *Clin Cancer Res.* 2015. 21(8):1810–8.
19. Vogelstein B, et al. *Science.* 2013. 339(6127):1546–58.
20. Welch DR. *Cancer Res.* 2016. 76(1):4–6.
21. Hanahan D, et al. *Cancer Cell.* 2012. 21(3):309–22.
22. Dvorak HF. *N Engl J Med.* 1986. 315(26):1650–9.
23. Wyld L, et al. *Nat Rev Clin Oncol.* 2015. 12(2):115–24.
24. DeVita VT, Jr., et al. *Cancer Res.* 2008. 68(21):8643–53.
25. Kim W, et al. *Cells.* 2019. 8(9):1105.
26. Thariat J, et al. *Nat Rev Clin Oncol.* 2013. 10(1):52–60.
27. Risbridger GP, et al. *Nat Rev Cancer.* 2010. 10(3):205–12.
28. Khalil DN, et al. *Nat Rev Clin Oncol.* 2016. 13(6):394.
29. Collins FS, et al. *N Engl J Med.* 2015. 372(9):793–5.
30. Capdeville R, et al. *Nat Rev Drug Discov.* 2002. 1(7):493–502.
31. Joyner MJ, et al. *JAMA.* 2015. 314(10):999–1000.
32. Hojman P, et al. *Cell Metab.* 2018. 27(1):10–21.
33. Jones LW, et al. *J Appl Physiol (1985).* 2012. 113(2):263–72.
34. Betof AS, et al. *J Natl Cancer Inst.* 2015. 107(5):dvj040.
35. Research WCRFAIfC. 2018.
36. Gandini S, et al. *Eur J Cancer (Oxf, Engl: 1990).* 2005. 41(1):45–60.
37. Friedenreich CM, et al. *Clin Cancer Res.* 2016. 22(19):4766–75.
38. Cormie P, et al. *Epidemiol Rev.* 2017. 39(1):71–92.
39. Spei ME, et al. *Breast.* 2019. 44:144–52.
40. Lauby-Secretan B, et al. *N Engl J Med.* 2016. 375(8):794–8.

40a. Colbert et al. Med Sci Sports Exerc 2009. 41(8):1597-605
40b. McClellan et al, Int J Oncol 2014. 45(2):861–8.
 41. Puterman E, et al. PLoS One. 2010. 5(5):e10837.
 42. Sjogren P, et al. Br J Sports Med. 2014. 48(19):1407–9.
 43. Nomikos NN, et al. Front Physiol. 2018. 9:1798.
 44. Tomasetti C, et al. Science. 2017. 355(6331):1330–4.
 45. McTiernan A. Nat Rev Cancer. 2008. 8(3):205–11.
 46. Pages F, et al. Oncogene. 2010. 29(8):1093–102.
 47. Zhou R, et al. Cancer Immunol Immunother. 2019. 68(3):433–42.
 48. Sharma P, et al. Science. 2015. 348(6230):56–61.
 49. Pedersen L, et al. Cell Metab. 2016. 23(3):554–62.
 50. Rundqvist H, et al. Elife. 2020. 9:e59996.
 51. Rusch HPK, BE. Am Assoc Cancer Res. 1944. 4(2):116–8.
 52. Leung PS, et al. J Appl Physiol (1985). 2004. 96(2):450–4.
 53. Barnard RJ, et al. Eur J Cancer Prev. 2007. 16(5):415–21.
 54. Knuppel A, et al. Cancer Res. 2020. 80(18):4014–21.
 55. Shanmugalingam T, et al. Cancer Med. 2016. 5(11):3353–67.
 56. Higgins KA, et al. Cancer Am Cancer Soc. 2014. 120(21):3302–10.
 57. Hojman P, et al. Am J Physiol Endocrinol Metab. 2011. 301(3):E504–10.
 58. Hanahan D, et al. Cell. 2011. 144(5):646–74.
 59. Wu NZ, et al. Cancer Res. 1992. 52(15):4265–8.
 60. Peake JM, et al. J Appl Physiol (1985). 2017. 122(5):1077–87.
 61. Goh J, et al. Front Endocrinol (Lausanne). 2016. 7:65.
 62. Kruger K, et al. Brain Behav Immun. 2008. 22(3):324–38.
 63. Abdalla DR, et al. Eur J Cancer Prev. 2013. 22(3):251–8.
 64. Martinez-Outschoorn UE, et al. Nat Rev Clin Oncol. 2017. 14(1):11–31.
 65. Vaupel P, et al. Cancer Metastasis Rev. 2007. 26(2):225–39.
 66. Pavlova NN, et al. Cell Metab. 2016. 23(1):27–47.
 67. Aveseh M, et al. J Physiol. 2015. 593(12):2635–48.
 68. Wang ZY, et al. Breast Cancer Res Treat. 2012. 131(3):791–800.
 69. Hussien R, et al. Physiol Genomics. 2011. 43(5):255–64.
 70. Vaupel P, et al. Antioxid Redox Signal. 2007. 9(8):1221–35.
 71. DeClerck K, et al. Front Biosci (Landmark Ed). 2010. 15:213–25.
 72. Nordsmark M, et al. Radiother Oncol. 2005. 77(1):18–24.
 73. Carmeliet P, et al. Nat Rev Drug Discov. 2011. 10(6):417–27.
 74. McCullough DJ, et al. J Natl Cancer Inst. 2014. 106(4):dju036.
 75. Loges S, et al. Cancer Cell. 2009. 15(3):167–70.
 76. Ebos JM, et al. Proc Natl Acad Sci U S A. 2007. 104(43):17069–74.
 77. Psychogios N, et al. PLoS One. 2011. 6(2):e16957.
 78. Keshishian H, et al. Mol Cell Proteomics. 2015. 14(9):2375–93.
 79. Schadler KL, et al. Oncotarget. 2016. 7(40):65429–40.
 80. Christensen JF, et al. Compr Physiol. 2018. 9(1):165–205.
 81. Ashcraft KA, et al. Cancer Res. 2016. 76(14):4032–50.
 82. Ju J, et al. BMC Cancer. 2008. 8:316.
 83. Kiserud CE, et al. Eur J Cancer. 2010. 46(9):1632–9.
 84. Patnaik JL, et al. Breast Cancer Res. 2011. 13(3):R64.
 85. Scott JM, et al. JAMA Oncol. 2018. 4(10):1352–8.
 86. Morishita S, et al. Integr Cancer Ther. 2020. 19:1534735420917462.
 87. Buffart LM, et al. Cancer Treat Rev. 2017. 52:91–104.
 88. Reinertsen KV, et al. J Cancer Surviv. 2010. 4(4):405–14.
 89. Smeland KB, et al. Bone Marrow Transplant. 2019. 54(4):607–10.
 90. Mustian KM, et al. JAMA Oncol. 2017. 3(7):961–8.

91. Jones LW, et al. *J Clin Oncol.* 2012. 30(20):2530–7.
92. Hurria A, et al. *Am Soc Clin Oncol Educ Book.* 2016. 35:e516–22.
93. Jarden M, et al. *Bone Marrow Transplant.* 2007. 40(8):793–800.
94. Lipshultz SE, et al. *Circulation.* 2013. 128(17):1927–95.
95. Adams MJ, et al. *J Clin Oncol.* 2004. 22(15):3139–48.
96. Stenehjem JS, et al. *Br J Cancer.* 2016. 115(2):178–87.
97. Wood WA, et al. *Bone Marrow Transplant.* 2013. 48(10):1342–9.
98. Lakoski SG, et al. *JAMA Oncol.* 2015. 1(2):231–7.
99. Collins J, et al. *BMJ Open.* 2014. 4(1):e003697.
100. Joglekar S, et al. *J Surg Oncol.* 2015. 111(6):771–5.
101. Pin F, et al. *Curr Opin Support Palliat Care.* 2018. 12(4):420–6.
102. Miyamoto Y, et al. *PLoS One.* 2015. 10(6):e0129742.
103. Sakai H, et al. *Toxicol Appl Pharmacol.* 2014. 278(2):190–9.
104. Rosenblatt JD, et al. *J Appl Physiol (1985).* 1992. 73(6):2538–43.
105. Hanson ED, et al. *J Clin Endocrinol Metab.* 2017. 102(3):1076–83.
106. van Londen GJ, et al. *Crit Rev Oncol Hematol.* 2008. 68(2):172–7.
107. Ozbakir B, et al. *J Control Release.* 2014. 190:624–36.
108. Lin KT, et al. *Steroids.* 2016. 111:84–8.
109. Frost RA, et al. *Endocrinol Metab Clin North Am.* 2012. 41(2):297–322.
110. Foletta VC, et al. *Pflügers Archiv – Eur J Physiol.* 2011. 461(3):325–35.
111. Sandri M, et al. *Cell.* 2004. 117(3):399–412.
112. Gupta A, et al. *Indian J Endocrinol Metab.* 2013. 17(5):913–6.
113. Christensen JF, et al. *Br J Cancer.* 2014. 111(1):8–16.
114. Mijwel S, et al. *FASEB J.* 2018. 32(10):5495–505.
115. Ni J, et al. *Cancer Manag Res.* 2020. 12:5597–605.
116. Advani SM, et al. *BMC Cancer.* 2018. 18(1):1174.
117. Friesen DE, et al. *Theor Biol Med Model.* 2015. 12:17.
118. Baracos VE, et al. *Ann Palliat Med.* 2019. 8(1):3–12.
119. Baracos VE, et al. *Nat Rev Dis Primers.* 2018. 4:17105.
120. Braun TP, et al. *J Exp Med.* 2011. 208(12):2449–63.
121. Surtees R, et al. *J Inherit Metab Dis.* 1989. 12(Suppl 1):42–54.
122. Guertin DA, et al. *Cancer Cell.* 2007. 12(1):9–22.
123. Grande AJ, et al. *Cochrane Database Syst Rev.* 2021. 3:CD010804.

Satellite cells and exercise

Neil R.W. Martin and Adam P. Sharples

LEARNING OUTCOMES

At the end of this chapter, you should be able to:

1. Define what satellite cells are and explain their function during muscle adaptation.
2. Describe the molecular regulation of myogenesis.
3. Explain how satellite cells respond both acutely and chronically to resistance and endurance exercise.
4. Evaluate the importance of satellite cells in muscle hypertrophy.
5. Describe the differences in satellite cell function as a result of ageing.

INTRODUCTION

Skeletal muscle fibres are some of the largest cells in the body. Some muscles contain hundreds and possibly even thousands of nuclei along their length, which provide the RNA and protein required to sustain the structure and function of such large cells. Skeletal muscle however is **terminally differentiated**, which means that unlike many other tissues in the body, myonuclei cannot re-enter the cell cycle and proliferate to support the repair and growth of the muscle. Nonetheless, skeletal muscle possesses incredible regenerative capacity, which is attributable to muscle specific progenitor cells termed **satellite cells**.

In **Chapter 2 and 8**, we outlined a short history as to the identification of satellite cells. Briefly, the existence of satellite cells was first reported in 1961 by two separate laboratories studying frog skeletal muscle using high-powered electron microscopes (1, 2). The scientists observed a population of cells which contained a single nucleus, had very little cytoplasm and were found to possess high levels of DNA which was not being actively transcribed, and contained underdeveloped and few cellular organelles. The cells were positioned on the periphery of muscle fibres, between the muscle sarcolemma and the basal lamina (a thin sheet of connective tissue surrounding muscle fibres) and based on this location Mauro termed them satellite cells (1). We know today that satellite cells play a crucial role in skeletal muscle repair and regeneration processes following trauma, injury and exercise.

In this chapter, we will first provide an overview of basic satellite cell biology, where much of what we understand about satellite cells and their function has been attained using in vitro cell cultures, studies in animals and the extraction of intact muscle fibres where satellite cells remain present. We then discuss satellite cell responses to exercise in humans and broadly separate these discussions into studies of resistance and aerobic exercise where the use of immunohistochemistry to label satellite cells from muscle biopsy samples has been imperative to improving understanding in this area. For thorough explanations of these procedures, readers are referred back to **Chapter 2**. We also highlight the role of

DOI: 10.4324/9781315110752-13

satellite cells in muscle fibre hypertrophy that was covered in **Chapter 8**, and go into more depth into this area, which has been a topic of intense focus and controversy over the last decade. Finally, we discuss how ageing affects satellite cells and how the environment in which satellite cells reside may play an important role in regulating their activity.

SATELLITE CELL FUNCTION

Satellite cells adopt a quiescent state on the periphery of skeletal muscle fibres. In this quiescent state, they do not actively transcribe genes and have very few organelles (3). However, quiescent satellite cells can be activated, at which point they re-enter the cell cycle and proliferate extensively, giving rise to progeny called **myoblasts**. Myoblasts, in turn, can differentiate and either fuse together to form myotubes (immature muscle fibres) or fuse with existing muscle fibres (**Figure 13.1**). Satellite cells are therefore myogenic precursors and myonuclear donors, which are known to be important in the following processes:

(i) Skeletal muscle regeneration: The production of new muscle fibres as a result of trauma (hyperplasia) and/or death of existing muscle fibres.

When severely injured, muscle fibres undergo degeneration and cell death which is followed by expansion of satellite cell derived myoblasts and their subsequent fusion to give rise to new (regenerated) multinucleate myotubes inside the basal lamina of the original fibre (4, 5). An absolute requirement of satellite cells in this process has been demonstrated by the failure of muscle to regenerate following chemical injury when the satellite cell population is abolished in vivo (6–8).

Regeneration is a rare event in healthy people, and exercise alone does not seem to cause severe muscle injury or fibre death to warrant regeneration (9). However, unaccustomed electrical stimulation of human muscle, along with chemical injury or non-physiological muscle loading (in rodents) do stimulate regeneration and have been used extensively to inform scientists about satellite cell biology.

(ii) Skeletal muscle repair: The process of restoring the morphological and functional defects caused by damage to a skeletal muscle fibre.

Satellite cells are also required to repair more moderate muscle damage via fusion with the damaged fibre. In mice, when the ability of satellite cells to fuse (differentiate) is blocked, excessive muscle damage and fibrosis is detected in the early phases of exercise training resulting in exercise intolerance (10). However, when satellite cells are able to function normally and act as myonuclear donors, muscle damage and fibrosis are reduced.

(iii) Skeletal muscle growth: An increase in size of a whole muscle which may occur through the processes of regeneration (hyperplasia) or hypertrophy.

Growth of skeletal muscle is typically accompanied by an increase in the number of myonuclei (11), suggesting an important role for satellite cells in regulating muscle growth. Whilst, in rodents at least, satellite cells can contribute to muscle growth through muscle fibre regeneration, the requirement of satellite cells for growth of existing muscle fibres (hypertrophy) is less clear and will be discussed later in this chapter.

MOLECULAR REGULATION OF MYOGENESIS

Quiescent satellite cells can be identified through expression of the paired box transcription factor Pax7. In mice, loss of Pax7 completely abolishes the satellite cell pool (12), illustrating its importance for satellite cell specification. The transition of quiescent satellite cells through activation, proliferation and differentiation to form muscle fibres is termed **myogenesis**. Myogenesis is a highly orchestrated process, characterised by and reliant upon temporal changes in genes termed **myogenic regulator factors**, or **MRFs**, that were briefly discussed in **Chapters 2 and 8.**

Briefly here, upon activation, satellite cells start to proliferate and are termed myoblasts. Myoblasts maintain Pax7 expression alongside upregulation of the MRFs, Myf-5 and MyoD (13). The relative importance of these transcription factors for normal myogenesis is evidenced by studies in which myoblasts lack either the Myf-5 or MyoD gene. Myf-5$^{-/-}$ myoblasts undergo less proliferation and favour early differentiation and muscle growth (14, 15), whereas MyoD$^{-/-}$ myoblasts continue to proliferate and fail to differentiate normally (16, 17). After multiple rounds of proliferation, the majority of myoblasts undergo terminal differentiation, resulting in their fusion to existing muscle fibres or new myotube formation (**Figure 13.1**). This process is characterised by the loss of Pax7 and Myf-5 expression, persistent MyoD levels and enhanced expression of Myogenin and MRF4 (18). Indeed, MyoD and Myogenin act as transcription factors for a large number of structural and contractile skeletal muscle genes such as actin and myosin, and therefore expression of these factors in differentiating myoblasts is required for correct muscle structure and function (19, 20).

Importantly, muscle damage or injury at one location on a muscle fibre can cause satellite cell activation along the length of a fibre (21). These activated satellite cells can then move (migrate) to the site of

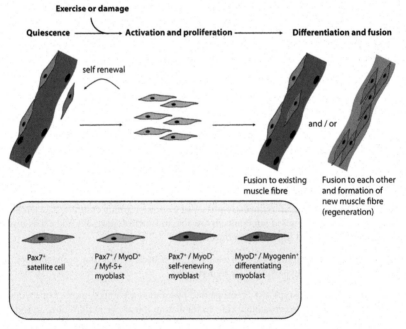

Figure 13.1 Molecular and cellular regulation of myogenesis. Satellite cells lie in a quiescent state on the periphery of skeletal muscle fibres, wedged between the sarcolemma and basal lamina and are characterised by the expression of the Pax7 gene. Exercise or muscle damage activates satellite cells, and they begin to proliferate and express the MRFs; MyoD and Myf-5. Activated myoblasts subsequently differentiate by upregulating Myogenin (and latterly MRF4) and downregulating Pax7. Differentiating myoblasts can either fuse together to form new muscle fibres (regeneration) or fuse to existing muscle fibres to help repair damage or facilitate muscle growth. A subpopulation of activated myoblasts lose MyoD expression prior to differentiation, and these myoblasts return to quiescence to restore the satellite cell pool.

muscle damage and fuse with the muscle fibre. By fluorescently labelling satellite cells and imaging them under a microscope, one study found that satellite cells can also migrate between muscle fibres to contribute new myonuclei to growing/regenerating muscle (22). The migration of satellite cells appears to be regulated on a molecular level by numerous signalling pathways, including HGF-MAPK (see 'regulation of satellite cell activation' below), PI3 kinase and mTOR (23, 24).

SATELLITE CELLS ARE SKELETAL MUSCLE STEM CELLS

To be considered a stem cell, a cell must be able to give rise to a differentiated cell type and self-renew. Self-renewal is the process whereby activated myoblasts return to a being a quiescent satellite cell on the fibre periphery, where they lie in waiting for their use in the future. Therefore, self-renewal maintains satellite cell number in the muscle which is often termed the 'stem/satellite cell pool'. In 2005, an important study in the field showed definitively that satellite cells were indeed skeletal muscle stem cells (25). Individual muscle fibres containing as little as seven satellite cells were grafted (inserted) into the muscles of mice that had previously been exposed to γ-irradiation, which eliminates satellite cells from the host muscle. Three weeks later, the researchers found that more than 100 new muscle fibres, each containing thousands of myonuclei, had been generated from the small number of original satellite cells on the single muscle fibre graft. The grafted satellite cells also contributed new nuclei to damaged muscle fibres from the host mice. Importantly, when the new fibres were removed, it was observed that there was a 10-fold expansion of the satellite cell pool. This clearly illustrates that satellite cells have the ability to proliferate extensively, and then either differentiate to form new fibres or return to quiescence (self-renew), thus satisfying the criteria of a stem cell (25).

Self-renewal of satellite cells relies on alterations in the expression of MRFs. Activation of satellite cells results in co-expression of Pax7 and MyoD. However, these activated satellite cells can progress down two potential routes: they can either downregulate Pax7 and upregulate Myogenin and differentiate (as described above), or downregulate MyoD expression and return to quiescence (26) (**Figure 13.1**).

How do dividing satellite cells decide whether to progress to differentiation or to self-renew? This is an area of active research, but it is known that the decision relies upon asymmetric cell division, which is the process by which two daughter cells adopt differing cellular fates. A group of proteins known as the Par complex are well established regulators of asymmetric cell division (27). In a satellite cell undergoing cell division (mitosis), the Par complex is localised to one side of the cell and activates p38 Mitogen-Activated Protein Kinase (p38 MAPK) signalling (28). MAPK signalling induces cell proliferation (29, 30) and turns on MyoD within the cell, such that one daughter cell becomes a bona fide myoblast, whilst the satellite cell without active p38 MAPK and MyoD returns to quiescence.

The Notch signalling pathway also plays a role in determining cell fate. Activation of the Notch pathway promotes myoblast proliferation and overexpression of the transmembrane receptor Notch-1 results in higher levels of Pax 7 in myoblasts and enhances satellite cell self-renewal (31, 32). By contrast, the Notch signalling inhibitor, Numb, promotes Myf-5 expression, myoblast differentiation and myotube formation (31). Numb is inherited asymmetrically when myoblast divide (31), potentially directing one daughter cell towards differentiation and another towards self-renewal. Numb localisation is also under the control of the Par complex (27), demonstrating how integrated cellular signalling can control satellite cell proliferation, differentiation and the return to quiescence.

SATELLITE CELL RESPONSE TO EXERCISE

As satellite cells are important for regeneration, repair and growth of skeletal muscle, it is unsurprising that over the past 20 years, the relationships between satellite cells and exercise, and their contribution to exercise adaptation have been studied extensively.

Resistance exercise

Following an acute bout of resistance exercise in humans, satellite cells are activated from their quiescent state and co-express Pax7 and MyoD (33, 34) which occurs by 3 hours following exercise (33). Thereafter, extensive increases in total satellite cell numbers are observed by 24 hours (35–37), consistent with results from studies that have observed satellite cell doubling time (the time for each satellite cell to replicate) is less than 24 hours (38, 39).

Chronic resistance training lasting between 8- and 16-weeks results in an elevation in the total number of satellite cells associated with muscle fibres in young and older men and women (40–45), which appears to largely be driven by the proliferation of satellite cells in the early phases of a resistance exercise programme (46). Expansion of the satellite cell pool appears to be most pronounced in type II muscle fibres (42, 44–46) and the increase in muscle size with training is strongly related to the growth of the satellite cell pool (43, 46, 47). This suggests that the expansion of the satellite cell population is an important regulatory mechanism for muscle hypertrophy following resistance training.

In keeping with our understanding of the basic biology of satellite cells, after the activation and proliferation of satellite cells following resistance exercise in humans, some satellite cells lose Pax7 expression (33, 34), differentiate and fuse with existing muscle fibres to contribute new myonuclei and increase the overall myonuclear number per muscle fibre (43, 44, 47, 48). This increase in myonuclear number affords muscle a greater capacity for transcription of RNA and translation of new protein and subsequent muscle growth. However, some studies have failed to observe myonuclear accretion following resistance training (41, 46), and it has been suggested that muscle fibres must hypertrophy beyond a given size before satellite cells contribute new nuclei to support the further growth of the muscle. This concept is referred to as the **myonuclear domain ceiling size theory** (41) and will be discussed further in relation to muscle hypertrophy in the sections below.

Finally, there is evidence that the activation and proliferation of satellite cells following acute resistance exercise are enhanced after undertaking resistance training (49), suggestive of increased satellite cell sensitivity to exercise. It is likely that this improved satellite cell response post-training reflects a requirement for satellite cells in hypertrophy, since the same group failed to observe the same phenomenon in a shorter training study where participants' gains in muscle mass were minimal (46). Whilst the mechanisms behind this training-induced satellite cell sensitisation are not clear; it may relate to increased capillarisation of the muscle following training. Indeed, there appears to be some degree of communication between endothelial cells (resident in capillaries) and satellite cells (50) and following training satellite cells are found in closer proximity to capillaries (49), potentially making them more able to respond to an acute exercise stimulus. Satellite cell communication with endothelial cells is discussed later in this chapter.

Aerobic exercise

There is less information available relating to the satellite cell response to endurance training. Studies in humans and rodents have demonstrated that the number of satellite cells per muscle fibre is increased following endurance training (45, 51–53) without the addition of new myonuclei to the muscle. This can be explained by the fact that endurance exercise promotes activated satellite cells to self-renew (return to quiescence) rather than differentiate and this appears to be related to reduced ATP demand and oxygen consumption in the satellite cells (54). The increase in satellite cell density following traditional endurance training appears to be specific to type I muscle fibres (55, 56) or to type I/IIa hybrid fibres following high intensity interval training (57).

Expansion of the satellite cell pool following endurance exercise may be dependent on exercise intensity. In older men an acute bout of high intensity interval cycling led to an increase in MyoD+ satellite cells (activated satellite cells) which was comparable with resistance exercise, whereas 30 minutes of cycling at a lower intensity of 60% of maximal heart rate had a more modest effect (34). Furthermore, in rodents, ten weeks of low intensity treadmill running failed to increase muscle satellite cell content, whereas robust increases in satellite cell numbers were observed in rats exercised at a high intensity, irrespective of the duration of each exercise bout (58). There was also no increase in satellite cell numbers in human skeletal muscle following six months of exercise training four times per week at a moderate intensity (59). Collectively, these observations suggest that intensity and not duration is the key determinant of satellite cell activation and expansion following endurance exercise.

Because high intensity endurance exercise is often associated with hypertrophy, the increase in satellite cell numbers may help to support the growth of the muscle. However, the fact that satellite cells are activated (34, 60) and their numbers can increase in the absence of hypertrophy (57, 58, 61, 62) suggests that satellite cells may play a role in more global skeletal muscle adaptation and health. For example, satellite cell expansion following endurance exercise primes skeletal muscle to respond to future damage, resulting in improved skeletal muscle regeneration, repair and growth (54, 62) with lower levels of inflammation (54).

Nonetheless, satellite cells do not appear to be important for the cellular adaptations to endurance exercise. When satellite cells are depleted from adult mice prior to eight weeks of voluntary wheel running, increases in vascularisation, mitochondrial density and a shift towards oxidative metabolism are not altered (63). However, mice lacking satellite cells exhibit a general loss of motor control and run less than normal mice, possibly due to atrophy of and excessive extracellular matrix (ECM) deposition around muscle spindles (63), which are important in providing sensory feedback on muscle length and contraction velocity.

Overall, endurance exercise can increase the number of satellite cells associated with skeletal muscle fibres in an intensity-dependent manner. Prior endurance training may enhance satellite cell function and enhance the satellite cell pool, thereby improving the efficiency of skeletal muscle in responding to subsequent bouts of more damaging exercise or muscle injury.

SATELLITE CELLS AND SKELETAL MUSCLE HYPERTROPHY

Within skeletal muscle fibres, each myonucleus controls the growth of a given volume of cytoplasm (64–66), which is called the myonuclear domain. Myonuclei are not only required for transcription of mRNA, but also for ribosome biogenesis which is important for supporting protein synthesis and muscle growth. Whether myonuclear domain size is dynamic and can expand as muscle fibres grow or remains fixed, requiring satellite cells to provide new myonuclei during muscle hypertrophy, is controversial (67, 68). However, recent research has suggested, that myonuclei display a 'reserve' capacity. When myonuclear numbers are reduced by 25% or even 55% in the neonatal period there appears to be little impact on muscle function because the existing myonuclei can increase their capacity to produce more mRNA per nucleus. However, the muscles with fewer myonuclei do tend to be smaller in size and when myonuclear number is reduced by 75% the muscle is abnormal and not functional (69). This supports the notion that if you have fewer myonuclei then muscles tend to be smaller and if you have more myonuclei then the muscle is larger. Some studies have also shown that satellite cells are not required for hypertrophy during adulthood per se because muscle fibres can grow and myonuclear domains expand considerably in the absence of satellite cells in response to growth promoting pharmacological interventions (10, 70–72), perhaps in part due to this reserve capacity described above. Therefore, as introduced in Chapter **8**, the necessity for satellite cells and additional myonuclei for muscle hypertrophy in response to mechanical loading and exercise is a point of much debate amongst scientists working in this field.

Can myonuclear domains expand during load-induced hypertrophy?

By functionally overloading the plantaris muscle of rats through the surgical ablation of the synergistic gastrocnemius and soleus muscles, robust muscle growth has been observed alongside an increase in myonuclear number (73, 74). This, in turn, results in a constant myonuclear domain size and points towards a requirement for satellite cells to support hypertrophy. Further evidence for this comes from cross sectional data from humans showing a positive correlation between muscle fibre cross sectional area and myonuclear number (75). However, studies which have examined myonuclear accretion over a time course of hypertrophy induced through synergist ablation/denervation have yielded contradictory results, with evidence that myonuclear addition to muscle fibres occurs prior to (76) and after (77) muscle fibre hypertrophy in rodents.

Experiments in humans suggest that moderate hypertrophy can occur without satellite cell-mediated myonuclear addition (41, 46), leading to the proposal that a myonuclear domain ceiling size exists (41). According to this model and shown in **Figure 13.2**, each myonucleus can regulate a muscle volume of approximately 2,000 μm^2, and hypertrophy of a muscle fibre to exceed this myonuclear domain size would require the addition of new nuclei from the fusion of myoblasts (progeny of satellite cells). In support of this theory, in mice, it has been observed that myonuclei can increase their transcriptional activity by up to 7-fold which is sufficient to support short-term muscle growth without the requirement for satellite cells (78). In addition, more extreme muscle hypertrophy following resistance training in humans is associated with an increased number of myonuclei, whereas more moderate increases in fibre size do not demonstrate this association (43, 48, 79).

Requirement of satellite cells for load-induced hypertrophy

The gold standard experiment to address whether satellite cells are necessary for muscle hypertrophy would require the complete removal of satellite cells from adult muscle before testing whether it can

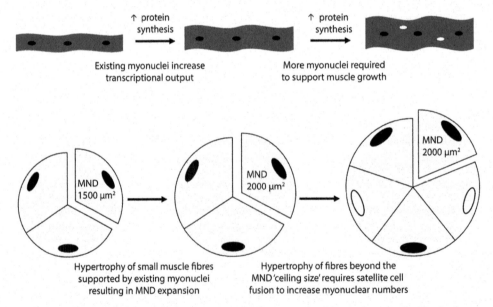

Figure 13.2 Myonuclear domains in skeletal muscle hypertrophy in response to exercise. Due to the multinucleate nature of muscle fibres, each myonucleus transcriptionally regulates a given volume of sarcoplasm, known as the myonuclear domain (MND). Small muscle fibres can hypertrophy without satellite cell-mediated myonuclear addition because each myonucleus can increase its transcriptional output, principally supporting ribosomal biogenesis and enabling effective protein synthesis. A hypothetical MND 'ceiling size' of ~ 2,000 μm^2 has been postulated, and therefore hypertrophy of muscle fibres beyond this ceiling requires satellite cell fusion to muscle fibres and additional myonuclei to support the additional protein synthetic requirements of the muscle fibre. Redrawn from Bamman MM, et al. (165).

grow in response to overload. Initial attempts at this experiment used ionising radiation to eliminate satellite cells from skeletal muscle of rats. The results showed that in the absence of satellite cells skeletal muscle fibres failed to hypertrophy in response to functional overload mediated through synergist ablation (80–82). These authors therefore concluded that satellite cells are indeed necessary for hypertrophy. However, the conclusions drawn from such experiments have been questioned, since the use of irradiation cannot specifically ablate satellite cells (68) and may have wider effects.

Recently, scientists have revisited the question of myonuclear accretion for hypertrophy using genetic mouse models of satellite cell ablation or impaired satellite cell fusion. In mice in which Pax7+ cells are conditionally eliminated from muscle during adulthood, one research group have observed that muscle hypertrophy was unaffected by the loss of satellite cells and that myonuclear domains can expand after two weeks of functional overload (83). However, muscle growth over longer periods perhaps cannot be sustained without satellite cells and muscles accumulate excessive ECM (fibrosis), illustrating a requirement of satellite cells for normal muscle growth (84, 85). Using an identical mouse model, other researchers have found a complete lack of muscle fibre growth following functional overload in the absence of satellite cells (86). In addition, in a mouse model in which satellite cells and their progeny are rendered incompetent of fusion (Myomaker^SCKO^), it has also been observed that muscles are resistant to hypertrophy following synergist ablation (87). Whilst there is not complete agreement between these studies, which may be due to small differences in the methods and analyses used, there is nonetheless evidence that satellite cells are required to sustain muscle fibre hypertrophy in response to muscle loading.

Whilst functional muscle overload using synergist ablation in rodents is known to induce robust hypertrophy, it does not represent normal muscle growth and induces damage to muscles which, in turn, requires satellite cell-mediated repair and regeneration irrespective of hypertrophy. Therefore, more recently, exercise training models which elicit muscle growth and hypertrophy in mice have been used to investigate satellite cell requirements for physiological hypertrophy. Eight weeks of progressive weighted wheel running in satellite cell ablated adult mice resulted in blunted muscle growth and hypertrophy compared to wild-type mice, along with dysregulated gene networks; suggestive of a role for satellite cells in wider muscle adaptation to exercise (88). Similarly, increases in muscle mass, myonuclear content and fibre hypertrophy were completely blocked in Myomaker^SCKO^ mice (10). Collectively, these experiments suggest that addition of new myonclei by satellite cells is a requirement for physiologically normal load/exercise induced muscle hypertrophy and adaptation. Finally, in Chapter **8**, we also discussed that most of the experiments demonstrating no requirement for satellite cells during hypertrophy come from rodent studies using synergistic ablation. Where it is perhaps important to highlight that human muscle fibres are larger than rodent muscle fibres, and that data in humans suggests growth of larger fibres compared with smaller fibres seem to be more dependent on satellite cells. This is supported by data suggesting hypertrophy of smaller fibres correlates with myonuclear domain whereas hypertrophy of larger fibres correlates with myonuclear number (described in Chapter **8** 'Satellite cells in human hypertrophy in response to resistance exercise').

In summary, there is still controversy as to whether satellite cell-mediated myonuclear accretion is obligatory *per se* for muscle fibre hypertrophy in response to loading. However, there is substantial evidence that hypertrophy in more physiological models is limited in the absence of myonuclear addition or a functional satellite cell pool or in larger human fibres. It is possible that requirements for satellite cells in the early phases of muscle loading may represent a need for satellite cells in muscle fibre repair, and/or to support appropriate ECM deposition by non-myogenic cells, and that myonuclear addition thereafter is required to support and maintain hypertrophy.

REGULATORS OF SATELLITE CELL ACTIVATION

Precisely how exercise/muscle contraction results in satellite cell activation is poorly understood. However, the fact that unilateral leg exercise fails to increase satellite cell activation in the contralateral leg (89) suggests that local changes in the muscle or the muscle niche are responsible for satellite cell

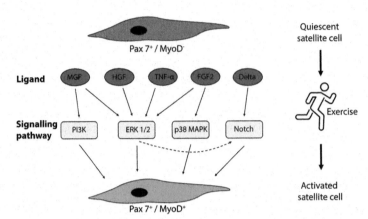

Figure 13.3 Regulation of satellite cell activation following exercise. Satellite cells lie quiescent in resting muscle, but following exercise are awoken from this dormant state and become activated and proliferative. The precise mechanisms which regulate satellite cell activation are not completely understood; however, several cytokines and proteins secreted from the exercised muscle have been implicated in the process which can activate signalling pathways known to induce myogenic gene expression and/or entry into the cell cycle. MGF = Mechano growth factor; HGF =Hepatocyte growth factor; TNF-α = Tumour necrosis factor-α; FGF2 = Fibroblast growth factor-2; PI3K = Phospho inositide-3 kinase; ERK 1/2 = Extracellular signal regulated kinase 1 and 2; p38 MAPK = p38 mitogen-activated protein kinase.

activation following exercise, rather than factors transported to muscle through the blood. Indeed, there are a number of growth factors and cytokines released from muscle tissue which may work in an autocrine/paracrine manner to awaken satellite cells from their quiescent state following exercise (**Figure 13.3**), the best characterised of which are hepatocyte growth factor (HGF) and insulin-like growth factor I (IGF-I).

Hepatocyte growth factor and nitric oxide

HGF is present in the ECM surrounding skeletal muscle fibres (90,91), and satellite cells themselves express HGF and its receptor c-met (90, 92–94). HGF binding to c-met is known to activate multiple signalling pathways which result in entry of cells into the cell cycle (95), and studies on isolated satellite cells and whole muscle extracts have demonstrated that HGF is able to activate satellite cells from quiescence and induce proliferation (90,92). Importantly, there is evidence that HGF may be crucial for exercise mediated satellite cell activation because mechanical loading of quiescent satellite cells in vitro induces their activation and proliferation in a HGF dependent manner (94), whereby loading releases HGF from the matrix, enabling its binding to c-met on satellite cells. In humans, an acute bout of eccentric exercise increased HGF levels within muscle biopsy samples, and HGF levels were augmented in the blood (36) suggesting release of HGF from damaged muscle. There were also increased levels of proteins involved in the cleavage of an inactive form of HGF (pro-HGF) to its active form (36). Collectively, these observations make HGF a strong candidate as a stimulus for satellite cell activation following exercise.

HGF release from the ECM requires the presence of nitric oxide (NO) (91,96,97). Exercise stimulates increases in the activity of nitric oxide synthase (NOS) (98), which, in turn, would result in higher NO levels in skeletal muscle, HGF release and satellite cell activation. Indeed, genetic loss of NOS in mice inhibits muscle regeneration following injury (99). The mechanism by which NO releases satellite cells from their matrix tethering is likely through a group of enzymes called matrix metalloproteinases

(MMPs). MMPs are endopeptidases which are involved in degradation and remodelling of the ECM, and MMP levels are increased by NO (100). Indeed, when MMPs are inhibited in isolated satellite cell cultures in vitro, HGF release and satellite cell activation by either mechanical loading or NO is prevented (101). Specifically, MMP2 appears to be required for HGF release and satellite cell proliferation (102). Overall, HGF release from the skeletal muscle ECM via the action of NO and MMP2 may be responsible for satellite cell activation following exercise/muscle loading.

Insulin-like growth factor I

Insulin-like growth factor I or IGF-I is an important mediator of tissue growth. IGF-I secreted by the liver in response to growth hormone acts in an endocrine fashion and drives post-natal growth and development of many organs, including the brain, spleen, kidney and bone (103). Skeletal muscle is also a source of IGF-I which can act in an autocrine/paracrine manner to induce muscle growth at the local level (104). In fact, in skeletal muscle, the IGF-I gene can be alternatively spliced during transcription to give rise (in humans) to three different IGF-I isoforms or splice variants: IGF-IEa (which is the same as the liver IGF-I), IGF-IEb and IGF-IEc (105). IGF-IEc has been termed mechano-growth factor (MGF) due to the fact that it is not detectable in resting muscle but is rapidly expressed at the mRNA level following mechanical stretch of rabbit muscle and eccentric exercise in humans (105).

Inhibition of IGF-I using a neutralising antibody, or through an inactive IGF-I receptor, worsens muscle regeneration following injury (106,107), and IGF-I is present on satellite cells 24 hours after damaging muscle contractions (108), suggesting a role for IGF-I in the satellite cell responses to exercise/damage. Interestingly, examination of the IGF-I splice variants shows that following exercise or injury, MGF mRNA increases within 24 hours, whereas IGF-IEa mRNA peaks between 3 and 10 days (108–110). Where, the MGF mRNA responses correlate well with Myf-5 levels, suggesting that MGF specifically may be important in satellite cell activation (108). Evidence from cell culture experiments supports this notion, as addition of MGF peptides to myoblasts cultured in vitro enhances proliferation and motility but impairs fusion into myotubes (111–113). However, it should be noted that MGF has also been shown to drive differentiation of myoblasts (114) or have no effect on proliferation or differentiation in vitro (115). Nonetheless, based on the majority of experimental evidence, following mechanical loading/exercise skeletal muscle MGF levels are enhanced, and through binding the IGF-I receptor on satellite cells, MGF activates signalling cascades which enhance Myf-5 gene expression (116), resulting in satellite cell activation and proliferation (117).

Other factors regulating satellite cell activation

Muscle damage/exercise results in an inflammatory environment in the satellite cell niche, principally due to release of inflammatory cytokines from the skeletal muscle fibre and from infiltrating pro-inflammatory macrophages, which are a type of white blood cell shown to be essential for muscle regeneration (118). Indeed, infusion or oral consumption of non-steroidal anti-inflammatory medication prevents the increase in satellite cell numbers normally observed after exercise in humans (119,120), clearly illustrating the importance of inflammation in the satellite cell response to exercise. Chronically elevated levels of the inflammatory cytokines Interleukin 6 (IL-6) and Tumour Necrosis Factor Alpha (TNF-α) can cause apoptosis/cell death, the breakdown of muscle protein and ultimately muscle atrophy. However, acute and transient increases in these cytokines may also be important for satellite cell function.

Theoretically, IL-6 may function to activate satellite cells through its binding to the IL-6 receptor (IL-6R), which, in turn, causes the phosphorylation (and activation) of STAT3 and its movement into the nucleus of a cell, where it acts to increase the transcription of numerous genes involved in the cell cycle and cell proliferation. In support of this, the IL-6R is present on satellite cells, and IL-6 and activated STAT3 are detected in satellite cells after, but not before, an acute bout of exercise (33, 121). An important paper in this research area showed that muscles from mice lacking IL-6 (IL-6$^{-/-}$) failed to grow effectively following synergist ablation, which was due to a failure of satellite cells to proliferate and differentiate to contribute new myonuclei to muscle fibres (122). Myoblasts from the same IL-6$^{-/-}$ mice also proliferated poorly when cultured in vitro (122). However, in the first few days after the onset of synergist ablation, satellite cells from IL-6$^{-/-}$ mice expressed MyoD (a marker of satellite cell activation) to the same extent as normal mice (122). This indicates that IL-6 is an important regulator of satellite cell proliferation but is perhaps not required for the initial activation of satellite cells from quiescence.

It is well established that the addition of TNF-α to in vitro cultures of myoblasts enhances their proliferation (123–125) and prevents their differentiation (126,127). Interestingly, at very low concentrations, TNF-α can also enhance differentiation and myotube formation (128). These pleiotropic effects can be reconciled by the fact that TNF-α activates a variety of signalling pathways; likely in a bi-phasic manner during myogenesis. For example, TNF-α activates MAPK. MAPK is an important signalling pathway which regulates myogenic gene expression (128,129) and TNF-α stimulation of Extracellular signal-regulated protein kinases (ERK1/2) and p38 MAPK are important in the regulation of myoblast proliferation and differentiation, respectively (117,130). TNF-α also activates inflammatory signalling through NF-κB signalling which prevents the induction of MyoD and Myogenin and thus prevents differentiation (127,131). Loss of the TNF receptor results in impaired muscle regeneration following chemical injury in mice (130) indicating its importance in satellite cell function in vivo, however whether TNF-α activates satellite cells following exercise is not yet fully elucidated.

Other proteins and growth factors have been demonstrated to play important roles in satellite cell activation. Myotube injury increases the expression of the protein Delta, which activates Notch signalling in satellite cells and causes their activation from quiescence (31, 132). The fact that inhibition of Notch signalling prevents satellite cell activation indicates that Delta may be important for satellite cell activation following exercise. Fibroblast growth factors (FGFs) may also be implicated in the activation of Notch signalling and satellite cell proliferation. Addition of several FGF family members to satellite cells cultured in vitro increases their proliferation (133–135) and inhibition of FGF-2 impairs muscle regeneration following injury in mice (107). FGF-2 binding to its receptor causes activation of the MAPK signalling pathway (30, 136) which, in turn, increases the expression of Delta and activates Notch (137). Incidentally, HGF and IGF-I (MGF) also activates MAPK signalling (95), and therefore may induce Notch to mediate its effects on satellite cell activation. Although there is some evidence of important roles for Delta/Notch and FGF-2 in satellite cell activation following exercise (137), direct links between these factors and satellite cell activation and proliferation following exercise are lacking.

SATELLITE CELL COMMUNICATION

As alluded to previously in this chapter, skeletal muscle communicates with satellite cells following exercise. For example, growth factors from the ECM and muscle fibres may regulate satellite cell activation, proliferation and differentiation. Furthermore, recent evidence suggests that satellite cells themselves can communicate with other cells in their environment (the satellite cell niche) to regulate muscle adaptation and satellite cell function.

Satellite cell communication with fibroblasts

Skeletal muscle fibroblasts are located in the interstitial space between muscle fibres and in close proximity to satellite cells. Fibroblasts are responsible for the secretion of ECM proteins such as collagens, which increases following exercise (138) and are important for correct skeletal muscle remodelling (7).

Genetic depletion of satellite cells in mice before functional overload (synergist ablation surgery) or chemical muscle injury results in an increase in fibroblast numbers and excessive production of ECM proteins (7, 85). Evidence that satellite cells directly communicate with fibroblasts comes from in vitro experiments. When myoblasts are cultured in a media which is subsequently transferred to fibroblasts, collagen synthesis is reduced in the fibroblasts (85), suggesting that myoblasts secrete a product which influences fibroblast synthesis of ECM proteins. Myoblasts can also release exosomes (a type of very small extracellular vesicle), which contain a microRNA called miR-206 that is transported into fibroblasts. miR-206 specifically represses the translation of a gene called Ribosomal binding protein-1 (Rrbp-1) which is important for collagen synthesis in fibroblasts (84). The levels of miR-206 are normally increased following muscle loading, which represses Rrbp-1 and prevents excessive ECM production by fibroblasts. However, when satellite cells are absent, miR-206 cannot be synthesised or transported to fibroblasts via exosomes, and the result is excessive collagen production and fibrosis (84). Overall, it is apparent that satellite cells communicate with fibroblasts via secretory factors and exosomes to regulate the production of ECM proteins following muscle loading.

Satellite cell communication with the muscle fibre

Under normal physiological circumstances, satellite cells fuse with muscle fibres to support muscle hypertrophy and regeneration following loading and/or exercise. However, a recent study conducted in mice and cultured cells, showed that satellite cells can transfer molecules to muscle fibres in the early stages of functional overload via exosomes (139). After muscle overload, muscle of mice lacking satellite cells have much higher expression of MMP-9 (a protein involved in ECM remodelling and muscle regeneration) compared to mice with functional satellite cells. In addition, culturing myotubes with exosomes derived from myoblasts reduces MMP-9 mRNA levels in the myotubes (139). These results indicate that satellite cells communicate with muscle fibres via exosomes to alter their gene expression during hypertrophy. As is the case for satellite cell communication with fibroblasts, this likely occurs via the delivery of microRNAs.

Satellite cell communication with endothelial cells

Endothelial cells are found on the inner lining of blood vessels, including capillaries in skeletal muscle. Satellite cells are located near to capillaries, and as discussed earlier in the chapter, exercise brings satellite cells and capillaries into closer proximity (49). The interaction between endothelial cells and satellite cells appears to be important for regeneration and communication between satellite cells and endothelial cells may be important for satellite cell function.

A recent study showed that when cultured together in vitro, release of the angiogenesis promoting protein Vascular Endothelial Growth Factor A (VEGFA) from myoblasts resulted in movement of endothelial cells towards myoblasts (140). When VEGFA was genetically deleted from satellite cells in mice, satellite cells and endothelial cells (capillaries) were found further apart, and this was associated with a loss of quiescent satellite cells (140). Endothelial cells were found to express the Notch ligand Delta-Like Protein 4 (Dll4), which through binding to Notch receptors on satellite cells prevents myoblast differentiation and therefore restored the satellite cell pool. Indeed, myoblasts that were cultured with endothelial cells had more Pax7+/MyoD− cells after myoblast differentiation (140), suggesting that more myoblasts return to quiescence in the presence of endothelial cells. People with more capillaries in their skeletal muscle are also able to activate and expand the satellite cell pool more effectively following exercise (141). Overall, this indicates that satellite cells communicate with endothelial cells to bring them into closer proximity, which enables improved satellite cell activation, expansion and restoration of quiescence following muscle damage.

AGEING, SATELLITE CELLS AND EXERCISE

Ageing and satellite cells

Analysis of skeletal muscle samples from rodents and humans has revealed that the number of satellite cells associated with myofibres is reduced by up to 50% with age (137,142). The reduction in satellite cell number is far more pronounced in type II muscle fibres (33, 143–145), which also appear to lose myonculei in older age (146). Since type II fibres are known to experience the most age-related muscle atrophy, reductions in satellite cell numbers and myonuclear content may play a crucial role in muscle wasting associated with older age, known as sarcopenia (147).

The regeneration of skeletal muscle in response to severe muscle damage is also impaired in older rodents (132,148,149), and is associated with reduced satellite cell proliferation, a blunted inflammatory response within the muscle and an increase in fibrosis (62, 132). However, when aged myoblasts from humans and rodents are cultured in vitro, their ability to proliferate and differentiate compared to young myoblasts is variable; with some studies showing age-related impairments (e.g. 150–152) and others finding no difference (132, 153, 154). Therefore, whilst the satellite cell itself may become dysfunctional in older age, the impaired regeneration observed in aged muscle may also be a result of (i) simply having fewer satellite cells able to mount a regenerative response, or perhaps more intriguingly, (ii) changes within the muscle niche that inhibit satellite cell regeneration.

Several experiments have provided evidence that the satellite cell niche (i.e. the muscle fibre itself) and the systemic environment contribute to the observed loss of satellite cell function in aged muscle. Transplantation of extensor digitorum longus muscles from old rats into young rats improves the recovery of muscle mass and force in the regenerating tissue compared with transplantation into an old host animal. Conversely, transplanting young muscle to an old host animal worsens regeneration (155). Heterochronic parabiosis studies, where the circulatory systems of young and old animals are surgically joined together, provide further evidence that the ageing environment is detrimental to satellite cell function. When old mice are joined with young mice prior to muscle injury, the old mice are able to mount a regenerative response which far exceeds the normal response of aged muscle (156). Conversely, young mice joined with old mice suffer more fibrosis during regeneration and reduced satellite cell proliferation (157). Furthermore, a recent study found that more than 50% of the genes that change in satellite cells with ageing can be returned to young levels when exposed to a young niche (158). Collectively, this suggests that exposure to a 'young' environment can help towards rejuventating aged satellite cells.

The aged niche (i.e. the muscle fibre) may impair satellite cell function via the dysregulated release a number of growth factors such as Delta, TGFβ or Wnt. Muscle injury leads to increased expression of the Notch-1 ligand Delta from the muscle fibre which promotes myoblast proliferation, however aged muscle fails to induce Delta upon injury (132, 137). Aged muscle fibres also produce more TGFβ than younger fibres (137, 159). TGFβ localises to the basal lamina (the residence of satellite cells) and causes phosphorylation of smad transcription factors which stimulate the cell to synthesise cell cycle inhibitors (159), preventing satellite cell activation. Finally, blood from old mice contains components which enhance Wnt signalling in satellite cells during regeneration (157). Wnt signalling is important for myoblast differentiation (160), but in quiescent satellite cells excessive Wnt results in the conversion of myoblasts to fibroblasts which, in turn, are responsible for the excessive collagen deposition and fibrosis observed in aged muscle during regeneration (157).

Reductions in satellite cell numbers in aged muscle may also be regulated by factors secreted by the muscle fibre. Aged muscle produces higher levels of FGF2 compared to young adult muscle (161) which promotes satellite cell proliferation and inhibits their self-renewal, resulting in depletion of the satellite cell population which ultimately blunts muscle regeneration (161). In addition, satellite cells may transition from a quiescent state into a senescent state (irreversible state of cell cycle arrest) in very old age (162) which renders them incapable of entering the myogenic cycle and self-renewing. This switch from quiescence to senescence impairs both satellite cell function and in the longer-term, cell numbers.

Overall, ageing is accompanied by a loss of satellite cell number and function which may be driven largely by changes in the environment to which they are exposed as well as intrinsic changes (e.g. induction of senescence) within satellite cells themselves. In particular, the aged skeletal muscle fibre appears to be a source of numerous factors which disrupt the normal myogenic programme, reduces the satellite cell pool and impairs the regenerative capacity of muscle.

Exercise and the ageing satellite cell

The satellite cell response to an acute bout of exercise is impaired in older people. Following both traditional resistance exercise (33, 163) and damaging eccentric exercise (35), the increase in satellite cell numbers is delayed and attenuated compared with young individuals. Fewer satellite cells express MyoD following resistance exercise in older individuals, suggesting that satellite cells fail to activate and induce the myogenic programme correctly. When measured according to fibre type, there is evidence that following exercise, satellite cells associated with type II fibres are particularly resistant to activation (33). In part, this may be due to an impaired MGF response to muscle loading in older individuals (164). This poor satellite cell response to muscle loading in type II fibres may explain, at least in part, specific loss of satellite cells in fast fibres with ageing.

Despite a poor satellite cell response to acute exercise, long-term exercise training increases the satellite cell content of muscle fibres in older individuals. Interestingly, 12 weeks of resistance training increased satellite cell numbers only in type II fibres (44, 146) which restored the diminished satellite cell pool associated with fast fibres. Similarly, resistance training results in the acquisition of new myonuclei specifically to type II muscle fibres (146), but this does not appear to occur when muscle hypertrophy is modest (45), further supporting the notion of a myonuclear domain ceiling size. Endurance exercise also results in preferential type II fibre satellite cell expansion (45), which differs from the response in younger individuals (see aerobic exercise section) and likely reflects the disparity in fibre type atrophy and satellite cell loss in type II fibres. This data shows that following long-term exercise, satellite cells from aged muscle retain the capacity to proliferate and differentiation to support muscle growth.

Evidence in animal studies suggests that exercise training improves the ability of aged satellite cells to regenerate muscle. One study demonstrated that 13 weeks of running exercise in rats increased satellite cell numbers, and those satellite cells were better able to differentiate in vitro, with fewer myoblasts converting into fibroblasts (52). More recently, 8 weeks of exercise pre-conditioning improved satellite cell numbers and regeneration following chemical muscle injury, with a more robust inflammatory response to injury and less fibrosis following regeneration (62). Whether exercise improves satellite cell function through alterations to the cells themselves (intrinsic) or to the niche (extrinsic) remains unknown.

Overall, satellite cells in aged skeletal muscle respond poorly to an acute bout of exercise, but longer-term training can increase satellite cell numbers and improve their ability to function and differentiate. Thus, suppression of satellite cell numbers and function with advancing age can be restored by exercise training.

Summary

In this chapter, we have provided an overview of satellite cells and their importance in skeletal muscle response and adaptation to exercise. Satellite cells are stem cells which lie in a quiescent state on the periphery of muscle fibres. Both resistance and endurance exercise provide a stimulus for satellite cell activation, whereby they enter the cell cycle and begin to proliferate, and a number of proteins and molecular pathways may be implicated in this process. Some satellite cells return to quiescence after activation therefore increasing satellite cell numbers, whereas others fuse with muscle fibres to provide additional myonuclei, a response which appears to be more specific to resistance exercise which promotes muscle growth. Whether muscle hypertrophy specifically requires the addition of new myonuclei by satellite cells is a point of much debate amongst scientists in the field. But under physiological conditions, new nuclei are added to muscle during growth and sustained hypertrophy is compromised in their absence. Satellite cells are also important for regulating ECM secretion as well as their own activation status, via communication with other cells within their niche. Ageing results in a loss of satellite cells and impaired function, which is likely a result of significant alterations in their environment. Finally, the deterioration of satellite cells in ageing can be improved by long-term exercise.

REVIEW QUESTIONS

• Describe what satellite cells are and explain their function during muscle adaptation.

• Describe the molecular regulation of myogenesis.

• Explain and discuss how satellite cells respond both acutely and chronically to resistance and endurance exercise.

• Discuss the important (or not important) role of satellite cells in muscle hypertrophy.

• Describe and discuss the differences in satellite cell function as a result of ageing and exercise.

FURTHER READING

Snijders T, Nederveen JP, McKay BR, Joanisse S, Verdijk LB, van Loon LJC, Parise G (2015) Satellite cells in human skeletal muscle plasticity *Frontiers in Physiology* 6:283.

REFERENCES

1. Mauro A. *J Biophys Biochem Cytol.* 1961. 9:493–8.
2. Katz B. *Philos Trans R Soc Lond B Biol Sci.* 1961. 243(703):221–40.
3. Schultz E, et al. *J Exp Zool.* 1978. 206(3):451–6.
4. Bischoff R. *Anat Rec.* 1975. 182(2):215–35.
5. Konigsberg UR, et al. *Dev Biol.* 1975. 45(2):260–75.
6. Lepper C, et al. *Development.* 2011. 138(17):3639–46.
7. Murphy MM, et al. *Development.* 2011. 138(17):3625–37.
8. Sambasivan R, et al. *Development.* 2011. 138(17):3647–56.
9. Crameri RM, et al. *J Physiol.* 2007. 583(1):365–80.
10. Goh Q, et al. *Elife.* 2019. 8:1–19.
11. Allen DL, et al. *J Appl Physiol.* 1995. 78(5):1969–76.
12. Seale P, et al. *Cell.* 2000. 102:777–86.
13. Yin H, et al. *Physiol Rev.* 2013. 93(1):23–67.
14. Gayraud-Morel B, et al. *Dev Biol.* 2007. 312(1):13–28.
15. Ustanina S, et al. *Stem Cells.* 2007. 25(8):2006–16.
16. Yablonka-Reuveni Z, et al. *Dev Biol.* 1999. 210(2):440–55.
17. Sabourin LA, et al. *J Cell Biol.* 1999. 144(4):631–43.
18. Cornelison DDW, et al. *Dev Biol.* 1997. 191(2):270–83.
19. Berkes CA, et al. *Semin Cell Dev Biol.* 2005. 16(4–5):585–95.
20. Penn BH, et al. *Genes Dev.* 2004. 18(19):2348–53.
21. Schultz E, et al. *Muscle Nerve.* 1985. 8(3):217–22.
22. Jockusch H, et al. *J Cell Sci.* 2003. 116(8):1611–6.
23. Brown AD, et al. *Biogerontology.* 2017. 18(6):947–64.
24. González MN, et al. *Skelet Muscle.* 2017. 7(1):1–13.
25. Collins CA, et al. *Cell.* 2005. 122(2):289–301.
26. Zammit PS, et al. *J Cell Biol.* 2004. 166(3):347–57.
27. Neumüller RA, et al. *Genes Dev.* 2009. 23(23):2675–99.
28. Troy A, et al. *Cell Stem Cell.* 2012. 11(4):541–53.
29. Bennett AM, et al. *Science.* 1997. 278(5341):1288–91.
30. Jones NC, et al. *J Cell Biol.* 2005. 169(1):105–16.

31. Conboy IM, et al. *Dev Cell.* 2002. 3(3):397–409.
32. Wen Y, et al. Mol Cell Biol. 2012. 32(12):2300–11.
33. McKay BR, et al. *FASEB J.* 2012. 26(6):2509–21.
34. Nederveen JP, et al. *Acta Physiol.* 2015. 215(4):177–90.
35. Dreyer HC, et al. *Muscle Nerve.* 2006. 33(2):242–53.
36. O'Reilly C, et al. *Muscle Nerve.* 2008. 38(5):1434–42.
37. McKay BR, et al. *PLoS One.* 2009. 4(6):e6027.
38. Siegel AL, et al. *Skelet Muscle.* 2011. 1(1):1–7.
39. Zammit PS, et al. *Exp Cell Res.* 2002. 281(1):39–49.
40. Roth SM, et al. *Journals Gerontol – Ser A Biol Sci Med Sci.* 2001. 56(6):B240–7.
41. Kadi F, et al. *J Physiol.* 2004. 5583:1005–12.
42. Mackey AL, et al. *J Physiol.* 2011. 589(22):5503–15.
43. Petrella JK, et al. *J Appl Physiol.* 2008. 104(6):1736–42.
44. Verdijk LB, et al. *J Gerontol – Ser A Biol Sci Med Sci.* 2009. 64(3):332–9.
45. Verney J, et al. *Muscle Nerve.* 2008. 38(3):1147–54.
46. Damas F., et al. *PLoS One.* 2018. 13(1):1–12.
47. Bellamy LM. et al. *PLoS One.* 2014. 9(10):17–21.
48. Petrella JK. et al. *Am J Physiol Endocrinol Metab.* 2006. 291(5):E937–46.
49. Nederveen JP. et al. *Am J Physiol – Regul Integr Comp Physiol.* 2017. 312(1):R85–92.
50. Bellamy LM. et al. *Am J Physiol – Cell Physiol.* 2010. 299(6):1402–8.
51. Charifi N, et al. *Muscle Nerve.* 2003. 28(1):87–92.
52. Shefer G, et al. *PLoS One.* 2010. 5(10):e13307.
53. Shefer G, et al. *FEBS J.* 2013. 280(17):4063–73.
54. Abreu P, et al. *J Cachexia Sarcopenia Muscle.* 2020. 11(6):1661–76.
55. Fry CS, et al. *J Physiol.* 2014. 592(12):2625–35.
56. Murach KA, et al. *Physiol Rep.* 2016. 4(18):1–10.
57. Joanisse S, et al. *FASEB J.* 2013. 27(11):4596–605.
58. Kurosaka M, et al. *Acta Physiol.* 2012. 205(1):159–66.
59. Snijders TIM, et al. *Muscle and Nerve.* 2011. 43(3):393–401.
60. Joanisse S, et al. *Am J Physiol – Regul Integr Comp Physiol.* 2015. 309(9):R1101–11.
61. Kurosaka, et al. *J Sport Sci Med.* 2009. 8(1):51–7.
62. Joanisse S, et al. *FASEB J.* 2016. 30(9):3256–68.
63. Jackson, JR, et al. *Skelet Muscle.* 2015. 5(1):1–17.
64. Ralston E, et al. *J Cell Biol.* 1992;119(5):1063–8.
65. Dix DJ, et al. *J Histochem Cytochem.* 1988. 36(12):1519–26.
66. Pavlath GK, et al. *Nature.* 1989. 337(6207):570–3.
67. O'Connor RS, et al. *J Appl Physiol.* 2007. 103(3):1099–102.
68. McCarthy JJ. *J Appl Physiol.* 2007. 103(3):1100–102
69. Cramer A., et al. *Nat Commun.* 2020. 11(1):6287.
70. Englund DA, et al. *Am J Physiol – Cell Physiol.* 2019. 317(4):C719–24.
71. Amthor H, et al. *Proc Natl Acad Sci U S A.* 2009. 106(18):7479–84.
72. Lee SJ, et al. *Proc Natl Acad Sci U S A.* 2012. 109(35):E2353–60.
73. Roy RR, et al. *J Appl Physiol.* 1999;87(2), 634–42.
74. McCall GE, et al. *J Appl Physiol.* 1998;84(4):1407–12.
75. Kadi F, et al. *Histochem Cell Biol.* 1999;111(3):189–95.
76. Bruusgaard JC, et al. *Proc Natl Acad Sci.* 2010. 107(34):15111–6.
77. van der Meer SFT, et al. *Ann Anat.* 2011. 193(1):56–63.
78. Kirby TJ, et al. *Mol Biol Cell.* 2016. 27(5):788–98.
79. Stec MJ, et al. *Am J Physiol – Endocrinol Metab.* 2016. 310(8):E652–61.
80. Rosenblatt JD, et al. *J Appl Physiol.* 1992. 73(6):2538–43.
81. Rosenblatt JD, et al. *Muscle Nerve.* 1994. 17(6):608–13.
82. Adams GR, et al. *Am J Physiol – Cell Physiol.* 2002. 283(4 52–4):1182–95.

83. Mccarthy JJ, et al. *Development*. 2011. 138(17):3657–66.
84. Fry CS, et al. *Cell Stem Cell*. 2017. 20(1):56–69.
85. Fry CS, et al. *FASEB J*. 2014. 28(4):1654–65.
86. Egner IM, et al. *Development*. 2016. 143(16):2898–906.
87. Goh Q, et al. *Elife*. 2017. 6:1–19.
88. Englund DA, et al. *Function*. 2020. 2(1):1–18.
89. Crameri RM, et al. *J Physiol*. 2004. 5581:333–40.
90. Tatsumi R, et al. *Dev Biol*. 1998. 194(1):114–28.
91. Tatsumi R, et al. *Muscle Nerve*. 2004. 30(5):654–8.
92. Allen RE, et al. *J Cell Physiol*. 1995. 165(2):307–12.
93. Sheehan SM, et al. Muscle and Nerve. 2000. 23(2):239–45.
94. Tatsumi R, et al. *Exp Cell Res*. 2001. 267(1):107–14.
95. Furge KA, et al. *Oncogene*. 2000. 19(49):5582–9.
96. Tatsumi R, et al. *Mol Biol Cell*. 2002. 13(8):2909–18.
97. Tatsumi R, et al. *Am J Physiol – Cell Physiol*. 2006. 290(6).
98. Roberts CK, et al. *Am J Physiol – Endocrinol Metab*. 1999. 277(2 40–2).
99. Rigamonti E, et al. *J Immunol*. 2013. 190(4):1767–77.
100. Ridnour LA, et al. *Proc Natl Acad Sci U S A*. 2007. 104(43):16898–903.
101. Yamada M, et al. Muscle and Nerve. 2006. 34(3):313–9.
102. Yamada M, et al. *Int J Biochem Cell Biol*. 2008. 40(10):2183–91.
103. Stewart CEH, et al. *Physiol Rev*. 1996. 76(4):1005–26.
104. Adams GR. *J Appl Physiol*. 2002 .93(3):1159–67.
105. Yang S, et al. *J Muscle Res Cell Motil*. 1996. 17(4):487–95.
106. Heron-Milhavet L, et al. *J Cell Physiol*. 2010. 2 25(1):1–6.
107. Lefaucheur JP, et al. *J Neuroimmunol*. 1995. 57(1–2):85–91.
108. McKay BR, et al. *J Physiol*. 2008. 586(22):5549–60.
109. Hill M, et al. *J Physiol*. 2003. 549(2):409–18.
110. Hill M, et al. *J Anat*. 2003. 203(1):89–99.
111. Ates K, et al. *FEBS Lett*. 2007. 581(14):2727–32.
112. Mills P, et al. *Exp Cell Res*. 2007. 313(3):527–37.
113. Yang SY, et al. *FEBS Lett*. 2002. 522(1–3):156–60.
114. Kandalla PK, et al. *Mech Ageing Dev*. 2011. 132(4):154–62.
115. Fornaro M, et al. *Am J Physiol – Endocrinol Metab*. 2014. 306(2):150–6.
116. Perez-Ruiz A, et al. *Cell Signal*. 2007. 19(8):1671–80.
117. Coolican SA, et al. *J Biol Chem*. 1997. 272(10):6653–62.
118. Lescaudron L, et al. 1999. 9(2):72–80.
119. Mackey AL, et al. *J Appl Physiol*. 2007. 103(2):425–31.
120. Mikkelsen UR, et al. *J Appl Physiol*. 2009. 107(5):1600–11.
121. Toth KG, et al. *PLoS One*. 2011. 6(3):e17392.
122. Serrano AL, et al. *Cell Metab*. 2008;7(1):33–44.
123. Meadows KA ,et al. *J Cell Physiol*. 2000. 183(3):330–7.
124. Li YP. *Am J Physiol – Cell Physiol*. 2003. 285(2 54–2):370–6.
125. Otis JS, et al. *PLoS One*. 2014. 9(3):1–10.
126. Foulstone EJ, et al. *J Cell Physiol*. 2001. 189(2):207–15.
127. Guttridge DC, et al. *Science*. 2000. 289(5488):2363–5.
128. Chen SE, et al. *Am J Physiol – Cell Physiol*. 2007. 292(5):C1660–71.
129. Wu Z, et al. *Mol Cell Biol*. 2000. 20(11):3951–64.
130. Chen SE, et al. *Am J Physiol – Cell Physiol*. 2005. 289(5 58–5):1179–87.
131. Langen RCJ, et al. *FASEB J*. 2004. 18(2):227–37.
132. Conboy IH, et al. *Science*. 2003. 302(5650):1575–7.
133. Kastner S, et al. *J Histochem Cytochem*. 2000. 48(8):1079–96.
134. Sheehan SM et al. *J Cell Physiol*. 1999. 181(3):499–506.

135. Yablonka-Reuveni Z, et al. *Basic Appl Myol*. 1997. 7(3–4):189–202.
136. Fedorov Y V et al. *Cell*. 2001. 152(6):1301–5.
137. Carlson ME et al. EMBO Mol Med. 2009. 1(8–9):381–91.
138. Koskinen SOA et al. *Am J Physiol – Regul Integr Comp Physiol*. 2001. 280(5 49–5):1292–300.
139. Murach KA, et al. *Function*. 2020. 1(1):1–15.
140. Verma M et al. *Cell Stem Cell*. 2018. 23(4):530–43.
141. Nederveen JP, et al. *J Physiol*. 2018. 596(6):1063–78.
142. Renault V, et al. *Aging Cell*. 2002. 1(2):132–9.
143. Shefer G, et al. *Dev Biol*. 2006. 294(1):50–66.
144. Verdijk, et al. *Am J Physiol – Endocrinol Metab*. 2007. 292(1):151–7.
145. Mackey, et al. *Acta Physiol*. 2014. 210(3):612–27.
146. Verdijk, LB et al. *Age (Omaha)*. 2014. 36(2):545–57.
147. Verdijk, LB et al. *J Am Geriatr Soc*. 2010. 58(11):2069–75.
148. Sadeh, M. *J Neurol Sci*. 1988. 87(1). 67–74.
149. Zacks SI et al. *Muscle Nerve*. 1982. 5(2):152–61.
150. Beccafico S et al. *Age (Omaha)*. 2011. 33(4):523–41.
151. Chargé SBP, et al. *Am J Physiol – Cell Physiol*. 2002. 283(4 52–4):1228–41.
152. Lorenzon P, et al. *Exp Gerontol*. 2004. 39(10):1545–54.
153. Alsharidah M, et al. *Aging Cell*. 2013. 12(3):333–44.
154. Renault V et al. *Exp Gerontol*. 2000. pp. 711–9.
155. Carlson BM, et al. *Am J Physiol – Cell Physiol*. 1989. 256(6).
156. Conboy IM, et al. *Nature*. 2005. 433(7027):760–4.
157. Brack AS et al. *Science*. 2007. 317(5839):807–10.
158. Lazure F et al. *bioRxiv*. 2021. 2021. 05.25.445621.
159. Carlson ME, et al. *Nature*. 2008. 454(7203):528–32.
160. Brack AS, et al. *Cell Stem Cell*. 2008. 2(1):50–9.
161. Chakkalakal JV, et al. *Nature*. 2012. 490(7420):355–60.
162. Sousa-Victor P, et al. *Nature*. 2014. 506(7488):316–21.
163. Snijders T et al. *Age (Omaha)*. 2014. 36(4):9699.
164. Owino V et al. *FEBS Lett*. 2001. 505(2):259–63.
165. Bamman MM, et al. *Cold Spring Harb Perspec Med*. 2018. 8:a029751.

Index

Note: **Bold** page numbers refer to tables and *italic* page numbers refer to figures.